Igal Galili

Scientific Knowledge as a Culture

The Pleasure of Understanding

 Springer

Igal Galili
Faculty of Mathematics & Natural Sciences
The Hebrew University of Jerusalem
Jerusalem, Israel

ISSN 2520-8594   ISSN 2520-8608   (electronic)
Science: Philosophy, History and Education
ISBN 978-3-030-80203-5   ISBN 978-3-030-80201-1   (eBook)
https://doi.org/10.1007/978-3-030-80201-1

This Springer imprint is published by the registered company Springer Nature Switzerland AG
The registered company address is: Gewerbestrasse 11, 6330 Cham, Switzerland

Science: Philosophy, History and Education

Efforts to leverage the history and philosophy of science (HPS) to improve science education have a long and productive history. This book series serves as a venue for science education and HPS scholars to continue this tradition. Thoughtful consideration of the synergistic relationships among HPS and science education can improve science teaching and learning, science education policy and outreach, and the teaching and learning of HPS. Science education efforts to improve teaching and learning about the nature of science (NOS) should obviously be informed by HPS scholarship, but HPS also offers much for improving science teaching more generally. For instance, HPS plays an important role in teaching and learning for authentic conceptual understanding; making clear how such understanding requires*all* students to at times abandon everyday reasoning when learning particular science ideas. Science education scholarship, in turn, can assist efforts among HPS scholars to promote public engagement with science and HPS. Recent emphasis on engineering and technology in science education calls for historians and philosophers in those disciplines to contribute their expertise in promoting a robust STEM education and avoiding undesired and unanticipated problems. This series is directed toward publishing authoritative books that overtly address how history and philosophy can and should inform science education. This series complements the journal *Science & Education* (http://www.springer.com/journal/11191). Questions regarding author ideas for book proposals should be directed to the Series Editor, Michael Clough (mclough@tamu.edu). Book proposals are to submitted to the Publishing Editor: Claudia Acuna (Claudia.Acuna@springer.com).

More information about this series at https://link.springer.com/bookseries/13387

To all those (including my children, Yudit and Guy) who wanted to know about science and enjoy its understanding but were denied such opportunity by the curricula that have betrayed the spirit of science, making it formal, disciplinary, designed to support a profession, a craft, and ignoring its culture.

Foreword

There is a well-known story of the "two cultures." In 1959, C. P. Snow's Rede lecture brought to prominence the discussion about whether our societies, our education systems, and our intellectual lives are characterized by a split between two cultures: the arts or humanities and the sciences. The books in this series of course provide strong counterarguments against such a division: history and philosophy of science are not only highly relevant to science but they can enrich it. More broadly the two cultures are not distinct but very much interrelated and interacting. This is also the message that Igal Galili aims to convey with the present book. Focusing on physics, the author discusses concepts such as motion, image and vision, inertial force, and weight. He also introduces a new educational paradigm of discipline-culture, which he illustrates through a comparative presentation of the scientific contributions of Leonardo and Galileo. As the author notes in his introduction "…making scientific knowledge cultural in both internal and external senses, that is within the disciplinary knowledge and out of it (fine arts), will bring to the learner much intellectual enjoyment of genuine understanding, satisfaction together with general literacy. It will hopefully motivate the reader for further learning, appreciation and love of science." I am sure that the readers who follow this series will be rewarded by reading the present book as much as I did.

University of Geneva
Geneva, Switzerland

Kostas Kampourakis

About the Book

This is an outstanding book. It is among the very best science education books to be published in the past many decades. The subtitle of the book is: 'The Pleasure of Understanding'. Equally true, the 'pleasure of reading' will be experienced by all who turn its pages. Its historic and conceptual canvas is grand, from Aristotle to Aquinas, through Galileo, Descartes and Newton, to Einstein, Bohr and contemporary physicists. Each of the highlighted episodes in the contested history of science is painted in sufficient detail to absorb and satisfy all readers.

Galili is a physicist with enviable competence and erudition in the history and philosophy of science. In each chapter, he demonstrates that the unfolding understanding of the physics – of free fall, of projectiles, of optics, of collisions, of dynamics – cannot be separated from the ontological and epistemological commitments of the investigating natural philosopher. Sometimes the commitments are explicit, other times implicit. Physics cannot be separated from philosophy: it both draws from philosophy and informs philosophy. Good teachers can draw out this interconnection.

Galili is also a physics educator who believes that the teaching of physics has to be informed by the history and philosophy of science. Many educators share this commitment to the importance of HPS for ST. For Galili, this commitment is spelt out in detail across the widest range of pedagogical and curricular questions including, of course, elaboration of the nature of science or NOS. They are organized within a novel paradigm of the *discipline-culture* argued by the author as a fundamental tool of science education.

Galili is also a wonderful communicator. The book is richly illustrated with a hundred or more diagrams and illustrations, and scores of carefully selected pictures and portraits. There is an intrinsic linking of science with culture and art. Art historians will find many illuminating discussions in the book. For Galili, science itself is a culture and needs to be appreciated as such, specifically, as a *discipline-culture*.

University of New South Wales
Australia

Michael R. Matthews

Preface

Science education in schools has become highly disciplinary and minimally engaging for many students. Through the eyes of educational administrators, there are two types of individuals, two breeds of students: those who are good at science and those who are not. C. P. Snow, in 1960, emblemized the situation in the metaphor of two cultures: science and the arts (natural science versus humanities, "physicists" versus "poets"). This vision contradicts the basic idea of the harmony of human nature as defined in Classical Greece—*Paideia*. It has been preserved throughout history in the liberal arts tradition of education. A new question is how this ancient tradition matches with the new requirement of literacy in science by the broad public in a society saturated with technology and science-based activities.

Greek God Janus had two faces to symbolize the fundamental dichotomy of reality as perceived by humans (Earth-Heavens, inside-outside, etc.). Today, this image may represent the spilt of education into humanities and natural sciences – a painful cultural schism. (Photo by the author from the Vatican Museum)

The new challenge requires a wider platform which should surpass strict formalism and aim at the holistic knowledge of science, its nature and meaning, and include aspects of liberal arts beyond disciplinary training. The common split of school education to humanities and natural sciences implies a fundamental division in the cultural baggage of society, causing tension and lacking common ground.

Commonly, a smaller group of students chooses physics; others prefer humanities and remain practically illiterate in science. All rely on "common sense" and receiving knowledge "later," after school, when required. A scientifically literate worldview among school students has become less common, which is often justified by the *practical* needs. The disciplinary curriculum suitable for a small group of students in effect threatens our future as missing much of our cultural heritage. This is a problem for a society praising plurality.

As a way to resolve the problem of the curricular polarization, science educators argue for the History and Philosophy of Science (HPS) in science teaching.[1] It was argued that being educated must include a representative image of science, its nature, beyond the ability to solve simple problems or retrieve several formulas. HPS should enter the basic curriculum together with languages, history, and mathematics. Though launched in the rather distant past,[2] the need of HPS is far from being in consensus with educators. The complexity stems from the fact that HPS presents an ocean of subjects and products, which may provide controversial and chaotic information about science. The selection of some elements and the ignoring of others seem arbitrary; therefore, some argue refraining from involving HPS in science education.[3] Those who do not agree justify HPS involvement through "general" interest, natural curiosity to scientific products in a great span of areas: space voyages, climate change, genetics, medicine, dispute with religion, ethical codes, and moral values. Researchers reported that some students' intuitive conceptions could be similar to some historical ones (conceptual recapitulation), suggesting their consideration in teaching, building on the productive cognitive resonance in students.[4] In any case, the investigation of what, how, and when to use HPS continues.

This book presents a new approach to the benefits of historical conceptions. In so doing, it chooses a special conception—Discipline-Culture (DC). DC provides a specific organization of scientific knowledge as theory based, while theories are structured in nucleus, body, and periphery. While the *nucleus* and *body* represent a scientific theory in the traditional disciplinary sense, the *periphery* converts a discipline to a discipline-culture. It reveals the dialogic nature of scientific knowledge and specifies the limits of validity of each theory.

Since the training of pre-service teachers commonly lacks courses in science history, equally absent in the requirements of physics departments, we have produced new learning materials. Instead of addressing isolated historical cases, we reconstructed the historical consolidation of a particular concept or conception. The units, *excursus*, revived the diachronic discourse of scholars regarding ontological and epistemological aspects while adding curricular comments. The first part of the book illustrates this genre, addressing several important physics concepts. Excursuses encourage construction of cultural content knowledge of the subject matter.

The second part of the book further elaborates the new approach, illustrates and expands it. The new paradigm depicts scientific knowledge as a specific

---

[1] Gauld (1991, 2014), Matthews (1994/2015, 2009, 2014), Holton (2003).

[2] Mach (1883/1919/1989).

[3] Brush (1974).

[4] For example, Wandersee (1986, 1990), Galili & Hazan (2000).

culture—the culture of rules.[5] According to Kant, *culture* and *discipline* (keeping rules) are in opposition.[6] The DC approach fuses them in one structure (Chap. 6) which implies the vision of history of science, curriculum, and typology of students. This is applied to the optics curriculum in Chap. 7. While a *discipline* is a tool for *indoctrination* proclaiming a certain picture of the world (and ignoring others), a *discipline-culture* promotes *enculturation* by exposing the conceptual dialogue of different pictures, the construction of Cultural Content Knowledge (CCK). "Culture," as opposed to "discipline," is understood in the sense of plurality of inter-pretations, beliefs, and arguments regardless their status of being true or erroneous.

In anthropology, *culture* signifies the whole compendium of life habits and tradi-tions ascribed to a certain group of people. Multiplicity is inherent in culture, not only in the number of items but also in their wholeness, their style, in their shared flavor. Even with regard to a homogeneous group, its image as a culture emerges only in the eyes of an *external* observer. Equally in science, culture does not appear before the existence of another perspective, a comparison must be present. This is culture in the *internal* sense. Yet, *culture* in everyday use possesses a different (*external*) meaning. It often signifies activities and knowledge beyond that which is professional, of a particular individual. It refers to the spiritual life and intellectual satisfaction, such as dealing with history, art, music, and literature for their own sake, for pleasure.

In effect, the DC framework specifies the role of HPS in education. HPS materi-als are chosen (in the way which could) to support conceptual comparison, known for its power in the clarification of meaning. Chapter 8 illustrates this perspective through a comparative analysis of the scientific contributions of two brilliant minds, Leonardo and Galileo, who adopted very different manners of knowledge exploration.

Cultural disciplinary knowledge contributes to the refinement of the features of scientific knowledge and method. Chapter 9 presents a revision of the nature of sci-ence (NOS) as discussed among researchers of science education. The chapter ana-lyzes questions often faced by practicing teachers of science/physics and often answered through mere intuition. The "consensus view" as sometimes stated,[7] regarding NOS, was reconsidered and deconstructed. The new claims emphasized the genus of science as objective knowledge and a theory of reality.

Finally, the style of liberal teaching suggests the involvement of artistic images for knowledge representation. Chapter 10 expands this perspective in addressing issues of scientific content and the nature of science. This approach creates an affec-tive intellectual appeal rare in the formal teaching of science. It intends to trigger and enhance the pleasure of understanding science through aesthetics and artistic metaphors.

---

[5] Why culture? The notion of *culture* originally presumed multiplication, for instance, agriculture, the *cultivating* of crops. Similarly, in the biological laboratory, a colony of microbes is *cultured* in a Petri dish.

[6] Kant (1781/1952).

[7] Lederman et al. (2015).

The presented excursuses, in their original form, were produced within the HIPST educational project in 2010. Three (Chaps. 1, 2, and 4) were then developed in close collaboration with my colleague, discussant and friend Dr. Michael Tseitlin. In preparing this book, all the excursuses have been rewritten and expanded, drawing on the new studies in HPS and science education research.

I greatly appreciate the tremendous work invested by Susan Torfstein, to whom I am deeply obliged for the pleasure of appreciating her skill of translation from English to English, and for revealing to me previously unknown points of logic, and lack of logic, in this language, often well-hidden from non-native speakers. I also appreciate the help of Michael Morris who helped me in the language revision of Chaps. 6 and 7.

It is my pleasure and honor to express my heartfelt appreciation to Professor Michael Matthews, whom I have had the privilege, honor, and real pleasure of cooperating with for many years. Professor Matthews has played a pivotal role in establishing an international community of science educators who share the conviction of the fundamental role of history and philosophy of science in science education. Many people are obliged to him both for his wisdom and his generous support. He was very kind to read the text of this book while offering stylistic, historical, and philosophical comments.

The last, but not the least, pleasure is to express my gratitude to Professor Kostas Kampourakis. His friendly attitude, encouragement, and academic fairness as the Editor of *Science & Education* journal as well as the Springer book series in Science Education Research were decisive in the writing and publishing of this book.

It is simply impossible to mention all those colleagues and students to whom I feel gratitude for their collaboration over the years of my work in different countries around the globe. They have all contributed to the knowledge which I have employed in this book. In a way, book creation imitates the creation of science knowledge itself in being a cumulative social product possessing validity.

References

Brush, S. G. (1974). Should the history of science be rated X? *Science, 183*, 1164–1172.

Galili, I. & Hazan, A. (2000b). The influence of historically oriented course on students' content knowledge in optics evaluated by means of facets-schemes analysis. *Physics Education Research, American Journal of Physics*, 68(7), S3–S15.

Gauld, C. F. (1991). History of science, individual development and science teaching, *Research in Science Education, 21*, 133–140.

Gauld, C.F. (2014). Using History to Teach Mechanics. In M. R. Matthews (ed.), *International handbook of research in history, philosophy and science teaching* (pp. 57–95). Springer.

Holton, G. (2003). The project physics course, then and now. *Science & Education, 12*(8), 779–786.

Kant, E. (1781/1952). *The critique of pure reason*. In Britannica Great Books (Vol. 42). The University of Chicago Press.

Lederman, N. G., Schwartz, R. & Abd-El-Khalick, F. (2015). Conceptualizing the construct of NOS. In R. Gunstone (Ed.), *Encyclopedia of science education* (pp. 694–698). Springer.

Mach, E. (1883/1919/1989). *The science of mechanics, a critical and historical account of its development*. Open Court.

Matthews, M. R. (1994/2015). *Science teaching. The contribution of history and philosophy of science*. Routledge

Matthews, M. R. (2009). Teaching the philosophical and worldview components of science in science. *Science & Education 18*, 697–728.

Matthews, M. R. (Ed.) (2014). *International handbook of research in history, philosophy and science teaching*. Springer.

Wandersee, J. H. (1986). Can the history of science help science educators anticipate students' misconceptions? *Journal of Research in Science Teaching, 23*(7), 581–597.

Wandersee, J. H. (1990). On the value and use of the history of science in teaching today's science: Constructing historical vignettes. In D.E. Herget (ed.), *More History and Philosophy of Science in Science Teaching* (pp. 278–283). Florida State University.

The Hebrew University of Jerusalem   Igal Galili
2021

The original version of this book was revised. The correction to this book is available at https://doi.org/10.1007/978-3-030-80201-1_12

Contents

Part II   Perspectives

Part I
Conceptual Excursus

Chapter 1
Understanding Classical Mechanics: A Dialogue with the Cartesian Theory of Motion

And the demonstrations are so certain that, even if experience seemed to show us the contrary, we would nevertheless be obliged to place more faith in our reason than in our senses.

Descartes

© Springer Nature Switzerland AG 2021
I. Galili, *Scientific Knowledge as a Culture*, Science: Philosophy, History and Education, https://doi.org/10.1007/978-3-030-80201-1_1

Abstract We make an excursus to the history of mechanics and consider the set of laws of motion established by Rene Descartes in the seventeenth century, followed by his inferred rules of collisions of material bodies. This knowledge preceded the mechanics of Newton who was inspired by Descartes' theory. Observing the history in retrospect, the laws of Descartes may be seen as including important ideas—the uniform rectilinear motion as a natural state and the concept of quantity of motion (momentum). Descartes stated conservation of momentum in interaction (collisions); however, because Descartes' momentum was defined incorrectly, his account of collisions was incorrect, violating Galileo's principle of relativity which implies rest-uniform motion equivalence. We analyze Descartes' laws of motion and each of his rules for elastic collisions, revealing the specific logic of Descartes and show how the following scholars—Wallis, Wren, Huygens, and Newton—refined some and refuted other of his statements regarding the conservation of momentum. Finally, we compare the laws of motion obtained by Descartes with those of Newton. The critique of Descartes' theory is both ontological and epistemological. It emphasizes the inappropriate neglecting of empirical verification in favor of principles postulated as certain and convincing.

Cartesian Laws of Nature

The most influential work of classical mechanics is undoubtedly Isaac Newton's *Principia* [*The Mathematical Principles of the Natural Philosophy*]. So influential was it that we often ignore the fact that it lacked much of what we find in school textbooks of mechanics nowadays due to the central role of Newton's contribution to what we call the theory of classical mechanics.

In fact, not only did Newton's work not cover the whole of classical mechanics, it was also not the first attempt to produce a comprehensive coverage of mechanics as a theory. The very name of Newton's work suggests that it should be understood to be part of a dialogue with an earlier work known under a similar title—*Principia* [*The Principles of Philosophy*] published by Rene Descartes in 1644.[1]

Descartes

A better understanding of this dialogue would lead a reader to a better understanding of classical mechanics. Therefore, we are going to revive this discourse by considering Descartes' *Laws of Nature* and their implications in order to compare them to those we know from Newton.

Descartes' *laws* of motion, and the *rules* governing *collisions* between material bodies, are presented in the second part of his *Principia*,[2] which is where we now begin with Descartes' own words.

---

[1] Rene Descartes (1596–1650)—French philosopher, mathematician, physicist, and physiologist, the founder of the modern philosophy of rationalism, contributed to the foundation of modern science.

[2] Descartes (1644/1983, p. 59).

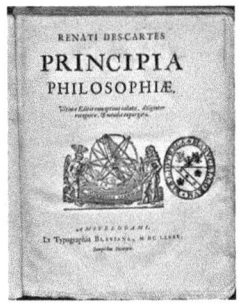

Title Page of Descartes'
Principia, 1685

37. The first law of nature: that each thing, as far as is in its power, always remains in the same state; and that consequently, when it is once moved, it always continues to move.

Furthermore, from this same immutability of God, we can obtain knowledge of the rules or laws of nature, which are the secondary and particular causes of the diverse movements which we notice in individual bodies. The first of these laws is that each thing, provided that it is simple and undivided, always remains in the same state as far as is in its power, and never changes except by external causes. Thus, if some part of matter is square, we are easily convinced that it will always remain square unless some external intervention changes its shape. Similarly, if it is at rest, we do not believe that it will ever begin to move unless driven to do so by some external cause. Nor, if it is moving, is there any significant reason to think that it will ever cease to move of its own accord and without some other thing which impedes it.

We must therefore conclude that whatever is moving always continues to move as far as is in its power. However, because we inhabit the earth, which is so constituted that all movements which occur near to it cease in a short while (and frequently from causes which are concealed from our senses), we often judged, from the beginning of our life, that those movements which thus ceased for reasons unknown to us, did so of their own accord. Indeed, because experience seems to have proved it to us on many occasions, we are still inclined to believe that all movements cease by virtue of their own nature, or that bodies have a tendency toward rest. Yet this is assuredly in complete contradiction with the laws of nature; for rest is the opposite of movement, and nothing moves by virtue of its own nature toward its opposite or its own destruction.

What Descartes is saying here is that an external influence is required to change the *state* of a body. Descartes considers motion to be a state rather than a process. Just as a solid body's shape, such as that of a metallic ball, remains the same unless externally manipulated, so the states of rest or motion of a body remain the same, unless influenced. The reason for this was, in his view, the "immutability of God." Descartes introduces here what we now call *inertial motion* by nature: a core principle of classical mechanics. Yet, he acknowledges that things do not seem to behave this way as bodies do tend to stop. This he explained as being a result of the influence of "causes that are hidden from our senses."

Importantly, Descartes considers the state of motion as being opposite to the state of rest—"rest is contrary to motion." This echoes the obsolete Aristotelian understanding which led Aristotle to suggest a mover was required to sustain motion. Descartes upgraded this view by considering motion as a state which changes under external influence, open or hidden. He, however, ignored the Galilean claim of bodies being indifferent to homogeneous motion during which they behave as if at rest.

Classical mechanics states that one cannot distinguish in any physical sense between an object travelling in a straight line with different constant velocities. The magnitude of velocity has no intrinsic implications for the body and is only meaningful in relation to another body. Accordingly, a body may be considered simultaneously to be in a state of motion and a state of rest with respect to different frames of reference. This lack of any impact on a moving body regardless of the magnitude of a constant velocity represents Galileo's principle.[3] In contrast, Descartes clearly argues that motion and rest are opposites and that each of them is preserved, unless subject to an external influence.

Descartes's generalized claim that "nothing can be moved to its contrary, or to its own destruction, by its own nature" is not supported by empirical evidence. Descartes immediately applied this law to the oldest problem in physics—the explanation of projectiles. He asked[4]:

38. Why bodies which have been thrown continue to move after they leave the hand.

and answered:

> For there is no other reason why things which have been thrown should continue to move for some time after they have· left the hand which threw them except that, {in accordance with the laws of nature}, having once begun to move, they continue to do so until they are slowed down by encounter with other bodies. It is obvious, moreover, that they are always gradually slowed down, either by the air itself or by some other fluid bodies through which they are moving, and that, as a result, their movement cannot last for long.

Descartes deduced a simple answer to the question so difficult for Aristotle. Aristotle suggested a mechanism of air turbulence—*antiperistasis*—which provides continuous pressure on the thrown stone by the surrounding streams of air.[5] This mechanism was criticized and refuted by later scholars by counterexamples of continuous motion of a spinning top, or a wheel, or an arrow, sharp at both ends. Descartes' new principle of "preserved motion" removed the need for antiperistasis. The air acts causing retardation, and eventual stopping the projectile.

Although working after Galileo and knowing about his account of projectiles, Descartes did not depict such motion, did not resolve it into horizontal and vertical components thus losing the ability to predict trajectory, velocity, and acceleration. He addressed the motion holistically ignoring any quantitative features, as required by physics.

Instead, Descartes proceeded to the second law[6]:

39. The second law of nature: that all movement is, of itself, along straight lines; and consequently, bodies which are moving in a circle always tend to move away from the center of the circle which they are describing.

> The second law of nature is that any part of matter, considered apart, never tends to continue to be moved along any oblique lines, but only along straight lines, even if many are often forced to deflect due to the collision of others, and, as has been said shortly before, in any motion a circle is somehow made from all the matter moved at the same time. The cause of this rule is the same as that of the one preceding, namely the immutability and simplicity of the operation by which God conserves motion in matter. For He does not conserve it other than precisely the way it is in the moment of time in which He conserves, with no rela-

---

[3] In his *Discourses on Two New Sciences* (1638), Galileo considered rest as a state of motion with infinitely low speed. This way rest lost its special meaning.

[4] Descartes (1644/1983, p. 59).

[5] See Chap. 2. How could it be that the same air supports and resists motion? Descartes mentioned birds, seemingly to illustrate the antiperistasis controversy. Birds fly against the air resistance, but without air they cannot fly at all. So, air helps and resists. Yet, the motion of a stone is similar to that of a bird.

[6] Descartes (1644/1983, p. 60).

tion to what perhaps was shortly before. Although no motion occurs instantaneously, it is nevertheless manifest that everything that is moved, in the single instants that can be designated while it is moved, is determined to continue its motion toward some direction along a straight line, and never along any curved line.

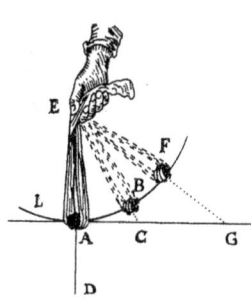

For example, stone A, rotated in sling EA around circle ABF, at the instant in which it is at point A is determined to motion in some direction, namely along a straight line toward C, such that the straight line AC is tangent to the circle. But one cannot arrange that it be determined to any curved motion; for, even if it previously came from L to A along a curved line, nevertheless nothing of this curvity can be understood to remain in it when it is at point A. This is also confirmed by experience, because if it then left the sling it would not continue to be moved toward B, but toward C. From which it follows that every body that is moved circularly, perpetually tends to recede from the center of the circle it describes. We experience this by tactile sense in a

Illustration from Descartes' *stone that we move in a circle with a sling. . .*
Principia, 1644, p. 56

Here, Descartes has refined his first law of motion preservation: it is not that every motion is preserved, rotation, for example, is not. Only rectilinear motion is preserved. Descartes did not mention "uniform" motion, but he did explicitly reject the idea of preserved curved motion. Galileo still thought about horizontal motion as that which moved along the Earth's surface,[7] "equidistant from this same common center" as being naturally preserved.[8] By rejecting Galileo's idea, Descartes provided rectilinear motion with a special status: only rectilinear motion is preserved and may be considered inertial.

Again, as in his first law, he deduced this idea from God's disposition: *the immutability and simplicity of the operation by which God conserves motion in matter.* Yet, if rectilinear uniform motion is immutable, why should circular motion not be similar in the eyes of God? Descartes remains silent on this but rather appeals to experience, as illustrated by a stone rotated on a sling.

One may ask whether Descartes' second law is different from the first law of Newton, commonly taught in physics classes. Indeed, they are rather similar but do not coincide. Newton is more specific regarding the state of motion representing its basic nature of being not only strait but also uniform in speed. Addressing a body compelled to move in a curved path, Descartes asserts that such a body possesses a *tendency* to proceed straight, in the tangential direction. This tendency manifests itself the very moment that the agent enforcing the curved motion ceases its action. Considering the tendency was adopted by Newton who stated in his *Principia*[9]:

Every body perseveres in its state of resting or moving uniformly straight on, except inasmuch as it is not compelled by impressed forces to change that state.

Newton asserts the *tendency* to preserve the state of *rest* or *uniform* forward motion. It is realized *inasmuch as the external force is not present.* Like Descartes,

---

[7] Galilei (1613/1957, pp. 113–114).
[8] Galilei (1638/1914, p. 181).
[9] Newton (1687/1999, p. 416).

Newton illustrates through addressing motion under active forces—projectiles, spinning tops, and the motion of planets.[10]

Finally, in his third law, Descartes addressed the influence of interactions between bodies in motion. Interestingly, Descartes considers only contact interaction between bodies by means of collision. He wrote:[11]

> **40. The third law: that a body, upon coming in contact with a stronger one, loses none of its motion; but that, upon coming in contact with a weaker one, it loses as much as it transfers to that weaker body.**

We may schematically represent the stated law of collisions by the following diagrams (Figs. 1.1a, b).

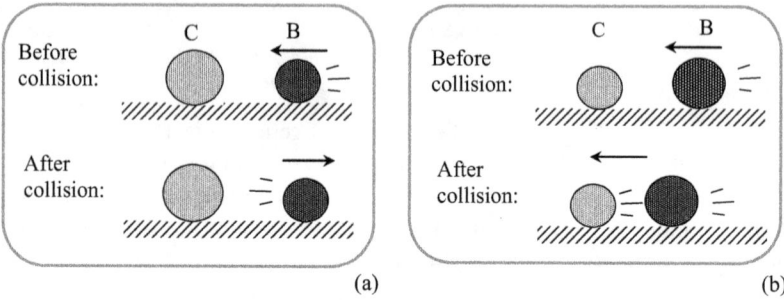

(a)   (b)

Fig. 1.1 The third law of nature

Descartes explained:

> *This is the third law of nature: when a moving body [B] meets another [C], if it has less force to continue to move in a straight line than the other has to resist it [Fig. 1.1a], it is turned aside in another direction, retaining its quantity of motion and changing only the direction of that motion. If, however, it has more force [Fig. 1.1b]; it moves the other body with it, and loses as much of its motion as it gives to that other.*

Descartes stated here the Law of Conservation of the Quantity of Motion, or Momentum Conservation, as we refer to it now. The significant ambiguity here is because Descartes split the law into two cases: collisions with "weaker" and "stronger" bodies, which is an obscure and erroneous separation. The correct conservation lacks such separation. But if to separate, what happens in the case of the collision of equally "strong" bodies? Descartes ignored this case, addressing the meaning of "weaker" and "stronger" in the rules of collisions which followed.

The third law addressed the collision of hard bodies.[12] Descartes correctly separated two essentially different cases of *hard* and *yielding* (soft) bodies. Each requires a different account. He comments:[13]

---

[10] Galili and Tseitlin (2003).

[11] Descartes (1644/1983, p. 61).

[12] Descartes (1644/1983, p. 62).

[13] Descartes (1644/1983, pp. 61–62).

Thus, we know from experience that when any hard bodies which have been set in motion strike an unyielding body, they do not on that account cease moving, but are driven back in the opposite direction; on the other hand, however, when they strike a yielding body to which they can easily transfer all their motion, they immediately come to rest.

This comment regarding collision with soft bodies is apparently wrong: soft (inelastic) collision does not necessarily imply that the colliding bodies must stop. Descartes seemingly had in mind the extreme case of a hard ball meeting a "soft" *medium*, such as sand. Descartes adds the area of validity of his law:

All the individual causes of the changes which occur in [the motion of] bodies are included under this third law, or at least those causes which are physical; for I am not here enquiring into what kind of power the minds of men or Angels may perhaps have to move bodies.

By excluding objects of an animated and unnatural kind, Descartes defines the domain of physics—inanimate Nature. Descartes proceeds with an explanation and justification of the two parts of his third law separately[14]:

41. The proof of the first part of this law.

[that a body coming in contact with a stronger one, loses none of its motion]
    The first part of this law is proved by the fact that there is a difference between motion considered in itself, and its determination in some direction; this difference makes it possible for the determination to be changed while the quantity of motion remains intact. For, as has been stated above, each thing which is not complex but simple, as motion is, always continues to exist as long as it is not destroyed by any external cause. And in an encounter with an unyielding body, there certainly appears a cause which prevents the movement of the body which strikes the other from maintaining its determination in the same direction. However, there is no cause which would remove or decrease the motion itself, {since none is taken from it by this body or any other cause and} since movement is not contrary to movement. From which it follows that its motion must not be diminished.

Descartes could imagine a collision with a much bigger, hard object (a wall). He clearly separates between motion (its amount) and its determination (its direction) and, incorrectly, treats each one of them independently. Descartes states that the collision of *hard* bodies may influence the direction of the motion, but not its amount, or in his language: *movement is not contrary to movement*. In reality, however, we observe a wide variety of *redistribution* of the amount of motion during collisions.

In the elastic collision, when the two objects initially move in different directions, one cannot correctly account for the conservation of motion without direction. In other words, one cannot render the conservation of motion solely in terms of positive numbers, which implies the motion described in magnitude and direction. Introduction of momentum as a vector removes the need for "weaker" and "stronger" identification. For us (but apparently, not for Descartes!) motion is only a *relative* quantity whose amount depends upon the chosen frame reference. That implies equivalence of motion and rest—Galileo's principle of relativity.

---

[14] Descartes (1644/1983, p. 62).

Descartes proceeded:[15]

42. The proof of the second part.

[...but that, upon coming in contact with a weaker one, it loses as much as it transfers to that weaker body]

Similarly, the second part is proved by the immutability of God's manner of working in always uninterruptedly maintaining the world by the same action by which He created it. From the fact that all places are full of bodies and that, nevertheless, the movement of each of these bodies tends in a straight line; it is obvious that when God first created the world, He not only moved its parts in various ways, but also simultaneously caused some of the parts to push others and to transfer their motion to these others. So in now maintaining the world by the same action and with the same laws with which He created it, He conserves motion; not always contained in the same parts of matter, but transferred from some parts to others depending on the ways in which they come in contact. Thus, this continuous changing in created things is an argument for the immutability of God.

Here, the law states the conservation of motion in the world but allowed distribution of motion among all bodies through collisions to produce all possible changes. The imagined picture is seemingly the collision of a "big" moving body with a small one, obtaining the impulse to move.

Descartes draws on the idea of conservation of motion from the immutable nature of the world's creator. This metaphysical claim is a postulated belief, and it represents the *rational philosophy* of Descartes. Such reasoning was criticized throughout history.[16] From his position in the seventeenth century, Descartes did not identify any natural reason for motion (momentum) conservation.

Emma Noether

The physical explanation of momentum conservation was found later: initially by Newton, but in a more fundamental way, by an outstanding scholar, Amalie Emmy Noether, in the beginning of the twentieth century.[17] She demonstrated that the conservation of a physical quantity is caused by the certain continuous symmetry possessed by a physical system. In particular, the symmetry of translation in space (space homogeneity) implies the conservation of momentum.

Newton had a difficult time avoiding metaphysical ideas, especially when he could not provide any mechanism for the considered phenomenon. Thus, he described gravitation, but avoided providing any explanation as to how it "worked," how it acts through space, at a distance. Nevertheless, based on the postulated gravitation, Newton provided highly successful accounts for numerous natural phenomena.

---

[15] Descartes (1644/1983, p. 62).

[16] German physicist, Ernst Mach, rewrote classical mechanics in 1893 in order to remove metaphysical claims, those which were not a subject of empirical verification. This new philosophical approach is identified as *logical positivism*.

[17] Amalie Noether (1882–1935), an outstanding mathematician of Jewish origin, faced great difficulties to be accepted and then to work at the University of Gottingen in Germany. In 1933, she was expelled from the university and flew to the USA.

The Account for Collisions

Having presented the general laws, Descartes proceeded to an implication—a set of rules "in order to determine, from the preceding laws, how individual bodies increase or decrease their movements or turn aside in different directions because of encounters with other bodies."[18]

46. The first rule.

Illustration from Descartes'
Principia, 1644, p. 60

First, if these two bodies, for example B and C, were completely equal in size and were moving at equal speeds, B from right to left, and C toward B in a straight line from left to right; when they collided, they would spring back and subsequently continue to move, B toward the right and C toward the left, without having lost any of their speed.

The first rule may be schematically represented by Fig. 1.2, in which magnitudes of speed and mass are introduced for illustration.

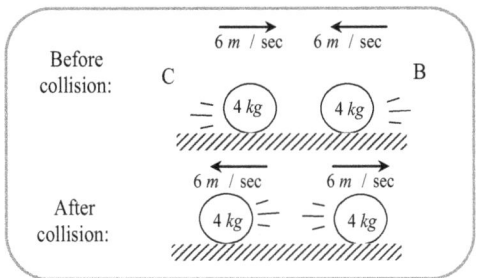

Fig. 1.2 The first rule of collision

Taken in isolation, this rule of Descartes is correct in its claim. However, to discover the problem, it is instructive to consider, what could Descartes' arguments be for supporting this rule? Is it consistent with the stated laws?

Descartes proceeded:[19]

47. The second.

Second, if B were slightly larger than C, and everything else were as previously described, then only C would spring back, and both would move toward the left at the same speed. {For B, having more force than C, could not be obliged by C to spring back}

---

[18] Descartes (1644/1983, p. 64).
[19] Descartes (1644/1983, p. 65).

This rule may be schematically represented by Fig. 1.3, in which magnitudes of speed and mass have been exemplified.

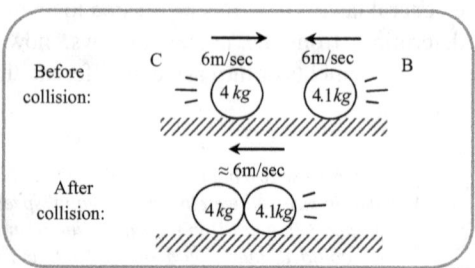

Fig. 1.3 The second rule of collision

Descartes predicted that after such a collision the two hard bodies would both move to the left. That is to say that a total "victory" is assured for B, however small its advantage in mass is. This is not what happens in reality.

Indeed, in the case of two equal bodies (the first rule), Descartes stated that each returned backwards with equal speed. Practically, there are no absolutely equal bodies (having an equal number of atoms!): one body is always "slightly" bigger than the other. Therefore, if the second rule were correct, we would never observe cases of bodies receding after collision. This conclusion falsifies the rule.

What was wrong? This failure reveals the falsity of definition of the quantity of motion. The slightest preference of mass causes a drastic change of output: B totally "overcomes" the motion of C and both proceed to the left with almost the same speed. This cannot be true. The failure is due to the defective definition: the quantity of motion must be dependent on the velocity (which is directional and hence a vector quality), not the speed (which is unidirectional and hence a scalar quality). The new definition would cause a different result, close to the case of equal bodies: a slight change in mass—a slight change in the output of the collision, which is reasonable and draws on continuity.

Descartes proceeded:[20]

48. The third.

Third, if the two bodies were equal in size, but if B were moving slightly more rapidly than C; after their collision not only would {C alone spring back and} both continue their movement toward the left, {that is, in the direction from which C came}, but also one half of B's additional speed would be transferred from it to C, {since B could not move more rapidly than C which would be ahead of it}. For example, if B had initially been travelling at six degrees of speed [toward the left], and C at a speed of only four [toward the right], {B would transfer to C one of its two additional degrees of speed, and} both would subsequently move toward the left at five degrees of speed.

---

[20] Descartes (1644/1983, p. 65).

The third rule of collisions may be schematically represented by the diagram in Fig. 1.4, using the magnitudes as suggested by Descartes.

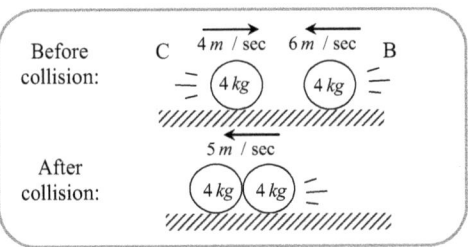

Fig. 1.4 The third rule of collision

This is again an incorrect statement. Descartes provides exact numbers to illustrate his account. One may argue by continuity: surely there should not be any drastic change from the first rule when B moves only slightly faster. If the first rule holds, after the collision, the bodies should recede from each other with only slightly altered speeds (B a little slower, C a little faster). This is what happens in reality.

Let us reconstruct the logic here. If the quantity of motion is defined by the product of mass and speed mv (not direction), then according to his theory, the higher quantity of motion of body B must win out and both will move to the left. Now, to preserve the quantity of motion, both bodies must move together with the mid speed, which is 5 m/s (masses are equal) to preserve the quantity of motion:

$$4 \times 4 + 4 \times 6 = (4+4) \times 5$$

Descartes proceeded:[21]

49. The fourth.
  Fourth, if the body C were entirely at rest, {that is, if it not only had no apparent motion but also were not surrounded by air or any other fluid (which makes the hard bodies immersed in such a fluid very easily movable, as I shall show)}, and if C were slightly larger than B; the latter could never {have the force to} move C, no matter how great the speed at which B might approach C. Rather, B would be driven back by C in the opposite direction: because {for B to move C, C would have to be driven as rapidly as B subsequently moves and} a body which is at rest puts up more resistance to high speed than to low speed; and this resistance increases in proportion to the difference in the speeds. Consequently, there would always be more force in C to resist than in B to drive, because C is larger.

---
[21] Descartes (1644/1983, p. 66).

The fourth rule of collisions may be schematically represented by Fig. 1.5, in which magnitudes of speed and mass are introduced for illustration.

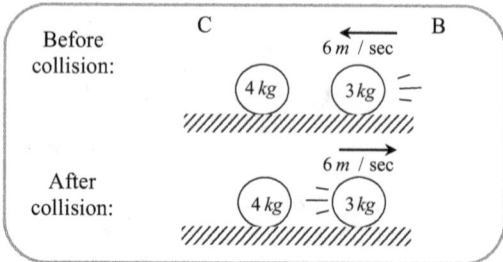

Fig. 1.5 The fourth rule of collision

Again, this understanding of Descartes is amazingly at odds with everyday experience. According to classical mechanics, if C remained at rest and B was reflected back, the total momentum would simply not be conserved. The problem is apparent (but not to Descartes).

Apart from having an incorrect scalar nature of motion, Descartes also erroneously believed that motion and rest were essentially different states—this was a great misconception introduced by the physics of Aristotle and overcome 2,000 years later by the principles of Galileo, all before Descartes wrote his rule!

Descartes ascribes to the greater mass at *rest* a resisting force superior to the striking force of the *moving* body, regardless of its speed. This violates the action-reaction symmetry of classical mechanics (Newton's third law). The scenario of one force overcoming the other in interaction is in accord with Aristotelian physics. According to Descartes, the resisting force is always greater in magnitude, increases with the increase of the speed of the colliding body B, and always overcomes it, while preserving the state of rest of body C, despite the impact. Descartes' imagination suggested a greater force to stop the faster body. The quantity of motion, as defined by Descartes, is preserved when body B returns back with the same speed.

Descartes proceeded:[22]

50. The fifth.

Fifth, if the body C were at rest and {even very slightly} smaller than B; then, no matter how slowly B might advance toward C, it would move C with it by transferring to C as much of its motion as would permit the two to travel subsequently at the same speed. Thus, if B were twice as large as C, it would transfer to C {only} one third of its quantity of motion; because that one third would move the body C at the same speed as the remaining two thirds would move the body B which {we are supposing} is twice as large as C. Therefore, after B had collided with C, its speed would be reduced by one third; that is to say, B would then need as much time to travel a distance of two feet as it previously did to travel a distance of three feet. Similarly, if B were three times as large as C, it would transfer to C one quarter of its motion; and so on.

[22] Descartes (1644/1983, p. 67).

The fifth rule of collisions may be schematically represented by Fig. 1.6, using the magnitudes as suggested by Descartes to illustrate the rule.

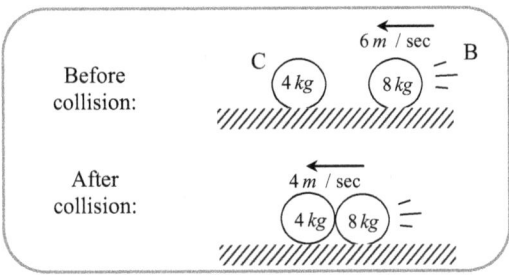

Fig. 1.6 The fifth rule of collision

In this case too, the *greater body* wins and imposes its motion on the *smaller body*. Descartes preserves the conservation of quantity of motion (mv), which, as he states, is in one direction after the collision. Using the numbers shown on the figure, before the collision, we arrive at $(mv)_{tot} = 8 \times 6$, and after: $(mv)_{tot} = (4 + 8) \times 4$, which is the same magnitude.

However, Descartes' claim is erroneous, despite the conservation of motion. In classical mechanics, to account for elastic collision, one should conserve not only momentum but also the kinetic energy, which was not known to Descartes. Application of both conservation laws —those of kinetic energy and momentum— provides a very different result: body B will continue at the speed of 2, while body C will move at the speed of 8, in the same direction.[23] It is this result which is confirmed by experiment.

Importantly, however, if one applies the relativity principle (ignored by Descartes), the fifth rule contradicts the fourth rule. Indeed, consider an observer sitting on B. The observer would perceive B to be at rest, while C approaches B. According to the fourth rule, body B should remain at rest, and body C would be repelled in contrast to the fifth rule which predicts that the two bodies would move together after the collision.

Descartes proceeded:[24]

51. The sixth.
Sixth, if the body C were at rest and exactly equal in size to body B, which was moving toward it; necessarily, C would be to some extent driven forward by B and would to some extent drive B back in the opposite direction. Thus, if B were to approach C with four degrees of speed, it would {have to} communicate one degree to C, and be driven back in the opposite direction with the remaining three.

---

[23] Not only will the momentum conserve: $8 \times 6 = 8 \times 2 + 4 \times 8$, but also the kinetic energy of the bodies will: $8 \times 6^2 = 8 \times 2^2 + 4 \times 8^2$.

[24] Descartes (1644/1983, p. 67).

The sixth rule of collision may be schematically represented by Fig. 1.7a.

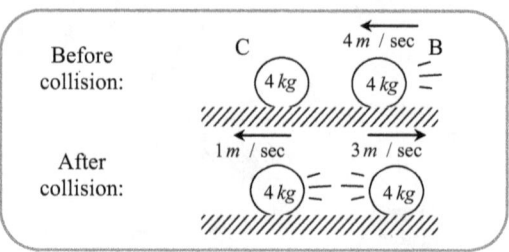

Fig. 1.7a The sixth rule of collision

This rule may be quite puzzling but can be interpreted[25] as a combination of two previous rules. Since C is neither larger nor smaller than B, the rule "should" fit both the tendencies stated in two previous rules. The first will imply C reacting as if it were smaller than B, and so resulting in both bodies moving with speed to the left (fifth rule). However, at the same time, there will be an equal tendency for C to behave as if it were larger, and hence, it should acquire no speed at all (fourth rule). Taking the average of these two tendencies provides the result predicted by Descartes: for body C the speed will be: $\dfrac{2+0}{2}=1$. Similarly, for body B it will be: $\dfrac{4+2}{2}=3$.

It is clear that Descartes conserves *his* "momentum"—the product of mass and *speed*—and not *ours*—the product of mass and *velocity*. According to his conservation:

$$4 \times 4 = 4 \times 3 + 4 \times 1 \,(\text{directions ignored!})$$

Again, Descartes violates the principle of relativity here: For an observer moving at speed 2 in the direction of body B, the case of the sixth rule is equivalent to the case of the first rule. And the first rule implies that when two equal masses collide elastically, they recede from each other at the same speed. Viewed in this way, the case of the sixth rule would suggest that B stops and C moves with the initial velocity of B. This does not match Descartes' statement in this case.

It is quite amazing how one could ignore a common situation in a game involving colliding hard balls (popular in France) in which one ball thrown toward another of the same size at rest stops and the other continues with the same speed (Fig. 1.7b).

---

[25] Descartes (1644/1983, p. 68), Banham (2009).

Fig. 1.7b Games with equal balls collisions popular in France

Despite this well-known situation,[26] Descartes preferred to follow his understanding, using it as a principle that is superior to any empirical evidence—a blindly rationalist philosophy.

Descartes proceeded:[27]

52. The seventh.

Finally, if B and C were travelling in the same direction, C more slowly than B, so that B (which would be following C) would eventually strike it; and if C were larger than B but B's speed exceeded C's by a greater extent than C's size exceeded B's: then B would transfer to C as much of its speed as would be required to permit them both to travel subsequently at the same speed and in the same direction (

Fig. 1.8a). However, if, on the contrary, B's speed exceeded C's by a smaller extent than C's size exceeded B's; B would be driven back in the opposite direction, and would retain all its movement (Fig. 1.8b).

The seventh rule of collisions may be schematically represented by Fig. 1.8.

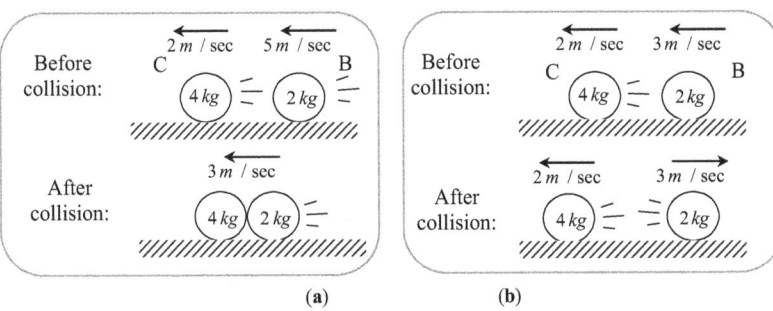

Fig. 1.8 The seventh rule of collision

---

[26] See below how Jan Marci in Prague, at about the same time, provided an impressive empirical demonstration of such a collision.

[27] Descartes (1644/1983, p. 68).

This rule addresses a general case of collision between two hard bodies moving in the same direction with different speeds. Here, Descartes fully reveals what he considers to be a decisive factor determining which of two scenarios will happen. In order to "win" (i.e., to be "stronger" or possess "more force"), a body needs to have a bigger quantity of motion. In case (a), B wins and impels C to change its motion (comparing the quantities of motion gives: $10 > 8$). The balance of motion looks, then, like this:

$$2 \times 5 + 4 \times 2 = 4 \times 3 + 2 \times 3$$

In case (b), however, C is stronger and wins (comparing the quantities of motion gives: $6 < 8$). This implies that C does not change its state of motion. B is compelled to change its direction, but not its motion (speed). Altogether, the quantity of motion, for Descartes, did not change:

$$4 \times 2 + 2 \times 3 = 4 \times 2 + 2 \times 3$$

The two cases (a) and (b) are completely different and do not match the continuity of the dependence on physical parameters. From the perspective of the theory of classical mechanics, as known to us, this rule of Descartes violates both the principle of conservation of momentum and the conservation of kinetic energy.

Contemporaries of Descartes quickly realized that most of the rules Descartes suggested to account for collisions failed to match real situations. Descartes also understood that his rules would be criticized for this disparity. However, this did not convince him to change his mind: he fully trusted reasoning by deduction from the principles that he considered to be true. If the result contradicts the empirical evidence, so be it, this must mean that something is preventing the principles from their realization. Thus, after outlining the rules, Descartes addressed their possible discrepancy with reality. He wrote:[28]

> **53. That the application of these rules is difficult, because each body is always surrounded by many contiguous ones.**
>
> *{Indeed, experience often seems to contradict the rules I have just explained}. However, because there cannot be any bodies in the world which are thus separated from all others, and because we seldom encounter bodies which are perfectly solid; it is very difficult to perform the calculation to determine to what extent the movement of each body may be changed by collision with others. Since, {before we can judge whether these rules are observed here or not}, we must simultaneously calculate the effects of all those bodies which surround the bodies in question and which affect their motion. These effects differ greatly, depending on whether the surrounding bodies are solid or fluid; and it is therefore necessary that we should immediately enquire into the difference between solid and fluid bodies.*

Descartes argued that the discrepancies are due to the complexity of reality in contrast to the ideal condition—a totally empty space, whereas the world is filled with continuous medium. Yet, the truth is that even in a vacuum, the rules of Descartes do not hold, while the principle of relativity, the conservation of *vectorial* momentum, and the conservation of kinetic energy in the elastic collisions are fully

---

[28] Descartes (1644/1983, p. 69).

supported experimentally. To overcome the barrier mentioned by Descartes, physicists use experiments with controlled parameters, and it is possible to monitor the influence of the medium by decreasing the impeding factors. The principles of mechanics were checked by approaching the ideal case, in which case, the observed behavior of real bodies never approached the predictions of Descartes (let alone the first rule, of course).

An alternative account of collisions was developed very soon after his *Principles* was published. The process of refutation started from establishing the empirical rules governing collisions. Eventually, these rules led to the new theory, the Newtonian theory, free from the metaphysical ideas employed by Descartes.

Discourse on Collisions Continues

Descartes' seminal work illustrates the interests of the seventeenth-century natural philosophers in constructing a mechanical picture of the universe. The interaction of bodies was central to that understanding, and collisions were considered as a building block of interaction.

Unlike the program of earlier natural philosophy, and rather in contrast to Descartes, several researchers first tried to obtain reliable empirically based quantitative rules that govern natural phenomena, whether or not they could explain them theoretically. It was Galileo's approach: first, we should reveal *how*, and then, understand *why*. Galileo was renowned for demonstrating experimentally that when things fall freely, they all move at a constant acceleration—g. Others proceeded toward "why."[29] Obtaining similarly accurate knowledge regarding collisions became a goal for several researchers of mechanics.

In 1668, the newly founded Royal Society in London issued a call for a study which could provide a reliable account of collisions. Three of the responses follow.

There was, however, an obstacle to performing such an exploration: how could one reliably measure velocities before and after collision? Galileo inferred the magnitude of velocities by measuring the distance travelled by moving bodies along an inclined plane and measuring time using a water dropper as a clock. This method is apparently not appropriate in an investigation of collisions which required a more precise technique.

An elegant solution to this problem was suggested in the studies of Edme Mariotte[30] published in Paris, in 1677.

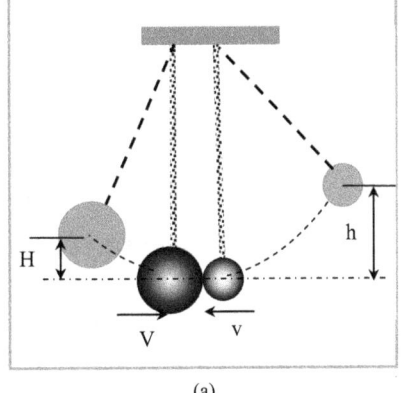

(a)

---

[29] Descartes' opposite approach failed, but not entirely, not in a more comprehensive sense. Galileo's approach is equally limited. Which way to go in science? We will return to this methodological aspect below.

[30] Edme Mariotte (1620–1684)—French priest and physicist, member of the Academie des Sciences.

Fig. 1.9 (a) Schematic representation of the apparatus suggested by Mariotte to investigate regularity in collisions of bodies. (b) The title page of the book *Works* by Mariotte published in 1717

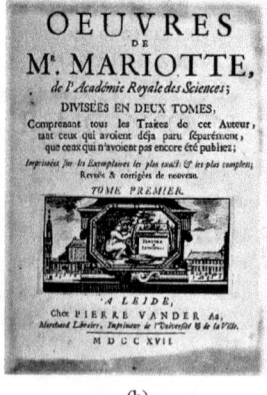

(b)

Mariotte

Mariotte suggested measuring velocities of colliding bodies (m, M) by making them bobs of pendulums (Fig. 1.9). It was known from Galileo that the height of elevation of a body thrown upward is a function of its initial velocity alone, and that the relationship between velocity and height is quadratic, or conversely a square root proportion:

$$h \propto v^2 \left(\text{or } v = \sqrt{2gh} \right).$$

This relationship allowed researchers to easily measure the velocity of the colliding bodies by measuring the heights to which they rose following the impact.

However, this was not enough by itself. For the cases in which one of the bodies cannot be converted into a pendulum, Mariotte described an apparatus we call today

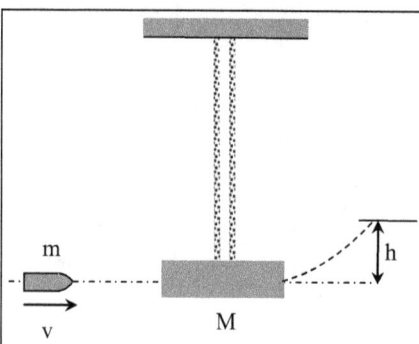

Fig. 1.10 Schematic representation of ballistic pendulum

ballistic pendulum (Fig. 1.10). In it, the pendulum bob is a massive cylinder (M) which is struck by a small body (a bullet, m) approaching at high speed and penetrating into the cylinder. Again, the height that the cylinder rose (h) was used to infer the velocity of the bullet that caused the collision. Again, it appears that there is a quadratic,[31] or square root, proportion between the speed of the bullet and the height of pendulum rise:

$$h \propto v^2 \left(v = \sqrt{2\frac{M+m}{m}gh} \right).$$

The First Step: Inelastic Collision

The success of the ballistic pendulum brings us to the fact that the first success in the experimental account of collisions was achieved by addressing the entirely *inelastic* collision that was ignored by Descartes. In 1668, the new results were

---

[31] The physical account is now changed from the elastic collision of Fig. 1.9. This type of collision is in the following section.

submitted to the Royal Society by John Wallis (1616–1703),[32] the eminent English mathematician and physicist. He also employed the concept of quantity of motion. Yet, unlike Descartes, Wallis assigned to it positive and negative values: both velocities were considered positive when the two bodies moved in the same direction, but one was considered positive and one negative, when the colliding bodies moved in opposite directions.

Wallis

Wallis defined *perfectly hard* bodies as those that did not "yield" (deform) on impact, *elastic* bodies, which yielded on impact, but spontaneously regained their original shape (as a spring), and *soft* bodies, those that deformed on impact and did not recover, remaining deformed. The latter may stick together in the collision and proceed afterwards to move as a single body.[33] It is regarding this case that Wallis established the rule for the resultant velocity (u), shared by the bodies with masses M and m[34] that initially moved with velocities V and v (Fig. 1.11):

$$u = \frac{MV \pm mv}{M + m}$$

This result displayed the conservation of the quantity of motion (momentum) before and after the impact:[35]

$$MV \pm mv = u(M + m)$$

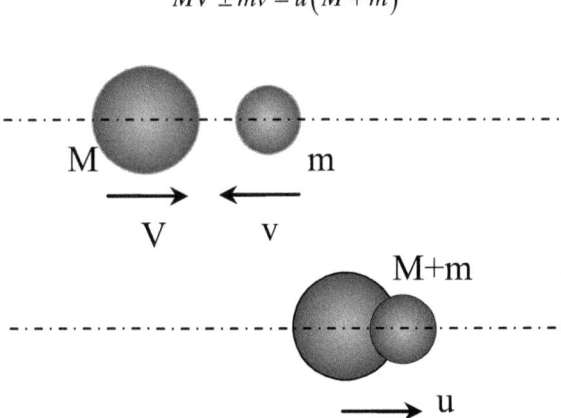

Fig. 1.11 Schematic representation of an inelastic collision of two balls

This was a significant progression since, for Wallis, the quantity of motion became an *algebraic quantity*—a step toward the vector quantity as we use it today.

---

[32] Wallis (1670), Mach (1883/1919).

[33] Contrary to Descartes' predictions, this never happens to elastic bodies.

[34] In fact, Wallis used weights and not masses to formulate his results. Only Newton separated the concepts of weight and mass. However, technically, this does not invalidate the empirical results of Wallis.

[35] Dugas (1988, pp. 172–175).

Moreover, the area of validity of momentum conservation expanded. Wallis formulated the rule in terms of "weights," instead of the "bigger" and "smaller" of Descartes. Several years later Newton distinguished *mass* and *weight* as different concepts and used the term *inertial mass* for the quantity of matter.

The Second Step: Elastic Collision

Wallis also addressed the elastic collisions of hard or elastic bodies. However, he only described a single case of such collisions: two equal bodies approaching each other from opposite directions (Fig. 1.12). In such a case, the two bodies rebound and exchange velocities: Descartes' first rule of collision was confirmed. Further progress in the empirical account of the elastic collisions with different velocities was due to Christopher Wren (1632–1723)[36] in London, in 1669.[37] Their experiments (of Wallis and Wren) were replicated and extended by Mariotte in France, in 1677.

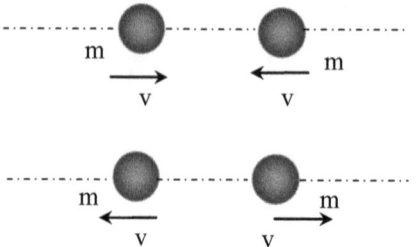

Wren

Fig. 1.12 Schematic representation of an elastic collision of two identical balls

However, it became clear that the empirical account was not sufficient. Descartes' theoretical approach remained in the background. The scholar who surpassed Descartes in this direction was Christian Huygens.[38] He applied Galileo's newly introduced *principle of relativity* to the rule which governs elastic collisions.

Huygens started from the same known case: two equal bodies approach each other at equal speeds.[39] Following the collision, they recede at the same speeds and exchange velocities. This result was postulated by Huygens and called by him *hypothesis*.

Huygens

---

[36] Christopher Wren was celebrated as the architect of London of that period, with its wide-scaled reconstruction. Being a cultured person with broad intellectual interests, he also contributed to physics.

[37] Wren, Ch. (1669), Mach (1883/1919, pp. 314–317).

[38] Christian Huygens—an outstanding physicist, one of the founders of modern science and classical mechanics in the seventeenth century.

[39] Huygens (1700/1977), Mach (1883/1919, pp. 314–317).

Fig. 1.13 A drawing from Huygens' treatise *De Motu Corporum Ex Percussione* of 1700

Then Huygens postulated that all the claims regarding collisions should be considered relatively: the motion of one body should be addressed with respect to another considered as being at rest.[40] This was a breakthrough. Based on this, Huygens returned to the case of the two equal bodies colliding with equal speeds and imagined that event on a boat, from the viewpoint of a sailor (Fig. 1.13). For the sailor, the bodies rebound at the same speeds. Huygens then asked about the same event as it appeared to a person standing on the bank. If the boat moves at the same speed v, Huygens stated, the person on the bank observes the collision of a body at rest with another body moving at double speed $2v$ (not simple to imagine as it addresses a single instant). The result of the collision was that the body at rest moved at the velocity $2v$ and the colliding body stopped.

Huygens deduced his *Proposition I*:

> *If a body is at rest and an equal body collides with it, after the impact the second body will be at rest and the first will have acquired the velocity that the other had before the impact.*[41]

It was a theoretical(!) refutation of the third rule of Descartes, as presented above. Huygens further developed the initial case of symmetrical collision to a more general case of unequal bodies of a certain ratio between their velocities and masses. He arrived at the following proposition (*Proposition VIII*).[42]

> *If two bodies moving in opposite directions with speeds inversely proportional to their magnitudes collide with each other, each one rebounds with the speed that it had before the impact.*

This more general result theoretically refuted Descartes' sixth rule and matched the established empirical fact.

From our point of view, it is easy to see that this claim is coherent with the principle of momentum conservation. Indeed, if M and m are the masses of the colliding bodies which approach each other with velocities u and v (Fig. 1.14), and, as stated in the proposition, their masses relate as:

$$\frac{M}{m} = -\frac{v}{u} \quad \text{or} \quad Mu + mv = 0$$

---

[40] It might look like a trivial statement to us; however, it is this unconditional relativity of motion that does away with the medieval concept of *impetus* often understood as a sort of absolute feature implanted into a moving body—a "charge" of motion.

[41] Dugas (1988, p. 177). Quotations from Huygens' *De Motu corporum ex percussione* (1700).

[42] Dugas (1988, p. 178).

Then, the general claim of momentum conservation:

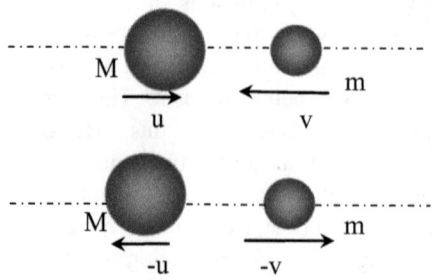

$$Mu + mv = Mu_1 + mv_1$$

is satisfied exactly, where the velocities u_1 and v_1 after collision are related as follows:

$$u_1 = -u \quad \text{and} \quad v_1 = -v,$$

as can be checked by direct substitution. Like Wren, who confirmed this rule experimentally, Huygens termed such velocities—*proper velocities*.[43]

Fig. 1.14 Schematic representation of an elastic central collision between two unequal balls

Huygens' *Proposition XI* is of special interest, for a collision between hard bodies (elastic collision):[44]

In the mutual impact of two bodies, the sum of the products of the masses into the squares of the respective velocities is the same before and after impact.

If we rewrite this new claim in the modern symbolic form, we arrive at:

$$Mv^2 + mu^2 = Mv_1^2 + mu_1^2$$

In fact, this is the claim of conservation of the quantity, later introduced by Leibniz[45] and termed vis viva (the living force), for the collision of two hard (elastic) bodies. Furthermore, this was an important step in the long road of physics toward the principle of energy conservation. Therefore, the treatment of elastic collisions actually led to the concept which is now at the foundation of physics—*energy*. In this context, it was *mechanical energy*.

As was already established by Wren empirically, a collision of any two bodies could be characterized by the change of their relative velocities in the impact, regardless of their masses. If the initial velocities of the bodies were, say, u and v, and the terminal velocities were, say, u_1 and v_1, then we should consider the ratio:[46]

$$e = \frac{v_1 - u_1}{u - v}$$

Clearly, then, when the bodies collide softly, $v_1 = u_1$ and the ratio nullifies ($e = 0$)—the bodies stick together. The experiment showed that the magnitude of e could at most approach one ($e = 1$), for the case of very hard metal or bone balls.

---

[43] For the way in which Huygens himself demonstrated this proposition, see Dugas (1988, p. 179).
[44] Wolf (1968, p. 233).
[45] Leibniz (1695/1968, pp. 29–31).
[46] Taylor (1941).

This is the case of *elastic collision*. Therefore, the rule obtained for the elastic collision of two bodies was:

In an elastic collision, the relative velocity of the two colliding bodies reverses:

$$u - v = -\left(u_1 - v_1\right)$$

For all intermediate cases of collision, the value of e (justifiably termed as the *coefficient of restitution*) must lie in the interval between 0 and 1:

$$0 < e < 1$$

Let us obtain Huygens' result: Consider the elastic collision, then:

$$v_1 - u_1 = u - v \quad \text{or} \quad v + v_1 = u + u_1$$

At the same time, the conservation of momentum implies:

$$Mu + mv = Mu_1 + mv_1 \quad \text{or} \quad M\left(u_1 - u\right) = -m\left(v_1 - v\right)$$

Multiplication of the last two equations yields:

$$M\left(u_1^2 - u^2\right) = -m\left(v_1^2 - v^2\right)$$

Or, after rearranging the terms, one obtains:

$$Mu^2 + mv^2 = Mu_1^2 + mv_1^2$$

So, Huygens succeeded in arriving at the rule which characterizes elastic collision in addition to momentum conservation—the claim of conservation of the kinetic energy. The exact statement should, of course, include coefficient 1/2 in all terms, but this does not change the equation regarding the magnitudes of its terms.

Historically and philosophically, the importance of this result goes beyond the claim of a conservation of specific quantity. This result elegantly resolves the conceptual debate between Leibniz and Descartes regarding *the* quantity to be adopted as the "true" characteristic of motion. Descartes (as well as Newton) used the quantity of motion—mv—whereas Leibniz argued for vis viva—mv². The elastic collision demonstrates that both quantities—energy and momentum—are essential characteristics of motion.

Final Refutation: Newton

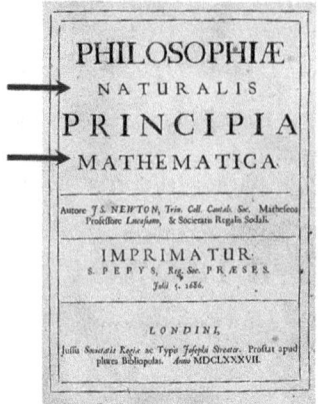

Fig. 1.15 The title page of Newtonian *Principia* (1687). If the words pointed to by arrows are dropped, one obtains the title Descartes' *Principles*

The final word in the debate on Descartes's mechanics of motion belongs, of course, to Newton. As a teenager at the University of Cambridge, he thoroughly studied every word in Descartes' *Principles of Philosophy*, copying them into his notebooks and making his own notes. Newton knew, of course, about the refutation of Descartes' mechanics through the accounts of collisions by Wallis, Wren, Huygens, and Mariotte.[47]

The qualitative approach of Descartes could not produce any demonstration of Kepler's laws—precise mathematical statements regarding the motion of planets. That, alongside the debates with Hooke in the Royal Society, stimulated Newton, bringing him to the fundamental treatise of physics culture— *The Mathematical Principles of Natural Philosophy*—Newton's answer to Descartes' *Principles of Philosophy*. Newton clearly began the debate from the title itself: **Mathematical** *Principles* instead of the *Principles*, and **Natural** *Philosophy* instead of *Philosophy* (Fig. 1.15).

Newton started his treatise with presenting his Laws of Nature, which came to replace Descartes' laws. We juxtapose them for comparison:

Laws of nature in Descartes' *Principles*		Laws of nature in Newton's *Principia*[48]	
	*The first law of nature: that any object, in and of itself, always perseveres in the same state; and thus, what is moved once always continues to be moved*	*Law I. Every body perseveres in its state of being at rest or of moving uniformly straight forward except insofar as it is not compelled to change its state by forces impressed*	
The second law of nature: that every motion of itself is rectilinear; and hence what is moved circularly tends always to recede from the center of the circle it describes		*Law II. A change in motion is proportional to the motive force impressed and takes place along the straight line in which that force is impressed*	
Third law: That a body, in colliding with another larger one, loses nothing of its motion; but, in colliding with a smaller one, loses as much as it transfers to that one		$$\Delta(mv) \propto F\Delta t$$ *Law III. To any action there is always an opposite and equal reaction; in other words, the actions of two bodies upon each other are always opposite in direction*	

---

[47] Newton (1687/1999, p. 427).

[48] Newton (1687/1999, pp. 416–417).

The changes introduced by Newton to the Laws of Descartes were essential.

1. Descartes' first law claims that a body left to itself remains in the same state. The important innovation was giving motion the status of a natural state. The form of motion preserved by a body left to itself—rectilinear motion—is stated in the second law of Descartes. In his second law, Descartes deprived circular motion of being a natural state—a historic change. In a sense, Newton's first law corresponds to the first two laws of Descartes. Newton's law relates the change of state of motion, explicitly specified (rest or uniform rectilinear motion), to the impressed force. While Descartes' law ignores forces and does not specify the reason for the state of change, Newton's first law states the tendency of a body to preserve its state *insofar as it is not compelled to change its state.* Unlike Descartes, for Newton, the states of uniform motion and rest were physically equivalent. Newton made this paradigmatic shift of mechanics from Aristotelian and medieval physics, and it became the central feature of classical physics.

2. In his second law, Newton specified the impact of the impressed force on the motion of a material body over a certain period of time—an integral law which refined the claim of the first law quantitatively.[49] It provided a tool for a quantitative account of motion of material objects under any force action. Descartes addressed the change of motion state only as following collision with other bodies. He deals with it in his third law.

3. The third law of Descartes was totally rejected as incorrect with its fundamental fallacy being in regarding the interaction of bodies as an asymmetrical process (in terms of "winners" and "losers"). In contrast, the third law of Newton stated a complete symmetry of action-reaction. The symmetry of action-reaction constituted a great advancement reached by Newton in contrast to all previous scholars. Unlike the first two laws, which were axioms, the third law of Newton was proved by him, drawing on the second law applied to a closed system of bodies.[50] The type of forces applied to bodies was not specified which allowed contact forces (as in collision) but also the forces acting over distance (as gravitation).

The conservation of the quantity of motion was essentially changed. The erroneous distinction between the magnitude and direction of the quantity of motion was replaced with a *vectorial* quantity (momentum). The statement of conservation of quantity of motion of a closed system became a direct inference of the second law, as was stated in Corollary III:[51]

> *The quantity of motion, which is determined by adding the motions made in one direction and subtracting the motions made in the opposite direction, is not changed by the action of bodies on one other.*

---

[49] After Newton, Euler modified the second law to its differential form ($F = ma$). The original form of the first law was also changed, firstly, to its simplified version of zero net force and, in the twentieth century, to the claim of a state of preservation in an inertial frame of reference (Chap. 2, Galili & Tseitlin, 2003).

[50] Newton (1687/1999, pp. 427–428).

[51] Newton (1687/1999, p. 420).

Unlike Descartes' laws, Newton's laws quantitatively corresponded to all available experience with natural phenomena and mechanical contrivances. These laws were, therefore, unanimously adopted as a fundamental theory of mechanics. All the empirical results of Wallis and Wren regarding collisions were incorporated into the theory of Newtonian mechanics.

Huygens' results regarding conservation of vis viva (kinetic energy) later entered the mainstream of physics. Although not introduced by Newton, the concept of kinetic energy presented its natural development contributing to a full account of motion and the change of its states under the impact of exerted forces.

Questions for Reflection

1. Present the account of collisions of hard bodies that was introduced by Descartes.
2. Critique the rules of collisions suggested by Descartes.
3. Justify the name for the coefficient $e = \dfrac{v_1 - u_1}{u - v}$ as the coefficient of *restitution*.
4. Exemplify the violation of the principle of relativity in Descartes' rules of collisions.
5. Exemplify the violation of the Newton's third law in Descartes' rules of collisions.
6. List and discuss the differences between Descartes and Newton's laws of motion.
7. Compare the title pages of treatises by Descartes and Newton. Justify the differences.

Philosophical Perspective

Descartes' Knowledge of Mechanics

Physics incorporates several fundamental theories, few in number, but very powerful, inclusive and useful, and each of them creates a unique picture of the world. Classical mechanics is one of them. To better understand these pictures, their features and scope, one should look into their creation and conceptual consolidation. In doing this, we, as readers in the future, have the benefit of hindsight, knowing much more than the pioneers knew about nature from the distant past. We possess a richer perspective enabling us to grasp the meaning of specific accounts of the phenomena which usually have been much refined and developed since then.

The world picture of Classical Mechanics established by Newton placed all material bodies in the all-inclusive empty space and time stream. These bodies influence each other through central forces causing changes of their states. Mechanics asks about the laws and principles which govern the movements and changes of objects, and how exactly bodies influence each other. Scholars considered two major options in this regard: *action at a distance* (objects influence each other through space) and *action at a contact* (objects influence each other by touch or collision).

We have considered one of the major fragments of the emerging mechanics—the account of collisions which served as a model of interactions of material bodies.[52] Within this progress of eliciting the rules of collisions, scholars revealed the conceptual underpinnings of a much wider scale—the principles which govern motion. In this regard, Descartes elicited the following:

1. The *rectilinear [uniform] motion* establishes a natural *state* of matter and is preserved (the inertial motion) unless other bodies cause its change.
2. Bodies interact through collisions in which the total *quantity of motion is conserved*.

Descartes claimed that these principles fit a general arrangement of reliable knowledge, which is (should be) hierarchical, that is, organized from claims that are more general to more specific. Here is how one may represent Descartes' arrangement of thought and knowledge (Fig. 1.16).[53]

As we know, Newton modified many statements of Descartes which showed that the proclaimed "distinct" and "clear" features of claims though very appropriate for formulation of features of Nature do not guarantee their correctness. In fact, Descartes appealed to common sense, as well as did Aristotle and… failed in many of his statements. Indeed, the equivalence of motion and rest stated by Newton is not obvious and contradicted the commonsense of Descartes, but this is what happened

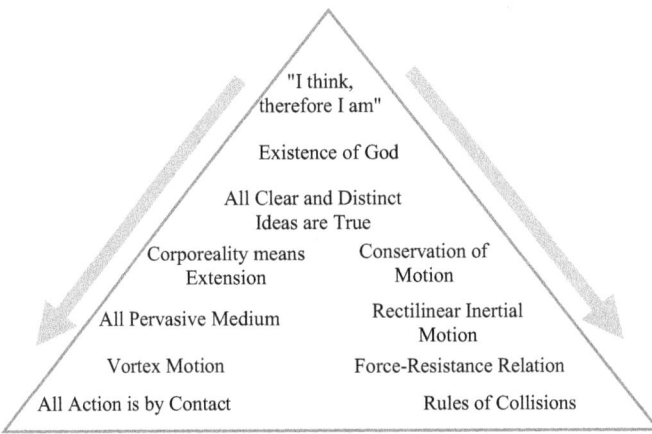

Fig. 1.16 Conceptual pyramid of Descartes. Arrows signify the direction of hierarchy from the more general

---

[52] Westfall (1989).

[53] Not all the items shown here were discussed. The template of this representation was adopted from Losee (1993, p. 76). We have included the elements discussed in this excursus.

to the truth of classical mechanics. Similarly, there is little obvious in the *quantity of motion* as the product of mass and velocity (instead of mass and speed in Descartes), or in the symmetrical nature of interaction (instead of "strong against weak" in Descartes), but they are correct even if not obvious, as may be required by experience.

Plato (Rafael, 1509)

Descartes was distinctive in combining his skills as a philosopher and as a physicist. He demonstrated this in his argument for the laws of motion. Descartes believed that philosophical principles can undermine the evidence of the observed and therefore should be fundamental. Physics must subdue these principles and deduce physical laws from them (Fig. 1.16). This approach is known as *rationalism*. Plato is considered its founder and Euclidean geometry—its illustration: several definitions and axioms serve the nucleus of a theory, while its body are the theorems developed through logical deduction. In his *Principles*, Descartes tried to act on this and deduce rules of collisions from the stated laws of motion. His failure illustrates the essential difference of physics theory from mathematical theory. Physics requires empirical evidence. In philosophy, this claim corresponds to *empiricism*, which might be related to Aristotle, another Greek philosopher whose scheme of knowledge production beyond a mere deduction took the form of the *Inductive-Deductive* cycle (Fig. 1.17). It became fundamental in philosophy, being refined through the history of science.

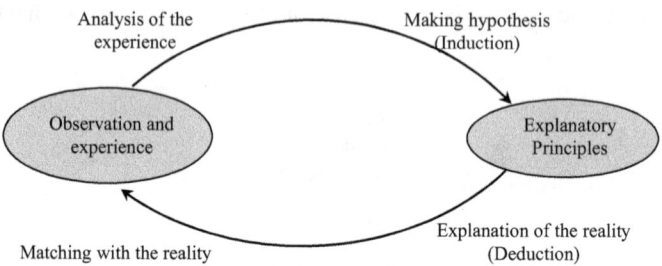

Fig. 1.17 Inductive-deductive cycle of the scientific method by Aristotle. Descartes neglected the element of "matching with the reality" and violated this procedure of knowledge creation

Aristotle (Rafael, 1509)

Ignoring empirical evidence did not mean that all claims of Descartes were false. Some of them were of crucial importance for classical mechanics: rectilinear motion as a state, conservation of the quantity of motion. Yet, ignoring empirical evidence caused their deficient form and clearly mistaken inferences which actually prevented their practical application beyond philosophical claims. His method was inferior even to Aristotle who saved place for self-correction of the produced claims.

Descartes specified his method of knowledge creation being explicit regarding its components[54]:

*I should have here shortly explained wherein all the science we now possess consists, and what are the degrees of wisdom at which we have arrived. The **first** degree contains only notions so clear of themselves that they can be acquired without meditation; the **second** comprehends all that the experience of the senses dictates; the **third**, that which the conversation of other men teaches us; to which may be added as the **fourth**, the reading, not of all books, but especially of such as have been written by persons capable of conveying proper instruction, for it is a species of conversation we hold with their authors...*

Descartes lists the four reliable ways to knowledge: (1) the clarity (obviousness) of notions, (2) the correspondence to sense experience, and (3, 4) awareness of the documented and ongoing discourse with other researchers. Nevertheless, to all these, Descartes preferred the fifth way which he saw as the best: revealing the *first causes and true principles* through illumination from which the specific claims are produced by deduction:[55]

Two considerations alone are sufficient to establish this--the first of which is, that these principles are very clear, and the second, that we can deduce all other truths from them; for it is only these two conditions that are required in true principles.

Descartes did not save words and did not exaggerate in modesty. Facing discrepancy between his claims and reality, he nevertheless insisted on the superiority of his philosophy to any practical aspect:[56]

And the demonstrations are so certain that, even if experience seemed to show us the contrary, we would nevertheless be obliged to place more faith in our reason than in our senses.

This approach brought Descartes to his erroneous rules of collisions which could be saved by his reference to the absence of vacuum and influences of other bodies. Many years after, Einstein formulated the epistemological dictum of modern science. In addressing the scientific revolution of the seventeenth century, Einstein wrote:[57]

But before mankind could be ripe for a science which takes in the whole of reality, a second fundamental truth was needed, which only became common property among philosophers with the advent of Kepler and Galileo. <u>Pure logical thinking cannot yield us any knowledge of the empirical world; all knowledge of reality starts from experience and ends in it.</u> Propositions arrived at by purely logical means are completely empty as regards reality. Because Galileo saw this, and particularly because he drummed it into the scientific world, he is the father of modern physics – indeed, of modern science altogether [the emphasize added].

These words of Einstein summarize the epistemological lesson that we may learn in studying Descartes in his exploration of the physics of motion and collisions.

---

[54] Descartes (1644/1983), Letter from the Author to the Translator of this Book, pp. XVII–XVIII.

[55] Descartes (1644/1983, p. XXI).

[56] Descartes (1644/1983, p. 69).

[57] Einstein (1934/2011).

Questions for Reflection

- Discuss the ways in which Descartes's mechanics was different from classical mechanics.
- Demonstrate the violations of rest-motion equivalence by Descartes (in the rules of collisions) from the perspective of the principle of relativity.
- Detect and discuss the ways by which Descartes' argumentation is different from those made common in science. What do we share with Descartes nevertheless?
- Discuss the advantages and disadvantages of the Aristotelian method of inductive-deductive circle in comparison to the modern scientific method.
- Are there any advantages in the rationalistic approach over the empiricist one and vice versa? What could be the reason for preferring one to the other? Exemplify your answer.

Educational Benefits

Descartes' theory of motion and collisions has been falsified by Newton's classical mechanics, and one may ask whether there is any value in addressing such a theory. A number of educational benefits for dealing with it in educational context may be mentioned.

- Firstly, Descartes' *Principles of Philosophy* is one of the most influential texts of Western culture. A few, easily understood, fragments of the text have been selected here and are parallel in content to the school physics curriculum. Through them, students may be introduced to the fundamentals of classical physics established in the scientific revolution.
- Within the perspective on physics knowledge as a culture, the theory of Descartes belongs to the periphery of classical mechanics.[58] Descartes' claims contrast the major ideas of Newtonian mechanics and reveal the *dialogical* nature of scientific knowledge, thereby facilitating its meaningful learning.
- Descartes demonstrated the rationalist epistemology. This approach does not fully represent the *scientific method*. By criticizing Descartes, we emphasize the importance of complementing the rational approach of Descartes and Huygens with the empiricist approach of Wallis and Wren as applied to the same subject of investigation.
- Among the conceptual shortcomings of Descartes was the scalar nature of momentum, mass-speed product. Revisiting this mistake emphasizes the correct knowledge: mass-velocity product.
- Descartes erroneously interpreted collision as a "competition" between the bodies to impel a change of motion on the other. This is a common misconception among students which could be treated through a critique of Descartes.
- Our excursus encourages distinguishing between *momentum*, always conserved in collisions, and *kinetic energy*, conserved only in elastic collisions.[59]

---

[58] Tseitlin and Galili (2005).

[59] Students' difficulties reported in Grimellini-Tomasini et al. (1993) and Sasson (2006).

- The critique of Descartes by Huygens essentially draws on the relativity principle of Galileo—a central principle of physics. In this, Huygens was ahead of his time. Students very seldom use the relativity principle. However, Huygens' account of collisions provides such an opportunity.
- Arguing for the deductive method is often exemplified by mathematical theories. Considering the attempt of Descartes to apply the same approach to physics phenomena illustrates the different natures of the physical and mathematical domains.
- Finally, considering Descartes' theory of collisions may illustrate the need for modesty in making scientific claims, even on the part of very smart people. Despite his great intellectual power, many of Descartes' physical claims were simply wrong. This may serve as a lesson to all students of science—and beyond.

The Media for Learning

The popular apparatus demonstrating the collision of hard objects is known as "Newton's cradle" (Fig. 1.18). It is comprised of several identical pendulums suspended in a line and touching each other. It is reminiscent of the Mariotte apparatus of two colliding balls used to measure the velocities of the colliding balls (Fig. 1.9). This simple device offers amusing opportunities for experimenting.

After several trials, one discovers the rule: the same number of balls that were initially raised up from one side of the line rise up at the opposite end, almost to the same height. This result is, of course, in accordance with the conservation of momentum and kinetic energy.

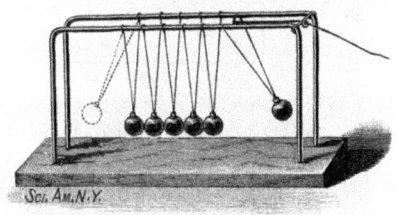

Fig. 1.18 Newton's cradle

The whole phenomenon seems to be similar to the ball falling from a height and colliding with an equal ball at rest: the striking ball stops and the struck ball proceeds with the velocity of the first. One might think that the balls in the middle, which remain at rest, do not participate at all and merely transfer the momentum. To discover the true scenario of this event, one might tie together the balls in the middle, those that seem not to be participating in the events. For example, if the line has six balls (Fig. 1.18) and we raise one ball on the left, a single ball will rise up at the right end of the line, whereas the first ball stops. If now we tie together the four balls in the middle, the result of the impact of the first ball with the line will be very different—all the balls will start to swing after the impact. Careful observation and additional thought may reveal that in the case of "free" suspended balls, a series of collisions takes place, each time between two adjacent balls. The collision event between adjacent balls travels very quickly—a series of subsequent collisions—along the line until the ball at the right end rises up. The succession of collisions is fast enough to be missed through mere observation, when not knowing what to look for.

The story of the Czech scientist Jan Marci (1595–1667) from Prague might be interesting. He presented the same phenomenon in an attractive context. Marci was among the first researchers who tried to study collisions preceding Wallis and Wren in the London of 1668. In his experiment, Marci fired a cannon ball that struck an identical ball placed on a stone table (Fig. 1.19). After the impact, the first ball stopped and the second ball started to move ahead in the direction of the original shot. Although this demonstration was nothing else but a repetition of the collision of pendulum bobs in the cradle, the context of a cannon ball that com-

Jan Marci

pletely stopped(!) and other ball that "did not move" took off, greatly impressed the observers by its anti-intuitive result.

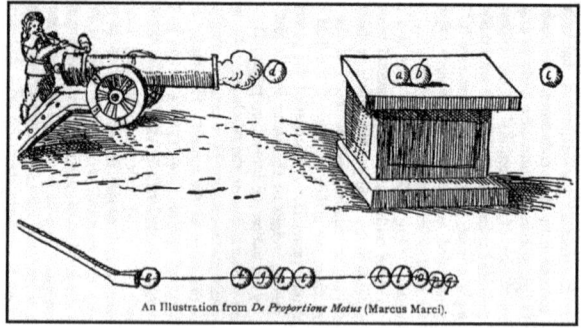

An Illustration from *De Proportione Motus* (Marcus Marci).

Fig. 1.19 Illustration from *De Proportione Motus* by Marcus Marci. A cannon ball strikes an identical ball at rest. Marci notified the positions of the two balls before (*d*, *b*), during (*a*, *b*), and after collision (*a*, *c*). In the lower drawing, Marci apparently addressed the experiment of a ball striking a line of identical balls (From Mach, 1883/1919)

Another interesting story related to collisions deals with the stirrup used by a horse rider. A stirrup is a light frame or ring attached to the saddle which holds the foot of a rider. Leaving aside the purpose of a stirrup for guidance and maneuvering the horse, one may proceed to the role of the stirrup in medieval warfare, in particular, the role of the stirrup in the assault of the mounted knight who tries to knock his rival off his horse with a lance during a joust or contest (Fig. 1.20a).

(a)

(b)

Fig. 1.20 (a) Jousting—horse riders competition in which they clashed striking each other with a pole. The illustration of the sixteenth century by Paulus Hector Mair—*De arte athletica*. (b) Stirrup (c) Battle of Hastings in 1066, as depicted on the medieval Bayeux Tapestry

(c)

To appreciate the advantage of the rider using paired stirrups, one should compare the collision of a rider without stirrups (allowing the sliding of the rider over the back of the horse) with a rider using them (preventing such sliding). The analysis of the collision draws on the conservation of momentum. The implication of this understanding for history may shed light, for instance, on the victory in the famous Battle of Hastings in 1066 (Fig. 1.20c) that changed the history of England.[60] One may realize the reason for the effectiveness of the cavalry of the Normans, who, unlike their rivals, used stirrups. This made the rider "one body" with his horse, providing a crucial advantage over the much lighter rider not firmly connected to his horse.

Historians argue that stirrup was one of the basic tools used to create and spread modern civilization, similar to the impact of the wheel or lever (Fig. 1.21).

(a)   (b)

Fig. 1.21 (**a**) Ancient statue of Roman Emperor (Marcus Aurelius—second century) in Rome—no stirrup; (**b**) statue by Verrocchio of an officer (Bartolommeo Colleoni—1488) in Venice—stirrup is shown (Photo credit: Naya Carlo, 1816–1882)

---

[60] Hannam (2009).

The tomb of Descartes (with detail of the inscription), in the Abbey of Saint-Germain-des-Prés, Paris

Newton's tomb in Westminster Abbey in London

Picture Credits

- Trinity College. Photo by Rafa Esteve. CCAttribution-Share Alike 4.0 International license. https://commons.wikimedia.org/w/index.php?curid=51077381
- Portrait of Descartes after Frans Hals. Photo by André Hatala. Public Domain. https://commons.wikimedia.org/w/index.php?curid=2774313
- Portrait of John Wallis by Godfrey Kneller, the National Portrait Gallery, London. Public Domain. https://commons.wikimedia.org/w/index.php?curid=6365975
- Portrait of Christopher Wren by Godfrey Kneller, the National Portrait Gallery, London. Public Domain. https://commons.wikimedia.org/w/index.php?curid=203683
- Portrait of Christiaan Huygens by Caspar Netscher, Kunstmuseum Den Haag. Public Domain. https://commons.wikimedia.org/w/index.php?curid=44047
- Portrait of Isaak Newton by Godfrey Kneller, 1689. Public Domain. https://commons.wikimedia.org/wiki/File:Sir_Isaac_Newton_(1643-1727).jpg
- Title page of Descartes' Principia, 1685, The BEIC digital library. Public Domain. https://commons.wikimedia.org/w/index.php?curid=50359310
- Illustrations from the book by Descartes, 1644, Louis Elzevir, Amsterdam, pp. 56, 60. Public Domain. https://www.wdl.org/en/item/3157/view/1/5/
- Photo of Emma Noether Free use https://commons.wikimedia.org/w/index.php?curid=42894188
- Fig. 1.7b Games with equal balls. Photo by the author.
- Fig. 1.7b Games with equal balls collisions popular in France. Photo by Darksidex—Flickr, CC BY-SA 2.0 International license. https://commons.wikimedia.org/w/index.php?curid=1686057
- Title page of Marriotte's book, 1717, by Edme Mariotte. Public Domain. https://commons.wikimedia.org/w/index.php?curid=2565164
- Portrait of Edme Mariotte by unknown author. Public Domain. https://commons.wikimedia.org/wiki/File:Edme_Mariotte.png

- Fig. 1.13 Illustrative drawing by Huygens from *Mechanics* of Mach, 1919, p. 314
- Fig. 1.15 The front page of Newtonian *Principia*, 1687. Public Domain. https://commons.wikimedia.org/w/index.php?curid=2681838
- Portraits of Plato and Aristotle by Rafael, 1511. Fragments from the fresco in the Vatican. Public Domain. https://commons.wikimedia.org/wiki/File:Sanzio_01_Plato_Aristotle.jpg
- Fig. 1.18 Illustration from S. Chapman, the *American Journal of Physics*, 1941, Vol. 9, pp. 357–369.
- Portrait of Jan Marci by unknown author, illustration from E. Mach (1883/1919), Mechanics, p. 306 Public Domain.
- Fig. 1.19 Illustration from E. Mach (1883/1919), Mechanics, p. 307. Public Domain.
- Fig. 1.20a Illustration from the book, P. H. Mair, De arte athletica II, 1540. Public Domain. https://commons.wikimedia.org/w/index.php?curid=15849932
- Fig. 1.21b Stirrup from a monument in Florence. Photo by the author. Public domain.
- Fig. 1.20c Bayeux Tapestry, 1066. Public Domain. https://commons.wikimedia.org/w/index.php?curid=684174
- Fig. 1.21a Monument of second century by anonymous author, Rome. Photo by Jean-Pol Grandmont. Public Domain. https://commons.wikimedia.org/w/index.php?curid=25778597
- Fig. 1.21b Monument of 1488 by Verrocchio, Venice. Photo by Carlo Naya, nineteenth century. Public Domain. https://commons.wikimedia.org/w/index.php?curid=5835633
- The memorial of Descartes in the Abbey of Saint-Germain-des-Prés, Paris. Photo by I, PHGCOM, CC Attribution-Share Alike 3.0 International license. https://commons.wikimedia.org/wiki/File:DescartesAshes.jpg
- Photo of Newton's tomb in Westminster Abbey in London. Public Domain. https://he.m.wikipedia.org/wiki/%D7%A7%D7%95%D7%91%D7%A5:Isaac_Newton_grave_in_Westminster_Abbey.jpg

References

Banham, G. (2009). *Descartes' Kinematics. Parallax, 15*(2), 69–82.

Descartes, R. (1644/1983). *Principles of philosophy*. D. Reidel.

Dugas, R. (1988). *A history of mechanics*. Dover.

Einstein, A. (1934/2011). Address at Columbia University, New York, January 15. In I. A. Einstein (Ed.), *Essays in science*. Open Road Integrated Media.

Galilei, G. (1613/1957). Second letter on sunspots. In S. Drake (Ed.), *Discoveries and opinions of Galileo*. Doubleday.

Galilei, G. (1638/1914). *Dialogue concerning two new sciences*. Dover.

Galili, I., & Tseitlin, M. (2003). Newton's first law: Text, translations, interpretations, and physics education. *Science & Education, 12*(1), 45–73.

Grimellini-Tomasini, N., Pecori-Balandi, B., Pacca, J. L. A., & Villani, A. (1993). Understanding Conservation Laws in Mechanics: Students' Conceptual Change in Learning about Collisions. *Science Education, 77*(2), 169–189.

Hannam, J. (2009). *Gods philosophers. How the medieval world laid the foundations of modern science*. Icon Books.

Huygens. (1700/1977). De Motu Corporum Ex Percussione. *Isis, 68*(4), 574–597.

Leibniz, G. W. (1695/1968). *Discourse on metaphysics*. The Open Court.

Losee, J. (1993). *A historical introduction to the philosophy of science*. Oxford University Press.

Mach, E. (1883/1919/1989). *The science of mechanics, a critical and historical account of its development*. Open Court.

Newton, I. (1687/1999). *The principia. Mathematical principles of natural philosophy* (B. Cohen & A. Whitman, Trans.). University of California Press.

Sasson, H. (2006). *Misconceptions and difficulties of high school students in understanding the concepts of impulse and momentum*. M.Sc. Thesis. The Hebrew University of Jerusalem.

Taylor, L. W. (1941). *Physics. The pioneer science*. Dover.

Tseitlin, M., & Galili, I. (2005). Teaching physics in looking for its self: From a physics-discipline to a physics-culture. *Science & Education, 14*(3–5), 235–261.

Wallis, J. (1670). *Mechanica Sive de Motu*. Typis Gulielmi Godbid.

Westfall, R. S. (1989). *Mechanical science in the construction of modern science* (pp. 50–64). Cambridge University Press.

Wolf, A. (1968). *A history of science, technology and philosophy in the 16th & 17th centuries*. Smith.

Wren, C. (1669). Lex Naturee de Collisions Corporum. *Philosophical Transactions, 3*, 867–868.

Chapter 2
De Motu: The History of the Understanding of Motion, from Aristotle to Newton

And the Lord said unto Cain, Where is Abel thy brother? And he said, I know not: Am I my brother's keeper? ... And the Lord put a mark on Cain, lest any who came upon him should kill him.

Genesis 4: 9, 14–16

Abstract This excursus reviews the conceptual basis of the classical theory of motion from Aristotle in Hellenic science, through the medieval theory of impetus, to the scientific revolution of the seventeenth century. The concept of impetus became central in physics after Aristotle and served as a mediator between Aristotelian and Newtonian mechanics. Familiarization with older theories provides the latter with both conceptual and cultural perspectives and encourages organizing knowledge while preserving a scientific discourse of ideas in science. Pioneers in the theory of motion were philosophers and enthusiasts of exploration of reality in an objective sense. Our depiction and analysis address not only the subject matter but also the employed epistemology, the type of evidence practiced, and the reasoning used—the method of science. We argue that understanding physical ideas is reached through comparison with older ideas, through acquaintance with the intellectual products of bright minds from the past who suggested other theories of motion and essentially contributed to scientific progress.

Theory of Motion

Why this epigraph?... Why these images?...

Science is a theory of nature. All natural objects are in motion, and so motion has become a subject of a special theory—*mechanics*. Unlike other theories, the title "mechanics" does not coincide with its subject—*motion*. This is because this fact was not fully understood—that mechanics is a theory of motion—until the seventeenth century when the understanding of motion became the core of the scientific revolution. Our epigraph suggests that this misunderstanding was then epitomized forever, and therefore students learn about motion in mechanics. Theories of motion cover the whole history of science from its start in the Hellenic physics of Aristotle[1] to the quantum and relativistic mechanics of the modern world, from the twentieth century on.[2]

---

[1] *Nike of Samothrace* by an anonymous Greek sculptor around the second century BC—this image of motion was produced at the time of the first scientific account of motion in the Hellenic culture.

[2] *Unique Forms of Continuity in Space* by Umberto Boccioni (1913)—the image was produced at the time of creation of quantum and relativist mechanics in modern times.

The theory of classical mechanics was established by Isaack Newton in his treatise *Philosophiæ Naturalis Principia Mathematica—Mathematical Principles of Natural Philosophy*—published in 1687. It started with stating the laws of motion. When we all learn the first law, we read[3]:

> *Law I. Every body perseveres in its state of being at rest or of moving uniformly straight forward except insofar as it is not compelled to change its state by forces impressed.*

We see here Newton addressing a body which "perseveres in its being *at rest or of moving uniformly straight forward.*" Newton juxtaposed the two states *at rest* and *moving uniformly straightforward* separated by *or* as if being not exactly the same. In the rest of this fundamental treatise, as well as in every modern textbook, we learn that the two aforementioned states are equivalent in physics: velocity is a relative entity. This implies that Newton did not need to specify *at rest*. However, he did. Why did he?

The answer is in the history of science. From the very beginning of physics in the fourth century BC, and up to the scientific revolution of the seventeenth century, scholars distinguished between *rest* and *motion* as essentially different concepts. In keeping with this background, one may interpret the text of the first law as a vestige of a certain commitment (the "Mark of Cain") and Newton's voice as one in the discourse regarding the nature of motion. Newton introduced his readers to the concept that was about to change. We are going to briefly reconstruct the major points of this discourse in order to understand the conceptual fundamentals of classical mechanics, which have since become the basis of every physics curriculum.

The First Theory of Motion: Aristotle

The first scientific theory of motion was produced by Aristotle[4] in his *Physics* (fourth century BC). This step was a great accomplishment, and all subsequent theories have drawn on it since this was where both subject matter and the conceptual agenda of the subject were determined. Aristotle defined *change* in general to be a subject for a physics inquiry.

Motion—the change of location in time, or, *locomotion*, in terms of Aristotle—was one of a small number of possible types of change taking place in the world. This part of physics was later termed—*mechanics*—for its treatment of the first simple devices—

Aristotle

machines. The mechanics domain focused on the motion of bodies, leaving aside all their specific characteristics (shape, quality, etc.). Yet, even under this apparently simplifying assumption, it was not a simple task to describe motion and to explain it.

---

[3] Newton (1687/1999, p. 416).

[4] Aristotle (384–322 BC)—one of the brightest minds ever, the Greek philosopher who made the fundamental contribution to the creation of science as the objective theory regarding Nature as a system of real objects organized according to a certain objective order—Cosmos.

Explanation in Science

It was important to Aristotle that in order to know and explain any investigation with regard to nature, the investigation should be organized within a general frame and the goal sought within a system of well-defined principles. He asked himself: "What does it mean to know something?"[5]

> Knowledge is the object of our inquiry, and people do not think they know a thing till they have grasped the 'why' (which is to grasp its primary cause). So, clearly, we too must do this as regards both coming to be and passing away and every kind of physical change, in order that, knowing their principles, we may try to refer to these principles each of our problems.

So, to explain a thing, one needs to find out and demonstrate the *causes* for it, from the knowledge of general principles. Firstly, he mentioned that awareness of the material that the object is constructed from is important to the understanding of the object itself. For example, if the object of our interest is a statue and it is made of marble, this explains why it is perceived by us as beautiful—the *material cause* of that feature. We find marble to be intrinsically beautiful. Then he stated that it is clear that the shape of things is also of importance. Thus, it is simple to agree that the shape of the statue is also the cause of our considering the statue to be beautiful. This constitutes the *formal cause* explaining this fact.

In the next form of explanation, Aristotle pointed to the "primary source" of the subject and the considered change. In the case of sculpture, this could be the special skill of the artisan who produced such an artifact. This special knowledge, or ability to create, served as the *effective cause* for the statue to be considered beautiful.

Finally, Aristotle mentioned that the goal for which the object was created was also important for its understanding, since it explains much of its nature. Knowledge relating to the purpose or function of the object also helps to explain the object. We acquire this knowledge when trying to answer how this or that feature fits into the considered goal, the overall system. This is the *teleological cause*. In the case of a statue, we may better understand the specific features of the statue when we understand its specific function, for example, if we know that the statue will be used to decorate a temple. So now we have four causes, and complete knowledge.

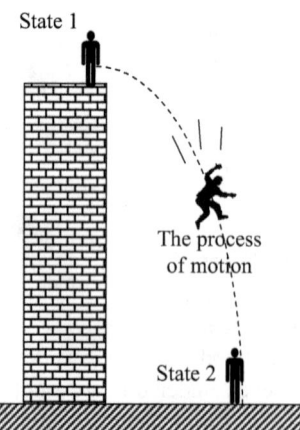

State 1

The process
of motion

State 2

Fig. 2.1 Motion as a process of transition between two states of rest

Explanation of Motion

Following this research agenda, Aristotle addressed the reality of motion. Motion was opposed to rest, the basic state of the existence of things. Motion

---

[5] Aristotle (1952a, Book II, part 3).

was perceived as a state of change, in particular, the change of location. It was therefore called *locomotion*. The state of rest, thought Aristotle, being natural, does not need any cause. Motion was considered as a process, transition from one state to another, the realization or actualization of a variety of causes (Fig. 2.1).

Aristotle distinguished two major types of locomotion: *natural* and *violent*. Each deserves separate treatment.

Natural Motion

Natural motion was related to the order of matter in the world of inanimate objects. Firstly, all material objects were considered to be comprised of four elements, which were naturally arranged in space in order of their weight or levity: from heavier down, becoming closer to the center of the universe, to lighter—upwards from the center. The elements were earth, water, air, and fire. Beyond the fire, celestial spheres start, being comprised of the fifth element—ether (Fig. 2.2).

When this order is violated by any object composed of different elements, a cause appears for its natural motion in order to restore the violated order, and the body seeks the compromise place subordinate to the natural environment.

Thus, it was understood that motion may take place for the *material* cause (such as when air is placed in water), the *formal* cause (such as when fire penetrates obstacles as it rises), and the *teleological* cause (to restore the violated order). Clearly, natural motion could be upward (the object is lighter than the environment) and downward (the object is heavier than the environment). In the case of a body comprised of several elements, the net intention of the components in comparison with the environment determines the resulting natural movement.

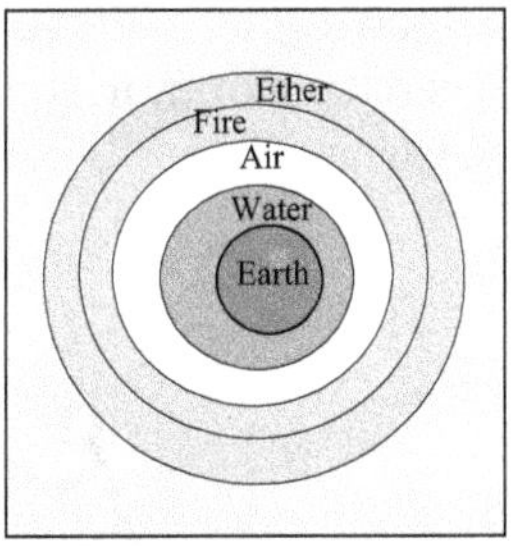

Fig. 2.2 The natural order of elements in the world in accordance to Aristotle

Weight, or gravity, as a faculty of an object, was determined by Aristotle in this context—as the tendency to restore the natural order, the measure of the intention to occupy the appropriate radial location in the universe. Absolute gravity was ascribed to earth and relative gravity—to water. As was common in the way the Greeks conceived reality in opposites, weight was introduced together with its opposite: levity—the tendency of light objects to ascend away from the center. Thus, absolute levity was ascribed to fire and relative levity—to air.

When motion starts, the moving object may change its place with different rapidity. The swiftness of the body is proportional to its weight and inversely proportional to the resistance of the medium. Aristotle did not introduce velocity or speed, neither did he present dependences of concepts in mathematical form. However, we may express such dependence of natural motion in the modern way, through providing a symbolic form of the functional dependence:

$$v \propto \frac{W}{R} \tag{2.1}$$

Here, v stands for velocity, W for the weight of the falling body, and R for the media resistance. The *falling* of things was considered paradigmatic of natural motion, its salient manifestation. The dependence of falling on weight and the resistance of the environment as represented by Eq. (2.1) make sense in everyday experience. Looking around, one sees a metallic ball falls faster than a leaf and the speed of falling decreases if one drops the metallic ball in water as compared to its falling through air. All such situations, at face value, support the account of falling by Aristotle.

Violent Motion

Any movement different from spontaneous vertical motion was considered *violent*. Unlike the natural motion, violent motion required an efficient cause—*pushing* or *pulling* the object. The origin of such effort was conceived as an agent or engine, external to the body and serving as its *mover*. The existence of an outside mover presents the principle of violent motion in Aristotelian dynamics.

Any lifting of heavy objects was considered as violent motion. Suppose two men lift a heavy barrel (Fig. 2.3a). The efficient cause of this motion is the men pushing the barrel (Fig. 2.3b). Addressing violent motion, Aristotle emphasized the external agent, the mover, in contrast with the natural motion which happens without force and agent.

Fig. 2.3 (**a**) Any lifting presents a violent motion. (**b**) To move an object, one needs a mover—the agent applying force and pushing the barrel along the inclined plane

As with regard to the natural motion, one may be interested how quickly the body changes its location—moves (we use for this velocity). With regard to violent motion, Aristotle related the swiftness of the body in direct proportion to the strength of the applied force, and in inverse proportion to the resistance of the medium or the weight of the body. We may express this reality by means that Aristotle never used:

$$v \propto \frac{F}{R} \text{ or } v \propto \frac{F}{W} \tag{2.2}$$

Here, v stands for velocity, F for the applied force, W for weight, and R for the resistance. In contrast with natural motion, the weight of the object resists its violent motion, as in moving object horizontally. These formulas hold only when F is bigger than *R* (or *W*). Only then, the force *overcomes* the resistance and causes motion. If not so, $F \leq R$ (or *W*), the body remains at rest ($v = 0$).

These formulas make sense in many everyday situations. It is not difficult to illustrate them. If one pushes the object harder, it moves faster. Consider an object on a rough surface. We may push or pull it. Our lesser effort will cause lesser speed than when we apply a greater effort causing a faster motion. Finally, if one applies the same force to bigger or smaller bodies, their speeds are in inverse proportion to their weights; to move a smaller object is easier.

Though often looking appropriate, this account was immediately countered by a challenging situation—the movement of projectiles. Where is the agent pushing the thrown stone after it loses its contact with the hand? What serves as a mover in this case? It was a great challenge to the Aristotelian theory of motion.

The Separation of Two Worlds

Observation of the world around brought Aristotle to the idea that it is organized in two areas, each in a different way, in particular with respect to motion. One area, our regular environment, extends from the center of the world to the orbit of the Moon (the sublunary world). In this area, the natural motion is vertical, that is, in the up-down direction, radially from the center of the Earth.

The second area is the celestial area, beyond the Moon (the supralunar world). There, in the heavens, which are comprised of ether, planets, and stars, move as being at their natural state.[6] The only motion conceived as possible to eternally preserve this was circular and uniform. Being natural excluded the need for any engine.

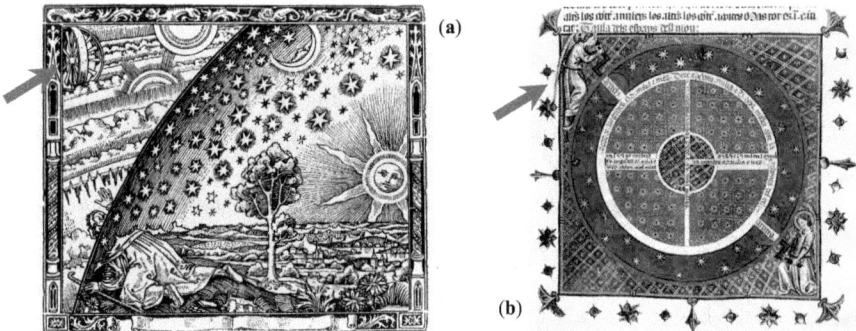

Fig. 2.4 Two worlds as depicted in the old engravings. In both, the moving mechanical agents are shown (arrows) to set in motion the celestial spheres. Though in the spirit of Aristotelian understanding, it was actually an expansion of the account for motion of the common objects to the celestial realm. ((**a**) Anonymous picture from Flammarion (Paris, 1888) https://en.wikipedia.org/wiki/Cosmos. (**b**) Illustration from the fourteenth-century manuscript in British Library https://en.wikipedia.org/wiki/Dynamics_of_the_celestial_spheres#/media/File:Angelic_movers.jpg)

---

[6] For a long time, this motion was related to a mysterious "intelligence"—a remnant of the mythology period of knowledge in Greece and beyond. Only in the seventeenth-century motion was recognized as a natural state (Descartes).

In medieval science, this Aristotelian view was already modified through the introduction of an engine which also caused the planets to move (Fig. 2.4). This was, in fact, the first step toward unifying the two worlds in one universe governed by the same laws, which happened in the scientific revolution of the seventeenth century.

The Motion of Projectiles

Let us consider now how the motion of a thrown stone is sustained as suggested by Aristotle. What could be the force pushing the stone after it leaves the hand? Aristotle's answer was quite original—the air provides the efficient cause for this movement. How could this be? When the hand accelerates the stone, the stone splits the air and puts its layers in motion. These layers circulate around the stone to fill the vacancy behind it. The area behind the stone pushes the stone from behind, propelling it forward. This mechanism was called *antiperistasis*—which means that the air *against* which the body is moving actually supports this motion (Fig. 2.5).[7]

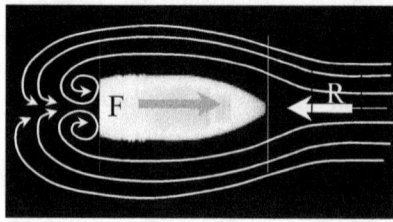

Fig. 2.5 Turbulent streams of air push, in the view of Aristotle, the projectile forward— *antiperistasis*

The result is rather strange: the air, which resists the motion through it (force R), also serves as the mover (applies force F), pushing the projectile forward. The mechanism fits in with the worldview of Aristotle that, without air, in a void, motion is not possible.[8]

> ... if the air were not endowed with this function, constrained movement would be impossible

Addressing the motion of a stone, Aristotle includes there two aspects: the intention of natural motion due to the weight downward (not a force) and the force exerted on the body:

> ...since movement is always due either to nature or to constraint, movement which is natural, as downward movement is to a stone, will be merely accelerated by an external force, while an unnatural movement will be due to the force alone. In either case the air is as it were instrumental to the force.

---

[7] In Greek, ἀντιπερίστασις is comprised of ἀντί ("against") and περίστασις ("standing around"). It is used in a more general sense too, such as when cold increases a body's temperature.
[8] Aristotle (1952b, Book III, Part 2).

Was this explanation convincing? Seemingly not. Although consistent with the theory, it spurred a desperate critique by other scholars who ultimately refuted the theory of Aristotle.

The Physics of Impetus

The explanation of projectiles through airflow, which both supports and resists (antiperistasis), already looked artificial to the contemporaries of Aristotle. The speculative mechanism looked too complex to a naïve observer. Indeed, if air maintains the movement of the thrown body why wind never causes stones to fly from the hand in the same way, people continued to seek for other ideas to explain projectiles.

Hipparchus

Hipparchus

Among the first who provided an alternative explanation for the projectile motion was the famous Greek astronomer Hipparchus (second century BC)—a brilliant mind of the Hellenistic culture.

Here is how he explained the movement of a tossed stone: The hand that throws a stone endows it with a power ("the throwing force") which counteracts gravity. We may call it *impetus*—the name that this force, or virtue, received later, in medieval Europe. As long as the "the throwing force" exceeds gravity, the body moves upward. The impetus naturally fades away and the body decelerates, stopping when the impetus and the weight equate with each other at the highest point of the trajectory.

According to Hipparchus,[9] after that moment, the weight prevails over the force implanted in the body, and it begins to fall while the accelerating force continuously diminishes. After this force is entirely dispensed, the body continues falling solely due to its weight, that is, in natural motion. We may represent this scenario graphically (Fig. 2.6) .

In the sixth century, another Hellenistic scholar John Philoponus (490–570), the early Christian commentator of Aristotle, returned to considering projectile motion. Similarly to Hipparchus, Philoponus proposed explaining the projectile motion by an unidentified "immaterial driving ability" imprinted on the thrown body. The process of launching the projectile looks, thus, as if charging a body with a form of

---

[9] Cohen and Drabkin (1948, p. 209).

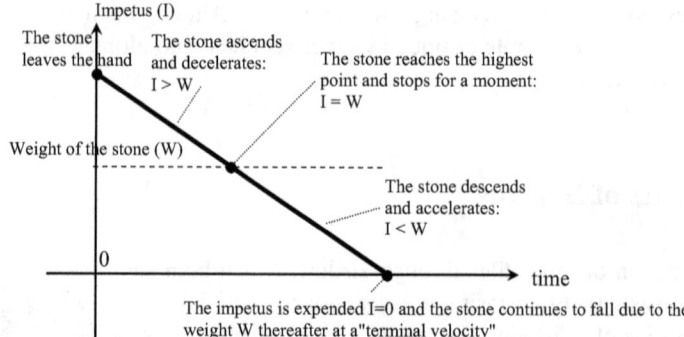

Fig. 2.6 The dependence on length of time of the impetus of the stone thrown upward

"motion charge," a specific virtue received by a body providing it with potentiality of motion .[10]

Philoponus understood that the situation when a body is simply dropped (zero initial velocity) apparently required a different treatment. In this case, he suggested an influx of impetus which streams through the support into the ball in order to prevent the falling of the object and explained the hand becoming tired. When a body is dropped, the impetus gradually disappears during falling. This explains the growing velocity at the beginning of falling.

Philoponus criticized the Aristotelian theory of motion in several ways. First, he addressed two heavy and unequal stones being dropped, one twice as heavy as the other. It is evident that the stones reach the ground almost together and not, as Aristotle would predict, the heavier stone—twice as fast. Furthermore, one may ask why we fail to throw light things (feathers) farther than heavier (stones). Philoponus repeated the evidence that wind does not cause stones to fly due to the antiperistasis. Not less puzzling seemed to be the motion of a thrown spear and a spinning top. The application of antiperistasis to such cases was apparently problematic.

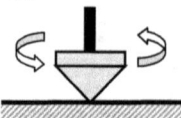

Indeed, how the air succeeds so effectively to push a very thin arrow after it leaves the string of a bow is a problematic question. It was well known that a spear (as well as an arrow) being sharp at both ends flies much faster and farther than an object with a flat back. One might think that the flat rear surface of a body would receive more impetus from the air streams of the antiperistasis and, therefore, would move faster and further. In fact, the opposite happens. An even more striking contradiction was presented by a spinning top. In this case, different parts of the moving body replace each other, leaving no place for the air to push. Why does a rotating spinning top continue to spin?

---

[10] As the electric charge which governs behavior of charged bodies, the introduced virtue (impetus) could stay and eventually dispends by the moving body while returning to its natural ("neutral") state—rest.

These questions could not be answered within the Aristotelian theory of motion, and they motivated the efforts of scholars to provide an alternative account of motion and its nature. The idea of the external mover looked unreliable. Instead, the agent of motion came to be considered something *internal*—the impetus.

Philoponus criticized the major law of motion by Aristotle (Eq. 2.1). This was especially important because Aristotle used this law in his demonstration of the absence of vacuum. He argued: if there were a void ($R = 0$), there would be an infinite speed—an obvious absurdity. Instead, Philoponus suggested a different dependence which could be expressed as following:

$$v \propto F - R \qquad (2.3)$$

This account implies that motion happens when the force F *overcomes* resistance R or gravity (in tossed bodies), that is, $F > R$ or $F > W$. Obviously, nothing dramatic happens when the resistance of the medium disappears ($R = 0$). Although Philoponus did not claim the existence of a void, he did not reject it either: zero resistance does not imply infinitely fast motion. The medium lost its essential role of sustaining motion (antiperistasis), remaining a mere impediment.

In this context, Philoponus started to think about the functional

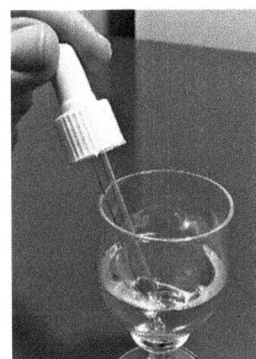

principle of the pipette (Fig. 2.7) (clepsydra—"water thief" in Greek)—a device in use in Greece to evacuate liquids from vessels without turning them on their side, as in, for example, taking wine from a deep barrel. Is there a vacuum there? The positive answer was given by Pascal much later. Galileo in the seventeenth century, addressing the sucking effect of a pipette, still explained it by nature "abhorring a vacuum."

Fig. 2.7 Pipette at work: After being collapsed through pressure, the small balloon recovers, being elastic, causing the liquid to rise and enter the balloon

Impetus in Medieval Physics

The scientific tradition of the Hellenic and Hellenistic cultures was preserved in the following cultural period of the Muslim world during the Middle Ages. Scientific development continued, drawing on the knowledge and discourse of the previous periods of science. Two scholars are known to us for addressing the idea of impetus: Ibn Sīnā (Latinized name Avicenna) (981–1037) —a Persian polymath, and Abu'l-Barakat al-Baghdadi (1080–1165)—a physicist and philosopher of Jewish descent from Baghdad. Aristotle continued to stimulate the critique of scholars seeking an alternative account, similar to Hipparchus and Philoponus.

Avicenna

Avicenna, in his own account for motion, introduced the entity of *inclination to motion* (*mayl*) transferred to the projectile from the thrower. He stated that the projectile motion in a vacuum would not cease. This inclination is dissipated due to the influence of the medium—the air resistance. Mayl of the moving body was defined as being proportional to weight (*W* stands for heaviness) and velocity (*v*), a clear precursor of the modern concept of momentum:

$$M \propto W \cdot v \tag{2.4}$$

Abul-Barakat al-Baghdadi also stated that the mover imparts a violent inclination (*mayl qasri*) on the moved, later diminishing in the course of motion. He described falling in terms of increasing impetus. Yet, Abul-Barakat went much farther. His account of motion was much more precise: he introduced acceleration as the rate of change of velocity. Quite amazingly, he stated that a constant force produced not a constant velocity as according to Aristotle (Eq. 2.2), but a constant acceleration.[11] In that, he, in the twelfth century, anticipated Newton's theory of the seventeenth century.

In Europe, the progress of mechanics was instigated in the fourteenth century by a French cleric and professor at the University of Paris, Jean Buridan (1295–1358), who mastered the theory of Aristotle and knew the works of Avicenna and Abul-Barakat.

Buridan adopted and applied the notion of impetus—the motion-maintaining faculty—in his argument against the Aristotelian view of motion.

Buridan suggested his explanation for the motion of the thrown-up body and the fact that this body accelerates when falling. He did this in his commentaries on Aristotle's *Physics*.[12] In his view, air cannot be a driver of motion. Rather, it impedes motion. He repeated the argument of the lack of spontaneous motion against the wind of the same strength in the absence of a thrower. Why is a thrower's hand essential? Also, how can we explain the movement preserved in a spinning potter's wheel when the pressure ceases? Air does not move against something, but only with it.

Buridan reminded us that, when people drag a boat against the river current and suddenly cease to pull the rope, the boat does not immediately go backward with the stream but keeps moving forward for a while with decreasing speed until it stops

and then starts to move backward—with the stream. Buridan stated, if Aristotle were correct, then [13]:

. . . you would throw a feather farther than a stone and something less heavy farther than something heavier, assuming equal magnitudes and shapes. Experiences show this to be false

---

[11] Pines (1986, p. 203).

[12] Buridan (1509/1959, in Clagett, 1959, pp. 532–540).

[13] Buridan (1509/1959) in Clagett (1959, p. 534).

Instead of Aristotle's mechanism of motion support, Buridan offered his account: the throwing hand transmits an "impetus" (pressure, power, desire, ability to move) into the body, which propels it[14]:

> *Thus, we can and ought to say - in the stone or other projectile there is impressed something which is the motive force (virtus motiva) of the projectile. And this is evidently better than falling back on the statement that the air continues to move that projectile. For the air appears rather to resist. Therefore, it seems to me that it ought to be said that the motor in moving a moving body impresses in it a certain impetus or a certain motive force (vis motiva) of the moving body, in the direction toward which the mover was moving the moving body, either up or down, or laterally, or circularly. And by amount the motor moves that moving body more swiftly, by the same amount it will impress in it a stronger impetus. It is by that impetus that the stone is moved after projector ceases to move. But that impetus is continually decreased by the resisting air and by the gravity of the stone, which inclines it in a direction contrary to that in which the impetus was naturally predisposed to move it. Thus, the movement of the stone continually becomes slower, and finally that impetus is so diminished or corrupted that the gravity of some stone wins out over it and moves the stone down to its natural place.*
>
> *This method, it appears to me, ought to be supported because the other methods do not appear to be true and also because all the appearances are in harmony with this method*

Although long and a little repetitive, this explanation of motion attracts by its simplicity. The impetus is similar to the momentum in classical mechanics. They both are in direct proportion to speed and, in a way, to the amount of matter. Buridan had no notion of mass and used the quantity of matter. Here is how Buridan justified the latter dependence[15]:

> *For if anyone seeks why I project a stone farther than a feather, and iron or lead fitted to my hand father than just as much wood, I answer that cause of this is that the reception of all forms and natural dispositions is in matter and by reason of matter. Hence by the amount more there is of matter, by that amount can the body receive more of that impetus and more intensely.*
>
> *Now in a dense and heavy body, other things being equal, there is more of prime matter than in rare and light one. Hence a dense and heavy body receives more of that impetus and more intensely, just as iron can receive more validity than wood or water of the same quantity. Moreover, a feather receives such an impetus so weakly that such an impetus is immediately destroyed by the resisting air. And so also if light wood and heavy iron of the same volume and of the same shape are moved equally fast by a projector, the iron will be moved farther because there is impressed in it a more intense impetus, which is not so quickly corrupted as the lesser impetus would be corrupted. This also is the reason why it is more difficult to bring to rest a large smith's mill which is moving swiftly than a small one, evidently because in the large one, other things being equal, there is more impetus.*

Buridan addressed jumping[16]:

> *[The impetus then explains why] one who wishes to jump a long distance drops back a way in order to run faster, so that by running he might acquire an impetus which would carry him a longer distance in the jump. Whence the person so running and jumping does not feel the air moving him, but feels the air in front strongly resisting him.*

A Greek athlete jumping to a distance at Olympics game

[14] Clagett (1959, pp. 534–535).

[15] Clagett (1959, p. 535).

[16] Clagett (1959, p. 536) Illustrative drawing from Gardiner (1910, p. 286, Fig. 66).

Impetus and Falling

Although the major effort was to explain *violent* motion, medieval scholars did not forget to apply the same idea of impetus to *natural* motion, too. Here is the account on falling by Buridan[17]:

> ...it follows that one must imagine that a heavy body not only acquires motion unto itself from its principle mover, i.e. its gravity, but that it also acquires unto itself certain impetus with that motion. This impetus has the power of moving the heavy body in conjunction with the permanent natural gravity. And because that impetus is acquired in common with the motion, hence the swifter motion is, the greater and stronger the impetus is. So, therefore, from the beginning the heavy body is moved by its natural gravity only; hence it is moved slowly. Afterwards it is moved by that same gravity and by the impetus acquired at the same time; consequently, it is moved more swiftly. And because the movement becomes swifter, therefore the impetus also becomes greater and stronger, and thus the heavy body is moved by its natural gravity and by that greater impetus simultaneously and so it will again be moved faster; it will always and continuously be accelerated to the end.

Buridan continued[18]:

> From this theory also appears the cause of why the natural motion of a heavy body downward is continually accelerated. For from the beginning only the gravity was moving it. Therefore, it moved more slowly, but in moving, it impressed in the heavy body an impetus. This impetus now together with its gravity moves it. Therefore, the motion becomes faster; and by the amount, it is faster, so the impetus becomes more intense. Therefore, the movement evidently becomes continually faster.

This was a new mechanism: gravity caused impetus and together they cause falling. It was an expanding of the Aristotelian law according to which speed is in direct proportion to the cause of motion. In falling, impetus joins gravity (weight) in causing motion. The process is repeated in a *self-enhancing* manner which explains the *accelerated* motion.

Thus, the context of *natural* motion—falling—brought scholars to the new feature of impetus. While in the course of *violent* motion, the impetus was normally dispensed, during the falling the impetus is continuously produced by the *natural gravity*. Importantly, the created impetus was identified with the *accidental gravity*[19] (or *accidental weight*) of the falling object. In this speculative account, the motion of the falling was identified as accelerated. It was also an extension of the concept of weight: *natural weight* at the beginning and additional *accidental weight* in the course of motion. Both weights matched operational logic. They corresponded to the effort which one would need to prevent or stop the falling: obviously, to support a body at rest is easier than to stop it while falling, and much easier than to stop it at rest, than when moving at a high speed.

---

[17] Clagett (1959, pp. 551–552).
[18] Clagett (1959, pp. 535–536).
[19] That is, obtained in particular circumstances.

The Birth of the Pendulum

One of the central tools of scientific investigation used by the scholars of medieval physics was *thought experiment*.[20] *Thought experiment* presumes a theoretical application of certain principles, conceptions to account for an imaginary situation, performing its analysis and making inferences—all purely theoretical. One such thought experiment became especially famous and well-known to numerous scholars. The reader was asked to imagine a tunnel through the Earth along its diameter (a complete fantasy, of course), and a person throwing a small object into the tunnel (Fig. 2.8). The question was, what would happen?

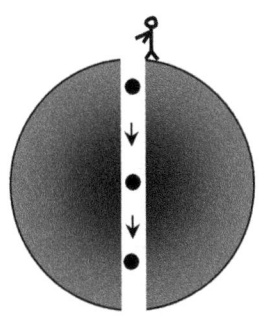

Fig. 2.8 Dropping an object toward the Earth's center

In accordance with Aristotle's theory, the object should fall toward the center of the Earth, which occupied the center of the universe, and stop there. Adelard of Bath, a well-known scholar of the twelfth century, that is, 1500 years after Aristotle, taught exactly this understanding.[21] However, the scholars of the fourteenth century were already thinking differently. Here is what Albert of Saxony (1316–1390), one of the pupils of Buridan and a well-known scholar himself, wrote [22]:

> ... it would be said also that if the earth were completely perforated, and through that hole a heavy body were descending quite rapidly toward the center, then when the center of gravity of the descending body was at the center of the world, that body would be moved on still further in the other direction, i.e., toward the heavens, because of the impetus in it not yet corrupted.

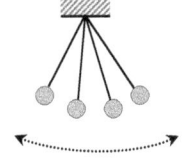

This way the medieval scholars of mechanics discovered a new type of motion: oscillation with respect to a certain point. Thus, Albert proceeded:

> And, in so ascending, when the impetus would be spent, it would conversely descend. And in such a descent it would again acquire unto itself a certain small impetus by which it would be moved again beyond the center. When this impetus was spent, it would descend again. And so it would be moved, oscillating [titubando] about the center until there no longer would be any such impetus in it, and then it would come to rest.

In accordance with the new theory, the following would happen. Due to the accumulated impetus, the body would not stop at the center of the Earth but continue to swing around it until, for some reason, the whole impetus would totally corrupt. It was common that scholars of that time did not consider it overly important to seek for empirical evidence of the prediction they made. In this case, however, the

---

[20] Galili (2009).

[21] Crombie (1959, p. 30).

[22] Albert of Saxony (1959) in Clagett (1959, p. 566).

comparison with the motion of a body suspended by a thread was offered. This analogy represents not only oscillations but also the eventually reached state of rest. A new object which played a prominent role in the following history of physical theories—the pendulum—was invented, and the oscillatory motion entered the collection of physics models. It was a great achievement.

Projectiles and Impetus

The concept of impetus modified the previous account of projectile motion which was comprised of *two* segments: violent and natural motions, as defined by Aristotle (Fig. 2.9a). The scholars of the fourteenth century produced a more sophisticated understanding, better corresponding to reality (Fig. 2.12). It now included *three* segments. At the beginning, the impetus totally prevails, and the launched ball moves straight ahead—the first segment of the trajectory. The second segment of the curve shows the gradual increase of the influence of gravity which diminishes the impetus. After the impetus is totally dispensed, the ball falls to the ground in a natural vertical motion, solely due to its gravity—the third segment of the trajectory (Fig. 2.9b).

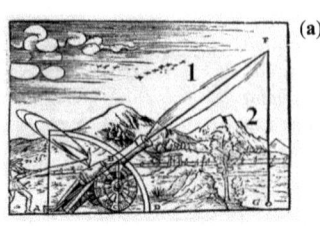
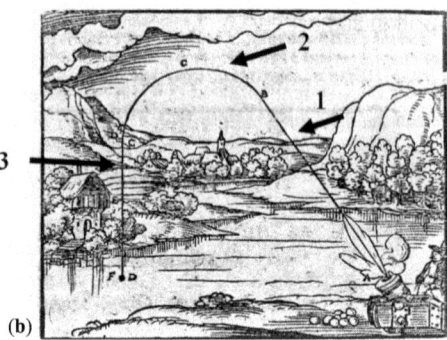

Fig. 2.9 (**a**) The trajectory of a projectile corresponding to the Aristotelian theory: the two segments are indicated: 1, violent motion; 2, natural motion. (The drawing from Daniel Santbech, 1533) (**b**) Ballistic trajectory according the impetus theory. Three segments on the trajectory are identified: 1, impetus prevailed; 2, impetus and gravity interplay; 3, gravity reigned (The drawing from Ryff, 1547)

Medieval Progress in Kinematics

From the beginning, the account for motion was rather "philosophical," that is, almost entirely qualitative. In accordance, the debates regarding motion remained imprecise, employing speculative, vaguely determined concepts, such as antiperistasis. For instance, Aristotle talked about the swiftness or rapidity of a moving body but did not introduce a quantitative account by means of speed—the ratio of the distance covered by the moving body and the time passed. In modern terms, this ratio is the *average speed*:

$$\text{average speed} = \frac{\text{distance}}{\text{elapsed time}} \tag{2.5}$$

Such an account was, however, not sufficient to effectively describe the reality.

The medieval scholar of the thirteenth century, Jordanus Nemorarius,[23] found a clever demonstration to show that falling water accelerates, that is, moves more and more swiftly. He pointed to the fact that a jet of water streaming downward becomes narrower and narrower (Fig. 2.10). Jordanus understood that this fact indicates a growing speed. It appears to be *the empirical, albeit purely qualitative, evidence for the accelerated motion of falling bodies.*[24]

Fig. 2.10 The stream of falling water becomes narrower and narrower

Gerard of Brussels,[25] the important Flemish physicist of the same time, also asked himself how to describe motion.[26] Suppose a body moves nonuniformly, swiftly, and slowly, in variation. Gerard introduced a *representative speed* of such motion: the speed of the imaginary uniform motion that would cover the same distance (S) during the same time interval (Δt). This is equivalent to what we mean by an average speed, $\bar{v}$:

$$S = \bar{v} \cdot \Delta t \tag{2.6}$$

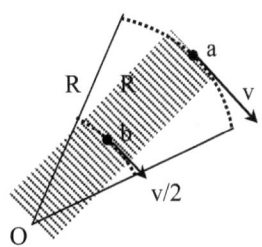

Fig. 2.11 Description of circular motion by Gerard

The idea of average speed was very important. It allowed the describing of the motion of a solid body as a whole. For example, Gerard proved that if point **a** (Fig. 2.11) describes a circle around the center **O**, then, point **b** at the middle of the radius moves forward at the speed of v/2. The area described by the radius to point **a** (the sector) at certain time interval would be equal to the area described by the radius if it were moving linearly with the speed of point **b** (the shaded rectangular). It was

---

[23] Not much is known about this scholar beyond his belonging to the Dominican order. Among his results was the account for the static equilibrium of two weights connected over the edge by means of a special principle. It was equivalent to the principle of virtual displacements—the introduction of the product of force and distance (work!).

[24] Much before, Strato, who followed Aristotle in leading his school, mentioned a similar evidence for the accelerated motion. A continuous stream of falling water from a considerable height splits into drops next to the ground (Barbour, 2001, p. 37).

[25] This thirteenth-century scholar is mainly known for his surviving work on kinematics.

[26] Gerard's *Liber de Motu* was the first treatise entirely devoted to kinematics.

a step toward the description of the rotational motion through the functional dependence of the linear velocity v of any point, the angular velocity ω of a spinning body, and the radius R of the point:

$$v = \omega \cdot R \tag{2.7}$$

The fourteenth century saw the extremely important development of the description of motion, later defined as *kinematics*. This advancement took place in Oxford, at Merton College, by the group of scholars known as the Merton School: Thomas Bradwardine, William Heytesbury, Richard Swineshead, and John Dumbldon. These scholars came to mechanics from the philosophical debates of scholastic philosophy[27] addressing the need to upgrade the qualitative accounts of motion by more precise, quantitative tools.[28]

Their approach implied the need to identify *intensive* as opposite to *extensive* qualities, *intension*, and *extension* in general sense. Such were, for example, the quantities of *mass* (extensive quantity) and *density* (intensive quantity), and the quantity of heat and the "effectiveness of heating" (in future—the *temperature*). With regard to motion, such quantities were *distance* covered and *speed*. Clearly, this problem was far from being trivial and led to the new idea of infinitesimals in the defining of *specific change*, dividing a small quantity by another small quantity, as required by intensive and extensive changes of qualities.

With regard to motion, Bradwardine defined uniform motion[29]:

> A motion is called uniform when equal distances are traversed in equal times with the same velocity.

The most important, however, was the introduction of the concept of *instantaneous* velocity. Heytesbury, suggested[30]:

> In a non-uniform motion, the velocity at a given instant of time may be conceived as the distance which the body would traverse if, in a certain interval of time, it was moved uniformly with the velocity it had at the given instant.

It indicated the enormous difficulty in defining instantaneous velocity—a purely imaginary quantity—while lacking calculus and the concept of a limit. Furthermore, Heytesbury defined the uniformly accelerated motion[31]:

> Any motion is uniformly accelerated when the velocity is increased with equal amounts in arbitrary, but equal, intervals of time.

---

[27] In particular, there was a discussion regarding adding different types of qualities.
[28] Clagett (1950).
[29] Pedersen and Phil (1974, p. 220); Dugas (1988, p. 67), Barbour (2001, pp. 193–194).
[30] Pedersen and Phil (1974, pp. 220–221).
[31] Pedersen and Phil (1974 p. 221).

The Merton scholars achieved several results of far-reaching importance for the theory of motion. The most famous among them regarded the distance traversed by the body in a uniformly accelerated motion[32]:

If the velocity of a body is increased uniformly from v_0 to v_1 during the time interval t, the distance traversed in that time will be:

$$s = \left[v_0 + \frac{v_1 - v_0}{2} \right] \cdot t \qquad (2.8)$$

For zero initial velocity, this formula provides $s = \frac{1}{2} v_1 \cdot t$ which together with the rule for velocity $v_1 = a \cdot t$ leads to the formula currently known to all students regarding the uniformly accelerated motion:

$$s = \frac{1}{2} a \cdot t^2 \qquad (2.9)$$

As we see, this result was known to scholars since the fourteenth century. The result (Eq. 2.8) matched the previous program of Gerard regarding average velocity $\bar{v}$ (Eq. 2.6). It is therefore called the *Mean Speed Rule*. Indeed:

$$s = \left[v_0 + \frac{v_1 - v_0}{2} \right] \cdot t = \frac{v_1 + v_0}{2} \cdot t = \bar{v} \cdot t \qquad (2.10)$$

with the mean speed of $\bar{v} = \frac{v_1 + v_0}{2}$.

However, the most obvious meaning of the formula (Eq. 2.10) was revealed by another bright mind from medieval science—Nicole Oresme from Paris.

Nicole Oresme

The greatest step in the theoretical analysis of motion belongs to the renowned medieval scholar of the fourteenth century from the University of Paris—Nicholas of Oresme, a pupil of Jean Buridan. He was the first to introduce a graphical representation of qualities as varying in time. This method visualized kinematics, the results achieved before Nicholas. It established the method highly important for all further progress in science.

Nicolas of Oresme

---

[32] Pedersen and Phil (1974, p. 222).

It is commonplace to represent the variation of velocity v in a graphical manner, assigning its magnitude to the vertical axis (latitude) placed against the horizontal axis (longitude) representing different instances of time. As a result, one obtains a graph—the image of velocity variation.

For the case of the uniform motion (v = const), the graph will be a horizontal line (Fig. 2.12a). Clearly, the area under the graph represents the distance covered by such a moving body (s = v·t). In the case of the nonuniform velocity, we obtain the graphical representation of motion in the form of Fig. 2.12b. And in the case of the "uniformly nonuniform" motion (constant acceleration), one receives the trapezium (Fig. 2.12c) for which the expression for the area below the graph reproduces the result (Eq. 2.10) obtained by the Merton School—Mean Speed Rule (Fig. 2.12d).[33]

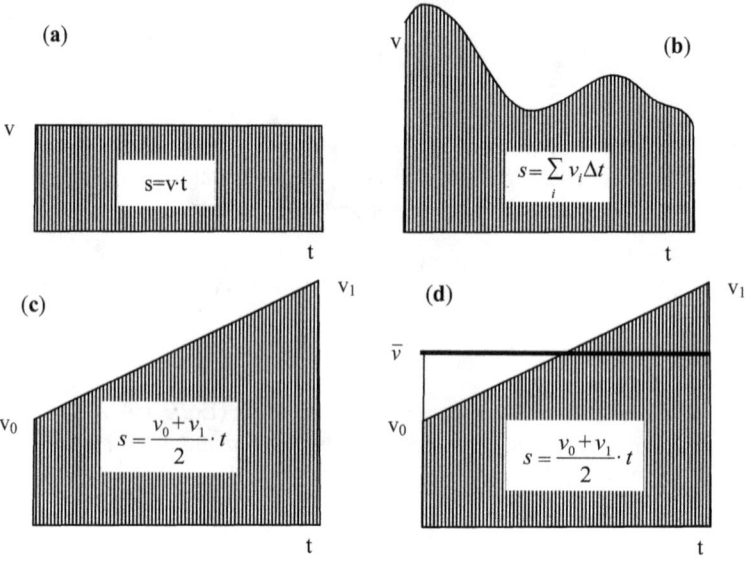

Fig. 2.12 Reconstructed graphical representation of velocity dependences in (**a**) uniform, (**b**) nonuniform, (**c**) uniformly nonuniform (constantly accelerated) motions. (**d**) The graphical representation of the Merton mean speed theorem regarding the distance traversed at a uniformly accelerated motion

Figure 2.13a reproduces a page from Nicholas of Oresme's manuscript where he presented his graphical method and applied it to various motions. Oresme thought in terms of vertical columns, narrow, but of finite width which actually replaced the motion with varying velocity with the motion with constant velocities for short intervals of time. It was actually the idea of the definite integral which was only introduced into mathematics in the seventeenth century by Newton and Leibniz, the newly invented calculus. They could see this treatment and adopt the idea.

---

[33] Dugas (1988, pp. 66–67), Lefetz (2014).

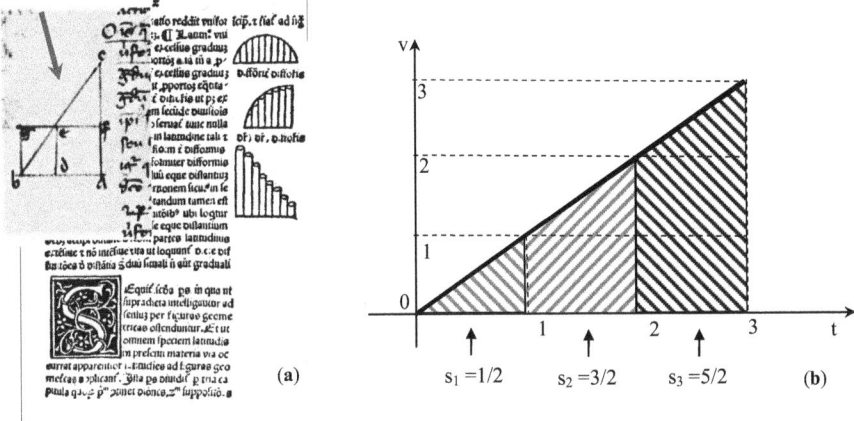

Fig. 2.13 (a) Fragments from the manuscript of Nicholas of Oresme. Mean speed theorem on the margins of the text. (b) Uniformly accelerated motion. Shaded areas allow comparison between the distances traversed in subsequent equal intervals

Equipped with the method of graphical representation, Oresme obtained the important characteristic of uniformly accelerated motion. Considering the graph of velocity versus time allowed him to obtain the ratio between the distances traversed by the body moving at a constant acceleration (Fig. 2.13b). Comparison between the marked areas of Fig. 2.13b leads to:

$$s_1 : s_2 : s_3 : s_4 : ... = 1 : 3 : 5 : 7 : ... \tag{2.11}$$

This result indicates that the ratio between the distances traversed by a body moving at constant acceleration (starting from a position of rest) is equal to that of odd numbers. Oresme's interesting result provides the feature by which the motion at a constant acceleration could be recognized and distinguished from every other motion.[34]

Two hundred years after Oresme, Galileo, in his book regarding the naturally accelerated falling body, drew the illustrated graph (Fig. 2.14a) and wrote[35]:

THEOREM I, PREPOSITION I

The time in which any space is traversed by a body starting from rest and uniformly accelerated is equal to the time in which that same space would be traversed by the same body moving at a uniform speed whose value is the mean of the highest speed and the speed just before acceleration began.

In this piece, we can recognize the ideas of Gerard of Brussels, scholars of the Merton School in Oxford and Oresme in Paris. Galileo seemingly adopted them to explain the motion of naturally falling bodies without any reference (compare Fig. 2.12d of Oresme with Fig. 2.14a of Galileo).

---

[34] Drake (1999, Vol. II, p. 249).

[35] Galilei (1638/1914, p. 173).

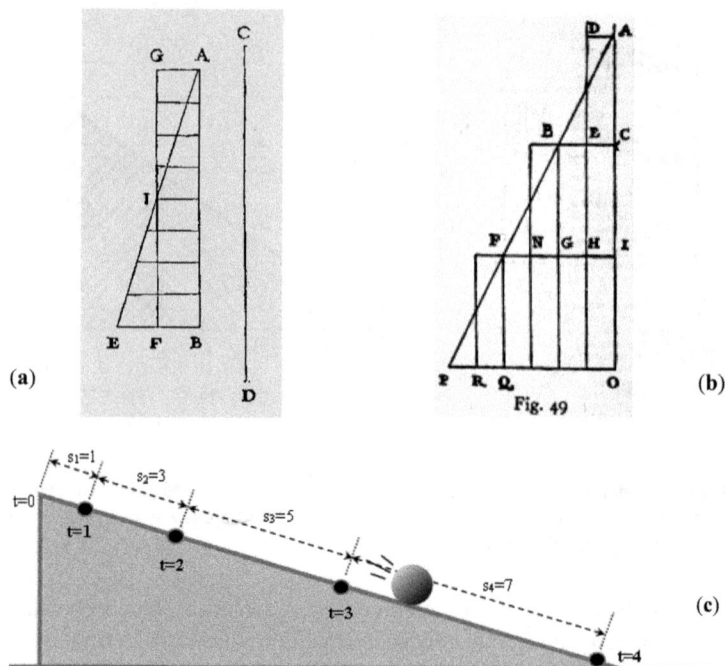

Fig. 2.14 (**a**) The graph of Galileo representing velocity change for a falling object in 1638. (**b**) Another Galileo's drawings in 1638. (**b**) Galileo's drawing supporting his proof that the measured distances traversed by a ball in equal time intervals "bear to one another the ratio of the series of odd numbers 1, 3, 5, 7...". (**c**) Schematic drawing representing the idea of Galileo's experiment with an inclined plane. Galileo placed tensed ropes (marked) on the way of the rolled down ball in such a way that the produced sounds by the ball passing over the ropes created a uniform sequence of ticks. This was a very accurate time intervals measurement given a special skill of hearing by Galileo

Galileo made direct use of the feature of (Eq. 2.11) when he experimentally (!) proved that the body, while descending along the inclined plane, moved at a constant acceleration (Figs. 2.14b, c).[36] It was a success of special importance, since he measured the directly accessible quantity—the distance and the rate of the ball encountering the ropes on its way along the edge. Physics teachers may miss much by a mere declaration of Galileo's results, skipping over his clever experimental setting and its historical background.

Medieval Progress in Dynamics

The scholars of Merton College in Oxford applied mathematical tools to describe the dynamics of motion, that is, to relate the motion to the active forces associated with that motion. Bradwardine in 1328, suggested his own theory that, similarly to

---

[36] Galilei (1638/1914, Corollary I, pp. 175–176).

Aristotle's (2–2) and Philoponus' (2–3) laws, related the velocity of a body to the forces exerted on it: the driving force F and the resistant force R.

The form of this dependence was rather difficult to express verbally, as was the norm in Merton debates. To simplify its presentation, we will introduce it with particular numbers. Suppose that forces F and R have the values: F = 3 and R = 1, and they cause motion (the Aristotelian understanding!) with velocity v. Then, in accordance with Bradwardine, the increased force of F = 9 and the same R = 1 would cause motion with double velocity 2v. In other words, double (or triple and so on) velocity is produced by squaring (or cubing, etc.) the ratio F/R, that is, $(F/R)^2$ (or $(F/R)^3$ and so forth).

In the general terms that we use today, Bradwardine's law stated that the velocity v was such a function φ of the ratio F/R that it satisfied the demand:

$$n \times v = \phi \left(\left[\frac{F}{R} \right]^n \right) \tag{2.12}$$

In other words, the rise of v by factor n is caused by the correspondent exponential rise of the ratio (F/R). In terms of the later, introduced in mathematics logarithmic function, this law obtains the following compact form:

$$v \propto \log\left(F / R\right) \tag{2.13}$$

The advantage of this form was that when F approached R in value, the velocity v smoothly approached zero, and the formula required F be greater than R ($F > R$) (again the Aristotelian conception). Importantly, it preserved the divergence of the resultant velocity when R is diminishing and approaching zero:

$$\text{when } R \to 0 \text{ then } v \to \infty \tag{2.14}$$

This behavior of (Eq. 2.14) is similar to the behavior of the Aristotelian law (Eq. 2.2) when when $R \to 0$, and it was this similarity to Aristotelian law that often attracted scholastic scientists who normally intended to *refine* Aristotle while *not refuting* his ideas.

Importantly, the scientists of that time (the fourteenth century) considered different dependences of motion characteristics, velocity, and acceleration, according to their matching certain theoretical principles, but they never tried to check it experimentally through any kind of data corresponding to the real situation. Perhaps they just could not do it due to the technological limitations of that time. This step lay in waiting for the illustrious time of the Renaissance in Italy.

Transition to Classical Physics

Galileo

Galileo Galilei (1564–1642) was the major hero of the coming change. At the beginning of his scientific activity, he fully adopted the impetus theory: impetus as an engine and cause of motion. In addressing Galileo's first work on motion—*De Motu*—Koyre[37] depicted Galileo's apologetic position regarding impetus, yet with a seed of criticism[38]:

> *[Galileo thought that] it is unthinkable and absurd not to admit that the cause or force which produces motion must necessarily spend and finally exhaust itself in this production. It can never remain unchanged for two consecutive moments, and therefore the movement that it produces must necessarily slow down and come to an end. Thus, it is a very important lesson from the young Galileo. He teaches us that impetus physics, though compatible with movement in a vacuum, is like that of Aristotle incompatible with the principle of inertia.*

The young Galileo stated that the idea of impetus as a charge for motion (a specific virtue of a moving body) contradicts the continuation of motion after the impetus is fully spent, implying that there is no place for inertia without the charge of motion. Eventually, in the course of his experiments, Galileo found a new result[39]:

> *The speeds acquired by one and the same body moving down planes of different inclinations are equal when the heights of these planes are equal.*

This means that the acquired velocity of the falling body depends on the vertical descent, the height of the fall, but does not depend on the distance actually traveled by the moving body (Fig. 2.15a).

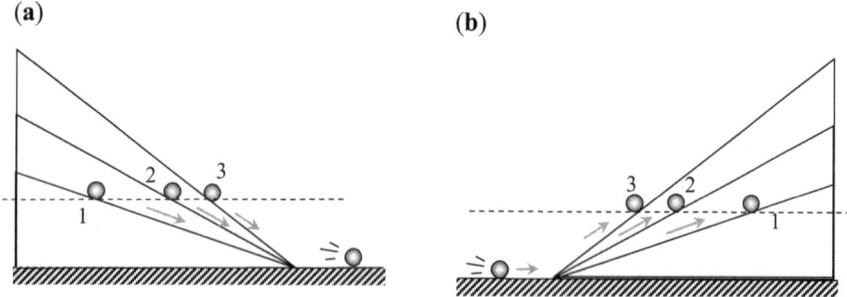

(a)   **(b)**

Fig. 2.15 (**a**) The ball acquires the same speed when released from the same height, regardless of the way it passed. (**b**) The ball climbs to the same height, regardless of the way it takes

---

[37] Alexandre Koyré (1892–1964) was a distinguished historian and philosopher of science who provided numerous interpretations of the events in the period to which he gave the name—the *Scientific Revolution of the seventeenth century.*

[38] Galilei (1590/1960), Koyré (1943).

[39] Galilei (1638/1914), p. 169.

Moreover, the same principle held regarding the elevation by the same body which could climb up along the planes of different slopes (Fig. 2.15b). If there were a "payment" for the motion itself, by means of impetus, this would not be the case (impetus seemingly depends on the height of elevation and not the distance passed).

Galileo demonstrated this rule quite convincingly with his famous constrained pendulum experiment (Fig. 2.16). He wrote[40]:

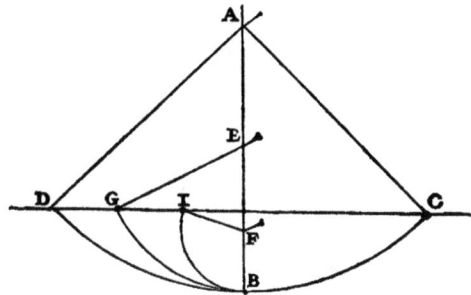

Fig. 2.16 Galileo's drawing of 1638. The constrained pendulum motion: released from point C, the bob of the pendulum stops at one of the opposite points—D, G, and I at the same level as D, C, regardless of the trajectory it might be compelled to take to pass the nail placed in its way in F or E

Imagine this page represents a vertical wall, with a nail driven into it; and from the nail let there be suspended a lead bullet of one or two ounces by means of a fine vertical thread, AB, say from four to six feet long, on this wall draw a horizontal line DC, at right angles to the vertical thread AB, which hangs about two finger-breadths in front of the wall. Now bring the thread AB with the attached ball into the position AC and set it free; first it will be observed to descend along the arc CBD, to pass the point B, and to travel along the arc BD, till it almost reaches the horizontal CD, a slight shortage being caused by the resistance of the air and the string; from this we may rightly infer that the ball in its descent through the arc CB acquired a momentum [impeto] on reaching B, which was just sufficient to carry it through a similar arc BD to the same height. Having repeated this experiment many times, let us now drive a nail into the wall close to the perpendicular AB, say at E or F, so that it projects out some five or six finger-breadths in order that the thread, again carrying the bullet through the arc CB, may strike upon the nail E when the bullet reaches B, and thus compel it to traverse the arc BG, described about E as center. From this we can see what can be done by the same momentum [impeto] which previously starting at the same point B, carried the same body through the arc BD to the horizontal CD. Now, gentlemen, you will observe with pleasure that the ball swings to the point G in the horizontal, and you would see the same thing happen if the obstacle were placed at some lower point, say at F, about which the ball would describe the arc BI, the rise of the ball always terminating exactly, on the line CD. But when the nail is placed so low that the remainder of the thread below it will not reach to the height CD (which would happen if the nail were placed nearer B than to the intersection of AB with the horizontal CD) then the thread leaps over the nail and twists itself about it.

---

[40] Galilei (1638/1914, pp. 170–171).

Galileo firmly stated: the quality possessed by the body in motion is determined by the height of its elevation or the height of falling. Therefore, to maintain the horizontal motion, that is to say, to move without any elevation, one does not need any "force," provided there is no friction with the air or the surface.

This was not only the claim for inertial motion that never stops, but at the same time, a refutation of dispensable impetus. If so, Galileo corrected terminology! Not the charge of motion, but instead, the characteristic of the intensity of motion was introduced. He started to use the notion of *momentum* instead of *impetus*. The latter, as well as the similar "force of motion," were left by Galileo in his *Dialogues* for use by Sagredo, the discussant who expressed naïve views on mechanics, and Simplicio, the discussant who represented Aristotelian views. Here is, for example, the piece in which Sagredo reproduces the medieval account for the motion of a tossed body[41]:

> *Sagredo: From these considerations it appears to me that we may obtain a proper solution of the problem discussed by philosophers, namely, what causes the acceleration in the natural motion of heavy bodies? Since, as it seems to me, the force impressed by the agent projecting the body upwards diminishes continuously, this force, so long as it was greater than the contrary force of gravitation, impelled the body upwards; when the two are in equilibrium the body ceases to rise and passes through the state of rest in which the impressed impetus is not destroyed, but only its excess over the weight of the body has been consumed – the excess which caused the body to rise. Then, as the diminution of the outside impetus continues, and gravitation gains the upper hand, the fall begins, but slowly at first, on account of the opposing impetus, a large portion of which still remains in the body; but as this continues to diminish it also continues to be more and more overcome by gravity, hence the continuous acceleration of motion.*

One may of course recognize in this piece the older view of Hipparchus (see above) and other scholars who, after Hipparchus, adopted that account for the motion of the upwardly thrown body.

In his attack on the concept of impetus, Galileo tried to show that the states of rest and motion were not so different *qualitatively*, and in fact, motion may continuously emerge from the rest (so they can be very close states…). This was a severe blow to the concept of impetus that implied the essential difference between the two: rest and motion[42]:

> *Sagredo. When I think of a heavy body falling from rest, that is, starting with zero speed and gaining speed in proportion to the time from the beginning of the motion; such a motion as would, for instance, in eight beats of the pulse acquire eight degrees of speed; having at die end of the fourth beat acquired four degrees; at the end of the second, two; at the end of the first, one: and since time is divisible without limit, it follows from all these considerations that if the earlier speed of a body is less than its present speed in a constant ratio, then there is no degree of speed however small (or, one may say, no degree of slowness however great) with which we may not find this body traveling after starting from infinite slowness, i.e., from rest.*

The *state of rest* appears here joined to the *state of motion* by continuity; rest is suggested as motion at an infinitely small speed. This approach signals the collapse of the qualitative opposition between the two concepts (rest and motion) that had survived in the minds of scholars for thousands of years. What, then, remained of

---

[41] Galilei (1638/1914, p. 168).
[42] Galilei (1638/1914, p. 162).

the idea of impetus? Does it exist in the moving body, at all? The rejection of the idea of impetus emerged as a radical conceptual change.

One may mention two more of Galileo's steps that contributed to the refutation of impetus and establishment of a new understanding of motion: the new account of falling bodies and the principle of relativity, that thereafter have borne the name of Galileo.

The Relativity Principle of Galileo and Inertial Motion

It seems natural that after comprehension of the fact that horizontal motion is not different from rest, and that horizontal uniform motion does not need any support (such as provided by impetus), one may pose a more general question: Is there any difference at all between the objects moving horizontally with different but constant velocities? Here, Galileo made the obvious but unprecedented conjecture that nothing changes with the whole reality, say, in a room, which moves at any constant velocity. "Nothing" means that all the objects, their behavior, and the regularities of physical reality are preserved. To represent this totality, Galileo depicts the new principle in a very casual manner through the words of his representative Salviati[43]:

Salviati: *Shut yourself up with some friend in the main cabin below decks on some large ship, and have with you there some flies, butterflies, and other small flying animals. Have a large bowl of water with some fish in it; hang up a bottle that empties drop by drop into a wide vessel beneath it. With the ship standing still, observe carefully how the little animals fly with equal speed to all sides of the cabin. The fish swim indifferently in all directions; the drops fall into the vessel beneath; and, in throwing something to your friend, you need throw it no more strongly in one direction than another, the distances being equal; jumping with your feet together, you pass equal spaces in every direction.*

When you have observed all these things carefully (though doubtless when the ship is standing still everything must happen in this way), have the ship proceed with any speed you like, so long as the motion is uniform and not fluctuating this way and that. You will discover not the least change in all the effects named, nor could you tell from any of them whether the ship was moving or standing still. In jumping, you will pass on the floor the same spaces as before, nor will you make larger jumps toward the stern than toward the prow even though the ship is moving quite rapidly, despite the fact that during the time that you are in the air the floor under you will be going in a direction opposite to your jump.

In throwing something to your companion, you will need no more force to get it to him whether he is in the direction of the bow or the stern, with yourself situated opposite. The droplets will fall as before into the vessel beneath without dropping toward the stern, although while the drops are in the air the

This picture of 1745 by Hogarth depicts the cabin of the captain (Lord George Graham). Galileo addressed situation in such a cabin claiming the indifference of the reality to the movement of the ship.

[43] Galilei (1632/1953, pp. 186–187, Second Day).

ship runs many spans. The fish in their water will swim toward the front of their bowl with no more effort than toward the back, and will go with equal ease to bait placed anywhere around the edges of the bowl. Finally, the butterflies and flies will continue their flights indifferently toward every side, nor will it ever happen that they are concentrated toward the stern, as if tired out from keeping up with the course of the ship, from which they will have been separated during long intervals by keeping themselves in the air. And if smoke is made by burning some incense, it will be seen going up in the form of a little cloud, remaining still and moving no more toward one side than the other. <u>The cause of all these correspondences of effects is the fact that the ship's motion is common to all the things contained in it, and to the air also.</u> That is why I said you should be below decks; for if this took place above in the open air, which would not follow the course of the ship, more or less noticeable differences would be seen in some of the effects noted. (Emphasis added)

This picture of 1745 by Hogarth depicts the cabin of the captain (Lord George Graham). Galileo addressed situation in such a cabin claiming the indifference of the reality to the movement of the ship.

Galileo deliberately depicts how the physical reality inside the ship moving at a constant speed is exactly the same as in the ship at rest. This claim of the complete equivalence of physical reality regardless of the uniform motion received the name of the *Relativity Principle of Galileo*. Yet, in 1632, this claim was not expressed using the concept of frames of reference (which was employed by Einstein[44]). Galileo explained it maintaining the spirit of medieval physics:

…The cause of all these correspondences of effects is the fact that the ship's motion is common to all the things contained in it, and to the air also…

To appreciate the conceptual development, we return to 1584. Giordano Bruno, while addressing the same settings—a coasting ship—explained why the stone dropped from the ship's mast falls on the desk just next to the mast, without any deviation (as Aristotelian physics would predict)[45]:

… we cannot draw any other explanation except that the things which are affixed to the ship, and belong to it in some such way, move with it: and the stone carries with itself the virtue [impetus] of the mover which moves with the ship. … From this it can evidently be seen that the ability to go straight comes … from the efficiency of the originally impressed virtue, on which depends the whole difference.

Following a similar explanation by Oresme, Bruno clearly employed the concept of impetus (impressed virtue) to explain the continuous motion shared by all objects on the moving ship. It was the medieval theory that replaced Aristotle's explanation of the same ship-situation within his proof of the motionless Earth.[46] Galileo dropped impetus and argued for shared ("common") motion. We know that this is not enough. In his next central treatise, *Discorsi*, in 1638, Galileo introduced inertial motion to depict a body remaining in motion in a horizontal direction which in the

---

[44] See also Chap. 4 for the first introduction of the concept of observer in physics by Huygens.

[45] Bruno, G. (1584). *La Cena de le Ceneri* (The Ash Wednesday Supper), quoted in De Angelis & Santo (2015). They also cited Oresme from his manuscript *Le livre du Ciel et du Monde* of 1377.

[46] Aristotle (1952b, Book II, Ch. 14, p. 388).

absence of obstacles and friction proceeds indefinitely—*inertial motion*. He dropped the concept of impetus possessed by a body in motion. Given that there was no manifestation of its existence, why should one ascribe it to the moving body at all? Bodies naturally move at any velocity without an engine inside them. This was a great conjecture, which has been maintained in physics ever since. There is no reason to assign impetus or any other special quality to bodies in order to account for motion.

Falling

So far, we have considered horizontal motion which, since Aristotle, had always been related to a *violent* process imposed on a body. Galileo's thoughts turned now to the *natural* motion manifested in the falling of bodies vertically toward the ground.

Although Aristotle's original claim—that objects fall with a swiftness proportional to their weight (Eq. 2.2)—often did not match empirical evidence, the observed swiftness of falling significantly varied from case to case and did not allow simple inference, leaving alone rejecting the radical change of Aristotle's general claim. Such refutation required serious investigation.

Galileo made two decisive contributions which practically dismissed the old view. Firstly, by performing precise experiments, he was able to demonstrate a quantitatively proven fact—the swiftness of falling is not in direct proportion to the weight of the fallen objects. Friction with the medium, water or air, may indeed mask the regularity, but the influence of friction decreases in objects of high density and size, such as iron balls. Thus, air friction cannot cause a decisive change in the falling of 2 balls, 1 of 100 pounds and 1 of 1 pound. Even if impeded by friction, the balls actually fell very close, one after the other, that is, clearly not fulfilling the prediction about the proportionality of their weights (Fig. 2.17).

Fig. 2.17 The famous Leaning Tower in Italian city of Pisa. Unfortunate circumstances made the tower an ideal place to perform the experiment of "free" falling objects dropped from a significant height. Galileo depicted his experiment of dropping two heavy cannon balls. Despite the significant difference in their weights they reached the ground at almost the same time

Secondly, Galileo brought to the fore the fact that Aristotle's claim regarding falling was inherently contradictive. Without a word of an alternative account of falling, Galileo demonstrated the inconsistency of the Aristotelian claim with regard to falling: that heavier things fall faster. For that, Galileo reproduced the argument suggested years before, in 1553, by his compatriot Gianbattista Benedetti (1530–1590), a mathematician from Venice (Fig. 2.18).

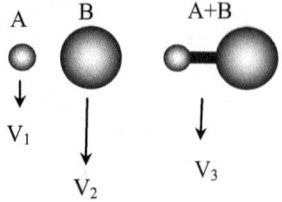

A   B   A+B

V_1

V_2

V_3

Fig. 2.18 The thought experiment by Benedetti: a small body A should fall at velocity V_1 smaller than V_2 of the bigger body B. Being connected, however, they had to fall at velocity V_3. The latter had to be less than V_2 (since A impedes the falling of B) but, at the same time, greater than V_2 (since A + B is bigger than B)—a contradiction

Here is how Galileo demonstrated the inherent contradiction of the Aristotelian theory[47]:

> *Salviati. If then we take two bodies whose natural speeds are different, it is clear that on uniting the two, the more rapid one will be partly retarded by the slower, and the slower will be somewhat hastened by the swifter. Do you not agree with me in this opinion?*
>   *Simplicio. You are unquestionably right.*
>   *Salviati. But if this is true, and if a large stone moves with a speed of, say, eight while a smaller moves with a speed of four, then when they are united, the system will move with a speed less than eight; but the two stones when tied together make a stone larger than that which before moved with a speed of eight. Hence the heavier body moves with less speed than the lighter; an effect which is contrary to your supposition. Thus, you see how, from your assumption that the heavier body moves more rapidly than the lighter one, I infer that the heavier body moves more slowly.*

Galileo realized that he could not argue by an alternative *theory*—he had none (one was later provided by Newton). Therefore, the best he could offer was the inconsistency of the old theory providing empirical evidence of the fact that bodies fall regardless of their material and size so far as the resistance of the medium decreases.

On the Motion of Projectiles

Following the two milestones of understanding motion: inertial horizontal and accelerated vertical (falling), Galileo achieved still another intellectual breakthrough—a complete account of projectile motion. This was presented on the *Fourth Day* of discussions described in his last big treatise—*Discourses Concerning Two New Sciences*—in 1638. For the first time in the history of physics, Galileo stated that the motion of projectiles is comprised of two component motions of a specific type: the horizontal motion—at a permanent velocity and the vertical motion—at a fixed acceleration. By combining the two motions, Galileo determined the precise type of the trajectory of any projectile, regardless of their initial condition(s)—the parabola (Fig. 2.19).[48]

---

[47] Galilei (1638/1914, p. 63).
[48] Galilei (1638/1914, p. 249).

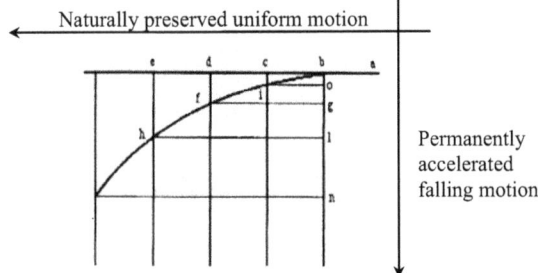

Fig. 2.19 Galileo's drawing (1638) representing the construction of a parabola as the trajectory corresponding to horizontal uniform and vertical constantly accelerated motions

In his quantitatively precise treatment, Galileo obtained theoretical proof of the fact that an initial projectile angle of inclination of 45° will provide the maximal distance reached by the projectile. In addition, a completely new result was discovered: two projectiles thrown at the angles which complement to 90°, traverse the same horizontal distance (Fig. 2.20).

In this way, the ancient problem of projectile motion received its full treatment. However, although complete, this was only a kinematic account. The question of *why* the particular types of motion take place in vertical and horizontal directions relative to the ground remained open until the dynamic (force-motion) explanation was provided by the theory of Newton.

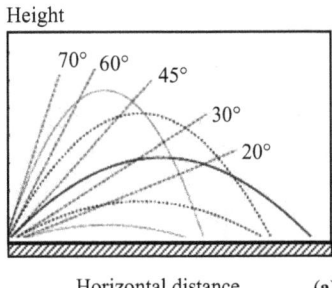

Fig. 2.20 (**a**) Trajectories of projectiles thrown at different angles to the horizon. Note the same distances corresponding to the launch at angles complementary to 90°. (**b**) Trajectories of water drops in the fountain (Sydney) demonstrate different paths of projectiles (yet, all parabolas) launched at different angles to the horizon

Descartes

Descartes

Rene Descartes (1596–1650), the renowned French phi-losopher, was familiar with Galileo's results, but he took another direction in considering the problem of motion. His approach was to construct a new theory based upon *correct* initial ideas, analytical logic, and mathematics. This methodology cared much less that the result obtained matched empirical evidence. The reason? Multiple inter-woven factors may mask the "true" phenomenon; there-fore, Descartes preferred correct logic. This method could bring only partial success, insofar as one intended to investigate and understand nature as it is.

Descartes formulated his vision in his treatise *The Principles of Philosophy* pub-lished in 1644. In it, he formulated the mechanical principles that govern nature, three laws on the motion of material bodies, and this work started from a great innovation. Descartes, unlike Aristotle, declared that motion is a natural state of matter. As such, it might require a cause for its change, but not for its being. Without such cause, movement just remains as it is[49]:

> *The First Law of Nature: that any object, in and of itself, always perseveres in the same state; and thus, what is moved once always continues to be moved.*

Descartes reasoned by the "immutability of God," which does not convince a contemporary learner, being an arbitrary claim. The type of motion was also not specified. However, in addressing motion as a state of a body (not a process), Descartes in fact introduced inertial motion, motion without a mover. This was a claim of great significance which was preserved as a principle of the new mechanics (of course, after adding motion to be rectilinear and uniform).

At the same time, Descartes continued to consider motion to be the state opposite to rest—the old Aristotelian conception contradicting the Galilean principle of rela-tivity. The rest-motion opposition matches naïve intuition, "commonsense." As was understood by Galileo, any observer moving uniformly at any velocity observes the same reality in the system moving with them and could be considered as being at a state of rest. Motion and rest "coexist" in the same body by mere relation to a different reference.

In the following, Descartes refined the type of inertial motion[50]:

> *The Second Law of Nature: that every motion of itself is rectilinear; and hence what is moved cir-cularly tends always to recede from the center of the circle it describes.*

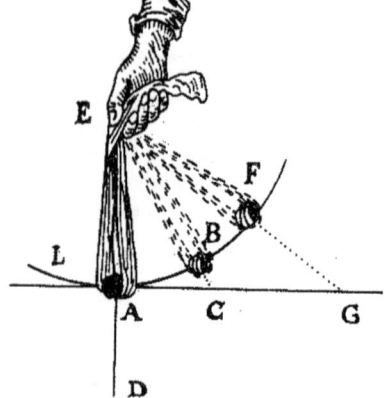

Fig. 2.21 Rene Descartes' drawing illus-trating the second law of nature

---

[49] Descartes (1644/1982, Part II).
[50] Descartes (1644/1982, Part II).

In this law, Descartes declared that the nature of motion preserved is being *recti-linear*. This law toppled the ancient status of the uniform *circular* motion as eternal, natural, and fundamental, being the motion of celestial bodies. Here too Descartes reasoned by "the immutability and simplicity" of the action by which God conserves motion in matter. Addressing conservation of the rectilinear motion Descartes refined[51]:

> *For He [God] does not conserve it other than precisely the way it is in the moment of time in which He conserves, with no relation to what perhaps was shortly before. Although no motion occurs instantaneously, it is nevertheless manifest that everything that is moved, in the single instants that can be designated while it is moved, is determined to continue its motion toward some direction along a straight line, and never along any curved line.*

In his reasoning, Descartes pointed to a very delicate characteristic of motion. Motion is not a collection of stills, but in any instant, the moving body possesses the specific quality manifested in the coming instant in its motion. The example is superior in exactness: the body in a sling left to itself at A "faces" two options: to continue along ACG (Fig. 2.21) or deviate along ABF. Nature reveals (Descartes pointed to experience!) that the rectilinear motion is the natural state and is thus maintained. The circular motion presents a changing of state and requires a cause (the sling and a hand)—a huge discovery[52]:

> *For example, stone A, rotated in sling EA around circle ABF, at the instant in which it is at point A is determined to motion in some direction, namely along a straight line toward C, such that the straight line AC is tangent to the circle. But one cannot arrange that it be determined to any curved motion; for, even if it previously came from L to A along a curved line, nevertheless nothing of this curvity can be understood to remain in it when it is at point A. This is also confirmed by experience, because if it then left the sling it would not continue to be moved toward B, but toward C. From which it follows that every body that is moved circularly, perpetually tends to recede from the center of the circle it describes. We experience this by tactile sense in a stone that we move in a circle with a sling. . .*

As in the first law, Descartes misses here the uniformity of the inertial motion. However, let us not forget that Galileo remained confused when he thought about *rotational* inertial motion.[53] Descartes explicitly rejected this idea in favor of the rectilinear motion, which is given a special status: only such motion is preserved as inertial. It was a high moment in the physics of motion.

Another aspect considers the acting "forces." Descartes used this term in a specific way and avoided using it here. His law addresses the tendency of the body being *compelled* to move on a circular path instead of proceeding in the tangential direction. Preservation of the circular motion is related to the agent imposing it— the sling. The moment it ceases to act, the stone moves naturally, that is, in a rectilinear motion. The Cartesian idea of *tendency* was adopted by Newton, who in the same spirit, stated in his *First Law* of motion (*Principia,* 1687):

---

[51] Descartes (1644/1982, Part II).

[52] Descartes (1644/1982, Part II).

[53] Galilei (1613/1957).

Every body perseveres in its state of resting or moving uniformly straight forward, except inasmuch as it is compelled to change its state by forces impressed.

As seen, Newton did *not* talk simply about the absence of forces but, like Descartes, rather about the *tendency* of preserving the state of rest or moving uniformly *except inasmuch as it is compelled by the external force.*[54]

Finally, Descartes formulated the law of interaction between bodies with relation to their motion. Importantly, the interaction between bodies could be, in his mind, solely by contact, in particular, through collisions[55]:

The Third Law: that a body that moves meets another, if it has less force to continue along a straight line than the other has to resist it, then it is turned aside in another direction, retaining its quantity of motion, and changing only the determination of motion. If, however it has greater force, then it moves the other body with it and loses as much of its motion as it gives to that other.

The two situations of the third law are visualized in Fig. 2.22.

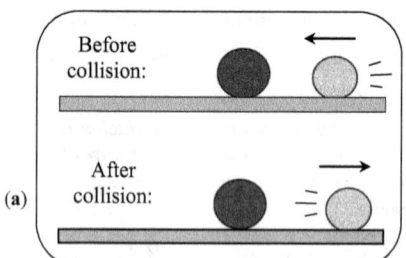 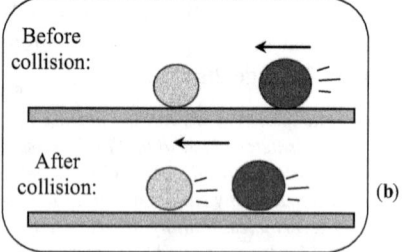

Fig. 2.22 The schematic representation of Descartes' law of collisions. (**a**) The moving body possesses a "lesser force" than the body it meets; (**b**) the moving body possesses a "greater force" than the body it meets

Descartes stated here one of the central principles of physics which has carried his name ever since—the conservation of the quantity of motion, or in our terms, momentum. However, he erroneously defined momentum, or the quantity of motion, as a scalar quantity. To represent this idea, we may write:

$$\sum_{in} mv_? = \sum_{out} mv_?$$

The uncertainty in the indexes is due to the separation of the law into two cases: a collision with a body possessing a smaller or bigger "force," which is obscure and erroneous. Descartes further refined this law in seven special cases: rules of collision.[56]

---

[54] Galili and Tseitlin (2003).

[55] Descartes (1644/1982).

[56] Chapter 1.

Consider the case of Fig. 2.22a. Descartes ascribes to the moving body "a lesser force" than in the body at rest. Descartes concluded that the body at rest will remain and that in motion will be pushed back with the same quantity of motion. This statement is simply wrong in a general sense; it approaches reality when a small body collides with a "wall" (a much more massive object). Under this specific assumption, the momentum of conservation appears to be a conservation of a scalar quantity mv. Perhaps, this observation brought Descartes to the erroneous generalization.[57]

Consider the case of Fig. 2.22b. Here the moving body possesses a "greater force" (quantity of motion) which overcomes the body at rest. An erroneous statement followed: Descartes predicted that the two bodies would move in the direction of motion after collision, and the speeds which would preserve the quantity of motion before the collision. Under this erroneous assumption, Descartes calculated the speeds of the two bodies, which did not match any reality.

In all six cases of asymmetrical parameters (masses or speeds), Descartes's claims were wrong. He was convinced, however, that he provided the true account of collisions of material bodies. In all the laws and rules, Descartes was consistent with his separation between rest and motion (violating Galileo's principle of relativity), with the asymmetry of the interaction of colliding bodies in accord with their "forces" (in contrast with Newton's third law), and the conservation of the scalar quantity of motion (instead of vector quantity). Descartes distinguished between the collisions of hard and soft bodies. This dichotomy has been preserved in physics. Although his rules addressed the collisions of hard bodies, Descartes commented regarding collisions of soft bodies. However, this too was wrong[58]:

> ... when they [moving bodies] meet a soft body, to which they can easily transfer all their motion, they immediately come to rest.

In reality, inelastic collisions do not necessarily cause the colliding bodies to stop. Seemingly, he kept in mind an analogy of the soft collision of a ball with a much more massive object.

On the Motion of Projectiles

Projectile motion was a traditional problem for those who addressed the physics of motion and so for Descartes. It is educative to realize how inferior Descartes' addressing the topic was in comparison with the triumph of Galileo[59]:

> Certainly, everyday experience of things that are thrown wholly confirms our rule. For there is no other reason why thrown [bodies] should continue in motion for any time after they have been separated from the thrower than that once moved they continue to be moved, until they are slowed by contrary bodies. And it is manifest that they usually are gradually retarded by the air, or some other fluid bodies in which they are moved, and hence their

[57] Descartes (1644/1982, p. 66).

[58] Descartes (1644/1982, p. 66).

[59] Descartes (1644/1982, p. 66).

*motion cannot last long. For we can experience air resisting the motions of other bodies by
our sense of touch if we strike it with a fan; the flight of birds also confirms the same thing.
And there is no other fluid which does not, even more manifestly than air, resist the motions
of projectiles.*

From his first law, Descartes deduced the difficulty treated by Aristotle with the
mechanism of *antiperistasis*; projectiles proceed with motion being placed in the
state of motion. This was a huge step, and it was quite different from Galileo's
understanding. Galileo took motion as a given. Without any question as to why,
Galileo provided a detailed account of motion, that is to say: trajectory, velocity
change, and acceleration, distance covered. Descartes stated a body moved in a
certain way but did not specify. He did not split projectile motion into two, horizontal inertial and vertical accelerated. Descartes and Galileo provided, thus, two complementary accounts: quantitative (empirical) and conceptual (rationalistic). Both
accounts are essential in science. They were further completed by Newton.

Final Synthesis in the Theory of Classical Mechanics: Newton

Ultimately, the new theory of motion was established due to the contribution of
Newton (1643–1727)—the outstanding English physicist and mathematician of the
seventeenth century. Newton founded the discipline we call today classical
mechanics.

As a teenage student in Cambridge, Newton studied thoroughly every word in
Descartes' *Principles of Philosophy*, copied them to his notebooks, and made his
comments. Newton studied the old views on motion gathered from the medieval
theory of motion as well as from the scholars of his era, those who shared his exploration of the seventeenth century: Galileo, Kepler, Descartes, Hooke, and Huygens.

Unlike Descartes, Newton's principle was not to leave
reality seeking a match with elaborated experience (such
as Kepler's precise *mathematical* statements regarding
the motion of planets) and experiment (such as Galileo's
heritage of the account of motion). Yet, his ambition was
to comprise the theory of the World more adequately and
better founded that that by Descartes.[60] Following a number of insensitive debates with Hooke in the Royal
Society, the new theory was comprised, in a rather short
time, from three revised and previously developed parts.
This treatise became fundamental for human culture—
The Mathematical Principles of Natural Philosophy, or
in short —*Principia* (1687). The debate with Descartes started from the title.

Newton

---

[60] Descartes' first version of *Principles* (1644) bore the title *Le Monde* (*The World*, 1633). In parallel, one of the three books which Newton wrote prior to the *Principia*, which was comprised of all
three books, was entitled *De Mundi Systemate* (*The System of the World*, 1728). The other two were
devoted to the theory of motion.

Newton modified Descartes' *Principles of Philosophy* to the **Mathematical Principles** of **Natural Philosophy**—two extra words representing another world-view and a different science.

From the outset, Newton presented his laws of motion,[61] which replaced all those introduced before him, from Aristotle to Descartes. In Table 2.1, they are given in comparison with those of Descartes.

Table 2.1 The laws of motion

Laws of nature in Descartes' principles	Laws of nature in Newton's *Principia*
The First Law of Nature: That any object, in and of itself, always perseveres in the same state; and thus what is moved once always continues to be moved	*Law I.* Every body perseveres in its state of being at rest or of moving uniformly straightforward, except insofar as it is compelled to change its state by forces impressed
The Second Law of Nature: That every motion of itself is rectilinear; and hence what is moved circularly tends always to recede from the center of the circle it describes	*Law II.* A change in motion is proportional to the motive force impressed and takes place along the straight line in which that force is impressed $\Delta(mv) \propto F\Delta t$
The Third Law: That a body, in colliding with another larger one, loses nothing of its motion; but, in colliding with a smaller one, loses as much as it transfers to that one $\sum_{in} mv_? = \sum_{out} mv_?$	*Law III.* To any action there is always an opposite and equal reaction; in other words, the actions of two bodies upon each other are always opposite in direction

What were the major changes introduced by Newton in the basic laws of motion? The changes were numerous and essential:

1. Both Descartes and Newton's first laws might seem rather similar. However, a closer look reveals essential differences.[62] Newton's law includes the relation of a body to the force impressed on it and describes the tendency of the body to preserve the state of uniform rectilinear motion or rest *at any moment*. The rectilinear uniform motion and rest are addressed as equivalent states, unlike the conception of Descartes. This equivalence of rest and motion presented a paradigmatic shift in physics knowledge and belongs to the central features of modern science.
2. The second and third laws of Descartes were totally modified. In a way, the second law of Descartes became a part of the first one, and a special case (circular motion) of the second of Newton's laws. Descartes' third law was replaced by a statement regarding any interaction between material bodies, not necessarily through collision (as in contact), but also at a distance (as in gravitation). Further, in Newton's laws, all interactions are stated to be through *forces*, replacing other confusing terms.

---

[61] Newton (1687/1999).
[62] Galili & Tseitlin (2003).

3. Newton essentially changed the statement regarding the quantity of motion with the principle of the quantity of motion sensitive to the direction of motion (a *vector*). Following this change, the conservation of momentum (Descartes' idea) became Corollary III of Newton's quantitatively precise second law[63]:

> *The quantity of motion, which is obtained by taking the sum of the motions directed towards the same parts, and the difference of those that are directed to contrary parts, suffers no change from the action of bodies among themselves*

This formulation also includes the essential change: the need for a *closed system* of bodies ("...*from the action of bodies among themselves")* in order to claim the conservation of the quantity of motion.

4. The fundamental fallacy of Descartes, who considered the interaction of bodies as asymmetrical processes (in terms of "winners" and "losers"), was corrected by Newton in his third law. Perhaps it was due to the debate with Descartes that Newton stated this claim as a separate law, although unlike the previous laws, presented as axioms (i.e., without demonstration), the third law was proved by Newton himself, based on the first law.[64] Interactions between any two bodies were stated to be reciprocal and symmetric: any acting force is accompanied by a correspondent force of reaction equal to it in magnitude.

5. Unlike Descartes's laws, Newton's laws matched experience *quantitatively* in a great number of physical situations, and therefore these laws were unanimously preferred over Descartes' laws by scholars everywhere, and very quickly. Empirical results, numerous observations, and predictions matched the Newtonian theory of mechanics with great accuracy.

Newton's theory established a firm basis for classical mechanics to replace the old theories of motion. From the seventeenth century and thereafter, classical mechanics have been firmly regarded as highly valid in the very broad area of reality around us.

The Three Laws of Newton

The laws of Newton as presented by him are rather different to those reproduced in contemporary textbooks. There is an important meaning to this difference.

1. The first law[65] is the most general law of motion in the view of Newton and so he put it as the first and principal law. It presents his words on the diachronic discourse on motion, starting from Aristotle and continuing for 2000 years, as we have traced here. It is a *differential* law, preferred by physicists, that is, it addresses the situation at each instant of time: each body preserved in its natural state of motion (rectilinear uniform) *to the extant it is not influenced* by an external force. Unlike that which is stated now by physics teachers, it is not a law

---

[63] Newton (1687/1999, p. 420).
[64] Newton (1687/1999, p. 428).
[65] Galili and Tseitlin (2003).

addressing the absence of forces.[66] Newton exemplifies his law by three physical situations, none of which is about the absence of force. They were projectile motions, the motion of a spinning top, and the motion of planets (Figs. 2.23). In each case, he pointed to the continuously changing states of motion.

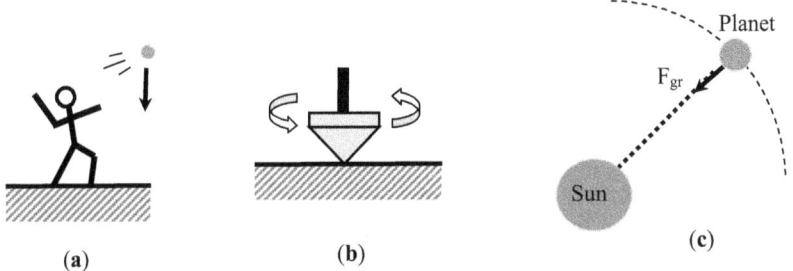

Fig. 2.23 The three cases Newton used to exemplify his first law of motion

2. The second law of motion presented by Newton also has no form commonly presented in textbooks. It is an integral law, addressing motion in a holistic manner. It is an integral law which summarizes the change of motion as being under the influence of a force during a certain period, and it states that the impact is the change of [amount of] motion (Descartes' term for momentum). The vector nature of the statement is presumed:

$$\Delta(mv) \propto F\Delta t \qquad (2.15)$$

This form creates a clear hierarchy between the two laws. The first law was the most fundamental law of motion. It was universally valid for all bodies in the universe, at each and every moment. It was followed by expanding to the motion described over time—the second law.

That was in the seventeenth century. What happened later, in the eighteenth century, was that the laws of motion were revised by another scholar—Leonhard Euler (1707–1783). He was an outstanding mathematician and apparently did not share Newton's worldview as a scientist-philosopher.[67] Therefore, in the view of Euler, the form of (Eq. 2.15) seemed rather vague and not sufficiently accurate. He developed it. Drawing on a simplified understanding of the first law,[68] he converted the second law into its *differential* form, thus destroying the grand design of Newton. Euler arrived, in a way, to an equivalent form:

Euler

---

[66] The lost meaning of the first law became apparent in its modern reformulation. For example, another great mathematician of the twentieth century reproduced the law as follows: "All bodies, if unimpeded, move at constant speed in a straight line, in technical terms, with uniform velocity" (Russell, 1959, p.190).

[67] Jammer (1961).

[68] Euler (1736/2008).

$$F = ma \text{ or } m = \frac{F}{a} \tag{2.16}$$

The transition from (Eq. 2.15) is straightforward, and it includes going to the limit of an infinitely short time period $\Delta t \to 0$. The meaning of the change escaped the attention of the consumers. Indeed, in solving any concrete problem one needs (Eq. 2.16). The first law possesses purely conceptual "philosophical" importance. Even after Euler, the second law of Newton was presented in its new "exact" form (Eq. 2.16) rather than (Eq. 2.15). Euler presented it as his own central law of dynamics without reference to Newton. Physicists adopted this form but named it after Newton, not Euler. Yet, due to Euler, the second law became differential. The immediate inference followed, in practice, the first law is a special case of the second. One may perceive that the great edifice of Newton's theory, in which the three laws were one whole complementing each other, was lost. To a great extent, the first law lost its Newtonian meaning, and the tradition of its modified understanding was established. In the twentieth century, the first law was often considered as a special case of the second[69] or a definition of force.[70] In light of the relativistic physics which introduced inertial observers, the first law was provided with the new meaning, often shared by many contemporary physicists[71]:

> ... (summarized here somewhat more clearly than by Newton [!]):
> Law #1. There exist reference frames (called inertial frames) relative to which every non-interacting particle moves with constant velocity.

This is the new meaning of the first law, which uses the vision of inertial frames of reference rather than the Newton reference.

3. The third law is the lowest in the hierarchy of the laws of motion. It refines the fundamental feature of interaction between each two material bodies and actually does not present a postulate being *proved* by Newton. Why, then, not to put it as a corollary? Seemingly, this was because Newton realized that in that claim (as, actually, in his two preceding laws), he was the first ever to state the symmetry of interaction. Nobody before him could imagine this symmetry preserved in the interaction between say, the Earth and an apple. Arbitrarily different objects apply the *same* force on each other. Is that not a paradox? Does it make any sense at all? In any case, it contradicts commonsense, but this aspect is already treated by the psychology of our conception, not by physics. In every generation after Newton, all students of physics struggled with this "absurdist" claim until being converted into people educated in physics.

Moreover, while Descartes understood interaction solely through contact and collision between bodies, the third law of Newton is silent on the issue. The new law included a new scenario never before considered in science—interaction at a distance. This heralded a new era in physics. Nobody before Newton claimed this kind of natural phenomenon.

---

[69] e.g., Taylor (1941, p. 130).

[70] e.g., Arons (1990, p. 52).

[71] e.g., Reif (1995, p. 95). This approach is often adopted in the advanced courses of physics.

Transformation of the First Law

It is of special interest to consider how the original form of the first law was correctly translated from the Latin original in the twentieth century[72]:

> *Law I. Every body perseveres in its state of being at rest or of moving uniformly straight forward except insofar as it is compelled to change its state by forces impressed.*

However, the first translation of the *Principia* from Latin was by Andrew Motte in 1729, 2 years after Newton's death. Motte was neither physicist nor philosopher. In that first English edition, Motte translated the first law as following[73]:

> *Law I. Every body <u>perseveres</u> in its state of rest, or of uniform motion in a right line, <u>unless</u> it is compelled to change that state by forces impressed thereon.*

In comparing the two translations, the reader discovers that the Latin equivalent of *except insofar as* was replaced with *unless*. Were they equivalent?—No... The new text implied the replacement of a gradual dependence of the state of motion on the impressed force with a clear case of a lack of external (impressed) force.

In the 1930s, Florian Cajori, a historian of mathematics, revised Motte's old translation again. This revision made another change to the first law. In the new form it became[74]:

> *Every body <u>continues</u> in its state of rest, or of uniform motion in a right line, <u>unless</u> it is compelled to change that state by forces impressed upon it*

In comparison with Motte's version, *continuous* replaced *perseveres*. Again, was this an equivalent change?—In no way. The new text excludes the resistance of a body to the impelled force (the factor which later became the inertial mass). In the absence of force, the body simply continued to be in the same state of motion. It is in this form that the first law was repeatedly used in a vast number of physics textbooks. Clearly, there was a major change from the original meaning—the external force and the resistance to the change of state were just removed from the context of the first law. This was the price of a lack of expertise, the error which was corrected much later and therefore had no impact on the established tradition in education.

Summary of the Theoretical Progress

We first may summarize the development of the classical theory of motion. A radical change of understanding took place regarding the natural state of a body. While the Aristotelian understanding identified rest as a natural state, after Descartes, Classical Physics states that uniform motion (constant velocity) presents a natural state (Fig. 2.24a). With regard to motion, while in the view of Aristotle motion was a process of transition from one state to another (Figs. 2.1 and 2.24b), in the

---

[72] The modern translations from Latin by physics experts: Krilov—Newton (1687/1989), and Cohen and Whitman—Newton (1687/1999). For discussion, see Galili & Tseitlin (2003).

[73] Newton (1729/1964, p. 23)

[74] Newton (1729/1952, p. 14).

Newtonian understanding, a body continuously changes its state under the influence of an external force (Fig. 2.24c).

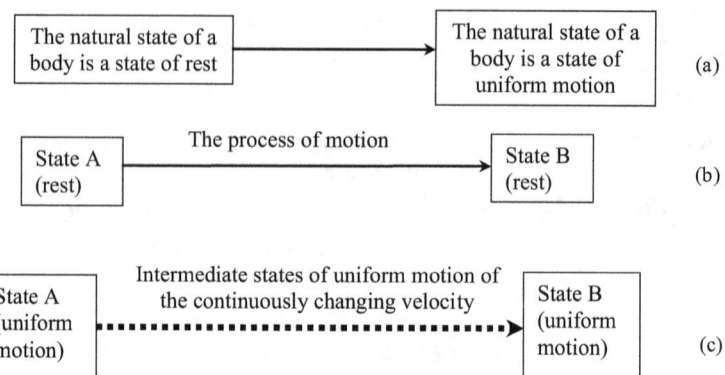

Fig. 2.24 The radical change in understanding of motion. (**a**) Understanding of the natural state of a body. (**b**) Understanding of Aristotle. (**c**) Understanding of Newton

The Classical Theory of Mechanics was formulated by Newton as based on three laws of motion. In the first law, Newton described the motion of a body as changing its state at any instant of time under the influence of external forces. The second law summarized the change of quantity of motion taking place under the influence of external force during a certain time interval. In the third law, Newton stated the interaction between any two bodies as reciprocal and equal.

Later, Euler transformed the second law into its differential form ($\mathbf{F} = \mathbf{ma}$) as it remained in use. As a direct result of this transformation, Newton's organization of the laws of motion significantly changed. The first law is often considered as solely addressing the case of zero net force exerted on the body, that is, the state of uniform motion ($\mathbf{F} = 0$, $\mathbf{V} = \text{const}$).

Note that the dichotomy of *natural* and *violent* motions was removed only from classical mechanics by Newton. Thus, the history of motion can be presented in the flowchart (Fig. 2.25) which mentions the ideas regarding motion from Aristotle to Newton.

As common in science, the discourse on motion continued throughout history among the scholars of different periods (Fig. 2.26) and culminated in the seventeenth century. Until then, this discourse produced three theories of mechanics—Aristotelian, Medieval, and Newtonian—but only that of Newton was proved to be correct.

Newton's theory provides a correct account of motion, valid for the objects in the wide scale of masses, distances, and time intervals including our regular environment, the subject of investigation from the dawn of science. A further stage in our understanding of motion was reached in the twentieth century in two theories—the *Theory of Relativity* and *Theory of Quantum*. Each of them established its own core principles, and each expanded the validity of mechanics to the different aspects of reality, beyond the area where classical mechanics holds true. They were later unified into the *Quantum Field Theory* which, however, did not include the modern theory of gravitation, the *Theory of General Relativity*.

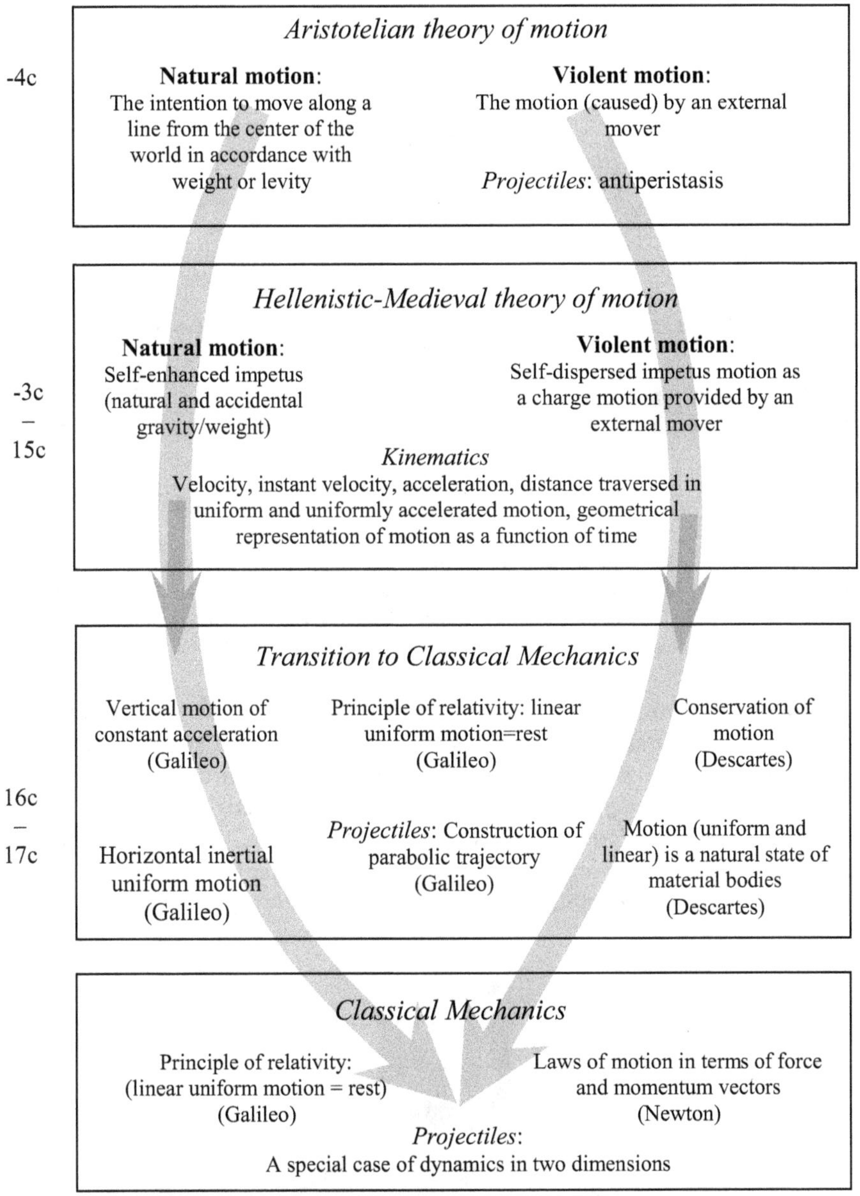

-4c

Aristotelian theory of motion

Natural motion:
The intention to move along a
line from the center of the
world in accordance with
weight or levity

Violent motion:
The motion (caused) by an external
mover

Projectiles: antiperistasis

-3c
–
15c

Hellenistic-Medieval theory of motion

Natural motion:
Self-enhanced impetus
(natural and accidental
gravity/weight)

Violent motion:
Self-dispersed impetus motion as
a charge motion provided by an
external mover

Kinematics
Velocity, instant velocity, acceleration, distance traversed in
uniform and uniformly accelerated motion, geometrical
representation of motion as a function of time

16c
–
17c

Transition to Classical Mechanics

Vertical motion of
constant acceleration
(Galileo)

Principle of relativity: linear
uniform motion=rest
(Galileo)

Conservation of
motion
(Descartes)

Horizontal inertial
uniform motion
(Galileo)

Projectiles: Construction of
parabolic trajectory
(Galileo)

Motion (uniform and
linear) is a natural state of
material bodies
(Descartes)

Classical Mechanics

Principle of relativity:
(linear uniform motion = rest)
(Galileo)

Laws of motion in terms of force
and momentum vectors
(Newton)

Projectiles:
A special case of dynamics in two dimensions

Fig. 2.25 The flowchart of the accounts of motion. The arrows signify the originally separated treatments of natural and violent motions unified only in the theory of Newton

The Theory of Relativity is valid *also* at very high velocities (close to the speed of light) and very big masses (stars, black holes), while the Quantum Theory expanded the validity of mechanics to very small masses of atomic scale (atoms, molecules, elementary particles). In the two realms where these modern theories are valid, classical mechanics fails to adequately account for natural phenomena. Yet,

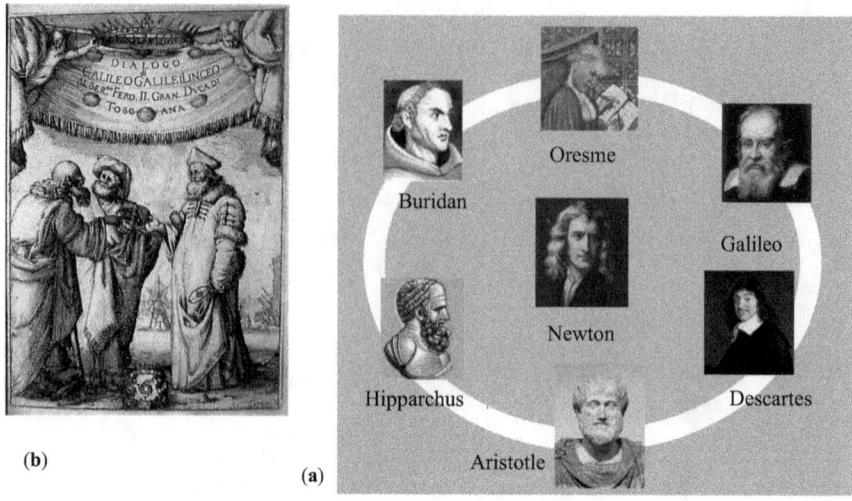

Fig. 2.26 (**a**) The participants of the historical discourse on motion which converged in the seventeenth century to the Newtonian understanding. (**b**) Another diachronic discourse on the system of the world was depicted as a conversation of Aristotle, Ptolemy, and Copernicus on the title page of Galileo's *Dialogue Concerning the Two Chief World Systems* published in 1632 in Florence

both theories—relativity and quantum—are able (in principle) to describe the realm of reality where classical mechanics is valid. Their accounts, however, are much more complicated and therefore classical mechanics remains in extremely wide use.

Thus, addressing the present account of motion, we may talk about all three theories of motion, those of Newton, Einstein, and Bohr (Fig. 2.27), as *correct*, each in its area of validity (these areas are not separate but overlap). It is not a mosaic (or patchwork) but a more complex relationship.

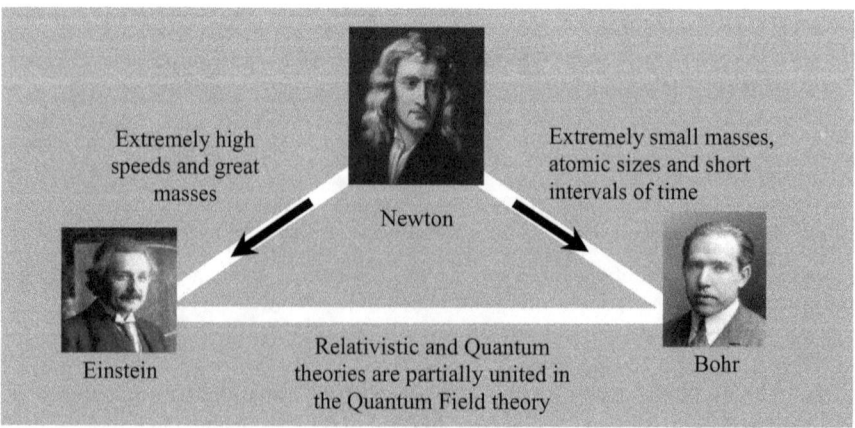

Fig. 2.27 Three correct theories of motion and their founders: classical, relativistic, and quantum. The image of Niels Bohr was chosen as being one of the founders of the Quantum Theory

Philosophical Considerations

Philosophy is a discipline different from science. It is important to know that, as well as that philosophy is of essential importance for science. Philosophy guides and organizes our knowledge, provides meaning to it, and answers the questions about what we do actually know. What is the status of this knowledge? What is the nature of the knowledge we possess? How do, how can we construct this knowledge? What is its validity? How do we prove its reliability? These questions are indispensable in physics, and they appear either spontaneously through our curiosity or intentionally, in an enlightened curriculum. One way or another, they seem unavoidable in any meaningful study of science education.

To illustrate the depth and specific power of philosophy, one may consider Zeno's[75] paradoxes of motion. There are several of them, but their logic is similar. Consider the motion of A toward B at rest. To reach B, A has to cover half of the distance. To cover the half, A has to cover the half of the half and so on. Since the division can be endless, A will never reach B, being occupied with an infinite number of missions. The inference becomes striking if one puts it in the figurative form of a story: Achilles (the fastest runner) will never reach a tortoise (apparently a slow mover); an arrow will never reach its goal and so on. Zeno inferred a challenging claim: if our logic is unbeatable, there is no motion at all, motion is an illusion... — not a weak statement... and the scientific answer to it was not immediate ...

Zeno of Elea

Fig. 2.28 This fresco (the fresco from the El Escorial palace in Spain is by B. Carducci or P. Tibaldi of 1588. https://commons.wikimedia.org/wiki/File: Zeno_of_Elea_Tibaldi_or_ Carducci_Escorial.jpg) depicts the philosopher Zeno with his students. Through an allegorical movement, the philosopher demonstrates the choice faced by anyone in choosing the way/door to truth or falsity. The fresco may even better illustrate the answer of Diogenes, who by a simple move demonstrated the actual falsehood of the logical claim that motion is a mere illusion

---

[75] Zeno of Elea (490–430 BC) was a pre-Socratic Greek philosopher from southern Italy and a member of the Eleatic School, founded by Parmenides.

One may say, OK, who cares? There must be some defect in Zeno's logical chain, but it is not our business as physicists. We, physicists, deal with the real nature, and there is motion there, it is given, and it is just obvious... ("No motion? Hm..." said another Greek philosopher, Diogenes, and simply walked away...) (Fig. 2.28). Indeed, the paradox revealed the weakness of ancient thinking in dealing with infinities.[76] Yet, Aristotle could not pass by... motion was among the central concepts of his *Physics*. He devoted much effort to treating the challenge of Zeno's paradoxes. His argument was not very persuasive, but it included several remarkably deep claims. Among them, "distance is not a collection of points," "time is not a collection of instances," "motion is not a collection of stills"—a huge advancement. His latter claim, for instance, was a forerunner of Descartes' understanding of the status of motion as a natural state, later adopted by Newton as fundamental. This initially philosophical debate was extremely important, instigating thought on the most fundamental concepts in physics and mathematics. It brought scholars to calculus, the concept of instant velocity as the limit of two infinitely small intervals of distance and time, so important to physics.

Philosophy suggests two basic aspects: *what* motion is (ontology) and *how* that knowledge was organized, produced, and validated (epistemology). We will view them in three basic periods of the science of motion—Hellenic (Aristotle's theory), Hellenistic-Medieval (the theory of impetus), and Modern (Newton's theory).

Motion as Actualization

As was described above, Aristotle distinguished between *natural* and *violent* motions in the sublunary world. The *natural motion* takes place without any agent and presents a spontaneous process of transition between the states of rest. This motion is vertical toward and from the center of the world. The effective cause for such motion was the violation of the order of the elements. When this happens (when a stone is raised), a certain potentiality of motion is created. It actualizes into reality—motion—if it does not meet the resistance of the environment. Thus, it is stated that *natural* motion is a process of the *actualization* of *potentiality*.

Violent motion can be considered from a similar perspective. An external mover possessing certain potentiality exerts force on a body causing motion while converting that potentiality into the reality of motion. The body leaves the state of rest and moves. Continuous motion presents continuous actualization. Aristotle's theory framed locomotion, natural or violent, in terms of the *actualization* of the potential to the actual.

This conceptualization (actualization of potentiality) remained valid throughout the whole history of science, including modern physics. It took different forms and

---

[76] Philosophers kept on thinking and the solution came in the Middle Ages (Oresme in Paris demonstrated the solution) and in modern times—in some cases, an infinite number of terms can be added to a finite number.

names in representing concepts. Potentiality was behind the impetus in the medieval conception of motion; it was behind the potential and potential energy in classical physics; it is in the probability of transition between the atomic states and superposition of states in quantum mechanics.[77] Actualization became a conceptual frame of physics processes.

Motion as "Charged" Activity

Greek natural philosophy was inherited by medieval science. Although scholars departed from or modified some elements of Aristotle's theory, they often considered those changes as refinements rather than replacements. The new understanding of motion was one such modification. In impetus physics, the cause of violent motion remained local (not distant), but the mover was placed inside the body—a specific faculty received from outside, causing motion. In a way, impetus served as a charge of motion or an internal engine. In the natural falling, the impetus increased itself and was dispensed when meeting any impediment or resistance to the medium.

With the introduction of impetus, the Aristotelian mechanism for moving projectiles (antiperistasis) was abandoned and the existence of a void became a logical possibility. Yet, interaction between bodies remained solely via actual contact. Impetus preserved the opposition of motion and rest. Moreover, though the scholars of this period, Buridan, Oresme and Nicolas of Cusa, realized the *relativity* of motion and questioned the unmovable Earth,[78] possession of a certain magnitude of impetus undermined the understanding of relativity, reducing it to a mere *kinematic* relativity. It differed from the modern understanding, known as Galileo's principle. The latter presumes *dynamic* relativity (total indifference of a body to uniform motion and the indistinguishability of physical phenomena to such motion), thus dismissing the absolute understanding of impetus as a certain entity.

Motion as a State

In classical physics, the uniform, rectilinear motion of a body is considered as a *natural state* which therefore does not require any cause or support. The mover as the cause of motion was dismissed. This understanding had its seeds in medieval thought but was first claimed by Descartes. Newton introduced the universal laws of motion of every kind. Falling, which before Newton did not require any cause but inherent gravity, was understood as being caused by the force of gravitation. Inertia

---

[77] We touch on this concept again with regard to forces in Chap. 5.

[78] e.g., Crombie (1959); Dijksterhuis (1986); Pedersen & Pihl (1974).

was refined to the concept of inertial mass—the property of a body to "resist" to *changing* the states of motion rather than just motion, as was understood before. Arbitrary motion was understood as a continuous transition from one state to another. The change of a state of motion was explained by the interaction with other bodies—the action of forces—including action at a distance (gravitation) across an empty space. Galileo recognized the relativity of motion in its physical sense—the equivalence of reality, the natural phenomena as observed by the observers in the systems moving at different constant velocities. The ontological development with respect to motion is summarized in Table 2.2.

Table 2.2 The ontological development of the conception of motion

Motion is the continuous change of place in time

Hellenic Science	*Hellenistic and Medieval Science*	*Classical Mechanics*
Rest is a natural state		▪ Rectilinear uniform motion presents a natural state
Natural motion is spontaneous, without external agent	▪ Natural motion is due to gravity (the weight of a body) and increasing impetus	▪ Motion (rectilinear uniform) is physically equivalent to rest or motion with any velocity (the relativity principle of Galileo)
Violent motion is due to the force exerted by an *externalmover* in contact with a body	▪ Violent motion is due to the *internalmover* – impetus, an imposed faculty of a body	▪ Change of motion state is due to the external force in contact or at a distance

Motion is a*process* of continuous transition between two natural states of rest Motion is a *continuous transition* of a body *through natural states*under the forceexerted on the body

According to Koyré,[79] two important features characterized the scientific revolution of the seventeenth century. First, the destruction of the Greek cosmos, i.e., the replacement of the finite and hierarchical order in the world of space, with the infinite universe, comprised of void and masses. Motion of masses was governed by forces and obeyed universal laws (Newton's paradigm). The second change was in the geometry of space, the substitution of the Aristotelian space of a plenum of radial geometry by the infinite space, a void, described by Euclidean geometry.

The classical theory of motion considered all bodies as placed in absolute space, running in absolute time. Newton considered motion in this space as absolute

---

[79] Koyré (1968).

motion. He admitted that in practice, one may not distinguish between absolute position and motion, and relative positions with respect to other bodies—apparent motion. Newton believed that he provided evidence of absolute space in his experiment with a rotating bucket of water.[80] The invalidity of this argument was shown later by Ernst Mach in his *The Science of Mechanics*[81] which heralded the age of modern physics. Mach pointed to Newton's neglecting of other objects in the environment (the Earth, for instance) as well as of the distant stars. It was this neglect that dismissed Newton's claim regarding absolute space and absolute motion.

Newton introduced gravitation as a force at a distance governing motion. In parallel to Newton's unifying natural and violent motion, in the new physics, Einstein demonstrated the physical equivalence of inertia and gravitation. He dismissed the separation between internal and external forces removing Newton's absolute space-time as a container of moved bodies. In the new mechanics, all motions and laws that govern them have to be formulated in relative terms (Mach's principle).

Epistemological Maturation in Considering Motion

In parallel with the ontology of motion, an essential change took place in the epistemology, in considering the way the knowledge was produced and validated. Unlike in ontology, the change of epistemology employed by scholars presented a cumulative and refining process of enrichment and maturation toward the method currently employed in science.

1. *Hellenic Science*

In the Hellenic culture, scholars sought for ideas and principles, standing behind principles and governing natural phenomena. Such was the early science—natural philosophy. Science distanced itself from human practice drawing on experiential knowledge. Hellenic science looked beyond the simple monitoring of natural things, trying to reveal the unchangeable behind the observed changes—the permanent essence of things and processes. By careful *contemplation*, an analysis of reality, scholars tried to elicit the nature of cosmos—the ordered, regulated world in which everything has its goal, place, and destination, the opposite of chaos. This was not an easy task in a world full of variety and changes. Yet, people believed that it was feasible, nature was considered as open and conceivable. Plato did not account for motion, perhaps because he could not offer a description in mathematical terms or symbols. Aristotle, his student, changed the approach and did create an all-inclusive theory of motion. He distinguished and separated between physics and mathematics and provided physics with solely qualitative explanations, pointing to the causes of

---

[80] Newton (1687/1999, pp. 412–413).

[81] Mach (1883/1919/1989, p. 233).

phenomena, analyzing the observed and making logical inferences, yielding claims and features of reality.

As a tool for this process, Aristotle introduced an inductive-deductive loop, implying circular refinement in a self-correcting process of conjectures and refutations in knowledge construction.[82] Thus, observing a stone sinking in water, flames of fire rising, or objects dragged on the ground Aristotle made generalizations (induction) regarding natural motion for all objects. Given the established principles, he made conjectures regarding other events (deduction). In case of failure, correction of the applied principles had to be considered, and the process continued.

Aristotle developed a speculative explanation of antiperistasis for projectile motion. The method drew heavily on careful observation and logical analysis but did not include experimentation. It rather drew on experience with real events, "natural experiment," and employing the established rules of logic. In such a process, less obvious causal connections were easily missed. With regard to motion, this method failed to account for some cases of motion later used to criticize the Aristotelian theory. Yet, the logical analysis using continuous induction-deduction loop was adopted as the scientific method, as well as the explanation for natural phenomena by reduction to the basic principles of a theory. These methods continued to be used in all the following developments.

2. *Medieval Science*

The scholars of the medieval period continued to develop the inductive-deductive method of Aristotle while addressing motion. This knowledge as an internally coherent all-inclusive theory remained of high value. The new idea of impetus allowed saving the problematic cases of projectiles, spinning top, etc. There was criticism too. Ockham criticized introducing a new faculty to a moving body—impetus—to explain its motion. In his view, a new concept was not more than a new word and should be avoided as superfluous. In a way, it was the next step toward recognizing motion as a natural state, made later by Descartes.[83] Avoiding unnecessary concepts, choosing the simplest (yet equally adequate) explanation received the name of *Ockham Razor* or the principle of *parsimony*. The problem was that despite its apparent intellectual attractiveness, the principle was vague in meaning: each scholar could define what is "simpler." For instance, both suggesting involvement of supernatural causes (God's design) and avoiding such causes can equally claim to be "simpler."

The specific of medieval scientific thought was its undisputable theological commitment. It essentially influenced the questions raised in the study of motion. For instance, scholars realized the shaky status of claiming the Earth to be unmovable and as occupying the center of the universe. Buridan, Oresme, and Nicolas of Cusa understood that relativity of motion implied that the Earth's motion, rotational and linear, is, at least, equally probable. Motion of the Earth might even be preferable

---

[82] Losee (1993, p. I-35, p. 6).

[83] Seemingly the first step was Aristotle's claim that motion should not be reduced to a collection of stills.

given the huge distance to the stars (thus rotating with enormous speed to close a cycle in 24 h). In such a case of uncertainty, they decided, the statement of the scriptures should be taken as decisive. Empirical testing was not seen. God remained the "designer" and "creator" of the world—Pantokrator.[84] The natural philosophers were theologians. Aristotle's theory of motion was adopted while providing theology with the strength of a scientific theory. Yet, the theological dictum of an appeal to the authority of the Bible as the ultimate judge was so strong that it undermined Aristotle's prescription to draw on the logical analysis of experience for verification and the adjustment of theoretical claims where required.

One of the central methodological features of science—mathematization—came to the fore with regard to motion. In fact, it received a certain freedom. Christian scholars allowed two ways of demonstration, by principles (physical theory) and by mathematics (modeling, in modern terms).[85] This allowed, for instance, tolerating Ptolemy's account for the motion of planets (as well as other astronomical accounts) despite its essential contradiction with some principles of Aristotelian theory and the silence of the Bible. Mathematical demonstration looked neither unique nor as obliging changes of conceptual nature.[86]

God the Geometer – in designing the World. Medieval illustration

Scholars made tremendous progress in the account of motion. They introduced *intensive* versus *extensive* qualities in kinematics.[87] Gerard of Brussels, Oresme of Paris, and the Merton School of Oxford employed new tools and approaches that allowed a *quantitative* account of motion: *instant and average velocity, acceleration, uniform*, and *uniformly deformed movements*. Oresme introduced the breakthrough methodology—a graphical representation of motion as a function of time. It was an essentially new language used by all in the following scientific revolution, often together with the formal results obtained by medieval scholars.[88] We teach them all in schools today. Such was, among others, the functional dependence of the distance traversed (mean speed theorem). It changed science forever.

3. *Modern Science*

The transition to classical mechanics included the epistemological revolution. The dialogue with the Aristotelian and medieval theories brought the new method to include previous methodology in a complementary synthesis of the rational and the empirical. Classical mechanics was rationally arranged as a hierarchical and

---

[84] The picture shows an illustration from a Biblical manuscript of the thirteenth century in France. https://en.wikipedia.org/wiki/Biblical_cosmology#/media/File:God_the_Geometer.jpg

[85] Pedersen & Pihl (1974, p. 268).

[86] In the Copernicus theory, the Sun is not in the center of the Universe, and epicycles were used as well. "Nobody" cared, it was only mathematics.

[87] Clagett (1959, pp. 414–416, 541–556), Pedersen & Pihl (1974, pp. 217–228).

[88] Hannam (2009), Galili (2011).

coherent knowledge system. The new science was not all inclusive as was that of Aristotle. Mechanics left electricity, magnetism, and thermodynamics for future treatment. The first concern was the motion of bodies.

The most important epistemological innovation was to *experiment with controlled parameters*. During Hellenic and Medieval periods, scholars often avoided artificial experimenting, preferring the natural addressing of reality. In contrast, the new science designed and investigated reality in the laboratory, making inferences to the reality outside.

Through using apparatus, scholars produced systems not readily available in nature. For example, the evidence for the law of inertia could be attained in the laboratory by reducing friction, often dominant in a regular environment. Francis Bacon, in the early seventeenth century, called this process an "interrogation" of nature. Researchers investigate nature in various deliberately organized settings. Such epistemology would seem erroneous to Aristotle and the medieval scholars, who avoided the changing of the natural conditions. Laboratory experimentation with monitoring of parameters became, and has remained, the central tool of science.

When possible, modern science investigates artificial "reality" and draws on an abundance of experiments. Such was the strategy of Galileo in his investigation of motion by means of the inclined plane and pendulum. Nevertheless, in no way does modern science exclude precise observations of nature for data accumulation, elaboration, and revealing law-type regularities. The laws of planetary motion were established by Kepler in 1605 through an analysis of the data collected by Tycho Brahe during years of painstaking observations. We may summarize the epistemological development of the theory of motion as it took place from Aristotle to Newton (Table 2.3). As already mentioned, the relationship between the different periods was not a simple replacement.

Table 2.3 Epistemological development of science with respect to motion (summary)

Hellenic science	*Medieval science*	*Classical mechanics*
Contemplation of nature seeking general regularities using inductive-deductive self-correcting loop	Logical analysis based on principles and rules	Controlled parameters, experimentapproach
	Appeal to authority and evidence	Rational based theory for the certain domain of reality
Establishing a unified coherent theory based on principles (experience based)	Subjugation to theology	Complementarymethod combining rational and empiricist approaches
Logical analysis seeking for the objective/natural causes of natural phenomena	Conceptual (physical principles) and mathematical accounts of reality not necessarily coherent	Coherent theory – experiment mathematical account

The Cumulative Nature of Science

Objective as a *collective* product, scientific knowledge remains vulnerable to sub-jective traits introduced by *individuals*—a continuous and sophisticated enterprise. For the limited abilities of each scholar, scientific knowledge cannot be but cumula-tive and discursive. The knowledge of motion illustrates this nature.

Aristotle debated the challenging claims of Zeno, Democritus, and others. In medieval science, being cumulative became the paradigmatic slogan of Bernard of Chartres, the twelfth-century French medieval scholar. His colleague John of Salisbury wrote in 1159[89]:

> *Bernard of Chartres used to say that we are like dwarfs on the shoulders of giants, so that we can see more than they, and things at a greater distance, not by virtue of any sharpness of sight on our part, or any physical distinction, but because we are carried high and raised up by their giant stature.*

Isaac Newton used the same metaphor in the letter to his rival Robert Hooke in 1676:

> *If I have seen a little further it is by standing on the shoulders of Giants.*

While presenting his account of motion, Aristotle reveals many other voices who all contributed to the consolidation of his claims. It became an epistemological norm; scholars use the results of their predecessors in both a positive and negative sense. With regard to Galileo's results on motion, Edward Grant asserted[90]:

> *Oresme's geometric proof and numerous arithmetic proofs of the mean speed theorem were widely disseminated in Europe during the fourteenth and fifteenth centuries and were espe-cially popular in Italy. Through printed editions of the late fifteenth and early sixteenth centuries, it is quite likely that Galileo became reasonably familiar with them. He made the mean speed theorem the first proposition of the Third Day of his Discourses on Two New Sciences where it served as a foundation of the new science of motion. Not only is Galileo's proof strikingly similar to Oresme's, but the accompanying geometric figure is virtually identical, despite a 90° reorientation that had already been made by some medieval authors.*

It could not be different: scientific knowledge is too complex and extended for an individual to create it alone. In a sense, scientific knowledge resides in numerous minds. Karl Popper introduced a powerful metaphor of the *third world*, which incorporates scientific knowledge[91]:

> *By world three I mean the world of the products of the human mind, such as languages; tales and stories and religious myths; scientific conjectures or theories, and mathematical constructions; songs and symphonies; paintings and sculptures. But also, aeroplanes and airports and other feats of engineering.*

---

[89] Crombie (1959, p. 27). See the allegorical representation of this metaphor in the medieval stained-glass windows of Chartres Cathedral in Chap. 10.

[90] Grant (1977, p. 58).

[91] Popper (1978).

At the same time, the theory of motion is not reached through a simple accretion. In the course of progress, people refined knowledge and changed the epistemological as well as ontological paradigms. It was done in a dialogue with previous knowledge and was inherently based upon it.

Educational Considerations

The new vision of the physics curriculum requires inclusion of the genesis of knowledge of motion. The arguments follow[92]:

1. The exposure of the discourse of motion creates an authentic image of physics as a domain of knowledge constructed in a conceptual dialogue. It provides a perspective often missed, even in those with intensive training in standard problem solving ("the forest is not seen for the numerous trees").
2. The alternative accounts of motion historically preceding the current understanding create a *contrast* background of classical mechanics, thus emphasizing its essential concepts.
3. The alternative accounts of motion create a *space of learning* in which the subject matter can be effectively learned though *conceptual variation*. These alternative accounts stimulate learning through comparison with the essential points of the learned, making the subject matter transparent rather than intuitive.
4. Old theories of motion belong to the cultural heritage and establish cultural knowledge. The discourse on motion spreads over different cultural periods. Appreciation of human culture stipulates liberal education and upgrades the common disciplinary training. It reveals the non-pragmatic aspects of scientific inquiry, universally valid for different societies.
5. The old theories of motion practiced different epistemologies. The contemporary epistemology is pluralistic, fusing rationalism with empiricism. Learning historical approaches and experiences, their successes and failures, effectively argues for plurality in inquiry-based learning.
6. The history of motion includes issues of a religious nature and displays different kinds of human inspiration. Aristotle, Hipparchus, Avicenna, Buridan, Oresme, Galileo, Descartes, and Newton were very different in this regard, whereas they all significantly contributed to the science of motion. Discussion of this matter can clarify the fusion of objective and subjective aspects in scientific knowledge.

---

[92] Chapters 6, 7 expand more on the subject.

Students' Conceptions

Much evidence has been accumulated regarding the naïve ideas of students reminding us of the claims of the Aristotelian theory of motion and of the theory of impetus.[93] Students' conceptions are context dependent. Considering dragging and pushing bodies, students often show the Aristotelian force-motion relationship, whereas in the context of motion (projectiles), the idea of impetus—the "charge" of motion—prevails. Impetus is often imagined in the role of fuel for motion.

Circular motion often provokes students' understanding in terms different from the Newtonian, but similar to those held by Kepler, Descartes, Huygens, and Borelli.[94] Students often consider the uniform circular motion of a body as a state of force competition: the moving force compensates the inertial force in the tangential direction and the centrifugal force on the body compensates for the centripetal force in the radial direction (Fig. 2.29). Such frequent misconceptions make relevant addressing the historical accounts of a circular motion of a similar kind. Learning about historical debate and the arguments used by the participants may help in the conceptual change of the learner through cognitive resonance with students' comprehension.

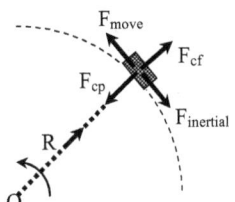

Fig. 2.29 Erroneous ensemble of forces exerted on the body in a circular motion. In 1666, Italian Giovanni Alfonso Borelli considered circular motion of planets as a kind of competition between centripetal and centrifugal forces (Koyré (1973, p. 509))—an erroneous view then and a frequent misconception of students today

Borelli

Media for Learning

To facilitate a shift in the perception of motion, we provide topics, questions, and tasks to support the learning:

- *The Aristotelian theory of motion* explains the motion of projectiles by the mechanism of antiperistasis. Consider a possible violation of the conservation laws of mechanics by this mechanism.
- *Theories of impetus* explained the motion of projectiles by providing them with a kind of charge of motion that changes their quality. Is an internal cause of motion possible from the perspective of classical mechanics?
- Discuss the following *thought experiment* which aims to dismiss the idea of impetus. Consider a coin tossed upward. We know how it moves: rises up, stops

---

[93] Whittaker (1983); McCloskey (1983, pp. 299–324).

[94] Galili & Bar (1992); Galili & Kaplan (2002).

 for an instant, and falls down. By throwing the coin upward, one violates the "world order," as Aristotle understood it. Hipparchus suggested the way to compensate for the tendency of heavy bodies to seek the center of the world: for the thrower to provide the body with *impetus*. We may now ask, what would happen if we compensated exactly for the gravity of the coin at the instant it reached the summit of its flight? Remaining uncompensated, the impetus in the body would immediately throw it upward due to the impetus it still holds. In contrast, if we follow the classical mechanics of Newton, the event should be different. If one switches off the gravity at the moment that the body is at the top of its flight, it would simply remain at rest and stay there, as long as there is no gravitational force.

We have, thus, two very different scenarios depending on which theory is correct—a crucial experiment: if the coin does not fly upward at the moment of compensation gravity, it would indicate that there is no impetus.[95]

The very possibility that the impetus theory can be falsified in an experiment is what makes this theory scientific in the sense Karl Popper called the Principle of Falsification. The teacher may initiate the discussion by asking:

1. Suppose that the impetus theory is correct. What would be the motion of the coin at the highest point of its flight if one supports it from below at that moment?
2. Suppose that classical mechanics is correct. What would happen with the coin if one provides such support at that moment?

   • One may say that impetus took the explanatory role we use to ascribe to force, momentum, and energy in classical mechanics. The following questions may indicate some limitations of the impetus theory, and hence, the advantages of classical mechanics.

1. How does the impetus theory explain that the body thrown up begins to fall down? In other words, why does the body not remain at the top of its trajectory?
2. How does classical mechanics answer the same question?
3. How does the impetus theory explain that the body thrown up begins to fall down and accelerate?
4. After the collision of two identical steel balls, they bounce away from each other. Can the impetus theory explain why the balls do not continue moving together after collision?
5. How does classical mechanics explain the same situation? Why do the balls not stick together after collision?
6. Compare Leonardo's sketch of a projectile with the modern stroboscopic photo of a similar motion (Fig. 2.30). Are the trajectories the same? What can be said

---

[95] For those who intend to perform such an experiment, we may mention that although we cannot switch off gravity, it may be compensated for by using another kind of force such as the electromagnetic force.

about the trajectory shown by Leonardo? Did he err in his depiction or another rationale? Explain your answer.

7. Consider the explanations of the Principle of Relativity by Galileo. Compare its explanations by Giordano Bruno and Galileo. Are they equivalent? In what way they reflect the difference between the understanding of motion by medieval and modern sciences?

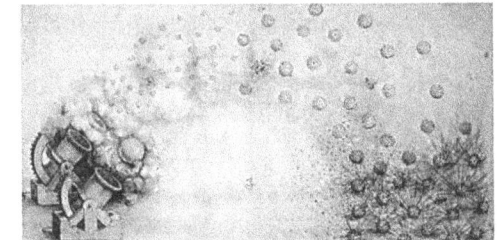 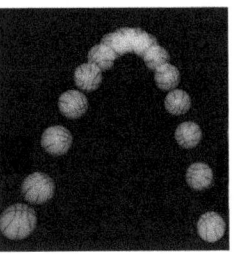

(a)   (b)

Fig. 2.30 (**a**) Drawing of projectiles by Leonardo da Vinci from the fifteenth century. (Da Vinci, Codex Atlanticus, fol. 24 r-a, Calvi (1956, pp. 284–285), http://codex-atlanticus.it/#/Detail?detail=31) (**b**) Modern stroboscopic photograph of a moving ball bouncing from the floor

Concluding Remarks

The title of this excursus—*De Motu* (*motion* in Latin)—was chosen for several reasons. In fact, *De Motu* became the name for a genre in physics discourse on motion in physics. As such, it was used in the multiple studies appearing at the time when the understanding of motion reached the establishment of classical mechanics: the late medieval period—the era of early modern science in Europe. This popularity signified the increased interest in the research of motion after a long time when the interest was to the problems in statics. Understanding of motion is fundamental in mechanics and represents its core. Yet, historically, mechanics received its name not as a theory of motion but as a theory of devices supporting manual labor (simple machines were the wheel, wedge, lever, screw, inclined plane, and pulley). In this sense, *de Motu* defines a specific domain of physics knowledge, with its focus on the account of motion by bodies considered as point masses (Fig. 2.31). The diagram is not all-inclusive and seeks only to define the topic area of *De Motu* by mentioning its close neighbors which joined classical mechanics after its great unification, drawing on the fundamental theory of Newton, introduced in 1687.

From the list of historical *De Motu* treatises with relevant content, we may mention the treatise by Galileo in which he joined the debate with other *De Motu*(s) including those published in his university, in Pisa. The work mainly represented a critique of Aristotelian views on motion, not rare in medieval science. The young

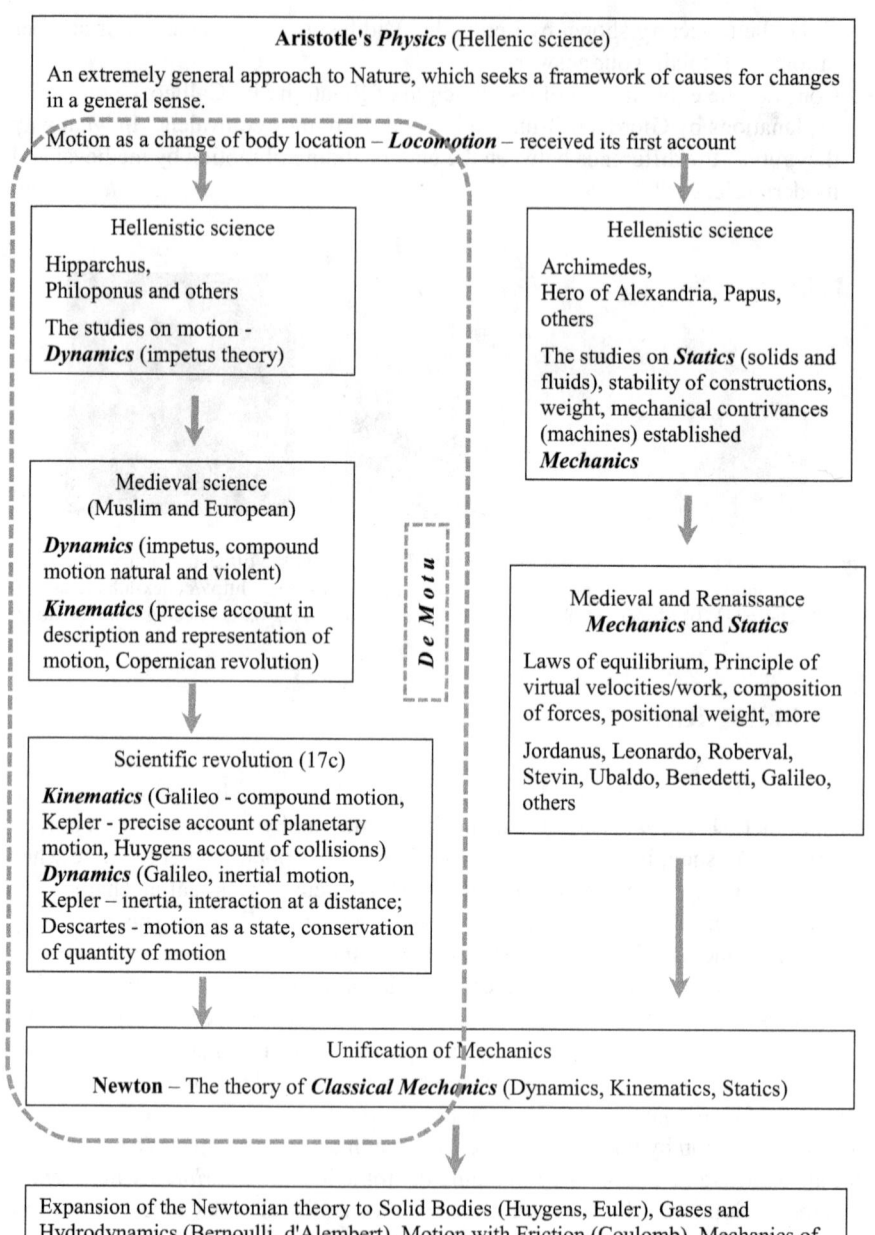

Aristotle's *Physics* (Hellenic science)

An extremely general approach to Nature, which seeks a framework of causes for changes in a general sense.

Motion as a change of body location – ***Locomotion*** – received its first account

Hellenistic science

Hipparchus,
Philoponus and others

The studies on motion -
Dynamics (impetus theory)

Hellenistic science

Archimedes,
Hero of Alexandria, Papus,
others

The studies on ***Statics*** (solids and fluids), stability of constructions, weight, mechanical contrivances (machines) established
Mechanics

Medieval science
(Muslim and European)

Dynamics (impetus, compound motion natural and violent)

Kinematics (precise account in description and representation of motion, Copernican revolution)

De Motu

Medieval and Renaissance
Mechanics and ***Statics***

Laws of equilibrium, Principle of virtual velocities/work, composition of forces, positional weight, more

Jordanus, Leonardo, Roberval, Stevin, Ubaldo, Benedetti, Galileo, others

Scientific revolution (17c)

Kinematics (Galileo - compound motion, Kepler - precise account of planetary motion, Huygens account of collisions)
Dynamics (Galileo, inertial motion, Kepler – inertia, interaction at a distance; Descartes - motion as a state, conservation of quantity of motion

Unification of Mechanics

Newton – The theory of ***Classical Mechanics*** (Dynamics, Kinematics, Statics)

Expansion of the Newtonian theory to Solid Bodies (Huygens, Euler), Gases and Hydrodynamics (Bernoulli, d'Alembert), Motion with Friction (Coulomb), Mechanics of systems of bodies and generalized forces– Analytical mechanics (Euler, Lagrange)

Fig. 2.31 Structure of classical mechanics for the purpose of definition of the topic domain in this excursus. The area of concern is shown by a dashed line

Galileo did not suggest a clear alternative. Seemingly dissatisfied with its content, Galileo did not publish this work.[96]

Another important *De Motu* was produced by Huygens.[97] In his study, Huygens derived the correct account of elastic collisions of bodies in which he arrived at the correct version of momentum (quantity of motion) conservation. In addition, he revealed conservation of the quantity named vis viva, which practically coincides with kinetic energy.

Newton's *De Motu* produced in 1684, prior to his major work, the *Principia*, included basic laws of motion, concept definitions, and a series of theorems regarding the motion of objects under the action of centripetal force. It is common to mention that these contents were incorporated in *the Principia*. This is, however, not precise. The published versions testify to continuous changes, in particular, with respect to formulation of laws and definition. We, thus, know that the content of *the Principia* presents a result of continuous painstaking labor during the years of seeking the form which satisfied the author.[98]

In 1721, Berkeley published his *De Motu* which presented a conceptual critique of Newton's theory of mechanics.[99] It was not a physical treatise since the author did not address any of the specific products of the Newtonian theory but attacked its philosophical foundations, the abstract concepts of space, force, time, motion, and velocity.[100] Berkeley considered their existence and the way in which we conceive them. Leaving alone the confusion of the existence of abstract notions by which we describe the real world (ontology which looks simple[101]), one may pay attention to his question of how we know about them, of whether we can rely on the mental (mathematical) manipulations involved in working with abstract spatial and temporal infinitesimals—the epistemology, which is not trivial.

Among the specific claims which are useful for us, we might emphasize the fact that such Newtonian concepts as absolute space and absolute time cannot be detected directly, and hence, they remain abstract products of mind. All our inferences regarding real objects, their location, and motion in space and time are only in relation to other objects such as distant stars. This was a true point to be noted and recognized. It was revived much later in the mechanical philosophy of Mach in the nineteenth century. He called these concepts metaphysical and tried to reduce physics knowledge solely to the concepts directly measured—positivistic philosophy. It was an important methodological clarification, by which science differs from non-science. It became a requirement of physics belonging to its nucleus.

---

[96] Drabkin (1960), Drake (1976), Camerota & Helbing (2000, pp. 319–365).

[97] Huygens (1700/1977). Mach (1883/1919/1989, pp. 314–317), Chapter I.

[98] Newton (1974, Vol. 6), Polak (1989, pp. 11–12).

[99] Berkeley (1721/1992, pp. 73–107).

[100] A comprehensive and elucidating analysis of Berkeley's *De Motu* was provided in Popper (1962, pp. 166–174).

[101] Popper resolves possible confusion of this kind by considering the *instrumentalism* of science (Popper, 1962, p. 443).

In a more general sense, *De Motu* of Berkeley can bring the students of science to a two-edged truth. We cannot know about Nature but by means of replacing reality by our mental theory of it. Nevertheless, our scientific knowledge about Nature is adequate, accurate, useful instrumentally, and penetrates the essence of Nature.[102]

We may mention other famous *De Motu*(s) which described investigations of motion beyond physics of plain objects. In his *De Motu Animalium*, Giovanni Alfonso Borelli (1608–1679), the Italian scholar considered the motion of a human body (as well as bodies of animals) as a mechanical system (Fig. 2.32a, b). Another *De Motu Cordis* (1628) reported the famous study by William Harvey (1578–1657), the English physician who revealed the circulation of blood due to the heart. It was considered as the pump pushing blood into vessels delivering it to the organs in a human body and from them back to the heart (Fig. 2.32c, d). Though different from mechanics, these studies shared with mechanics the interest in motion in its mechanical sense. Borelli considered the human organism as a machine, a skeleton moved by muscles, and Harvey—as a hydraulic system animated by heart.

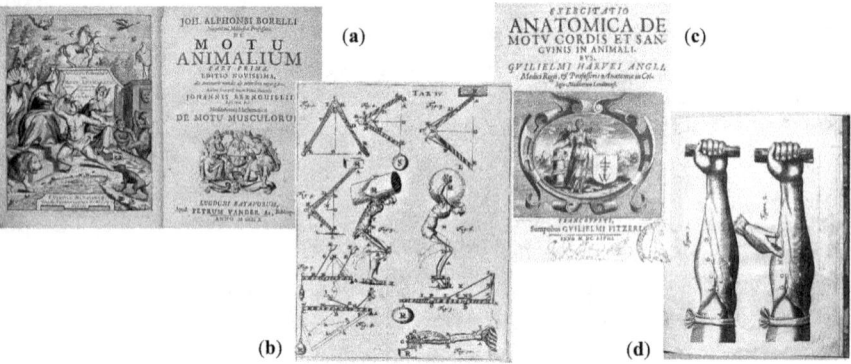

Fig. 2.32 (**a**) The front page of *De Motu Animalium* by Borelli. (**b**) An illustration from the 1710 book; (**c**) the front page of *De Motu Cordis* by Harvey; (**d**) an illustration from the 1653 book, p. 87

We thus defined *De Motu* as the area of physics treating the motion of material objects neglecting their structure. We traced the highlights of its understanding from Aristotle to Newton. This understanding framed classical mechanics. The story, however, continued within classical mechanics and beyond it. A new understanding of motion was introduced by the relativistic and quantum theories. They highly expanded the areas of validity regarding basic physical parameters—speed, mass, distance, and time. The *De Motu* of Newtonian physics remained valid in "regular" conditions and adequately depicts numerous activities including space navigation, but not others, related to the modern physics which provided advanced technology.

---

[102] See also a general discussion on this topic in Chap. 9.

Picture Credits

- Illustration figure from the book by Gardiner, E. N., 1910, *Greek Athletic Sports and Festivals*, p. 281, Fig. 52
- Photo of Shuttle launch. Picture credit: NASA. https://www.nasa.gov/multimedia/imagegallery/image_feature_1328.html
- The bust of Aristotle after Lysippos. Photo by Jastrow. Public Domain. https://commons.wikimedia.org/w/index.php?curid=1359807
- Portrait of Nicole Oresme. Public Domain. https://en.wikipedia.org/wiki/Nicole_Oresme#/media/File:Oresme.jpg
- Portrait of Galileo Galilei, 1636, by Justus Sustermans. Public Domain. https://commons.wikimedia.org/w/index.php?curid=230543
- Portrait of Descartes after Frans Hals. Photo by André Hatala. Public Domain. https://commons.wikimedia.org/w/index.php?curid=2774313
- Portrait of Isaak Newton by Godfrey Kneller, 1689. Public Domain. https://commons.wikimedia.org/wiki/File:Sir_Isaac_Newton_(1643-1727).jpg
- The statue of Nike of Samothrace by an anonymous Greek sculptor second century. BC. Photo by Lyokoï88. CC Attribution-Share Alike 4.0 International license. https://commons.wikimedia.org/w/index.php?curid=39152764
- The statue *Unique Forms of Continuity in Space* by Umberto Boccioni, 1913. Public Domain. https://commons.wikimedia.org/wiki/File:%27Unique_Forms_of_Continuity_in_Space%27,_1913_bronze_by_Umberto_Boccioni.jpg
- Fig. 2.4a Illustration from Camille Flammarion, Paris, 1888, Atmosphère: Météorologie Populaire, p. 163. Public Domain. https://commons.wikimedia.org/w/index.php?curid=318054
- Fig. 2.4b Cosmological diagram, the fourteenth century, scan by SteveMcCluskey, British Library, ms Yates Thompson 31, fol. 45. Public Domain. https://commons.wikimedia.org/w/index.php?curid=7755149.
- Portrait of Hipparchus, 190 BC. Public Domain. https://commons.wikimedia.org/w/index.php?curid=630396
- Sculpture portrait of Avicenna, Persian Scholar Pavilion in Vienna International Centre. Photo by Yamaha5. CCAttribution-Share Alike 3.0 Unported license. https://ru.m.wikipedia.org/wiki/%D0%A4%D0%B0%D0%B9%D0%BB:Persian_Scholar_pavilion_in_Viena_UN_(Avicenna).jpg
- Illustration from Gardiner, E. N. *Greek Athletic Sports and Festivals*, London: MacMillan, 1910, p. 286, Fig. 66
- Fig. 2.9a The drawing from D. Santbech, 1561. Public Domain. https://commons.wikimedia.org/w/index.php?curid=60700867
- Fig. 2.9b The drawing from Ryff, 1547, De Deutsche Fotothek. Public Domain. https://commons.wikimedia.org/w/index.php?curid=6476855
- Fig. 2.13a A page from Oresme, *Tractatus de latitudinibus formarum*, 1486. Public Domain. https://commons.wikimedia.org/wiki/File:Oresme,_Nicole_-_1486_-_Tractatus_de_latitudinibus_formarum_-_diagrams.jpg
- Fig. 2.14a Illustration from Galilei, 1638/1914, Fig. 47, p. 173
- Fig. 2.14b Illustration from Galilei, 1638/1914, Fig. 49, p. 176
- Fig. 2.16 Illustration from Galilei, 1638/1914, Fig. 46, p. 171
- Illustrative picture by W. Hogarth, the National Maritime Museum. Public Domain. https://commons.wikimedia.org/w/index.php?curid=24895095
- Fig. 2.19 Illustration from Galilei 1638/1914, p. 249
- Fig. 2.21 Illustration from Descartes, 1644, p. 56
- Portrait of Euler by J. Handmann, 1849, Kunstmuseum Basel. Public Domain. https://commons.wikimedia.org/w/index.php?curid=893656

- Portrait of Albert Einstein by Ferdinand Schmutzer. Public Domain. https://commons.wikimedia.org/w/index.php?curid=34239518
- Portrait of Niels Bohr, AB Lagrelius & Westphal, Nobel Prize biography, 1922, Public Domain. https://commons.wikimedia.org/w/index.php?curid=288274
- Portrait of Zeno of Elea by unidentified engraver, 1692. Public Domain. https://commons.wikimedia.org/w/index.php?curid=8762415
- Fig. 2.28 The fresco by Carducci, 1595, in El Escorial, Madrid. Public domain. https://commons.wikimedia.org/wiki/File:Zeno_of_Elea_Tibaldi_or_Carducci_Escorial.jpg
- Illustrative picture *God the Geometer*, 1230, by anonymous, Österreichische Nationalbibliothek, Vienna, Public domain. https://commons.wikimedia.org/wiki/File:God_the_Geometer.jpg
- Portrait of Giovanni Alfonso Borelli. Public Domain. https://commons.wikimedia.org/w/index.php?curid=110780
- Fig. 2.30a Drawing by Leonardo Da Vinci (1519), Codex Atlanticus, fol. 24 r-a. Public Domain. http://codex-atlanticus.it/#/Detail?detail=33; http://codex-atlanticus.it/#/Detail?detail=31

- Fig. 2.30b Picture by Michael Maggs, edit by Richard Bartz. CC Atribution-Share 3.0 Unported license. https://commons.wikimedia.org/w/index.php?curid=2880974
- Fig. 2.32a The front page of *De Motu Animalium,* Borelli, 1710. Public Domain. https://commons.wikimedia.org/wiki/File:Borelli_-_Motu_Animalium.jpg
- Fig. 2.32b An illustration from *De Motu Animalium*, Borelli, 1710. Public Domain. https://commons.wikimedia.org/wiki/File:Houghton_IC6_B6447_680db_-_De_motu_animalium,_TAB_IV.jpg
- Fig. 2.32c The front page of *De Motu Cordis,* Harvey, 1628, CC Atribution-Share 4.0 International license. https://commons.wikimedia.org/wiki/File:W._Harvey,_Exercitatio_anatomica,_title_page._Wellcome_L0021402.jpg
- Fig. 2.32d An illustration by unknown author, *De Motu Cordis,* Harvey, 1653, plate 26, p. 120. Public domain. https://commons.wikimedia.org/w/index.php?curid=4551254

References

Aristotle. (1952a). *Physics* (Vol. 1). Encyclopedia Britannica.
Aristotle. (1952b). *On the heavens* (Vol. 1). Encyclopedia Britannica.
Arons, A. (1990). *A guide to introductory physics teaching*. Wiley.
Barbour, J. B. (2001). *Discovery of mechanics*. Oxford University Press.
Berkeley, G. (1721/1992). *De Motu and the analyst*. : Springer.
Calvi, I. (1956). Military engineering and arms. In E. Vollmer (Ed.), *Leonardo da Vinci* (pp. 275–306). Reynal.
Camerota, M., & Helbing, M. (2000). Galileo and Pisan Aristotelianism: Galileo's "De Motu Antiquiora" and the Quaestiones de Motu Elementorum of the Pisan Professors. *Early Science and Medicine, 5*(4), 319–365.
Clagett, M. (1950). Richard Swineshead and late medieval physics: I. The intension and remission of qualities. *Osiris, 9,* 131–161.
Clagett, M. (1959). *The science of mechanics in the middle ages*. The University of Wisconsin Press.
Cohen, R. M., & Drabkin, E. I. (1948). *A source book in Greek science*. McGraw-Hill.
Crombie, A. C. (1959). *Medieval and early modern science*. Doubleday Anchor Books.
De Angelis, A., & Santo, C. E. (2015). The contribution of Giordano Bruno to the principle of relativity. *Journal of Astronomical History and Heritage, 18*(3), 241–248.
Descartes, R. (1644/1982). *Principles of philosophy*. D. Reidel.
Dijksterhius, E. J. (1986). *The mechanization of the world picture. Pythagoras to Newton*. Princeton University Press.

Drabkin, I. E. (1960). A note on Galileo's De Motu. *Isis, 51*(3), 271–277.

Drake, S. (1976). The evolution of De Motu. *Isis, 67*(2), 239–250.

Drake, S. (1999). Galileo: Scientific method and philosophy of science. In *Essays on Galileo and the history and philosophy of science* (Vol. I). University of Toronto Press.

Dugas, R. (1988). *A history of mechanics*. Dover.

Euler, L. (1736/2008). *Mechanics or the science of motion analytically demonstrated* (Vol. 1).

Galilei, G. (1590/1960). De Motu. In I. E. Drabkin & S. Drake (Eds.), *On motion & on mechanics; comprising*. The University of Wisconsin Press.

Galilei, G. (1613/1957). Second letter on sunspots. In S. Drake (Ed.), *Dicoveries and opinions of Galileo*. Doubleday, NY.

Galilei, G. (1632/1953). *Dialogue concerning the two chief world systems – Ptolemaic and Copernican*. University of California Press.

Galilei, G. (1638/1914). *Dialogue concerning two new sciences*. Dover.

Galili, I., & Bar, V. (1992). Motion implies force. Where to expect vestiges of the misconception? *International Journal of Science Education, 14*(1), 63–81.

Galili, I., & Kaplan, D. (2002). Students' interpretation of water surface orientation and inertial forces in physics curriculum. *Praxis der Naturwissenschaften Physik in der Schule, 51*(7), 2–11.

Galili, I., & Tseitlin, M. (2003). Newton's first law: Text, translations, interpretations, and physics education. *Science & Education, 12*(1), 45–73.

Galili, I. (2009). Thought experiment – Establishing conceptual meaning. *Science & Education, 18*(1), 1–23.

Galili, I. (2011). James Hannam: Gods philosophers. How the medieval world laid the foundations of modern science. *Science & Education, 21*(3), 415–422.

Gardiner, E. N. (1910). *Greek athletic sports and festivals*. Macmillan.

Grant, E. (1977). *Physical science in the middle ages*. Cambridge University Press.

Hannam, J. (2009). *Gods philosophers. How the medieval world laid the foundations of modern science*. Icon Books.

Huygens, C. (1700/1977). De Motu Corporum ex Percussione. *Isis, 68*(4), 574–597.

Jammer, M. (1961). *Concepts of mass in classical and modern physics*. Ambridge, Mass: Harvard University Press.

Koyré, A. (1943). Galileo and Plato. *Journal of the History of Ideas, 52*(4), 400–428.

Koyré, A. (1968). *Metaphysics and measurement: Essays in scientific revolution*. Cambridge, Mass.: Harvard University Press.

Koyré, A. (1973). *The astronomical revolution. Copernicus- Kepler – Borelli*. Cornell University Press.

Lefetz, L. (2014). *Oresme, or the quadruple root of the principle of relativity*. http://bergson-aujourd-hui.over-blog.com/2014/04/conclusion-oresme-ou-la-quadruple-racine-du-principe-de-relativite-10-10.html

Losee, J. (1993). *A historical introduction to the philosophy of science*. Oxford University Press.

Mach, E. (1883/1919/1989). *The science of mechanics, a critical and historical account of its development*. : Open Court.

McCloskey, M. (1983). (a) Naïve theories of motion. In D. Genter & A. L. Stevens (Eds.) *Mental models* (pp. 299–324). Hillsdale, NJ, Lawrence Erlbaum. (b) Intuitive physics. *Scientific American, 248*(4), 122–130.

Newton, I. (1687/1989). *Mathematical principles of natural philosophy*. Translated into Russian by A. N. Krilov. Moscow: Nauka.

Newton, I. (1687/1999). *The principia. Mathematical principles of natural philosophy*. Translated by B. Cohen & A. Whitman. Berkeley, CA: University of California Press.

Newton, I. (1729/1952). *Mathematical principles of natural philosophy*. Revised edition of Motte's translation by F. Cajori. Chicago: Britannica Great Books.

Newton, I. (1729/1964). *Mathematical principles of natural philosophy* translated by A. Motte. London: Middle-Temple-Gate in Fleetstreet; NY: Philosophical Library.

Newton, I. (1974). Whiteside, D. T. (ed.). *The mathematical papers of Isaac Newton. Vol. VI.* (1684–1691). Cambridge: The University Press.

Pedersen, O., & Pihl, M. (1974). *Early physics and astronomy.* McDonald & Janes.

Pines, S. (1986). *Studies in Arabic versions of Greek texts and in mediaeval science* (Vol. 2). Brill Publishers.

Polak, L. S. (1989). Introduction. In I. Newton (Ed.), *Mathematical principles of natural philosophy.* Nauka.

Popper, K. R. (1962). *Conjectures and refutations: The growth of scientific knowledge.* Basic Books.

Popper, K.R. (1978). Three worlds. The Tanner lecture on human values. The University of Michigan. http://www.tannerlectures.utah.edu/lectures/documents/popper80.pdf. Accessed 24 Sept 2015.

Reif, F. (1995). *Understanding basic mechanics.* Willey.

Russell, B. (1959). *Wisdom of the west.* Crescent Books.

Ryff, W. H. (1547). *Mathematischen und Mechanischen Kunst.* Johann Petreius.

Taylor, L. W. (1941). *Physics. The pioneer science.* Dover.

Whittaker, R. J. (1983). Aristotle is not dead: Student understanding of trajectory motion. *American Journal of Physics, 51*(4), 352–357.

Chapter 3
Optical Image and Vision:
From Pythagoras to Kepler

© Springer Nature Switzerland AG 2021, corrected publication 2022
I. Galili, *Scientific Knowledge as a Culture*, Science: Philosophy, History and
Education, https://doi.org/10.1007/978-3-030-80201-1_3

And God said, let us make man in our image, after our likeness: and let them have dominion over ... every creeping thing that creeps upon the earth.

Genesis 1:26

Thou shall not make thee [any] graven image, [or] any likeness [of any thing] that [is] in heaven above, or that [is] on the earth beneath, or that [is] in the waters beneath the earth.

Deuteronomy 5:8

Abstract This chapter outlines the consolidation of the concept of image in optics. We start from classical Greece where scientists suggested several theories for the understanding of vision. Early Hellenic theories provided nonmathematical, qualitative, holistic accounts. The later Hellenistic optics of Euclid refined the holistic conception but suggested the erroneous concept of "active vision" by means of visual rays. Alhazen, a distinguished Arab scholar during the medieval era, split an image into points, each "transmitted" to the eye by a single ray. Further progress in the Renaissance noted the similarity between the eye and the Camera Obscura, but scholars were puzzled by the inverted image in the eye. The solution was due to Kepler, in Germany, who introduced light flux to create an image and understood that human cognition provides the "inversion." The whole history of optical image—from holistic to light flux created and interpreted by the mind—reconstructs the theory of vision and sheds light on the nature of science. This is through a consideration of the scientific discourse of competing theories, a complex cumulative process, not a simple accretion, but a *conceptual change*. Research has revealed a similarity of students' accounts for vision with the old ideas, intromission and extramission, holistic and differential images, and the central role of light rays or light flux. This similarity suggests that historical debate may be included in teaching to remedy common misconceptions and upgrade the meaningful learning of optics. So we all were authoritatively informed by the Bible that each of us represents an image of someone, and secondly, we were instructed that we are forbidden from creating a reproduction of the original, that is, from producing an image of a living creature. This heavy/strict limitation, however, did not imply that we cannot investigate and reveal what an image is. Here, we will follow and clarify the way in which people conceived and perceived an optical image through vision. It turned out to be quite difficult for scholars to discover the nature of optical image. To understand this concept, we will familiarize ourselves with the dialogue of ideas carried out in science through a rather long history of development that lasted for more than 2,000 years.

Introduction

 Optics is the physical theory of *light and vision*. Reconstruction of an object's appearance by means of light constitutes optical image and presents the subject of optics. The idea of the *optical image* was central in people's comprehension of light and vision throughout the course of history. The nature of an image created by light, its origin, and the way it emerges occupied human curiosity long before science was established. Myths and tales were illustrated in numerous works of art, but the *scientific* inquiry into the optical image comprised a special story that we will briefly depict here. In it, light and vision were inherently interwoven from the dawn of science in classical Greece (considered then, and for a long time after, as Natural Philosophy). This interwoven concept of light and vision split much later, during the scientific revolution of the seventeenth century, into the separate studies of light, becoming physics, and vision, as in psychophysics. Here is how it was as historians tell us.[1]

Hellenic Science

Natural science was established in Classical Greece having evolved from the discourse of myths. The Hellenic scientists provided several competing accounts of vision. The first was introduced by Pythagoreans, Hippocrates of Chios (fifth c. BC), and Archytas of Tarentum (fourth c. BC). They explained vision as a form of radiation. *Opsis*, an *internal* "fire," was thought to emanate from an observer's eyes, reaching the observed objects and causing the observer to see them (Fig. 3.1). For the direction of activity in the process, historians identified this theory as *extramission*.

Pythagoras

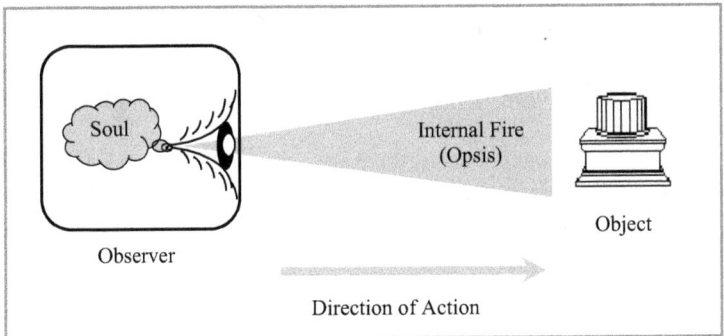

Fig. 3.1 Schematic representation of the Pythagorean extramission scenario of vision

---

[1] Lindberg (1976, 1978, 1985, 2002), Ronchi (1970, 1991), Magie (1969), Park (1997), Darrigol (2012).

The rationale of this conception was to consider vision as touch, in parallel with other senses: smell, hearing, touch, and taste (Fig. 3.2a). This understanding was seemingly shared with the much earlier culture of Ancient Egypt. Indeed, in the images of about 3500 years ago, the Sun was depicted as sending its "hands" (Fig. 3.2b). The touch manifested itself in heating that we perceive.

(a)   (b)

Fig. 3.2 (**a**) Is vision provided through *touch* like other senses? The drawing from the book of Descartes (1649, p. 25) explaining the sense of touch. (**b**) An Egyptian pharaoh, Akhenaten, from the fourteenth century BC worshiped Sun disc as the God Aten sending his hands down (pointed by arrow) to *touch* the world objects

In the Pythagorean theory of vision, the agent of touch was a special internal flux emanating from the observer's eyes—extramission. It seemed reasonable to ascribe the origin of vision activity to the observer who turned his head and eyes to the object, concentrated his attention to the object in order to see it. We "focus" our sight. Strangely enough, the adherents of "active vision" reasoned also by the outwardly curved, *convex* surface of the eyes: the organ of vision. It was in contrast with ears for hearing and nose for smelling, that possess a *concave* shape. Hence, eyes that are convex must radiate, not adsorb. The adherents of extramission  exemplified it with the night vision of animals, especially cats, which were known for their eyes sparkling in the dark.

Democritus

The opponents of this view were not convinced: if this theory is true, why do we not see at night? Or why do we see the distant stars immediately after opening our eyes? Both questions were difficult to answer.

The *second* theory was due to the school of Atomists of Democritus. They stated that each body around us produced a form of replica—*eidolon* (Fig. 3.3). The Eidolon left the surface layer of atoms of the object and traveled in all directions until it entered the eye of the observer, causing a vision of that object. In contrast to

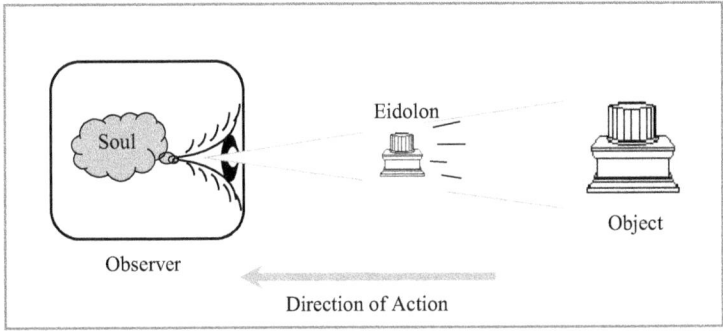

Fig. 3.3 Schematic representation of the Atomists' intromission theory of vision

the Pythagorean theory, this theory stated the process originated in the object and continued toward the observer. It, thus, was labeled as the *intromission* theory.

The opponents of the Atomists' theory questioned how the eidola of big objects, such as mountains, could possibly enter into a small eye. Or, why we do not see numerous images of things everywhere around, on walls, following the arrival of the eidola. The atomists left unanswered the question: Why can we not see at night?

Empedocles

The *third* theory was associated with Empedocles (490–430 BC), a distinguished philosopher from Elea—the Greek colony in Italy, and Plato (427–347 BC) in Athens. Their approach combined the two previous ideas, suggesting the understanding of vision as resulting from a meeting of fire fluxes of three kinds (Fig. 3.4). It was all about fire.

Firstly, the *pure fire* of ambient daylight, not mixed with other primary substances filling the space, originated in various light sources (Sun, Moon, etc.). In contrast with the flame of regular fire, the pure fire did not burn but provided eyes with an ability to see. Secondly, *internal fire*—the stream of visual flux, the entity contained in the eyeball was capable of emanating from the pupil toward the object observed. Thirdly, the *external fire* radiated by

Plato

the objects originated in their colors. These scholars believed that the external fire propagated toward the object and met there the stream of the internal fire of the eye. The collision of the two streams yielded the sensation of vision of the observed object.[2] In this theory, night vision was excluded by lacking the "pure fire" of the daytime. The radiation of external fire was induced by the pure fire. Thus, all three fires are necessary for the process of vision.

---

[2] Conford (1937), p. 152.

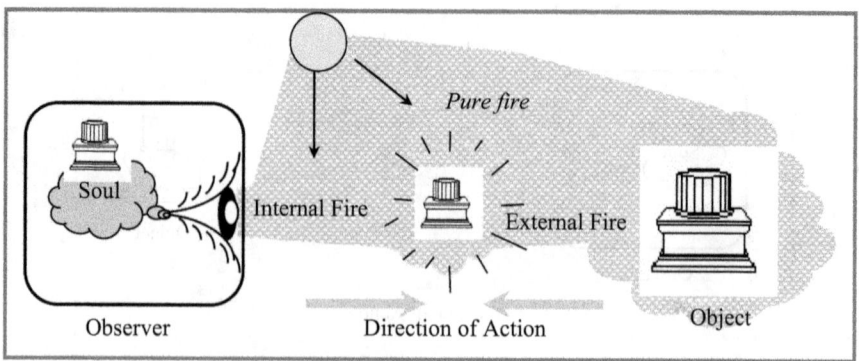

Fig. 3.4 Schematic representation of the Empedocles-Plato's hybrid theory of vision

The *fourth* theory was especially creative. It was suggested by
Aristotle (384–322 BC).[3] It was an intromission theory, that is, it
included the activity directed from outside into the eyes of the
observer.[4] Aristotle distinguished *color*—an inherent feature of
objects—from *light*, which he understood as a nonmaterial agent,
the trigger for the medium (air, glass, water) to become transparent
and allow the process of vision to take place (Fig. 3.5).

Aristotle

According to Aristotle, the *color* of the observed object causes
the state of tension in the surrounding transparent medium (air or
water). This compression in the medium expands between the object and the observer's eye. When it reaches the eye, it causes the perception of the image of the object.
It seemed plausible that the absence of color causes the object to be transparent, not
seen. The color relevant for vision covers the surface of objects.

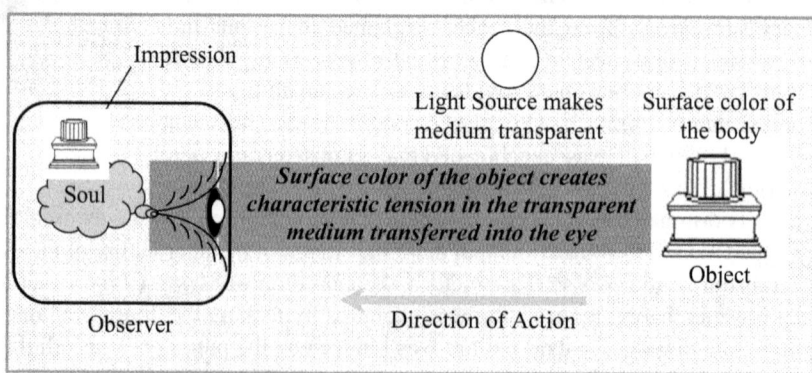

Fig. 3.5 Schematic representation of the Aristotle intromission theory of vision

---

[3] Aristotle (1952).

[4] Aristotle was, however, not consistent regarding vision. In explaining the rainbow, he employed
visual rays of the extramission theory. Perhaps the rainbow explanation was due to a student from
the school of Aristotle.

The perception of an image reaches the soul of the observer, causing the act of recognizing the image by cognition. On the whole, this theory could be considered as an intromission theory with a mediator between the object and the eye. Aristotle saw the evidence for the role of mediator in the fact that we cannot see an object when it is located too close to the eyes. To act, the medium needs a certain space in between.

All Hellenic optical theories understood the eye to be comprised of a kind of fluid. No structural elements of the eye were recognized.

Questions for Reflection

1. Discuss the arguments in favor of each of the four Hellenic theories of vision.
2. Suggest reservations with regard to each of the theories.
3. Was it possible for an individual living in Greece at that time to make a fair preference in favor of one of the theories and refute the others? Explain.

Hellenistic Science

The four Hellenic theories described various types of relationships between the observer, the medium, and the object seen. During the next period, that of the Hellenistic culture, scientists proceeded with their exploration of vision, seeking for a more detailed picture and a more refined mechanism of vision. This tendency brought new investigations and theories of more sophisticated descriptions of vision and the eye.

Hellenistic culture, in general, performed a kind of revolt against the philosophical preference of Hellenic science. Scholars were not satisfied with the holistic, qualitative account of vision and reconsidered it. Scientists of the Hellenistic world sought for more concrete information about the behavior of light and the vision process. This ambition required a more active exploration and a new methodology in order to perform a more precise elaboration of natural phenomena.[5]

Galen (Claudius Galenus) was a famous Greek physician in the Roman Empire of the second century AD. He accumulated vast experience in performing surgery on sick and injured people. Seemingly, he was the first to describe correctly the anatomic structure of the human eye and identified its major components: cornea, iris, pupil, crystalline lens (or humor), aqueous and vitreous humor, and retina (Fig. 3.6).

Galen

---

[5] The Hellenistic period provided an impressive array of scholars: Aristarchus, Euclid, Archimedes, Heron, Ptolemy, Hipparchus, Eratosthenes, and others. They all essentially contributed through scientific inquiry, analyzing, experimenting, modeling, measuring, inventing, and constructing (e.g., Russo, 2004) to more concrete knowledge, however, generally keeping within the Hellenic philosophical framework from the previous period.

Galen identified the humors that filled the eye (vitreous and aqueous) and ascribed to the lens a vital role in the vision process, as seemed to him evident from his surgical practice.

Among the most important findings was the connection of the retina to the brain. In this way, for the first time, the brain was included in the process of vision and was in fact recognized as the seat of vision perception.

In understanding the vision process, Galen adopted the new approach of the Stoic philosophers who ascribed a central importance to a special medium—*pneuma*—an all-pervasive active agent composed of air and fire that fills the whole universe.

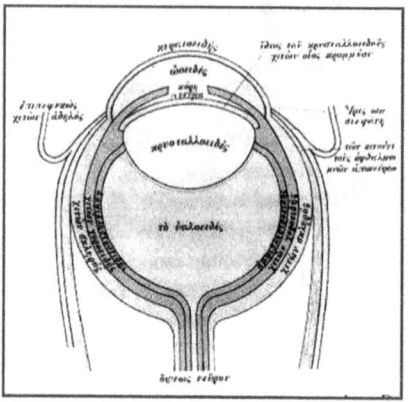

With regard to vision, *pneuma* was, in a way, similar to the Aristotelian theory in which the medium, that became transparent, also played a central role. However, according to Stoic physics, pneuma was a more active medium, itself possessing a feature of perception, a spirit originating in the observer. It acted in two ways, from and to, the observer.

Fig. 3.6 The structure of eye as established by Galen

According to Galen, visual pneuma emanates from the brain and comes to the eye through the visual nerve (Fig. 3.7). When visual pneuma touches the air, the air goes through a transformation and obtains the virtue of visual perception, which was similar to what happens with the sense of touch. The scenario included a mediator, *pneuma*, as in Aristotle's theory, but was closer to the extramission process. Galen rejected the intromission idea of an image of the observed object that physically approaches the eye. In considering the incompatible sizes of the pupil and the large objects observed, he asked: How could the image of a mountain enter the small opening of the observer's eye?

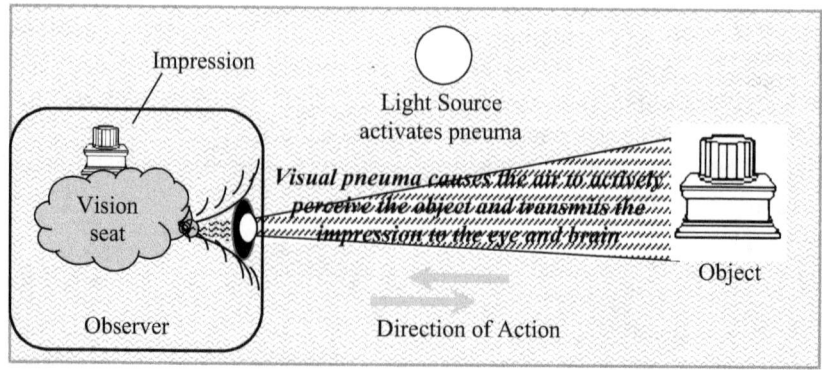

Fig. 3.7 Schematic representation of Galen's extramission theory of vision

Euclid—the renowned mathematician of the Hellenistic period, whose geometry we all still learn at school—made his own contribution to the understanding of light and vision. In a way, his contribution was revolutionary. Having adopted the extramission idea of vision flux, from the school of Pythagoras, he refined it by introducing the concept of *ray*. Euclid introduced two types of rays: visual and light. Both served as fundamental concepts for his theory. Light was considered as comprised of light rays and vision—as a process of searching the environment by the observer who emanates rays of vision.

Euclid

Euclid postulated that rays of light and rays of vision obeyed the same rules of behavior. In its structure, his theory explained that the experience of vision was structured in the same way as his famous geometry: fundamental postulates and definitions followed by their particular implications, derived statements. Euclid suggested that the observer scanned the environment with visual rays (Fig. 3.8). The ray concept structured the "internal fire" of Pythagoreans and provided a powerful tool for the graphical account of vision. In fact, it is that account by means of rays which established the theory of *perspective*—the flat, two-dimensional representation of the three-dimensional reality through drawing it. This approach explained, for example, why we see all distant objects smaller, and how one should draw the observed reality on paper in the way that the sketch being flat would, however, appear to the observer as a three-dimensional object.

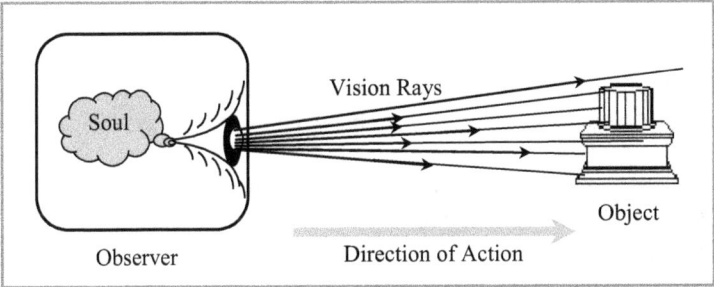

Fig. 3.8 Schematic representation of Euclid's refinement of the extramission theory of vision. The observer's eye scans the environment using visual rays

Other great minds of Hellenistic science—Heron and Ptolemy of Alexandria, and Archimedes of Syracuse—investigated the rules followed by light and vision rays in the phenomena of reflection and refraction. Heron and Archimedes formulated the law of specula reflection: the trajectory of light and vision includes (I) all three lines—the perpendicular to the reflecting surface at the point of reflection, the incident, and reflected rays—lay in the same plane and (II) the angle of incidence and the angle of reflection are equal (Fig. 3.9a). Heron based his formula on the postulate that the trajectory of light must take the shortest way, and Archimedes obtained the same result through the requirement of the reversibility of the light and vision path (being possible in both directions).

Ptolemy

Ptolemy (Claudius Ptolemaist), who adopted the idea of visual rays, was the first to experimentally investigate the rule under which vision (and light) rays behave at the border between two different transparent materials—*refraction*. This was the phenomenon of the change in the direction of a light/vision ray. Although Ptolemy failed to establish the correct mathematical dependence between the angle of incidence and the angle of refraction of the rays, he did observe that in the case of rather "small" angles, refraction happens, in keeping with the constant ratio of the angles, and that ratio depends on the medium.

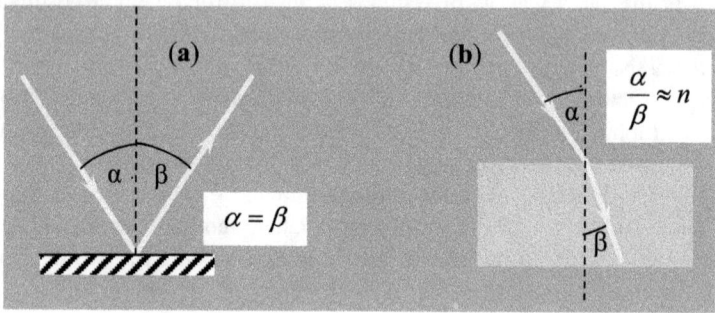

Fig. 3.9 The established regularity of vision and light rays at the processes of their reflection (**a**) and refraction (**b**)

However, although the behavior of the rays outside of an observer was investigated, nobody was able to describe what happened with the rays inside the eye that caused the image to be perceived.

Quite interestingly, when Ptolemy considered which of the two scenarios—intromission or extramission—takes place, he preferred the extramission theory (visual rays coming out of the eyes) because, as he thought, if the intromission theory were correct and eidolon left the observer body, he would receive it back as it rebounded from the mirror and would therefore see his own face. Intromission at that time was considered holistically, as a transfer of the whole replica of the object into the eye. This conception changed in the next historical period, that of medieval science.

Ptolemy speculated that vision rays should possess width. He suggested a sort of collection of "pencils" rather than rays of vision, thin but possessing a finite width. Hence, he argued, we perceive an image as a pattern of small spots from scanning by visual rays rather than a holistic entity.

Questions for Reflection

1. Compare the Hellenistic and the Hellenic theories described. What features were different? Formulate the rationale of these theories.

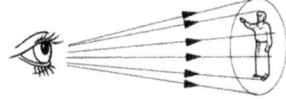

2. Discuss the justification for and reservations regarding each of the Hellenistic theories of vision.
3. Explain the reasoning by which Ptolemy refuted the intromission theory of light.

Medieval Muslim Science

Al-Kindi

During the ninth century, the center of scientific activity moved to the Muslim world. Europe was immersed in numerous local wars and suffered from instability. Remarkable progress in the account of vision and optical imagery took place in the Muslim world at that time.

Progress began with Al-Kindi, a distinguished Arab scholar from Baghdad in the ninth century, who stated an important principle regarding the behavior of light. He stated that light emanates from *each* point of a light source in all possible directions (Fig. 3.10a). This feature of light was seemingly not obvious in the past (Fig. 3.10b).

In fact, this was a recognition of the fact that an extended light source is always comprised of numerous smaller light sources, each radiating light in all directions.

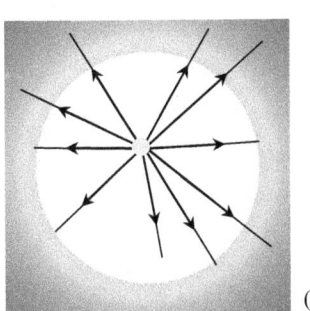

(a)   (b)

Fig. 3.10 (**a**) Illustration of the principle of Al-Kindi: light emanates from each point of the source in all directions. (**b**) Violation of the principle of Al-Kindi in the old Egyptian representation of the Sun

An important extension of this principle was made by another great Arab scholar—Abu Ali al-Hasan ibn al-Haytham (965–1039), or Alhazen, as he was known in Europe—in the eleventh century. He extended the same claim regarding *any* observed object (Fig. 3.11):

Each object, whether radiating or reflecting light, causes light radiation in all directions from each of its points.

This principle might look confusing as if contradicting the law of light reflection (Fig. 3.9a). The explanation stems from the appreciation of the microscopic roughness of the surface of regu-

Alhazen

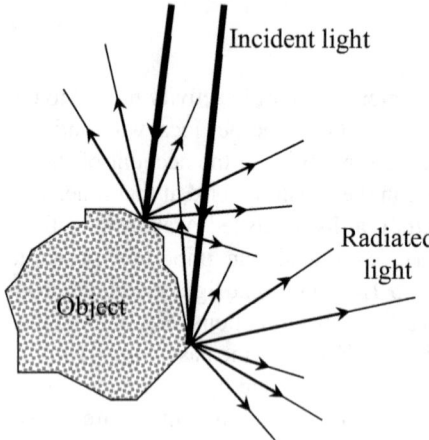

lar bodies that causes multiple reflections from each tiny area. Each such separate reflection obeys the law of specular reflection but, all together, light radiates in all directions (Fig. 3.11).

This understanding was very important for further progress in resolving the enigma of vision. Indeed, the principle of making relevant all directions for light reflection may be followed forward in time to the Huygens principle of light expansion (seventeenth century), Fresnel's treatment of light waves (nineteenth century), and even to the account of Feynman in modern physics (twentieth century) regarding the behavior of matter and light particles potentially expanding in all possible directions.

Fig. 3.11 Illustration of the principle of Al-Kindi—Alhazen: light is emanating from each point of any observed object in all directions

And what about vision? Firstly, Alhazen studied thoroughly the heritage of Hellenic and Hellenistic optical theories, with their wealth of ideas regarding the nature of light and vision. He adopted the concept of a light ray and considered light as an abundance of rays. This was helpful to his creating an *accurate* account of optical phenomena. Secondly, he dismissed the idea of *visual* rays and supported the intromission conception. He stated that our vision is due to the light entering the eyes of the observer.

He referred to the fact that looking straight into a strong source of light, such as the Sun, is very harmful to the eyes (never try it![6]). Furthermore, he pointed to the known effect of an after-image (if one looks at an object for about a minute and then closes his/her eyes, the observed image is still perceived). Alhazen considered it as evidence that light entered the eyes. Light travels from each observed object, and

---

[6] Learn from the unfortunate experience of Galileo who became blind after his long observation of the Sun.

this is what enables the object to be seen. The question remained, however, how exactly that happens.

To be convincing, Alhazen proceeded beyond a mere idea and suggested a mechanism of vision. In his analysis, Alhazen drew on Euclid, Heron, Archimedes, and others who all stated and used the rectilinear expansion of light.[7] Alhazen reconsidered the known phenomenon of image creation in the Camera Obscura. The phenomenon was known already to Aristotle, but Alhazen related it to the rectilinear expansion of light. In a darkened room with a small opening in one of its walls, one observes an image of the objects

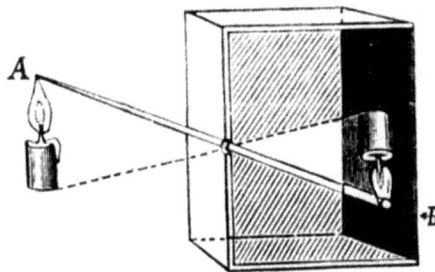

Fig. 3.12 Each point of the object generates a cone of light entering the Camera Obscura. So, the image observed is comprised of spots of light, each coming from a point on the object

outside (Fig. 3.12). Yet, for the first time in history, Alhazen *dissected* the process of the image creation. He drew on two prepositions:

1. Each point of any illuminated object sends light in *all* directions.
2. Light rays are straight.

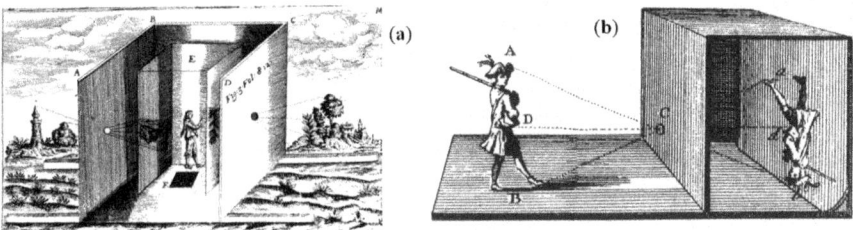

Fig. 3.13 (**a**) The image observed in the Camera Obscura is up-down inverted as it is seen in the illustration by Kircher in 1646. (**b**) The image has no depth and also right-left inversed. The unknown illustrator missed the latter feature in eighteenth century

Indeed, if a point on the object radiates light in all directions, only a fraction of this light, a cone, manages to enter through the hole of the camera (Fig. 3.12). This light creates a light spot on the opposite wall. The spots from the points of the object all together comprise the observed image. The new argument was that the image does not need to travel through space *as a whole*. It is enough that each point on the object surface will cause an illuminated spot. The spots create an illuminated area which reproduce the shape of the object. The spots vary in intensity. It is the information they carry. The emerged replica of the object by means of light constitutes the *optical image* (Fig. 3.13).

---

[7] Lindberg (1968).

There was, however, an immediate problem faced by Alhazen: the image in the Camera Obscura was upside down; that is, the upper and lower parts of the object (as well as the left and right ones) were inverted. Therefore, Alhazen seemingly thought the image we observe apparently needs something more to be addressed in order to understand its right-side up position.

There was another problem too. Since each point of the object sends many rays to the eye, how could it be that no rays cause a chaos of illumination pattern inside the eye? Indeed, we perceive clear and focused images of the objects around. Alhazen started to speculate regarding the kind of selection among the rays of light that "obviously" was responsible for this clarity. He knew about the refraction of light entering the transparent medium. Ptolemy of Alexandria had investigated that phenomenon. All the light is deflected in the second medium, except the light traveling perpendicular to the surface. Alhazen assumed this to be the form of selection among the rays. Light rays enter through the cornea (convex surface of eye) and continue penetrating inside. Only the ray arriving perpendicular to the surface proceeds and contributes to the image. Such rays reach the crystalline lens and create an image on its surface—an amazing modification of his understanding of Camera Obscura image. The image created on the lens was interpreted as similar to that emerged on the wall of the Camera Obscura. In Fig. 3.14, rays 1 and 2 map points a and b of the object to points a′ and b′ of the image. In summary:

In the view of Alhazen, during image creation, the points of the object are "reproduced" by the light rays which create the points of the image on the surface of the eye lens, point to point by means of a single light ray for each point.

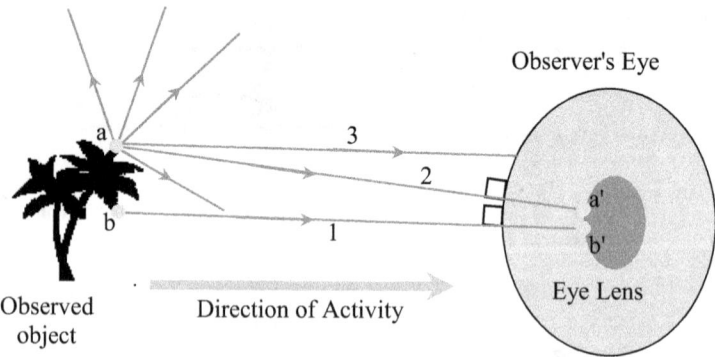

Fig. 3.14 Schematic representation of the mechanism of vision according to Alhazen. Ray selection: rays 1 and 2 are perpendicular to the eye surface and enter the eye. Ray 3 is oblique and so irrelevant for image construction

The question remains, however, why Alhazen stopped the image creation on the lens, making it a sensitive organ of human sight and did not proceed with the rays across the lens and arriving at the retina. Seemingly, one of the reasons, if not the only reason, was that the upside-down image observed after the convex lens, which seemed inappropriate. He saw this as a clear contradiction with the actual experience—the right-side up of images and removed this option.

The optical treatise of Alhazen: *Kitab al-manazir* [*The Book of Optics*; *De aspectibus* or *Perspectivae*] became famous in the Muslim world and in Europe. It reached the scholars of the Western Europe in the thirteenth century. The treatise of Alhazen was not only translated but actually stimulated scholars to produce their own commentaries and presentations of optics and to provide more experimental evidence. Such were the manuscripts of *Perspectiva* by the Polish scholar Witelo (Fig. 3.16) and *Perspectiva Communis* by the English cleric, Peckham who reproduced some of the described experiments of Alhazen. They disseminated the new knowledge of optics in Europe from the thirteenth up to the seventeenth century. Roger Bacon in Oxford could read *The Perspective* of Alhazen in Arabic and was inspired by its ideas in his own optical exploration.

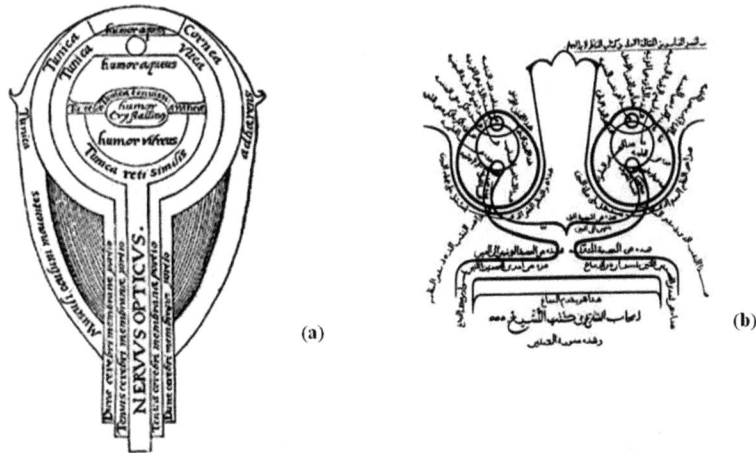

Fig. 3.15 The structure of the eye by Alhazen (**a**) as depicted in Witelo's text in (1572) (note the eye lens—"humor crystalling"—erroneously placed in the center of the eye). (**b**) The drawing in the original *Perspectiva* by Alhazen (1038)

The book of Witelo was especially popular, dominating the field in the universities of medieval Europe for about 400 years. Regarding the knowledge of the structure of the eye and the process of vision (Fig. 3.15a), Witelo's book copied Alhazen (Fig. 3.15b) who followed Galen of Hellenistic science. However, the light ray became an important theoretical concept of optics, a central tool in the account of vision, as well as of the explanation of light image in Camera Obscura.

After the thirteenth century, the main frontier of scientific activity moved to medieval Europe. Without listing all the reasons for this development, which continues to inspire analysis and discussion,[8] one may mention that this radical change

---

[8] e.g., https://www.dailysabah.com/feature/2016/01/29/why-the-islamic-world-fell-behind-in-science; and Overbye (2001). Another important trait was the asymmetry with regard to the authority of knowledge. The scholars of the Islamic period embraced the heritage of Hellenic and Hellenistic science, but later, they neither adopted nor followed European developments. For that, people there needed a special religious permission—a fatwa. This asymmetry vio-

happened as a result of a combination of several highly influential factors, social, political, and economic. It is sufficient here to mention one of the most significant factors—the absolutely devastating invasion of the Mongols in 1220 which crushed the major centers where scientific activities occurred. The collapse of the Islamic state in the Iberia peninsula in the fifteenth century caused a similar impact. In a sense, this scientific loss was a repetition of the no less devastating loss of the Academy in Alexandria (Mouseion, with its unique library) following the Islamic invasion of the seventh century. Now the roles were reversed. The Islamic culture was the victim.

Questions for Reflection

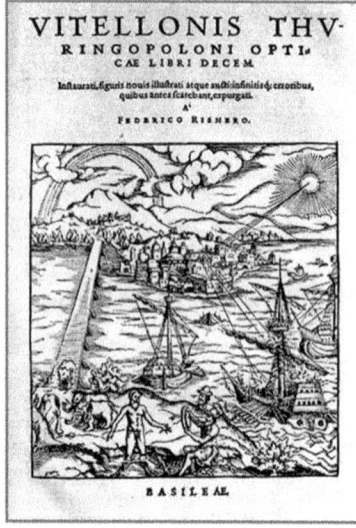

1. How did Alhazen reason for the intromission theory of vision?
2. Alhazen stated that light was reflected in all directions from any point of the illuminated object. Did this claim contradict the law of specula light reflection (the equality of the angles of incidence and reflection)?
3. What kind of light ray selection did Alhazen apply? What for?
4. Alhazen placed the image on the surface of the eye's lens. What could be the reason for that?
5. Could you imagine a different way to cause image inversion?
6. Identify the structural components of the human eye in Fig. 3.15a—the drawing by Witelo.

Fig. 3.16 The front page of Witelo's textbook *Perpectiva* as published in Basel in 1572. It presented Alhazen's optics treatise and introduced his ideas to the Europeans

Medieval Christian Science

Following a significant break in continuity, scientific progress in Europe was slow. Unlike the previous involvement of experimentation in Hellenistic and Muslim sciences, the major scientific effort in Europe was mainly of a qualitative and religious nature, with almost no empirical investigation. The first attempts at changing this situation, bringing optical research to the empirical agenda, were by Robert Grosseteste (1168–1253) and Roger Bacon (1214–1294)—natural philosophers in

---

lated the open nature of scientific progress and success. Voluntary ideological selection of knowledge according to the identity of its creators is incompatible with scientific progress and may lead to pseudoscience (Chap. 9).

Oxford. They stated the so-called prerogatives of experimental science[9] that required experimental investigation and testing of the scientific claims and using mathematics for that aim. Grosseteste wrote[10]:

The usefulness of considering lines, angles and figures is very great since it is impossible to grasp natural philosophy without them. They are absolutely important both in the universe as whole, and in its individual parts.

Within this framework, Grosseteste could empirically arrive at the understanding of light refraction by a spherical ball and light convergence in a focal point after the ball (Fig. 3.17a). The focal point was termed *combustion point*, which hints to the experience of heating objects placed at that point.

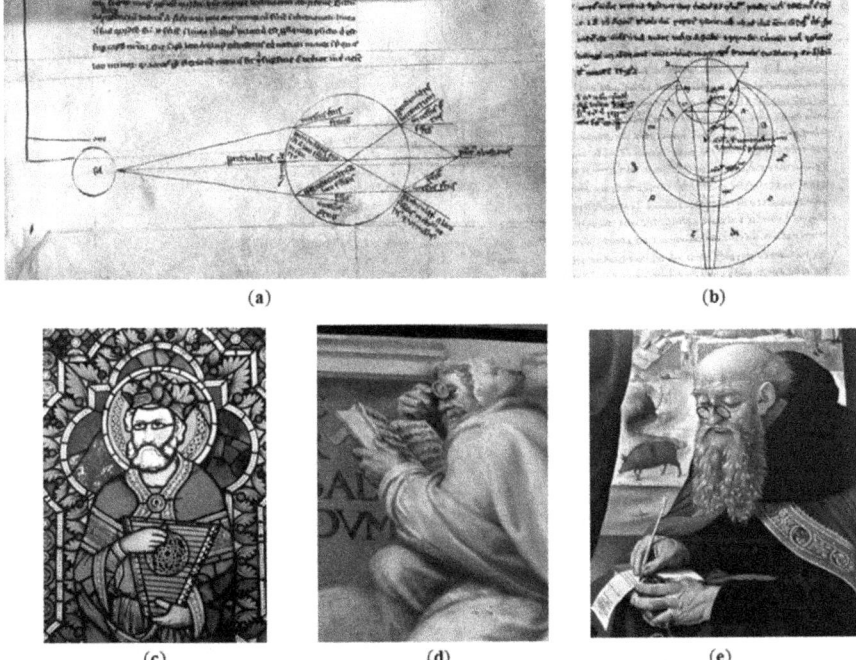

Fig. 3.17 (**a**) Sketch by Roger Bacon explaining lens functioning as causing refraction to light rays from the Sun and their convergence to the point called "point of burning", or "combustion point." (**b**) Drawing by Roger Bacon in which he described the structure of the human eye. (**c, d, e**) Artistic images documenting reading glasses in medieval life. Fragments from the stained glass in Florence (Santa Croce cathedral), the wall fresco in Parma Cathedral and the picture *Visitation* by Piero di Cosimo in Washington (National Gallery of Art). (Photos by the author)

---

[9] Losee (1993, p. 35).

[10] Pedersen & Pihl (1974, p. 196).

However, no further steps were made toward the understanding of an optical image, no relevant experimentation was carried out. Therefore, nobody in Grosseteste's time, including his follower Roger Bacon, could surpass the original schematic description of the structure of an eye in the investigation of image creation (Fig. 3.17b). Yet, Grosseteste stated the ability to improve vision in practice through using reading glasses—spectacles. As we know from art of that time (Fig. 3.17c–e), reading glasses were in quite regular use, while they remained without any scientific explanation. The understanding of optics by the Arab scholars remained superior until the scientific revolution of the seventeenth century when the new theory of vision was established.

Early Modern Science

During the Renaissance, scholars in Europe continued to puzzle over the enigma of vision. Alhazen sounded rather convincing, but his theory was still too qualitative, offering no numerical account. Alhazen's theory failed to make any practical implementation, for example, to explain how spectacles work. Spectacles or "reading glasses" were known for their ability to help with reading from the thirteenth century, but no theory could explain them. This pointed to the deficiency of the theory.

The major weakness of Alhazen's mechanism of vision was the unique role of the *relevant* rays (those at right angles with the eye surface). Indeed, what about the rays very close to the relevant ones? Can Nature be so selective in preferring one ray to its close neighbor? Such highly tuned discrimination did not look plausible. This unanswered question joined other unknowns in the periphery of that optical theory.

Optical theory continued to accumulate knowledge before the next breakthrough. Scholars tried to understand general features of vision. Among such features, perspective continuously attracted much interest. In fifteenth-century Renaissance Florence, the group of artists and architects known as the "Perspectivists," including Brunelleschi, Alberti, and others, revived the Euclidian theory of rays for promoting the abilities of architecture and painting (Fig. 3.18). In this activity, it was not important whether light came into the eye or emanated from it for the reversibility of light known after Archimedes. Nor was it important what happens inside the eye of the observer. Leon Battista Alberti (1404–1472) wrote, showing his knowledge of the old debate in optics[11]:

> Among the ancients, there was no little dispute whether these rays come from the eye or the plane. This dispute is quite useless for us. ... Nor is this the place to discuss whether vision, as it is called, resides at the juncture of the inner nerve or whether images are formed on the surface of the eye as on a living mirror.

---

[11] Alberti (1436/1970).

Alberti

Fig. 3.18 This flat artistic composition aims to generate a perception of depth in the observer in the Church of San Giacomo Maggiore in Bologna. (Photos by the author)

Leonardo da Vinci (1452–1519)—perhaps the most famous scholar and artist of the Renaissance—made a huge contribution to account for visual representation. The theory of perspective, introduced by Euclid in the third century BC, became central due to the interests of the artists of the Renaissance. In the hands of Leonardo, rays became a tool for estimating the intensity of light distribution in optical instruments (curved mirrors) and in shadowed areas. He traced numerous rays, initially equidistant after they reached the instrument, and evaluated light distribution in partially illuminated areas (penumbra) (Fig. 3.19).

Leonardo and his contemporaries continued to believe in mapping an observed object by means of light rays, which "transferred" the image into the eye. This

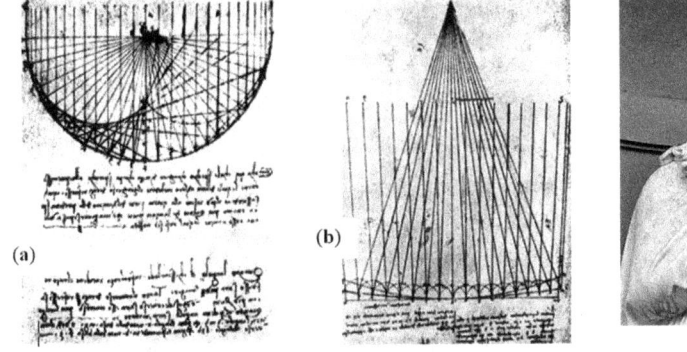

Leonardo da Vinci

Fig. 3.19 Two sketches of Leonardo show that he scrupulously followed numerous light rays in order to understand light distribution in mirrors (spherical, **a** and parabolic, **b**). Codex Arundel, fol. 87r

understanding we see in Leonardo's direct relating the appearance of objects to the opening of the eye:[12]

> *...the larger the pupil the larger will be the appearance of the object it sees.... All things seen will appear larger at midnight than at midday and larger in the morning than at midday. This takes place because the pupil of the eye is considerably smaller at midday than at any other time....at night it [pupil] sees things larger than by day.*

When one reads these words, it is difficult to avoid the impression that Leonardo still thought in terms of an image being transferred toward the observer, and the bigger the opening (the pupil), the bigger the image that can "enter" the eye. However, Leonardo did not suggest any selection among the multitude of rays emitted from the object and reaching the eye. He did not agree with Alhazen on the image location in the eye: it is not on the lens, which only refracts rays, but further inside the eye, on the extremity of the optical nerve—the retina.

It is interesting to follow the conceptual struggle of Leonardo to discover the mechanism of image creation. Image (in the terms of that time: *similitudine, spetie, impressione, forma, eidolon, simulacra*[13]) was transferred somehow through space, by rays. Leonardo knew about the Camera Obscura and the analogy of vision but wanted to clarify this theory through looking for the inversion of light rays when they enter a small opening in the pupil (Fig. 3.20a).

The inversion, which apparently caused the upside-down image, puzzled Leonardo, as it had Alhazen 400 years before. Leonardo, thus, suggested that the second crossing of the rays that took place inside the eye—by the eye lens—could cancel the first inversion on the way to the retina, causing the "correct," right-side-up image to appear there (Fig. 3.20b). Leonardo even imagined a device in which one may observe the image after a double inversion (Fig. 3.20c). Yet, it remained a pure guess (incorrect in modern understanding: there is no double crossing of light inside the eye) since Leonardo did not trace rays (as Grosseteste did before him, Fig. 3.17a) and did not specify the image creation.

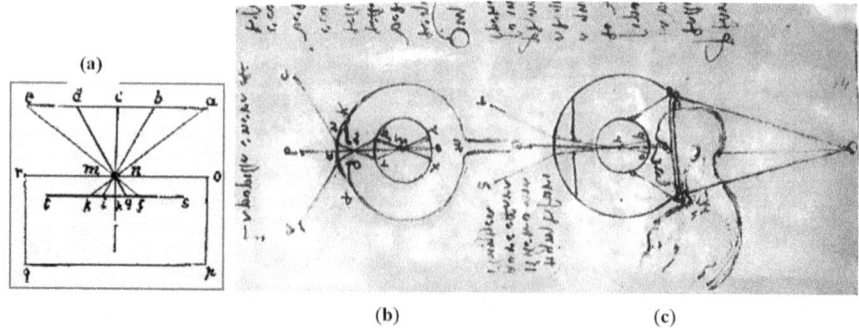

Fig. 3.20 (**a**) In this sketch of Camera Obscura, Leonardo explained the first inversion of the image in the eye as taking place in the pupil (Paris Manuscript D, 8r). (**b**) In this sketch, Leonardo showed his model of visual image with two consecutive crossing points of the rays on the way to final perception (Paris Manuscript D, 3v). (**c**) In this sketch, Leonardo depicted a device to observe the right-side-up image—the person represents the perception by the brain. (Paris Manuscript D, 3v)

---

[12] da Vinci (1955, p. 251)

[13] The abundance of synonyms seemingly indicated the lack of clarity regarding what the optical image is.

The ideas of Leonardo, however inspiring, did not have any impact on his contemporaries, perhaps, because he did not want that to happen. He encoded his writings, and they were never known by or debated with other scholars.  As a result, instead of stimulating further investigation, his ideas remained to impress the readers in the distant future and did not influence the progress in optical image understanding at that time.

It is informative to mention the views of two more heroes from the Italian science of the sixteenth century: Francesco Maurolico and Giovanni Batista Della Porta.

Maurolico

Maurolico (1494–1575), a professor in Messina, Sicily, understood that one should not overlook the crystalline lens of the eye. As a lens, it causes light rays to converge. This account helped to realize the influence of spectacles in supporting deficient far-sighted vision: the remedy is provided by using convex lenses which compensate for the insufficient power of the eye's lens. However, Maurolico did not introduce rays crossing after the lens—another way to avoid the upside-down inversion. "Properly arranged" after refraction, the rays reached their destination—the retina. In the view of Maurolico, the retina served as a screen (Fig. 3.21). In fact, Maurolico's picture maintained the medieval conception of vision by Alhazen: point-to-point mapping of the object to the image by single rays.

Even worse, Maurolico rejected the claim of Alhazen that only perpendicular rays construct the image and called this claim absurd. Yet, since he did not provide any alternative mechanism of image creation, his account strengthened the problem Alhazen raised pertaining to the mixing of the rays from different points. Regarding this, he simply mentioned that the central ray in the relevant "pyramid of rays" emanating from an object provides the major contribution to the image, while other rays play a minor role or yield "less certitude," which was quite an obscure claim. Lacking a consistent mechanism of vision, Maurolico, as Alhazen before him speculated regarding the crystalline humor. Alhazen considered it being an instrument of perception where the visual power resides and Maurolico—merely as a refracting device, in effect, directing the image to the retina.[14]

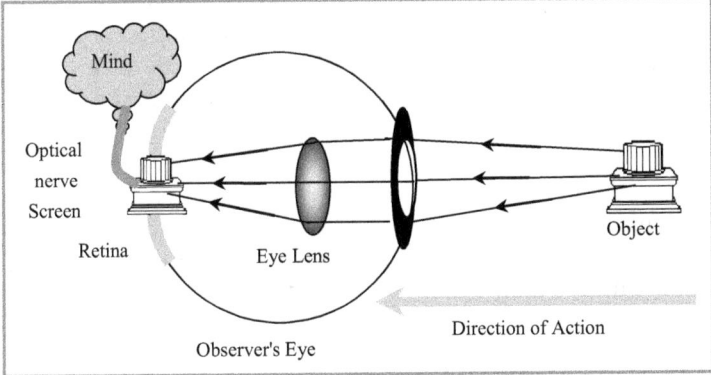

Fig. 3.21 Schematic representation of Maurolico's understanding of vision. Single rays projected from each point of the object to its image in the eye. The eye's lens causes light rays to reach the retina after they diverge from the object

---

[14] Lindberg (1976, p. 182).

We now mention another investigator from Renaissance Italy—Giambattista della Porta (1535–1615) who was a scholar and polymath—working mainly in Naples. His major interest was to discover new phenomena that he considered as revealing secrets of Nature but without providing a theoretical account, that is, a reliable explanation.

Within this activity, della Porta once found in 1586, that if one placed a convex lens in the opening of the *Camera Obscura*, the image in the camera would become much brighter and clearer. Not only this, while the image in the regular camera remained sharp in a wide range of distances between the wall and the opening, after a lens was put in the opening, a clear-edged image was observed only at a certain distance from the lens. No explanation followed, just a description of a new gadget in his book—the *Natural Magic*.[15]

We know today that by placing a convex lens into the opening of the Camera Obscura, della Porta converted it into a different optical instrument—a photo camera, in which a focused image can be obtained only at a certain distance from the lens. Indeed, a focused image is created at the specific distance from the convex lens. The bright image becomes blurry and unclear at other distances. For the objects very distant from the camera, a clear image is observed exactly at the focus distance of the lens known as the "combustion point" for those who used the lens to ignite fire by sunlight (Fig. 3.17a).

Although della Porta speculated that his invention might be meaningful in understanding vision, this did not shake the conception of visual image by Alhazen, which had dominated the science of optics for 600 years—a point-to-point mapping from the object to the image by means of light rays. The history of vision was on the verge of a radical conceptual change.

---

[15] Porta (1658/1957, p. 365).

Questions for Reflection

1. What problem did Leonardo try to resolve with regard to vision? What supposition guided Leonardo in his efforts to resolve the enigma of vision?
2. What was the similarity between Alhazen and Maurolico in their understanding of vision?
3. What was the contribution of della Porta to the understanding of vision?
4. Figure 3.22 is from an old book of optics and depicts human vision. To what conception of vision does it seem to fit? Is the choice univocal?
5. Should we consider della Porta to be a scientist? Discuss this question.

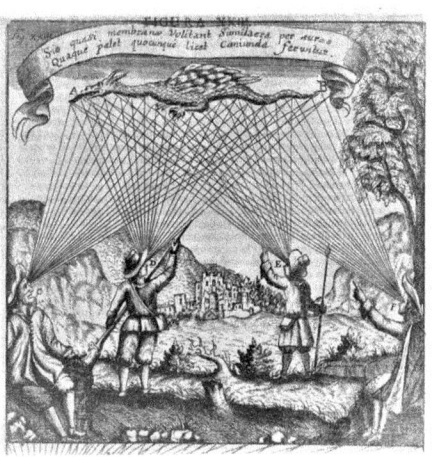

Fig. 3.22 A drawing depicting vision from the book by Johannes Zahn (1685)

Modern Science

Kepler

The major breakthrough in resolving the enigma of vision took place during the scientific revolution of the seventeenth century—a transition to Modern Science. Johannes Kepler (1571–1630), an outstanding scholar, physicist, and mathematician from Germany, was one of the pioneers in this process. He explained the course of action taking place in vision in an essentially new manner.

Kepler compared the construction of a visual image with the way in which an image is constructed by a convex lens (Fig. 3.22). Kepler reinterpreted the role of the light ray, considering it a mere representation of light flux without any real function in *transferring an optical image*.

It was a radical change. Instead of explaining image in terms of rays, one should think about vision in terms of light or, more exactly, light flux: "lines infinite in number issued from every point" in a visual field. Accordingly, each visible point of the object completely bathes the eye with light, creating a cone of light flux with the apex on this point and the base on the eye surface, at the pupil (Fig. 3.23).

In accordance with the Al-Kindi–Alhazen principle, light flux emanates from each point of the object and expands into space in all directions. When light flux meets the convex surface of the cornea, it enters the eye through its refracting layers, the optical system that causes the flux to converge. Due to the crystalline lens of the eye, the entering light flux converges precisely onto the retina, creating the

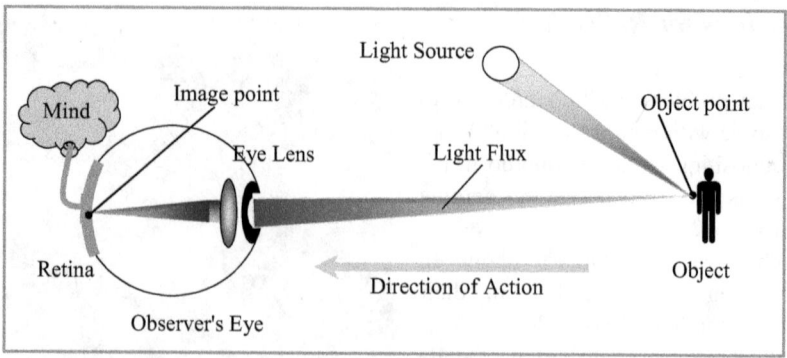

Fig. 3.23 Schematic representation of Kepler's solution to the problem of vision. The mind interprets the inversed image obtained on the retina of the observer's eye. The image is obtained through the point-to-point mapping of the object to the image by means of the first diverging and then converging light flux cones

corresponding image point. This is the basic mechanism of image creation. The cluster of image points, thus created, comprises an illumination pattern reproducing the appearance of the object—the optical image.

The conceptual change introduced by Kepler was the replacement of point-to-point mapping by means of *single rays* to the point-to-point mapping of the object by means of *light fluxes*, which are converged by the eye onto the retina.[16] This way Kepler dismissed Alhazen's idea of a sharp discrimination between the rays, perpendicular and oblique. He wrote[17]:

> *I refute Witelo by this very confusion of rays. For, as he says, oblique radiation too is seen insofar as oblique rays intersect perpendicular rays [within the eye]; therefore, the same point [of the eye] receives both oblique and perpendicular radiation. Consequently, two things [the oblique and perpendicular radiation] will be judged to be situated in the same place.*
>
> *The reception or sensing of the perpendiculars and the rays adjacent to them [should be] almost equal.*

Nevertheless, having explained the image, Kepler faced the same problem that for many years had challenged so many scientists: Alhazen, Leonardo, Maurolicus, and others. The image formed on the retina was inverted, exactly like the image in a photo-camera and a Camera Obscura. Yet, this did not trouble Kepler. He realized the existence of an additional post-imaging stage in the process of vision: the interpretation of the retina image by human consciousness:[18]

> *I say that vision occurs when the image of the whole hemisphere of the world that is before the eye... is fixed on the reddish white concave surface of the retina. How the image or picture is composed by the visual spirits that reside in the retina and the nerve, and whether it is made to appear before the soul or the tribunal of the visual faculty by a spirit within the*

---

[16] We write "converged by the eye" since the whole eye causes the light refraction, the eye lens performs only tuning to reach clarity of the image on the retina.

[17] Kepler (1604) *Ad Vitellionem paralipomena, quibus astronomiæ pars optica traditur*, in Lindberg (1976, p. 189).

[18] Kepler (1604) in Lindberg (1976, p. 203).

hollows of the brain, or whether the visual faculty, like a magistrate sent by the soul, goes forth from the administrative chamber of the brain into the optic nerve and the retina to meet this image, as though descending to a lower court – this I leave to be disputed by the physicists.

Unlike Alhazen, who followed light rays only up to the lens in order to prevent the image from being inversed, and unlike Leonardo, who looked for the additional intersection of rays in the eye to compensate for the first intersection and cause the image to be right-side-up, Kepler left the task of image perception on the retina to consciousness. It is in the human mind that the image is interpreted and "recognized" as right-side-up. In effect, Kepler ascribed to the consciousness the additional inversion of the image performed to support normal functioning of the whole organism (Fig. 3.24a). Descartes specified this understanding, pointing to the location of image interpretation in the brain (Fig. 3.24b). He believed that the inversion takes place in the pineal gland of the brain which is the seat of vision and common sense. Inside this gland, we can see the small right-side-up image (a, b, c) of the object ABC after the second inversion of the inversed image on the retina in both eyes.

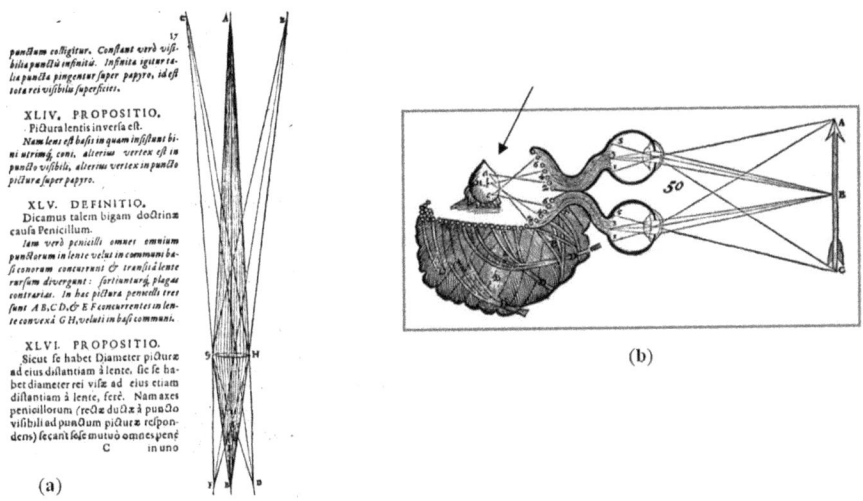

Fig. 3.24 (**a**) Page 17 from Kepler's *Dioptric* of 1611 (Kepler (1611, p. 17). Reproduced from the scan of the original, mentioned by Shapiro (2008)). The ray diagram shows the new understanding of image—creation of point images by the cones of light and their order inversion relative to the object. (**b**) Descartes refined the same understanding in his sketch in *L'homme, et la formation du foetus* (1677, p. 74). The sketch represents the mapping of points *A*, *B*, and *C* of an object by light flux and connection to the brain. Descartes showed the pineal gland (indicated by an arrow) which is the seat of vision and common sense. Inside this gland, we see the small right-side-up image of the object (*a, b, c*) after the second inversion inside the brain

Within this understanding, Kepler, in 1604, dismissed the whole idea of image transfer as a projection through space, the ancient Atomist idea of eidolon and the Aristotelian idea of transfer by disturbance in the medium. Kepler resolved the problem of vision mechanism by reducing it to the real image produced by the eye as a convex lens. In the new scenario, the optical image did not, even "potentially," exist anywhere between the object and the eye but is comprised of image points created on the retina.

Historical and Philosophical Summary

This excursus spans about 2000 years. Starting from the dawn of science in Classical Greece, it concludes at the beginning of the scientific revolution of the seventeenth century. The period sees several shifts in scientific understanding and scientific method, scientific ontology, and epistemology.

The process of understanding vision started with the debate between the intromission and extramission theories, the existence of visual rays versus the eidola of objects that enter into the eyes. The debate faded away only after Alhazen in the eleventh century; visual rays were dismissed.

The first theories of vision were within the agenda of Natural Philosophy—seeking universal regularities in the ordered world—cosmos—which were independent of anybody's will. The perspective of the stable objective organization of nature was a contrast to the myths of voluntary command by humans or superhumans on natural events. The claim of the objectively occurring process with regard to a natural phenomenon was revolutionary. Yet, Hellenic theories of vision only provided qualitative reasoning drawing on certain principles and formal logic. The introduced concepts were vague, poorly defined, and lacking any mechanism that could be detailed or confirmed. They introduced the ideas which directed exploration for years ahead. However, what is of great importance is to realize that Hellenic theories clearly considered Nature as being *open* for exploration and understanding. *Nature allowed its understanding!* This tremendous claim remained central to science in all its areas and for all the times to come, even if the philosophical and practical debate in all subjects became far more sophisticated. They laid the foundation stone of science structure.

During the following Hellenistic period, science became more pragmatic, physically concrete, mathematically elaborated, and less interested in philosophical claims which remained in use as a general framework—it was, in effect, a scientific revolution.[19] Such were the optical theories of Euclid, Galen, and Ptolemy. Euclid refined the Pythagorean flux of internal fire to visual rays, agents of sight, and established the theory of perspective. Euclid and Ptolemy stated the laws of visual rays' behavior—reflection and refraction (*Catoptrics* and *Dioptrics*[20]). Galen held the theory of a mysterious medium facilitating vision—*pneuma*—and upgraded theory of vision with an exact anatomical structure of the eye.

The scholars of medieval Muslim science drew on both Hellenic and Hellenistic heritage. Al-Kindi proceeded the trend of Ptolemy and adopted his perspective of extramission and empirical method. In contrast, Avicenna (Ibn Sina) and Averroes (Ibn Rushd) objected to the extramission theories of Euclid, Ptolemy, and Galen and remained close to understanding of vision by Aristotle. The greatest original

---

[19] Russo (2004).
[20] Cohen & Drabkin (1948, pp. 257–283).

progress, however, was due to Alhazen who demonstrated how science works. He synthesized the intromission idea from Aristotle, light rays from Euclid, refraction from Ptolemy, and the anatomy of the eye from Galen. The great synthesis enabled Alhazen to make significant progress. He explained the image in Camera Obscura and suggested a new mechanism of vision. This understanding he transferred to visual image. He argued for the correspondence between the image point in the eye and the certain point of the object created there by means of a single light ray. Alhazen imagined right-side-up images appearing on the surface of the eye's lens, and this implied considering the lens to be the sense organ of vision. His theory was not elaborated quantitatively, but it remained as the most advanced theory for about 600 years.

Kepler succeeded in the next breakthrough—he mapped the object to its image by means of the light flux emanating from each point of the object, entering the observer's eye and being converged by the eye to the corresponding point on the retina. It was suggested that the inverted image on the retina was subsequently interpreted by the observing mind as the realistic image of an object. These two steps were major discoveries by Kepler regarding the visual image. They demonstrated the nature of science in its ontological and epistemological sense.

The ontological change was in replacing of a single ray with light cone was an impressive invention in physics theory. It was reincarnation of the old Pythagorean idea of visual flux emanation from the eyes of an observer. It was later used also by Plato and Ptolemy to explain vision.

Taking the advantage of the time passed, we may identify the major steps of resolving the enigma of visual image a subject of scientific investigation for more than 2000 years (Figs. 3.25 and 3.26).

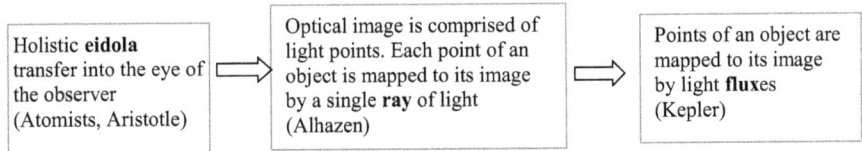

Fig. 3.25 Flowchart of the intromission theories of optical image in the history of physics

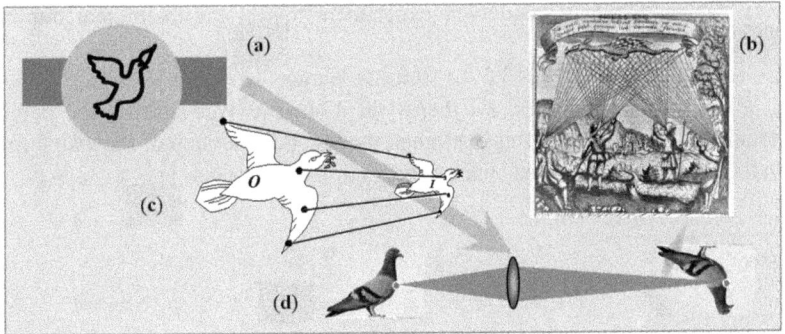

Fig. 3.26 Symbolic representation of the stages of understanding optical image: (**a**) holistic intromission, (**b**) extramission by vision rays, (**c**) differential intromission, point-to-point intromission by light rays, and (**d**) point-to-point intromission by light flux cones

The important epistemological change was overcoming intuition in understanding image creation. The reality may be more complex and may contradict intuition. For centuries, scholars (Alhazen, Leonardo, Maurolico, and others) rejected the inverted image in the eye. However, the inverted image in the eye appeared exactly to be the case calling for more careful use of intuition in physics.

Another remarkable ontological change happened to the concept of the *light ray*. In Hellenistic science, it was introduced as an "atom of light." In modern science, however, it is a mere notification tool of traveling light, lacking any physical existence. After dismissing the meaning of a ray as an object by Kepler, it was, however, revived in the optics of Newton as the "smallest amount of light that demonstrated the features of light behavior," the smallest beam which still obeyed optics laws. It presents a valid operational definition of light ray. Within Newton's speculation regarding light particles, light ray could be seen as the light particles' trajectory. In contrast, within the parallel theory of Huygens, the wave theory of life, the light ray signified the normal (perpendicular) to the wave front. In all cases, however, the light ray was no more than a representative tool, and so it has remained in modern physics which employs at least three theories of light: geometrical optics (the theory of light rays), physical optics (the theory of light waves), and quantum optics (the theory of photons).[21] Each theory is useful and practically valid within its area of validity, the magnitude of parameters: the dimensions of the environment in comparison with the wavelength, the intensity of light, the range of energy, and others. Kepler's understanding of the optical image remained a part of geometrical optics.

Questions for Refection

1. What was Kepler's mechanism of vision? In what way was it different from the theory of vision by Alhazen?
2. What was the novelty of Kepler's theory of vision? Why was it preferable to that of Alhazen?
3. Compare the structure of the eye by Kepler (Fig. 3.27a) with that of Alhazen (Fig. 3.15). What is different between the structures?
4. What was the problem of the inverted image? What were its implications? How was this resolved? Consider an experiment to test the claim that our mind inverts the visual image.
5. Consider the sketch (Fig. 3.27b) from Descartes (1677, pp. 74 and 76) treatise *L'homme, et la formation du foetus* mentioned above. Discuss the depicted details of the way Descartes explained the second inversion of the visual image in the brain (consider the separate image in left upper corner).

---

[21] Chapter 7

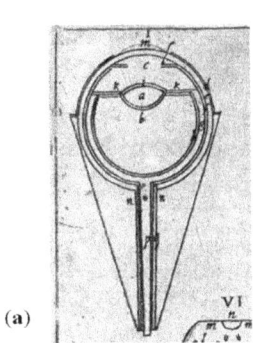 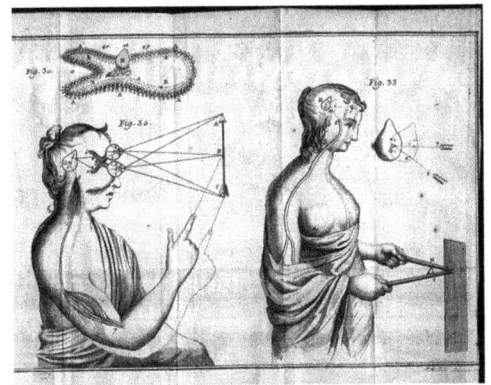

Fig. 3.27 (a) The structure of the human eye as used by Kepler. (b) Descartes' sketch explaining vision

6. Figure 3.27b includes two persons. What was the reason to include the figure on the right close to the first one? What human sense is illustrated by that figure?
7. Describe the changes in the concept of light rays throughout history.
8. Consider manifestations of the rationalist and empiricist approaches in each theory of optical imagery.
9. Why do we consider the old knowledge of optical image to be scientific?
10. Consider the changes of involvement mathematics in optics throughout the course of science history.
11. Identify the chain of claims regarding the optical image, ontological (What was stated regarding the physical process?) and epistemological (How was it reasoned?).

Educational Comments

The analysis of students' knowledge reveals a variety of conceptions that students develop regarding optical imagery. After being elicited in a number of studies,[22] they were organized in a hierarchical structure of scheme-facets of knowledge. It is worth comparing a few selected schemes of knowledge with the historical conceptions in optics (Table 3.1).[23]

---

[22] Galili & Hazan (2000a)
[23] Galili & Hazan (2000b)

Table 3.1 Conceptual parallelism in optics knowledge regarding image and vision

Historical conceptions of optics	*Students' conceptions in learning optics*
Pythagoras and Plato's conception of vision by emanating fire	Active vision scheme ("touching by sight")
Atomists' conception of *Eidola* traveling to observer eyes	Image holistic scheme (the image travels to the eye or screen/mirror where it is observed)
Euclidean (Ptolemy) visual and light rays	Light rays as the entity comprising light vision rays as the entity performing vision
Alhazen's conception of visual image mapping by means of light rays	Image projection scheme: Image is mapped point-to-point by means of single rays per point into the eye and on the mirror

The apparent conceptual similarity suggests the relevance of the historical material to the teaching-learning process. One may argue for the didactical benefit of this knowledge in the following aspects:

Cognitive Resonance

Within the constructivist theory of learning, *cognitive resonance* is considered as a phenomenon in conceptual construction performed by the leaner making sense of phenomena and being challenged and encouraged by teaching. The historical parallelism (Table 3.1) implies that addressing certain historical conceptions may cause *cognitive resonance*—an internally enhanced reaction to the considered conception, due to the conceptions that the students hold. This phenomenon might be interpreted using the idea of zone of proximal development[24] manifested by, among others, the selective attention of students and their preferences in individual learning. The exposure of historical theories in historical debates and the arguments for and against each of these theories may serve as an effective strategy in promoting conceptual change toward scientific conceptions.

Learning by Conceptual Variation

Recent studies in educational psychology stressed the advantage of learning by conceptual *variation*.[25] This approach suggests that in order to stimulate meaningful assimilation of certain concepts or conceptions by the learner, the educator should prepare the learning material which would present the target content in a certain conceptual variation. The learner is thus guided and encouraged to discern and

---

[24] Vygotsky (1934/1986).
[25] Marton et al. (2004).

adopt the target conception through comparison and teacher mediation. The more precisely the *contrast* is elaborated, the more appealing the teaching becomes, causing better results of the such arranged learning.

Fig. 3.28 Symbolic presentation of the diachronic dialogue on the nature of vision in science. The perspectives on optical image and vision mechanism, other than those by Kepler, present conceptual variations of his theory. All together, they create a space of leaning about the optical image and vision

The historical conceptions regarding optical imagery and the nature of vision provide the required conceptual variation in the form of an overview on the historical debate regarding different ideas and arguments (Fig. 3.28). Teaching enforcing comparison emphasizes the essential differences between the theories and leads to a genuine understanding of the goal conception. It is comparison that impels paying attention to the essential features of the subject of learning which otherwise could be very probably missed or considered obvious. For example, the scheme of the "direct" transfer of points of the object by single rays to the points of the image (Alhazen's conception) often seems plausible to students and emphasizes the innovation of Kepler's refutation of this idea—the image construction by light flux rather than by rays.

Construction of the Cultural Content Knowledge

The teaching of a certain disciplinary subject without any conceptual variation (in a generic curriculum) may represent a particular item of disciplinary knowledge but not fully represent science. Science presents a culture and as such, each of its items is immersed in a variety of ideas and interpretations. To represent this aspect of scientific knowledge, one needs an appropriate structure—discipline-culture.[26] In it, a fundamental scientific theory emerges as including the *nucleus* containing fundamental principles (ontological and epistemological), the *body*—applications of the principles (solved problems, affiliated empirical rules, explained phenomena, models, devices, etc.), and *periphery*—the knowledge elements contradicting the particular nucleus for any reason (erroneous or related to other theories).

Within this framework, the periphery of the geometrical optics includes the old conceptions of vision (visual rays, image transfer, etc.). In addition, the periphery incorporates the principles of the more advanced optical theories (the theory of light waves and the theory of photons). Together, they create the *space of learning* and encourage the construction of the *cultural content knowledge* of optics. Learning such organization promotes enculturation into the world of physics (principles, norms, models, laws, etc.) rather than its indoctrination. This presents a cultural education appropriate for the broad population of school students, above and beyond professional training.

Construction of Scientific Knowledge

The history of understanding the optical image and vision spreads over 2000 years. During this long period, science changed its features, passing through different cultural periods: Hellenic, Hellenistic, Muslim and European Medieval, and Modern. The excursus into the optical theories provides an opportunity to describe the *changing features of science*, the variance in its epistemology (the preferences of rational and empirical nature, mathematics, and experiment) as well as ontology (basic concepts and models of the content knowledge).

At the same time, alongside the changing subject matter, one may observe *the invariant features of scientific knowledge*, such as the commitment to seeking *objective* truth about Nature, revealing the principles and laws that govern Nature, and the contest and synthesis of theories. Thus, one may better appreciate the essential features of modern science—a well-balanced synthesis of rationalist and empiricist approaches and the essential use of a mathematical account in knowledge codification and monitoring.[27]

---

[26] Chapter 6.

[27] Galili & Hazan (2001).

Media for Learning

Here are a number of suggestions of educational topics to consider while teaching the science of optical images.

1. Alhazen treated light as a collection of light rays, and image as transferred by them to the eye. This idea became common knowledge in medieval Europe due to the effort of Witelo and Packham who wrote optical treatises mainly translating Alhazen's theory. Numerous artists adopted this conception in a religious context. Fra Angelico (1432), a monk of San Mark convent in Florence, was educated in contemporary optics in Western Europe. He portrayed the important religious event of the Annunciation where Mary of Nazareth received the divine message as transferring to her the holy image moving "on the rails" of rays (Fig. 3.29b)[28]

(a)

(b)

Fig. 3.29 Three ways of image creation. (**a**) The holistic transfer of the image in the *Annunciation* depicted in the Eastern canon (Fragment from the illustration by unknown author of 1323). (**b**) The image projection by light rays. The fragment from the *Annunciation* by Fra Angelico (1432); (**c**) the image construction by light cones in the geometrical optics of Kepler (Descartes' drawing of 1637)

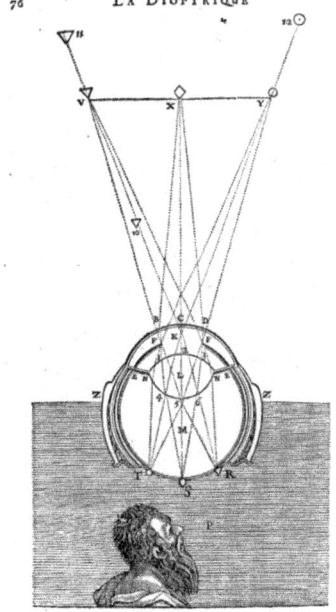
(c)

Those who lived in Eastern Europe prior to the period when the knowledge of Alhazen became known there depicted the same context in accordance with holistic transfer—the Atomists' conception of eidola (Fig. 3.29a). Compare the views on image transfer with Kepler's perspective as depicted later by Descartes in his *Dioptric* (Fig. 3.29c).

2. The context of the mirror image is effective in revealing students' views on image formation, its observation and the principle of light reflection from objects (Al-Kindi and Alhazen).

---

[28] Galili & Zinn (2007).

Fig. 3.30 (**a**) Multiple images can be used to illustrate the principle of Al-Kindi–Alhazen: light expansion in all directions from each point of the observed object in all possible directions. Multiple mirrors illustrate this reality. (**b**) A sundial in Geneva (Photos by the author)

Consider the image location in a mirror as dependent on the location of the object, the way the image is observed, the image existence without an observer looking into the mirror. It is especially elucidating to discuss the case of three images in the mirror corner, two parallel mirrors and multiple mirrors (Fig. 3.30a).

3. The image of the Sun created through a small opening (as in the Camera Obscura) was the popular way to observe the Sun throughout the history of scientific observations. It was safe and affordable, requiring mainly persistence and no sophisticated equipment. Following the motion of the Sun became a reliable tool of time measurement in churches, monasteries, and private homes. Today we see such sundials in public places. Consider such a device established in Geneva park (Fig. 3.30b). One may identify a small hole in the disc mounted on the metallic shaft. The image of the Sun emerges on a curved screen incorporating a time scale. Discuss the role of both shadow and image in using the sundial. The image changes its place during the day and during the year. It moves along a curved line (see the photo). Discuss the reasons for this.

4. In the old Byzantine water pool in Istanbul, one may observe the sculpture of the head of the mythological Medusa, from the time of Constantinople (Fig. 3.31). It is placed just above the water level in an upside-down orientation. Try to justify this orientation. To do this, one may draw on the myth that one could not look at Medusa's face *directly* for fear of being turned to stone on the spot.

Fig. 3.31 The head of Medusa as the basis of a column in a water pool in Istanbul. Why did the designer invert the position of Medusa's head? (Photo by the author)

Students often erroneously ascribe to the plane mirror the ability of inversion—the right to the left (instead of the inversion "away from the mirror"—"towards the mirror").[29] This discussion may include the observation of an identifying mark or sign on the front side of emergency cars written left-right inversed (Fig. 3.32). Many students believe that it is because the mirror, through which the driver in the car ahead of the ambulance observes the ambulance, "transforms right to left." Is it so?

Fig. 3.32 The inscription on the car AMBULANCE is right-left inversed. Why is this usually done with emergency cars? (Photo by the author)

---

[29] e.g., Galili et al. (1991); Galili & Zinn (2007).

5. Another culturally interesting story involves understanding of optical image in the plane mirror. One may wonder why Italian artist Giotto, in the thirteenth century, while depicting stigmatization of the St. Francis connected between the hands and feet of the two figures in a special manner: left to right both hands and feet (Fig. 3.33a). Other artists later commonly kept with a regular correspondence: left to the left (Fig. 3.33b).[30]

(a)   (b)

Fig. 3.33 (a) Stigmatization of St. Francis as depicted by Giotto, c.1299. The correspondence established between right and left hands and feet. This relationship hints to what happens with image in the mirror. (b) Stigmatization of St. Francis as depicted by unknown German artist in 1500. The correspondence established was between right and right, left and left as is between two people facing each other

The answer could be that Giotto intended to demonstrate the philosophy of St. Francis according to which each creature presents a "mirror image" of God. St. Francis was known under the nickname "the mirror of God." The story leads the learner to raise the question of what the mirror does to the image. Expanding this question to include comparison with different lenses and mirrors is recommended.

6. Leonardo was a pioneer of the scrupulous investigation of shadow images in order to facilitate representing variations of shading in producing images in paintings. These efforts included studies of traced light rays in various settings. His sketches help us interpret Leonardo's procedure of construction areas of partial shadow (penumbra) (Figs. 3.34). Explain the logic of penumbra area construction in painting.

7. Consider the problem of Aristotle regarding the image in the Camera Obscura. Aristotle points to the fact that when the opening is very small, the image of the Sun appeared inside the camera regardless of the shape of the opening, whereas

---

[30] Chapter 10, Galili (2014).

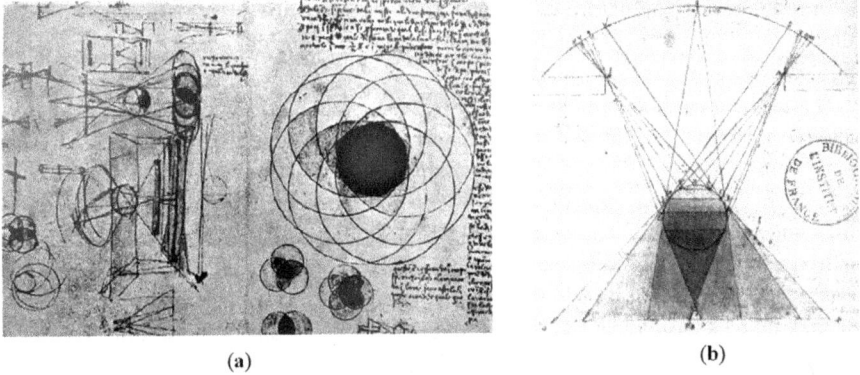

Fig. 3.34 Shadow construction by Leonardo (Da Vinci, Codex Atlanticus, fol. 187r, reproduced by Argentieri (1956, p. 413)). The activity supports interpreting Leonardo's constriction of the shadow image. (**a**) Construction of a shadow image in the case of a spherical source of light and a spherical obstacle; (**b**) construction of a shadow image in the case of two-point sources of light and a spherical obstacle

when the opening is big, the image inside the camera became an illuminated pattern in the shape of the opening regardless of the shape of the Sun. Explain the transition taking place between the two cases in the gradual increasing of the opening of the Camera.[31]

8. Consider the answer to each of the two following claims:

- If a tree falls in a forest and no one is there around, does it make a sound?
- You stand before a wall mirror and see your image in it. If you now close your eyes, is there an image there in the mirror?

Is there any difference between the two claims in terms of existence of the subject (ontological aspect)? How can you verify that your answer is correct (epistemological aspect)?

Concluding Remark

In closing this excursus to the history of optical image and its perception in vision, one may mention that regular teaching commonly deals with image concept as a given, without its definition, drawing merely on the intuitive conceiving image as a replica of an object created by light due to an optical tool. Vision is often mentioned in a physics class as a process in the eye similar to the photo camera—as if this is an equivalent case. The story discussed here shows that this approach, besides being ineffective,[32] misses an especially rich area of physics knowledge which may serve

---

[31] See Chap. 7 for more about the Camera Obscura.

[32] Vision is documented as a domain of greatly abundant misconceptions of students starting from elementary school and remaining in the following years thought school and beyond.

as introduction to the nature of science in school, revealing its method, multidisciplinary interaction, involvement of philosophy in empirical science, and implying high relevancy to students' learning.

If so, what could be an influential reason for neglecting the optical image especially with regard to its being treated solely technically (location, size, kind)? Seemingly, this policy reflects the vision of classical physics which tries its best to *exclude the observer*. Indeed, anything related to an observer is often considered as subjective, personal, related to feelings, views, etc. which is foreign, potentially deflecting attention and misleading understanding from the "pure" physical content. This is the situation in classical mechanics, thermodynamics, and electrodynamics. There, the observer often presents an obstacle and a factor to be overcome, and its influence to be corrected and dismissed. In elementary school(!), the physics teacher normally insists on the *heliocentric* world organization as the only true picture of the world, despite the apparent motion of the Sun. The physics teacher insists that the perception of *cold* and *hot* is only an illusion of tactile perception, the true characteristic is temperature. The physics teacher insists that there is nothing inside a mirror, only an illusion, which does not exist without an observer looking into the mirror. The physics teacher often insists that the astronauts in a satellite station possess weight, and their *weightlessness* is a mere illusion.[33] In all these examples, ignoring the observer presents a way to misunderstanding but brings a true relief for the teacher in class.

In optical comprehension of the observed, the situation is often the opposite, because the observer presents a part of an optical system. In this sense, classical optics can serve a unique springboard directly preparing students for learning quantum and relativistic theories in which, as in optics, the observer presents an integral part of the physical system considered.[34] It is with regard to the optical image that the students realize the importance of *who* and *how* observes the image, what is its nature (real or virtual), the involvement of the medium, and the role such external factors as cognition. All these aspects are left outside the technical account of a generic curriculum. For instance, the drawing of Descartes explaining optical image (Fig. 3.28c) explicitly shows the observer. It was not an artistic style; it was to emphasize the last phase in image observation discovered by Kepler—its inversion in the human mind. It implies optics textbooks to preserve this tradition of showing an observer in the ray diagrams.

This excursus argues for the discipline-cultural approach to teaching this topic. With such an approach, dealing with the history of vision is not merely a way to engage students, not simply an enjoyable enrichment by posing philosophical questions, but a part of the requirement of addressing the periphery of the theory employed. This presents a culture of physics itself. Not occasionally, this demand of *cultural content knowledge* coincides with the teaching by *conceptual variation* as a tool to stimulate *cognitive resonance*—the effective pedagogical strategy. In it, historical conceptions, the diachronic debate of ideas present indispensable resources.

---

[33] Chapter 5.

[34] The other case of a similar conceptual intention—the inertial force—will be considered in the next chapter, Chap. 4.

Dealing with image, interwoven with vision, is especially appropriate when dealing with often neglected addressing common sense and naïve intuition in science. It is the history of image understanding that provides a natural way to such questions as "How do we know that?", "What is your evidence?", "Is it a reliable reason?" In this sense, this excursus suggested an introduction to physics epistemology appropriate for middle school teaching.

Finally, with regard to the reservation that the discipline-cultural approach requires time lacking in the already overloaded science curriculum, one may realize that presenting a subject in its dialogical nature will not necessarily take more time. It is rather like adding sugar which does not increase the volume of the water in the glass but makes the drink different. Exposure of/to the dialogue dismisses indoctrination of a univocal truth making learning more effective, meaningful, and interesting. Such teaching often *saves* time for it reduces the rote learning unavoidable if the meaning remains unknown. Rather than simply refining the usual curriculum, considering image and vision in the cultural manner creates a new subject which enriches students intellectually.

Picture Credits

- Fig. 3.6 The structure of eye by Galen. Public domain History of Ophthalmology (mrcophth.com)
- Portrait of Euclid. Photo by M. A. Wilson, The College of Wooster. CC Attribution-Share Alike 4.0 International license. https://hif.wikipedia.org/wiki/Euclid#/media/file:Euclid_statue,_Oxford_University_Museum_of_Natural_History,_UK_-_20080315. jpg
- Portrait of Al-Kind, ninth c., Egyptian stamp. Public domain. https://commons.wikimedia.org/wiki/File:Al-Kindi_stamp,_Egypt_(1975).jpg
- Fig. 3.12 Camera Obscura. Illustration in the book of 1910. Public Domain. https://commons.wikimedia.org/wiki/File:Camera_obscura_1.jpg
- Fig. 3.13a Camera Obscura. Illustration from Kircher, *Ars Magna Lucis Et Umbra* (1645). Public domain. https://commons.wikimedia.org/w/index.php?curid=52598491
- Fig. 3.13b Eighteenth century, unknown author. Public Domain. https://commons.wikimedia.org/wiki/File:001_a01_camera_obscura_abrazolas.jpg
- Fig. 3.15a Illustration from Alhazen, *Perspectiva*, 1572, p. 6, in Latin. Public domain. https://books.google.co.il/books?id=oy9WAAAAcAAJ&printsec=frontcover&dq=inauthor:%22al-%E1%B8%A4asan+Ibn-al-%E1%B8%A4asan+Ibn-al-Hai%E1%B9%AFam%22&hl=en&sa=X&ved=2ahUKEwjJrcH56aDuAhVB6qQKHRgrBHgQ6AEwAHoECAQQAg#v=onepage&q&f=false
- Fig. 3.15b Illustration from Alhazen's *Perspectiva*, 1652, in Arabic. Public Domain. https://commons.wikimedia.org/w/index.php?curid=75144366
- Fig. 3.16 The title page of Witelo, *Perpectiva*, 1572, by unknown illustrator, Bavarian State Library. Public Domain. https://commons.wikimedia.org/w/index.php?curid=503708
- Fig. 3.17a Drawing by Roger Bacon, from Opus Maius, British museum. Public Domain. https://commons.wikimedia.org/w/index.php?curid=453736
- Fig. 3.17b Structure of the eye by Roger Bacon from Opus Maius, British museum. Public Domain. https://commons.wikimedia.org/w/index.php?curid=453045
- Fig. 3.19 Drawings of light reflection by Leonardo da Vinci. Codex Arundel, fol. 87r. Public domain. http://www.bl.uk/manuscripts/Viewer.aspx?ref=arundel_ms_263_f001r
- Statue of Leonardo da Vinci outside the Uffizi gallery, Florence. Public domain. https://commons.wikimedia.org/w/index.php?curid=408669
- Fig. 3.20a Sketch by Leonardo da Vinci, 1500. Public domain. https://upload.wikimedia.org/wikipedia/commons/8/8f/Da_vinci_-_camera_obscura_%28from_notebooks_71%29_0071-q75-644x596.jpg
- Fig. 3.20b, Sketches by Leonardo da Vinci—Leonardo da Vinci, Manuscript D, 1508-09, 3v. Bibliothèque de l' Institut, Paris. RMN. Public Domain. https://commons.wikimedia.org/w/index.php?curid=64303217
- Portrait of Maurolico by Bovis. CC Attribution 4.0 International license. https://commons.wikimedia.org/w/index.php?curid=35872950
- Front page of the book by della Porta. T. Young and S. Speed, London, 1658. Public Domain. https://commons.wikimedia.org/w/index.php?curid=4041273
- Fig. 3.22 Illustrative picture from J. Zahn, 1685. Public Domain. https://publicdomainreview. org/collection/images-from-johann-zahn-s-oculus-artificialis-1685
- Fig. 3.24a Illustration from *Dioptrice*, 1611. Republished in Kepler, Gesammelte Werke, 1941. ISBN 3-406-01644-8, Vol. 4, pp. 329–414.
- Fig. 3.24b Sketch by Descartes in L'Homme, 1677, p. 67. Public domain. https://archive.org/details/lhommeetlaformat00desc/page/64/mode/2up
- Fig. 3.27a Illustration from J. Kepler, 1604, p. 188. Paralipomena to Witelo & Optical Part of Astronomy. Francoevrtty Apud Claudium Marnium & Hxrcdcs Ioannis Aubrii
- Fig. 3.27b Sketch by Descartes, L'Homme, 1677, pp. 74, 76. https://archive.org/details/lhommeetlaformat00desc/page/76/mode/2up
- Fig. 3.29a Annunciation in Eastern canon, Évangile de Toros de Taron, 1323. Photo credit: Michel Bakni. Public Domain. https://commons.wikimedia.org/w/index.php?curid=96543029

- Fig. 3.29b Annunciation in Western canon by Fra Angelico, 1432, Museo del Prado. Public Domain.   https://commons.wikimedia.org/wiki/File:La_Anunciaci%C3%B3n,_by_Fra_ Angelico,_from_Prado_in_Google_Earth_-_main_panel.jpg
- Fig. 3.29c Sketch by Descartes in *La Dioptrique*, 1637, Fig. # 17, p. 35. Public domain. http:// classiques.uqac.ca/classiques/Descartes/dioptrique/dioptrique.html
- Fig. 3.33a Stigmatization of St. Francis, 1299, by Giotto di Bondone, in Louvre. Public Domain. https://commons.wikimedia.org/w/index.php?curid=51798726
- Fig. 3.33b Stigmatization of St. Francis, 1500, by unknown German master, Wallraf-Richartz-Museum. Public Domain. https://commons.wikimedia.org/wiki/File:16th-century_unknown_ painters_-_St_Francis_Altarpiece_(central_panel)_-_WGA23747.jpg
- Fig. 3.34a Studies on graduation of light and shadow by Leonardo Da Vinci. Codex Atlanticus, fol. 187r. Public domain. Reproduced by Argentieri, 1956, p. 413.
- Fig. 3.34b Studies on graduation of light and shadow by Leonardo Da Vinci, Ms. Ashburnham II, folio 13v. Public domain. https://commons.wikimedia.org/w/index.php?curid=59585

References

Alberti, L. B. (1436/1970). *On painting* (J. R. Spencer, Trans.). Yale University Press.
Argentieri, D. (1956). Leonardo's optics. In E. Vollmer (Ed.), *Leonardo da Vinci* (pp. 405–436). Reynal.
Aristotle. (1952). *On the soul* (Vol. 1). Encyclopedia Britannica.
Cohen, R. M., & Drabkin, E. I. (1948). *A source book in Greek science*. McGraw-Hill.
Conford, M. (1937). *Plato's cosmology. The Timaeus of Plato*. Bobbs-Merrill.
da Vinci, L. (1955). *Notebooks. Optics, Ch. IX*. Georger Braziller.
Darrigol, O. (2012). *A history of optics from Greek antiquity to the nineteenth century*. Oxford University Press.
Galili, I. (2014). Teaching optics: A historico-philosophical perspective. In M. R. Matthews (Ed.), *International handbook of research in history and philosophy for science and mathematics education* (pp. 97–128). Springer.
Galili, I., Goldberg, F., & Bendall, S. (1991). Some reflections on plane mirrors and images. *The Physics Teacher, 29*(7), 471–477.
Galili, I., & Hazan, A. (2000a). Learners' knowledge in optics: Interpretation, structure, and analysis. *International Journal in Science Education, 22*(1), 57–88.
Galili, I., & Hazan, A. (2000b). The influence of historically oriented course on students' content knowledge in optics evaluated by means of facets-schemes analysis. *Physics Education Research, American Journal of Physics, 68*(7), S3–S15.
Galili, I., & Hazan, A. (2001). The effect of a history-based course in optics on students views about science. *Science & Education, 10*(1–2), 7–32.
Galili, I., & Zinn, B. (2007). Physics and art – A cultural Symbiosis in physics education. *Science & Education, 16*(3–5), 441–460.
Kepler, J. (1611/2000). *Optics: Paralipomena to Witelo and the optical part of astronomy*. Green Lion Press.
Lindberg, D. C. (1968). The theory of pinhole images from antiquity to the thirteenth century. *Archive for History of Exact Sciences, 5*(2), 154–176.
Lindberg, D. C. (1976). *Theories of vision from Al-Kindi to Kepler*. The University of Chicago Press.
Lindberg, D. C. (1978). The science of optics. In D. C. Lindberg (Ed.), *Science in the Middle Ages* (pp. 338–368). The University of Chicago Press.
Lindberg, D. C. (1985). Laying the foundations of geometrical optics: Maurolico, Kepler, and the medieval tradition. In *Discourse of light from the middle ages to the enlightenment* (pp. 1–65). The University of California Los Angeles.

Lindberg, D. C. (2002). The Western reception of Arabic optics. In R. Rashed (Ed.), *Encyclopedia of the history of Arabic science* (Vol. 2, pp. 363–371). Routledge.

Losee, J. (1993). *A historical introduction to the philosophy of science*. Oxford University Press.

Magie, W. F. (1969). Light. In *A source book in physics* (pp. 265–386). Harvard University Press.

Marton, F., Runesson, U., & Tsui, A. B. M. (2004). The space of learning. In F. Marton & A. B. M. Tsui (Eds.), *Classroom discourse and the space of learning* (pp. 3–40). Lawrence Erlbaum.

Overbye, D. (2001, October 30). How Islam won, and lost, the lead in science. *New York Times.* https://www.nytimes.com/2001/10/30/science/how-islam-won-and-lost-the-lead-in-science.html

Park, D. (1997). *The fire within the eye. A historical essay on the nature and meaning of light.* Princeton University Press.

Pedersen, O., & Pihl, M. (1974). *Early physics and astronomy*. McDonald & Janes.

Porta della, J. B. (1658/1957). *Natural magic*. Basic Books.

Ronchi, V. (1970). *The nature of light: An historical survey*. Heinemann.

Ronchi, V. (1991). *Optics. The science of vision*. Dover.

Russo, L. (2004). *The forgotten revolution: How science was born in 300 B.C. and why it shad to be reborn*. Springer.

Shapiro, A. E. (2008). Kepler, optical imagery, and the camera obscura. *Early Science and Medicine, 13*(3), 270–312.

Vygotsky, L. (1934/1986). *Thought and language*. The MIT Press.

Zahn, J. (1685). *Oculus Artificialis Teledioptricus Sive Telescopium*. Johannis Ernesti Adelbulneri.

Chapter 4
Inertial Force: The Unifying Concept

The understanding of inertia only is sufficient to distinguish between those who learned physics and those who did not.

© Springer Nature Switzerland AG 2021, corrected publication 2022
I. Galili, *Scientific Knowledge as a Culture*, Science: Philosophy, History and
Education, https://doi.org/10.1007/978-3-030-80201-1_4

Abstract The concept of *inertial force* is one of the most important and the most difficult in classical dynamics. The complexity and obscurity come from both *inertia* and *force*. However, for that very reason, making sense of *inertial force* may shed light on much of the meaning of mechanics. In fact, *inertial force* designates a cluster of meanings and touches upon different aspects of science. This will emerge from the history brought here. Different meanings coexist today, thus causing confusion for the novice learner. The challenge and need is to address the concept of inertia and inertial force as the concepts unifying classical physics with the modern physics. Yet, even learned at introductory level, this topic provides benefit to the students through widening the picture of physics as continuously improving itself even in classical mechanics. In fact, without inertial force and the directly related *non-inertial observer*, classical mechanics cannot treat one of the most important aspects of movement—its relativity. This excursus reviews the development of *inertial force* as a concept and addresses the way it could be better represented in teaching.

Introduction

The term "inertia" is commonly used to refer to the phenomenon of a certain state or activity continuing even after its cause has ceased to act. Following Aristotle's natural philosophy, scientists seek for *causes* for everything that happens in the nature around. Since the scientific revolution of the seventeenth century, even in

The original version of this chapter was revised: The caption of Figure 4.18 has been corrected. The correction to this chapter is available at https://doi.org/10.1007/978-3-030-80201-1_12

nonscientific contexts, the word "cause" is often replaced by the word "force," as in the notions "the force of tradition," "the force of will," or "the motive force." Seemingly, nonscientists use the term "force of inertia" even more than scientists do. Why do we continue to do things when there is no reason for them anymore? Why do things go on moving without any push? Why is it difficult to put a heavy body into motion? Many people answer these questions with "because of inertia…" or "because of the force of inertia." However, scientists try to put an exact meaning to this concept. This meaning has gradually consolidated during the historical development of science. We will trace this process in order to understand the scientific meaning of "the force of inertia."

Kepler's Inertia: The Force of Sluggishness

Johannes Kepler (1571–1630)—a prominent German scientist— was among the main contributors to the scientific revolution of the seventeenth century. He worked in astronomy, optics, as well as mathematics, being among the founders of modern physics. He is especially famous for the formulation of his three laws of plane- tary motion, which he succeeded in eliciting from the rich astro- nomical data accumulated by his employer, the Danish astronomer Tycho Brahe.

Kepler

Having been educated in the tradition of Aristotelian physics, Kepler held many Aristotelian views, in particular on the nature of motion. However, like certain medieval thinkers, he already did not fully divide the universe into two realms of different orders—below and above the Moon (sublunary and supralunary).

Kepler, a contemporary of Galileo, did not stop at a mere description of the motion of celestial bodies and did not interpret their motion through ascribing to planets "souls" as the Greeks did, but continued to think about the cause of motion as a natural philosopher. Kepler regarded the motion of celestial bodies as *violent* (not natural[1]). In his mind, the Sun served as the mover for the motion of the planets by emanating a power or force (Fig. 4.1). If so, two implications followed. Firstly, within the Aristotelian framework of dynamics, the mover causes this motion—*vis motrix*. Secondly, there should be a resistance to motion. In the absence of medium, Kepler ascribed this special feature of resistance to the planets themselves—*inertia*[2]:

> If the matter of celestial bodies were not embowered with inertia, something similar to weight, no force would be needed for their movement from their place; the smallest motive force would suffice to impart to them an infinite velocity. Since, however, the periods of planetary revolutions take up definite times, some longer and others shorter, it is clear that matter must have inertia which accounts for these differences.

---

[1] See Chap. 2 regarding Aristotelian theory of motion.
[2] De causis planetarum in Keper's *Opera Omnia*, in Jammer (1961, p. 55).

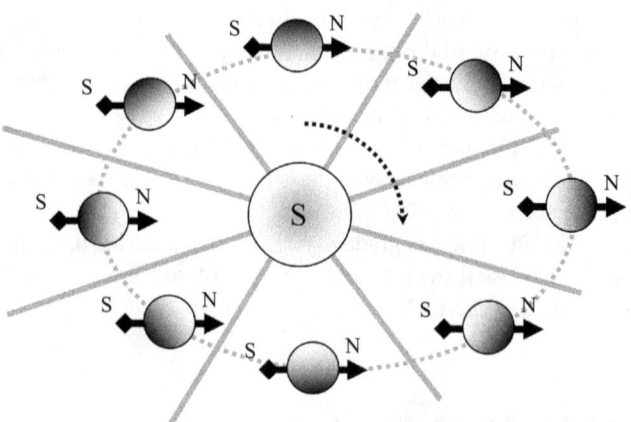

Fig. 4.1 This drawing is guided by Kepler's understanding of planet motion (Kepler 1621/1952, p. 899). The sun rotates and emanates "rays of power"—*vis motrix* (radial lines), which, while turning around the center, push the planet (like spokes of a wheel) along its path—the Aristotelian scenario of violent motion. The distortion of the circular trajectory of planets' motion is caused by magnetic interaction (magnetic axis and poles are shown): the magnetic axis of each planet remains about the same direction in space causing continuous change in facing the unipolar magnet of the Sun by the magnetic poles of a planet

This pure Aristotelian account implied, however, Kepler's far-reaching conclusion regarding the nature of the universe. Kepler broke with one of the basic presumptions of Aristotelian physics: the division of the universe into two areas of different orders or different laws. Kepler extended the physics of Aristotle, valid for the sublunary world, to the supralunary world.

However, Kepler immediately encountered a problem: Why were the periods of planetary motions different from one another? Indeed, if one thinks about the cosmos as the Greeks did, as an organism, one rotating body, all its components should share the same period of rotation (angular velocity). Why does the cosmos not actually behave this way? Kepler explained this by introducing a special feature of celestial matter, inertia, that distinguishes between the planets' ability to move under the action of the pushing force[3]:

> If this were the case, would not the planets make their periodic returns at the same time as the sun? Yes, if only this were the case. But it has been said before this, that besides this motor force of the sun there is also a <u>natural inertia</u> in the planets themselves with respect to movement: hence, by reason of their matter they are inclined to remain in their own place. Therefore, the motor power of the sun and the <u>powerlessness of material inertia</u> of the planet are at war with one another. Each has its share of velocity: the motor power moves the planet from its seat; the material inertia removes its own, i.e., the planetary, body somewhat from those bonds by which it was laid hold of by the sun (emphasis added).

Kepler mentioned that virtue was "something similar to weight." It was quite in accord with the Aristotelian view that the weight of regular bodies causes resistance to their violent motion, and that the heavier the body, the greater the "resistance to motion"[4]:

---

[3] Kepler (1621/1952, Book, IV).

[4] See the Aristotelian dynamics in Chap. 2.

$$v \propto \frac{F}{W} \tag{4.1}$$

(v, the speed of the body; F, the force of the mover; W, the weight of the body)

Kepler accepted, however, the premise that celestial bodies were by nature weightless. Accordingly, Kepler had to coin a new concept, *inertia*, to explain the resistance to motion and the finite speeds of the planets:

$$v \propto \frac{F}{I} \tag{4.2}$$

(v, the speed of the planet; F, the force of the Sun; I, inertia of the planet)

Why did Kepler not just allow the planets to have weight? Apparently, this was because in pre-Newtonian physics, ascribing weight to planets would imply their falling. Because the planets do not fall, Aristotle deprived them of weight. The element comprising celestial bodies—*ether*—was considered weightless. Kepler, therefore, sensed the need for a new concept in order to explain *resistance* to motion, *regardless of weight*.

It is noteworthy that our everyday use of the term inertia is often different. One may say: "I keep moving by inertia," meaning, "I do not push any more, but keep moving." It is spontaneous motion that one often ascribes to *inertia*. By contrast, Kepler introduced the concept to address the response of matter to the exerted force of its mover—the sun pushing the planets. For Aristotle, the source of any violent motion (the effective cause) was a *mover*, where there was no mover, there could be no violent motion. For Kepler, the source of motion was *vis motrix* (moving force), a sort of thrust by spokes or chains emanating from the Sun and acting on the planets. Inertia represented the inability of the planets to maintain motion, their inherent "laziness," the intrinsic tendency of the body to remain at rest. We read[5]:

> ...the motor power [potentia vectoria] of the sun and the powerlessness or material inertia of the planet [impotentia planetae] or its material inertia are at war with one another.

Preventing unsupported motion of the planets, Kepler's inertia would return them to rest if the force (*vis motrix*) between the Sun and the planet ceased to act. Inertia represented the sluggishness of the body which is in proportion to the amount of its matter:[5]

> Inertia or opposition to motion is a characteristic of matter; it is stronger, the greater the quantity of matter in a given volume.

One might think that by ascribing inertia to the planets, Kepler arrived at the meaning we usually associate with inertial mass in classical mechanics. This, however, was not the case. Newton fundamentally changed the Keplerian concept. Bernard Cohen, the renowned Newton expert, made a remarkable finding when he

---

[5] Kepler (1621/1952, Book, IV).

revealed Newton's marginal remark in his private copy of the first edition of the *Principia*. Newton wrote[6]:

> *I do not mean Kepler's force of inertia by which bodies tend toward rest, but the <u>force of remaining in the same state</u> whether of resting or of moving* (emphasis added).

However, before we get to Newton, it is worth taking a look at what Galileo, Kepler's great contemporary, had in mind in this regard.

Galileo's Inertia: Indifference

Although the name Galileo is regularly associated with "the law of inertia," his views on inertia were quite different from ours. In fact, Galileo never used the term *inertia* and never formulated the *law of inertia*.[7] However, he did address the notion of inertial *motion*, as we now call it, and so it is important to see how his thinking on the subject compares with other views.

Galileo's view on inertia changed with time, and it was not too well articulated. Coming, as he did, before the time of Newton, Galileo had no all-inclusive theory to account for motion nor did he formulate his own law-like framework succinctly. Instead, Galileo repeatedly expressed his views voluminously in nonformal dialogues, letters, essays, etc.

Galileo Galilei

In his famous *Dialogue Concerning the Two Chief World Systems* (1638), Galileo uses his creation Sagredo (an intelligent layman, initially neutral, representing an educated person applying common sense) to express the seemingly popular view (that also led Kepler, as we saw above)[8]:

> *I see in a movable body the natural inclination and <u>tendency</u> it has to an <u>opposite motion</u>. . . . I said <u>internal resistance</u>, because I believe that this is what you meant and not external resistances, which are many and accidental.* (emphasis added)

Galileo allows his other creation Salviati (who represented Galileo's own views) to clearly confirm and refine the point:

> *I wonder whether there is not in the movable body, besides a natural tendency in the opposite direction, another <u>intrinsic and natural property</u> which makes it <u>resist motion</u>* (emphasis added).

---

[6] Cohen (1971, pp. XVI, 28).

[7] Drake (1964).

[8] Galilei (1632/1953, p. 213).

Here, Galileo's thoughts appear similar to Kepler's understanding of inertia as the inherent *resistance to motion*. In another place, in his letter on sunspots (1613), Galileo expressed a wider view though still within the old perspective[9]:

I seem to have observed that physical bodies have physical <u>inclinations to some motion</u> (as heavy bodies downwards), which motion is exercised by them through an intrinsic property and <u>without need of a particular external mover</u>, whenever they are not impeded by some obstacle. And to some <u>other motion they have a repugnance</u> (as the same heavy bodies to motion upward), and therefore they never move in that manner unless thrown violently by an external mover. Finally, to some movements they [bodies] are <u>indifferent</u>, as are these same heavy bodies to horizontal motion, to which they have neither inclination (since it is not toward the center of the Earth) nor repugnance (since it does not carry them away from that center). And therefore, all external impediments removed, a heavy <u>body on a spherical surface concentric with the earth</u> will be indifferent to rest or to movements toward any part of the horizon. And it will <u>maintain itself</u> in that state in which it has once been placed; that is, if placed in a state of rest, it will conserve that; and if placed in movement toward the west (for example), it will maintain itself in that movement. Thus a ship, for instance, having once received some impetus through the tranquil sea, would move continually around our globe without ever stopping; and placed at rest it would perpetually remain at rest, if in the first case all extrinsic impediments could be removed, and in the second case no external cause of motion were added.

In this piece, Galileo kept with the common Aristotelian separation between natural and violent motion in falling bodies (like Sagredo),[10] but he also introduced the new idea. Galileo found the *third* type of motion in-between the two Aristotelian extremes: *natural* (downwards) and *violent* (upward). The third kind of motion, neither natural nor violent, but neutral, keeps a body at the same distance from the center of the Earth. A body on the surface of the Earth would remain, therefore, on a circular orbit around the center. In this motion, as inferred by Galileo, the body remains *indifferent* (to motion). Such a motion represents the idea of *inertial motion*, a motion without change and without support. In fact, this idea neatly combines Aristotelian and medieval (impetus) accounts of motion, in a way, the sublunary and supralunar orders.

The point to note here is that Galileo's version of inertial motion was *circular*. His loose phrasing confused generations of historians. Indeed, Galileo, in his *De Motu*, spoke about horizontal motion, leaving much space for questions[11]:

On a perfectly horizontal surface a ball would remain indifferent and questioning between motion and rest, so that the least force would be sufficient to move it, just as any little resistance, even that of the surrounding air, would be capable of holding it still. From this we may take the following conclusion as an indubitable axiom: That heavy bodies, all external and accidental impediments being removed, can be moved in the horizontal plane by any minimal force.

---

[9] Galilei (1613/1957, pp. 113–114).

[10] Galileo represents the medieval account of a tossed body motion (Galilei, 1638/1914, p. 165).

[11] Drabkin & Drake (1960, p. 171).

In light of the two views above, the first similar to Kepler (resistance to motion) and the second being his own idea (indifference to motion), it seems that Galileo, eventually, combined them. Stillman Drake further supported this view[12]:

> Galileo must have been teaching it in his private classes, for one of his pupils, Renedetto Castelli, who had left Padua some time before, wrote to Galileo in 1607 mentioning "your doctrine that although to start motion a mover is necessary, yet to continue it the absence of opposition is sufficient."

This view shows that Galileo reached, rather closely, the Newtonian understanding of motion. Nevertheless, Galileo remained distinct. Not only did Galileo believe in circular inertial motion, one of the cornerstones of the old physics refuted only by Descartes,[13] but a contemporary reader would also refine the idea of "resistance to motion," by replacing it with "resistance to the change of motion." Resistance to the change of motion combines *both* manifestations of inertia, resistance to the starting of motion and resistance to the stopping of it (changing it), or in other words, *preservation* of motion as a state. And so, we come to Newton.

Newton: Inertial Force and Inertia of Mass

Newton

Sir Isaac Newton (1642–1727) was a remarkable mathematician, physicist, and natural philosopher, one of the most brilliant scientists of all times. A Fellow of Trinity College at the University of Cambridge, he produced one of the most famous scientific books: *Mathematical Principles of Natural Philosophy*, known as the *Principia*, published in 1687.

The *Principia* introduced a new picture of the world in which material bodies interacted only by means of forces. Newton dismissed the distinction between natural and violent motions in the all-inclusive framework of mechanics—the Newtonian theory of motion.

Newton defined the inertial force at the very beginning of the book, where in a special chapter he introduced the basic concepts of his theory. In his third definition, he wrote[14]:

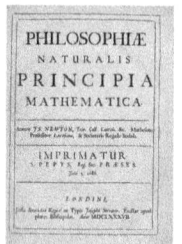

> Inherent force of matter [vis insita] is the power of resisting by which every body, so far as it is able perseveres in its state, either of resting or of moving uniformly straight forward.

He immediately explained:

> This [vis insita] force is always proportional to the body and does not differ in any way from the inertia of the mass except in the manner in which it is conceived. Because of the inertia of matter, every body is only with difficulty put out of its state either of resting or of moving.

---

[12] Drake (1964).

[13] See Chap. 1.

[14] Newton (1687/1999, p. 404).

Consequently, inherent force may also be called by the very significant name of force of inertia [vis inertia]. Moreover, a body exerts this force only during a change of its state, caused by another force, impressed upon it; and this exercise of force is, depending on the viewpoint, both resistance and impetus: resistance in so far as the body, in order to maintain its state, strives against the impressed force, and the impetus in so far as the same body, yielding only with difficulty to the force of a resisting obstacle, endeavors to change the state of that obstacle. Resistance is commonly attributed to resting bodies, and impetus to moving bodies; but motion and rest, in the popular sense of the terms, are distinguished from each other only by point of view, and bodies commonly regarded as being at rest, are not always truly at rest.

The force of inertia [*vis insita*] introduced by Newton was related to the first law of motion. Too often, this law is considered as a special case of the second law. This is due to the later inaccurate translation from Latin, which has only recently been changed.[15] Within the recent interpretation, it appeared that the second law, formulated by Newton as an integral statement, could be understood as a refinement of the first law formulated as a different statement. In this perspective, the first law should be rather stated as follows[16]:

Every body perseveres in its state of being at rest or of moving uniformly straight forward, except in so far it is compelled to change its state by forces impressed.

Careful interpretation of this text reveals that it talks about the situation with the active external force exerted on the body which tries to preserve its state of uniform motion (or rest). The text does not talk about the lack of force. The latter case is considered in the sense of a limit. In accordance, Newton exemplified the first law with three situations, the projectile, the spinning top, and the moving planet. Traditionally, however, physics teachers often keep with the simplified version of the first Newton's law addressing the case of zero external force, when the law becomes a trivial special case of the second law. The whole of the emphasis is put on the second law describing the general case. In it, Newton asserts that the external force determines the acceleration in direct proportion, while the body resists in proportion to the quantity of matter (inertial mass):

$$a = \frac{F}{m} \tag{4.3}$$

Important in understanding Newton is realizing his approach to reality. Newton always tried to address real situations, and such is the situation with active forces. This contrasts with Galileo who tried to establish the laws of nature in their "pure" form, that is, corresponding to ideal situations (such as the absence of friction). Therefore, the presentation of the first law in physics classes which is popular today is actually in the tradition of Galileo, rather than that of Newton.

Newton's revolutionary understanding was highly important for physics. In contrast to his predecessors, Newton understood that the states of rest and motion are fully equivalent! This change was striking, particularly in view of Aristotelian

---

[15] Galili & Tseitlin (2003).
[16] Newton (1687/1999, p. 416).

physics which claimed that a body is at some *state* when it rests, and while being in motion, it undergoes a *process* of change from the initial state to the final one.[17] Newton understood nonuniform motion as a continuous chain of states which replace each other in a sequence. *Vis insita* described the tendency of the body to preserve each single state in that sequence. By the end of the nineteenth century, for instance, in Mach's critical revision of Newton, the "force of inertia" did not appear any more.[18]

To clarify the development of inertia, we turn now to its relationship with the quantity of matter and mass. Newton opened his treatise with Definition I[19]:

> *DEFINITION I. Quantity of matter is the measure of matter that arises from its density and volume jointly.*

and followed explaining[20]:

> *I mean this quantity whenever I use the term body or mass in the following pages.*

This statement represents one of the most important and controversial ideas in the theory of motion. Newton tried first to clarify the fundamental notions of the physical account of motion. Material bodies were characterized by a variety of properties (form, weight, color, etc.). As we know *now*, the property known as mass is not equivalent to the quantity of matter, but it characterizes all material bodies. There is a reason why we say "the mass of the body" and not "the matter of the body" in order to characterize an object. "Mass" presents a quantified physical characteristic of a material object. We say "big mass" but not "big matter." Furthermore, mass presumes certain uniformity, whereas matter is known in its different kinds: water, air, wood, etc. We may say, that in a crowd, or "in a mass", people loose individuality and behave accordingly. We use the term "mass" to emphasize the absence of differences and to relate to a certain universal measure.

Newton did identify mass with the quantity of matter. At first glance, this may seem obvious: the bigger the mass—the larger the quantity of matter. However, a deeper thought may suggest otherwise. Consider how we know about the quantity of matter. Matter is not known to us in its elementary components, as an observable parameter. We estimate the quantity of matter by observing the properties of material bodies made of various materials (wood, metal, etc.). What is the fundamental characteristic of any object? It cannot be the number of its elementary components, atoms, because they cannot be directly observed as individual objects. For the same reason, the fundamental characteristic cannot be atomic volumes, as we do see and measure them directly.

For Newton, it was a fair assumption that the quantity of matter in two identical bodies is twice the quantity of one of them and so inertia (mass) should follow that fact. Therefore, the quantity of matter, he believed, must be primary and determine inertia. This approach placed belief over evidence. In fact, we measure inertia and

---

[17] Descartes was very close to Newton's understanding but not there (Chaps. 1 and 2).
[18] Mach (1883/1919, p. 241).
[19] Newton (1687/1999, p. 403).
[20] Newton (1687/1999, p. 404).

not the quantity of matter. It is inertia that determines for us the quantity of the matter and not the other way around.

Another fair assumption of Newton was that the quantity of matter (mass) determines gravitation. Indeed, the bigger the planet, the stronger it gravitates to its satellite. Thus, gravitation is determined by the quantity of matter (again). Newton, observing both relationships, quantity of matter—inertia and quantity of matter—gravitation, questioned the equivalence of inertial and gravitational masses and tried to check this empirically by experimenting with pendulums of different materials (i.e., different amounts of matter which determined both the acceleration and the gravitational force). He found the identity (universal proportionality) which greatly surprised him... Why? He did not see any reason for that but the (unnatural) design. The physical answer for that identity was given much later by Einstein in his theory.

For Newton, the quantity of matter seemed most fundamental, and it determined inertial mass as well as gravitational mass. The claim of fundamentality of quantity of matter prior to inertia and prior to gravitation may display Newton's perspective on physical theory, his epistemology (we will return to this).

Soon after, Euler in his *Mechanics* (see the following section) indicated that the quantity of matter was an obscure notion. In other words, the notion "quantity of matter" lacks operational definition, independent of inertia which is directly measured by the force required to accelerate the body to the given extent. Consequently, the definition "mass as a quantity of matter" was abandoned, whereas *inertial mass* remained the central concept determined in Newton's second law. Consequently, the *inertial mass* appeared in the definition of momentum—the quantity of motion. The *gravitational mass* was determined operationally in the law of universal gravitation and appeared to be equal to the inertial.

Inertia became the primary property of matter, and the law of inertia represented a fundamental ontological claim for classical mechanics. While for Descartes, bodies were defined by their impermeable boundary (geometry); for Newton, boundaries were secondary after inertia. Newton's explanation of definition I expanded the realm of the quantity of matter as fundamental to weight. Weight was determined by Newton as the gravitational force exerted on and by the body. He wrote[21]:

> ... It is this quantity [quantity of matter] that I mean hereafter everywhere under the name of body or mass. And the same is known by the weight of each body; for it is proportional to the weight, as I have found by experiments on pendulums, very accurately made, which shall be shown hereafter.

Newton showed the proportionality of *weight* to *inertial mass* and considered this discovery to be very important. Both were proportional to the quantity of matter.

In his Definition III, Newton introduced *internal force (vis insita)* proportional to the magnitude of a body, that is, its mass, as already stated in Definition I[22]:

> *DEFINITION III. The vis insita, or innate force of matter, is a power of resisting, by which every body, as much as in it lays, endeavours to persevere in its present state, whether it be of rest, or of moving uniformly forward in a right line.*

---

[21] Newton (1687/1999, p. 404).

[22] Newton (1729/1952, Definitions, p. 5).

> *This force is ever <u>proportional to the body</u> whose force it is; and differs nothing from the inactivity of the mass, but in our manner of conceiving it. A body, from the inactivity of matter, is not without difficulty put out of its state of rest or motion. Upon which account, this vis insita, may, by a most significant name, be called inertia [vis inertia], or force of inactivity. But a body exerts this force only, when another force, impressed upon it, endeavours to change its condition* (emphasis added).

As seen here, Newton did not distinguish between inertia and the force of inertia. This meaning did not preserve in physics. Why force? The answer is seen in the following definition of external (impressed) force (Definition IV) to which inertia provides resistance, a permanent partner: force to force. Newton mentioned inertial force as that which exists only at the time external force acts. It is by means of the inertial force that a body resists to the change of its state of motion[23]:

> *DEFINITION IV. An impressed force is an action exerted upon a body, in order to change its state, either of rest, or of moving uniformly forward in a right line.*
>
> *This force consists in the action only; and remains no longer in the body when the action is over. For a body maintains every new state it acquires, by its vis inertia only.*

This inertial force was very special element in the Newtonian picture of the world, the world of forces, and it became the subject of strong criticism soon after.

Euler: Inertia Is Not a Force

Leonhard Euler (1707–1783) was a famous Swiss mathematician and scientist who contributed greatly to a wide range of topics in mathematics and physics. As a young researcher, Euler produced his treatise, *Mechanics*, which in practice became the first textbook on classical mechanics, influencing the development of mathematics, physics, and philosophy. In it, Euler adopted the term "inertial force," although he did express his critical opinion of this concept in his preface[24]:

Euler

> *... a state of conservation is an essential property of all bodies, and all bodies, in as much as they are such, have the strength or facility to remain permanently in their state, which is called nothing other than the <u>force of inertia</u>. Indeed, calling the inertial effect a force for the source of this conservation is less than suitable, since <u>it is not a force of the same kind as the other forces</u> thus properly discussed, such as the force of gravity, nor can it be compared with these; into which error many are accustomed to fall, especially those involved in metaphysics, from the ambiguity of their deceptive discussion.* (emphasis added).

Accordingly, Euler defined inertial force as a faculty of a material body[25]:

> *The force of inertia [vis inertiae] in all bodies is that faculty of the body to maintain its state of rest or of continuing in its present state of [uniform] motion in a straight line.*

---

[23] Newton (1729/1952, Definitions, p. 6).

[24] Euler (1736/2008, Preface, p. 3).

[25] Euler (1736/2008, Ch. 1, Definition 9, p. 18).

This definition approaches the definition provided in text-books today. Perhaps because Euler wrote his *Mechanics* as a textbook (not as an original essay), he tended to provide more explanations than Newton did. In Proposition 17, Euler strength-ened the important feature stated by Newton less clearly[26]:

The force of inertia of any body is proportional to the quantity of matter, upon which it depends.

Euler, however, argued for this as follows[27]:

That [the force of inertia] therefore is to be estimated from the strength or the force applied to the body, with the aid of which it is to be disturbed from its state. Truly different bodies equally [disturbed] in their state are disturbed by forces, which are as the quantities of material contained in these. Therefore, the forces of inertia of these are [also] proportional to these forces.

This statement matches the form of the second Newton's law (F = ma) provided by Euler.[28] Indeed, if two bodies, of masses m_1 and m_2, are accelerated to the same extent ($a_1 = a_2$)—that is, they are disturbed from their state of motion to the same extent—then, the masses ("the quantities of matter") are in the same ratio as the external forces required:

$$\frac{F_1}{F_2} = \frac{m_1}{m_2} \qquad (4.4)$$

Euler considered this fact to serve as evidence that the two forces of inertia of the bodies are in the ratio of their quantities of matter since these inertia faculties caused the same change of motion.

In his later treatise, however, Euler returned to his criticism of the notion of "inertial force" and wrote[29]:

SCHOLIUM 95. Calling this property <u>inertia</u> indicates that character of bodies, whereby being at rest means that they will continue to be at rest, therefore as if they oppose not being in that state of motion themselves; but since bodies set up in motion themselves equally oppose all to be changed, either on account of the speed or direction, this name is not a bad choice taken to account for the conservation of the state of rest or of motion. Also, now and then, it is called the <u>force of inertia</u>, because the force is, in a way, resistance to the change of the state; but <u>if the force is defined by some cause, which is changing the state of the body, here it is not in the least acceptable with this meaning</u>; certainly, the reason for this strongly disagrees with that by which henceforth we show a force to be acting. Whereby, lest any confusion should arise, we omit the name force from this property of the body and we will call it by the simpler name of <u>inertia</u>.

Here, we read that Euler saw the force to be solely of active origin, that exerted *on* the body caused it to change its state, while inertia was due to the *innate* inert matter (conserving the state). Therefore, he concluded, the term *force of inertia* was mislead-ing and should be replaced by the term *inertia* to signify a faculty of a material body.

---

[26] Euler (1736/2008, Ch. 2, p. 59).

[27] Euler (1736/2008, Ch. 2, p. 59).

[28] See Chapter 2.

[29] Euler (1765/2009, Ch. 2, p. 60).

D'Alembert: Inertial Force as a Fictitious Force in Newton's Second Law

Another perspective on *inertial force* was subsequently introduced by the distinguished French scientist and philosopher Jean le Rond d'Alembert (1717–1783) whose approach was very different.[30] D'Alembert started with the adoption of the idea that the *force of inertia* is a property of bodies by which they preserve their state of rest or uniform rectilinear motion.

However, unlike Euler, for whom *inertia* was taken to be a fundamental property of matter, d'Alembert considered such a "property" to be obscure and wanted to define all forces, including inertia, in the same way: by their output—the change of motion.

D'Alembert

D'Alembert considered Newton's second law, in the form suggested by Euler: F = ma, to be the best framework to account for forces. He saw that law as defining "effective (or moving)" force.[31] D'Alembert had a rationalist intention to construct mechanics starting from universal principles. Accordingly, d'Alembert (under the influence of Archimedes' treatment of the lever) sought to base the physics of motion on the principle of static equilibrium.[32] This was what he did. A simple manipulation of the second law in the form of Euler yields:

$$\mathbf{F} - \mathbf{ma} = 0 \quad \mathbf{I} \equiv -\mathbf{ma} \quad \Rightarrow \quad \mathbf{F} + \mathbf{I} = 0 \tag{4.5}$$

Here, a new force **I** was defined —*the force of inertia*. Was it the Newtonian force of inertia? Yes and no. The Newtonian notion of inertia was defined qualitatively and thus obscurely (as a "tendency"). d'Alembert's new notion was given a precise mathematical formula. One may think about the force **I** as the "resistance to the change of motion," which nullifies the effect of the impressed external force. Together, the external agent and internal resistance create a sort of balance, equilibrium. We have arrived, thus, at the d'Alembert principle[33]:

Any system of forces is in equilibrium if we add to the impressed forces the force of inertia.

Lagrange even mentioned that d'Alembert reduced the theory of motion to statics.[34] This, however, was only an illusion. Dynamics was imbedded within the expression for the force **I** (in the concealed acceleration **a**). Many physicists (e.g.,

---

[30] d'Alembert (1758).

[31] Later this idea, to define force by means of Newton's second law, was adopted by other physicists and supporters of positivist philosophy, such as Mach, Hertz, and others.

[32] In the history of old physics, we observe the strongly embedded separation between the realm considering the state of rest (statics) and that considering motion (dynamics). The preference of statics was also for differences in practical needs. Yet, physics unified the two areas in one theory — classical mechanics.

[33] This is the modern formulation of the principle, e.g., Lanczos (1949/1970, p. 90).

[34] Lagrange (1788/1938).

Mach) saw this step as a manipulation, saying that d'Alembert did not say anything new: after all, no new physics can be obtained simply by transferring a term from one side of an equation to the other.

Indeed, the essential difference between the inertial force **I** and the external force **F** was that the latter was always interactive (in correspondence with Newton's third law) and the former was not. Therefore, for the physicists who adopted Newton's paradigm of mechanics (absolute and unique space-time, and central interactive forces between material particles), d'Alembert's force of inertia was a mere trick. Accordingly, it was given inferior status and a correspondingly dismissive name, *fictitious force*. Much later, during the twentieth century, it was understood that in a sense, dynamics may indeed be reduced to statics, but such a change implies the change of observer (or reference frame). This understanding, however, required a completely new picture of the world, foreign to the physicists in the eighteenth and nineteenth centuries. Yet, this claim is not precise. Quite surprisingly, the first steps toward this picture were already taken even before d'Alembert, by Christian Huygens, in the Netherlands in the seventeenth century.

New Vision: Newton versus Huygens

The influence of Newton on the framework of thought adopted by scientists was enormous. However, there was another brilliant mind active at the same time, Christian Huygens (1629–1695), a prominent Dutch physicist and philosopher who produced alternative ideas regarding the fundamental issues of physics, both in mechanics and optics. In optics, Huygens confronted and challenged Newton's ideas of particles traveling along light rays by introducing the wave paradigm: light as waves of elastic distortions in the ether.

Huygens

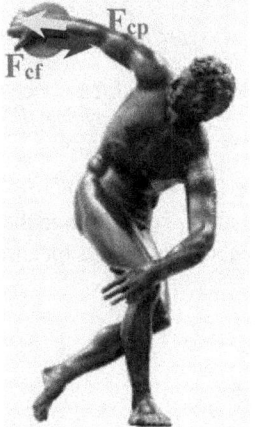

(a)

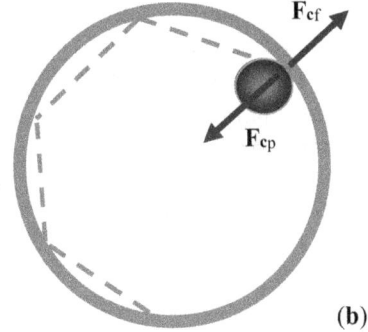

(b)

Fig. 4.2 (**a**) In Newton's interpretation, when the athlete whirls the disc while turning around, the force applied on the disc is *centripetal* and its counterforce acting on the hand is *centrifugal*. (**b**) The Newtonian pair action-reaction forces for the ball rolling on the circular ring

In our excurse, we touch on another fundamental idea interpreted differently by Newton and Huygens, the *centrifugal force*. Introduced by Huygens in 1673 in his treatise *De vis centrifuga* [*Concerning the centrifugal force*], centrifugal force was associated with circular motion. We present below, both Newton's and Huygens' approaches, one after the other. We start with Newton, although he actually followed Huygens by several years with his *Principia* being published in 1687.

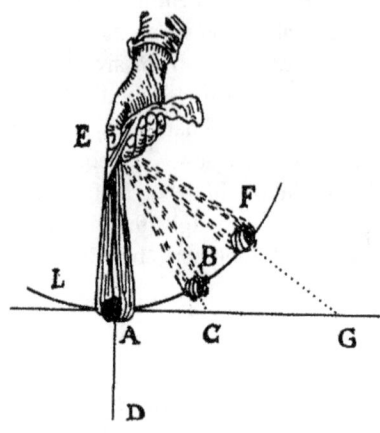

Considering circular motion, Newton first emphasized *centripetal* force—the force applied on a body, causing it to follow a circular path (such as a ball fastened to a rope) rather than an inertial rectilinear motion. *Centrifugal* force—the radial force outward from the center of rotation—was addressed by Newton as the force that the circling body exerted on the constraint, the other body that caused circular motion to the revolving body (Fig. 4.2a). In fact, in this way, Newton followed in Descartes' steps, who had analyzed circular motion before him.

Fig. 4.3 A stone in a sling rotated by a hand

Describing the circular motion of a stone in a sling, Descartes had explained that because of the tendency of a stone to proceed in a tangential direction, AG (Fig. 4.3), a centrifugal effect is experienced by the sling: a radial tension which stretches the sling outward, along EAD, manifesting the "inclination" (conatus, in Latin) of the body's tendency to move along the straight tangential path ACG.

Descartes

Newton adopted this view and, considering a body moving inside a solid circular frame along the sides of a polygon inscribed in a circle (Fig. 4.2b), he wrote[35]:

> And if a body, moved with a given velocity along the sides of the polygon, is reflected from the circle at the several angular points, the force, with which at every reflection it strikes the circle will be as its velocity: and therefore the sum of the forces, in a given time, will be as that velocity and the number of reflections jointly; … *This is the centrifugal force, with which the body urges the circle;* and the opposite force, with which the circle continually repels the body towards the center, is equal to this centrifugal force. (emphasis added)

This definition of centrifugal force matches the framework of Newton's paradigm of forces: they are central and interactive; that is, one always can arrange them in pairs obeying Newton's third law. Much later, Hertz,[36] elaborated on this meaning of the centrifugal force[37]:

---

[35] Newton (1687/1999, Book I, Section 2, Scholium, pp. 452-453).

[36] Heinrich Rudolf Hertz (1857–1894) was a prominent German physicist. In his *Principles of Mechanics Presented in a New Form* (1899), Hertz tried to create a new version of mechanics, reconsidering the concept of force.

[37] Hertz (1899/1956, p. 6).

We swing a stone attached to a string in a circle; we thereby consciously exert a force on the stone; this force constantly deviates the stone from a straight path, and if we alter this force, the mass of the stone or the length of the string, we discover that indeed the motion of the stone occurs at all times in agreement with Newton's second law. Now, the third law demands a force opposing that which is exerted by our hand on the stone. If we ask for this force, we obtain the answer familiar to everybody, that the stone reacts on the hand by virtue of the centrifugal force, and that this centrifugal force is indeed equal and opposite to the force exerted by us on the stone. ...

In summary, in Newton's mechanics, which became classical mechanics, centrifugal force was a considered a typical interactive force, the reaction to the centripetal force. So far, no relation to inertial force here. Not so for Huygens.

Huygens: The Centrifugal Force – The Force in View of the Rotating Observer

Huygens' view on the subject was highly original. As already mentioned,[38] in his argument with Descartes, Huygens managed to fully account for collisions of hard bodies as he actively used the just introduced relativity principle of Galileo. In 1659, before Newton introduced his laws of motion and theory of gravitation, Huygens used a similar approach in producing the concept of *centrifugal force* and determining its variables. Today we express this force by the well-known formula:

$$F = \frac{mv^2}{r} \tag{4.6}$$

Huygens never wrote this formula but described in separate statements the dependence of the centrifugal force on its parameters: weight, velocity, and radius of rotation, often through a comparison between two similar cases.

Moreover, before Newton's separation of weight and mass, Huygens used weight and not mass, as we do in the expression (4.6). Yet, Huygens did what Newton never did: he was the first to describe a physical situation from the point of view of the observer in a rotating system (Fig. 4.4). For that, he used a thought experiment[39]:

Let us imagine some very large wheel, such that it easily carries along with it a man standing on it near the circumference but so attached that he cannot be thrown off; let him hold in his hand a string with a lead shot attached to the other end of the string.

He made a statement regarding the tension in a string caused by the rotation and promised to clarify this situation from the chosen point of view:

The string will therefore be stretched by the force of revolution in the same way and with the same strength, whether it is so held or the same string is extended to the center at A and attached there. But the reason why it is stretched may now be more clearly perceived.

---

[38] Chapter 1.

[39] Huygens (1659/1703).

To appreciate the significance of Huygens' account, we should first present it in modern terms. For this, we should distinguish between the description of the situation made by observer N, outside of the rotating wheel, and that by observer H, standing on the rotating wheel, the one never used before in mechanics.

Observer N, the only one considered later by Newton and all his predecessors, detects the tension of the rope, designated as force **T**, and the gravitational force **mg**, acting on the mass m. These forces are sufficient to fully describe the rotation of the bob through applying the second Newton's law. As it is defined in modern physics, N presents *inertial observer*, the only kind of observer for whom Newton's laws hold.

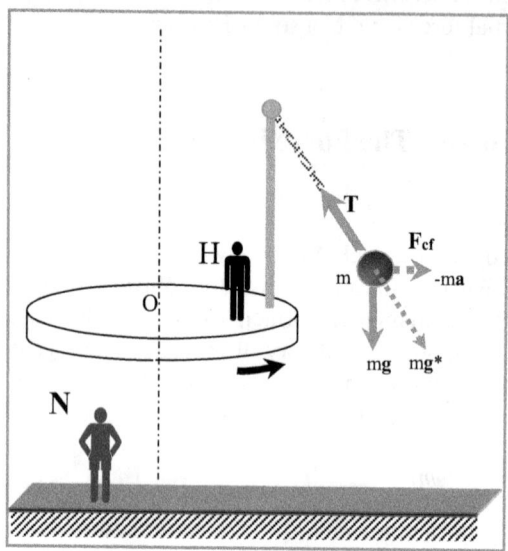

Fig. 4.4 Body m suspended on the rope is located on the rotating disc. The situation is similar to the one described by Huygens

Observer H, considered by Huygens, may not see mass m rotating (imaging him inside a rotating cabin). He rotates with the disc and may describe the environment inside the cabin. He may state that the ball is at rest, and hence, the forces acting on it are in equilibrium. The force of tension **T** and the gravitational force **mg** could not cancel each other and so could not keep the body at rest. Therefore, he infers regarding additional force, −ma, to reach equilibrium, zero net force.

The force ($\mathbf{F}_{cf}$) does not exist for N but is required by H. This force is termed today *inertial force* ($F_{in} = -ma$). Observer H (the one who introduced the force) is defined as a *non-inertial observer*, and the rotating cabin presents a *non-inertial frame of reference*.

From this perspective, we may better understand the meaning of d'Alembert's step of converting the dynamic problem to a static one. In fact, d'Alembert transformed the description of motion from the *inertial* observer N to the *non-inertial* observer H. Such a transition is accompanied by the appearance of the inertial force $F_{in} = -ma$ in the physical account of the situation in a rotating environment.

Huygens: The Inertial Force Is Similar to the Force of Gravity

Lacking Newton's theory of gravitation and Newton's laws of motion greatly limited Huygens' treatment of rotation. Nevertheless, he made an ingenious guess to show that the force of gravity is similar to the centrifugal force which he introduced. Were he to know the formulas that we use today:

$$F_{cf} = m\frac{v^2}{r} \text{ and } F_g = mg \tag{4.7}$$

he would have immediately seen that both forces are proportional to the mass of the object. In this sense, they are essentially similar (they possess the same form which is, using notations, $a_R = v^2/R$ and $a_g = g$:

$$F_{cf} = m \cdot a_R \text{ and } F_g = m \cdot a_g \tag{4.8}$$

This implies that observer H on the disc might be misled regarding the force of gravity, since he does not distinguish between the gravitational force and the centrifugal force. The weight of the object as considered by observer H—the heaviness of the suspended body at rest, or the net force needed to compensate T—is mg* (Fig. 4.3). Although observer N can clearly identify the gravitational weight (mg), he is also led to depart from the identity between weight (measured by how heavy the body is to support) and the gravitation—for him they are different forces. This indeed could be the point at which the concepts of gravitation and weight split from each other. However, although Huygens was so close to this inference, he did not make it. This claim was made by Einstein within his principle of equivalence.

Fig. 4.5a (a) Huygens' drawing illustrating the meaning of heaviness (gravitas) as a tendency to fall

Lacking Newton's laws of motion and being familiar with the works of Galileo and Descartes, Huygens drew on their results and conceptions. Unlike Newton, he described gravity using Descartes' notion of *conatus*—innate inclination (tendency) (Fig. 4.5a)[40]:

Heaviness is a tendency to fall [Gravitas est conatus descendendi]

Then, he considered a rotating body fastened with a rope. Here, Descartes specified the radial tension in the rope—*conatus*. The intention of Huygens was to show that there is no difference between this conatus while in rotation and the tension due to the gravity. How could he do it?

Huygens showed that if one disconnected the stone from the rope and the stone receded along the straight line of the tangent (in accordance with Descartes' law), then the distances from the center of the wheel would increase in the rate that was registered by Galileo in the free fall of objects. This proved, for Huygens, the identity of the nature of the rotational conatus (the centrifugal force on the body) and the natural gravity, considered as an intention to fall.

Here is the quote from Huygens' essay *De vi Centrifuga* (Fig. 4.5b)[41]:

Let BG be a wheel that rotates parallel to the horizon about center A. A small ball attached to the circumference, when it arrives at point B, has a tendency to proceed along the straight

---

[40] Huygens (1659/1703, p. 255).
[41] Huygens (1659/1703, p. 255).

line BH, which is tangent to the wheel at B. Now, if it were here separated from the wheel and flew off, it would stay on the straight path BH and would not leave unless it were pulled downward by the force of gravity or its course were impeded by collision with another body. At first glance it indeed seems difficult to grasp why the string AB is stretched that much when the ball tries to move along the straight line BH, which is perpendicular to AB. But everything will be made clear in the following way.

Let us imagine some very large wheel, such that it easily carries along with it a man standing on it near the circumference but so attached that he cannot be thrown off; let him hold in his hand a string with a lead shot attached to the other end of the string. The string will therefore be stretched by the force of revolution in the same way and with the same strength, whether it is so held or the same string is extended to the center at A and attached there. But the reason why it is stretched may now be more clearly perceived.

Take equal arcs BE, EF very small in comparison to the whole circumference, say hundredth parts or even smaller. Therefore, the man I spoke of [as] attached to the wheel traverses these arcs in equal times, but the lead would traverse, if it were set free, straight lines BC, CD corresponded to the said arcs, the endpoints of which [lines] would not, however, exactly fall on the straight lines drawn from center A through points E, F, but would lie off

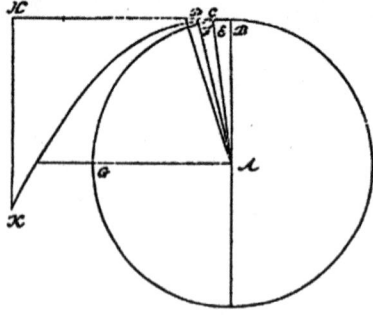

these lines a slight bit toward B. Now it is clear that, when the man arrives at E, the lead will be at C, if it was set free at point B, and when he arrives at F, it will be at D. Whence, we say correctly that this tendency is in the lead. But now if points C, D were on the straight lines AE, AF extended, it would be certain that the lead tended to recede from the man along the line drawn from the center through his position; and indeed, such that in the first part of the time it would move away from him by the distance EC, and in the second part of the time it would be distant by the space FD. But these distances EC, FD, etc. increase as the series of the squares from unity, 1, 4, 9, 16, etc. Now they agree with this series ever more exactly as the particles BE, EF are taken to be smaller, and hence at the very outset they may be considered as if they differed nothing.

Fig. 4.5b (**b**) Huygens' drawing illustrating the tendency of the rotation body to "fall" remaining on the tangent instead of rotation but being restrained by the wheel

Let us mention that Huygens did not show his calculation, perhaps to prevent the departure of the reader's attention from the thread of the proof. The distances in equal time intervals from the start of motion increase in accordance with the series: 1, 4, 9, 16, etc. This very same sequence was obtained by Galileo for the bodies in a free fall. Huygens proceeded keeping with weight as the tendency of the suspended body to fall and inferring:

Thus this tendency will clearly be similar to that which is felt when the ball is held suspended on a string, since then too it tends to recede along the line of the string with a similarly accelerated motion, i.e. such that in a first certain period of time it will traverse 1 interval, in two parts of time 4 intervals, in three 9, etc. ... it should turn out for those [bodies] that tend to move away from the center; their tendency has been clearly shown to be similar to the tendency arising from gravity.

Huygens' work on the centrifugal force is one of the most remarkable in the seventeenth-century mechanics. He preceded Newton's treatment of mechanics which, even as a mature theory, did not consider the issue of the observer at all. Later on, physicists expanded Newton's perspective and combined it with Galileo's principle, thus arriving at classical mechanics, stating its validity for all inertial observers. Huygens, in fact, tried to extend this idea even farther, to the view not reached before Einstein. Indeed, Newton's worldview of the "truly" static all-embracing universe placed in infinite and absolute space and time allowed for only one true observer of all—God (Fig. 4.6).

Fig. 4.6 The allegoric representation of the unique observer depicting the physical world. Traditionally people depicted God as *observing* from above the world He created. God symbolized the unique observer of the true picture of the world. This picture was created as a fresco by Michelangelo in 1508–1512 and placed by him, not by chance, on the ceiling of the most important hall in the Vatican—Cappella Sistine, where the divine supervision was especially required

Therefore, although much praised, even by Newton, Huygens' work was not truly appreciated by any of his contemporaries, seemingly not even by Huygens himself. Apparently, the time (the seventeenth century) was not yet ripe. At the beginning of the twentieth century, Albert Einstein put this subject at the center of a new physical theory. The inertial observers were basic in the special theory of relativity and the non-inertial observers—within his general theory of relativity. Huygens was the first to consider the rotating frame of reference and the non-inertial observer. Though he did not realize the grand scale of his step, in practice, he arrived at the idea of the *principle of equivalence* inherent in the general theory of relativity[42]:

> There is no experiment observers can perform to distinguish whether an acceleration arises because of a gravitational force or because their reference frame is accelerating

Furthermore, the equality of the inertial and gravitational masses declared by Newton (the inability to distinguish between them by any physical measurement) exactly fits Huygens' treatment.

Reality of Inertial Forces

After d'Alembert first defined inertial force as $F_I = -ma$, in the eighteenth century, and up until the twentieth century, this force was considered to be a *fictitious* force, a force that one may imagine, but can actually manage without, in classical physics. However, in practice, the situation was quite different. The account of reality for the

---

[42] Giancoli (1988, p. 155).

observer within a rotating system is simpler and more intuitively understood than in the ideal inertial system.

Surprisingly, the first one to break the Newtonian framework of treatment with *real* interactive (non-fictitious) forces was Newton himself! The significant example was his considering our spinning Earth. In his theoretical[43] *proving* that the shape of the rotating Earth was a spheroid oblate at its poles, Newton reasoned that the force equilibrium between the *gravitational forces* held the Earth together and the *centrifugal forces* pulled it apart.[44] This account, however, is only valid in the rotating frame of reference.

Fig. 4.7 Newton's treatment of the Earth's shape. Two columns of matter, one along the polar axis and one toward the equator should result in an equilibrium of centrifugal and gravitational forces

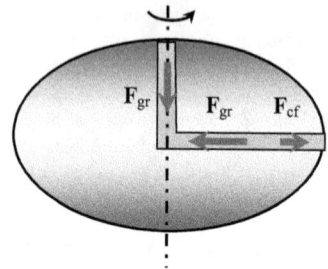

Newton equated the pressure of two columns of matter, one along the axis of rotation, from the center toward the pole, and the other outward toward the equator (Fig. 4.7). In this treatment, the centrifugal force is active as a real force. This way, Newton determined the ratio (R_P/R_E) to be 230 to 229 (the modern measurement of the equatorial radius—6378.14 km and the polar radius—6356.77 km, testify for the accuracy of his account to be about 0.1%).

However, Newton's theoretical abandoning of inertial forces was not shared by practitioners and engineers. During the steam revolution of the eighteenth century, James Watt developed the "centrifugal governor" to regulate the rate of rotation speed (Fig. 4.8a). In technical schools, it was explained in terms of simple equilibrium between elastic and centrifugal force $F_{cf} = m\omega^2 R$, often without even noting the non-inertial observer required for using centrifugal force. All steam machines were equipped with "Watt regulators."

During the nineteenth century, physicists explained several other natural phenomena in terms of the rotation of the Earth as observed by us on the surface of the rotating globe—a non-inertial system of reference. For moving objects, such descriptions introduced another inertial force: Coriolis force which is required to account for the behavior of an object moving between the locations with different radii of rotation. This happens when the object moves from the North to South and back.

---

[43] The empirical proof was due to the French astronomer Jean Richer who demonstrated that the beat of a pendulum is slower when approaching the equator rather than toward the poles. The opposite claim was due to another well-known physicist—Giovanni Cassini (1625–1712), the director in the observatory of Paris.

[44] Newton (1687/1999, Book III, Proposition XIX).

(a)   (b)

Fig. 4.8 (**a**) Centrifugal regulator of a steam machine (pointed by the arrow). Its simplest explanation uses inertial centrifugal force that requires a non-inertial observer in physics theory. (**b**) The
asymmetrical erosion of river banks is observed particularly if the river water flows along the
meridian of the Earth (Photos by the author)

Among such phenomena is the erosion of one of the banks of a river, especially
when its water stream is along the meridian line, to either the North or South
(Fig. 4.8b). A technological parallel of this phenomenon is the asymmetrical erosion of railway lines.

Perhaps the most notable use of the Coriolis forces was the explanation of the
famous pendulum of Foucault (nineteenth century)—"the demonstration of the
Earth's rotation," as it has been named in numerous textbooks since then (Fig. 4.9).

(a)   (b)

Fig. 4.9 (**a**) In 1851, the well-known French physicist Jean-Bernard-Leon Foucault suggested
using a pendulum of large dimensions to demonstrate the Earth's rotation (From the magazine *Le
Petit Parisien* of 1902). The demonstration was realized first in the Pantheon of Paris and since then
has been repeated in numerous places. (**b**) Foucault pendulum in the Cathedral of San Petronio in
Bologna (Photo by the author)

Generations of engineers have learnt about centrifugal and Coriolis forces[45] and make very practical and essential use of these concepts in the design of machines. Inertial forces present a commonplace in technical textbooks. They are treated as real forces complementary to the external ones (the d'Alembert principle).[46] Frames of reference and observers are not even mentioned. Here is what Sommerfeld wrote in his famous course on general physics regarding the inertial forces as real forces[47]:

> *Incidentally, the operation of railroads furnishes a very vivid example of the fact that the "fictitious" centrifugal force has a very real existence. On the curve, the rail bed is banked in such a way that the outer rail is higher than the inner. The difference in height is always such that for some mean velocity of the train the resultant of gravity and centrifugal force is perpendicular to the rail bed. This procedure eliminates not only the danger of overturning about the outer rail, but also a harmful unequal loading of the rails.*

Could one of the students seriously consider these forces as "fictitious"?

One more example of such use is that in the twentieth century, when people could directly see the giant cyclonic motion of the air streams in the atmosphere as well as the structure of distant galaxies, explanations using Coriolis force were extended to all those spectacular natural phenomena. Scarcely anybody ever thought to describe these phenomena using the Newtonian framework of solely interactive forces (Fig. 4.10).

(a)   (b)

Fig. 4.10 (**a**) Cyclonic motion of the atmosphere: its description by the observer on the Earth requires inertial force—Coriolis force. (**b**) The similarity with the structure of the spiral galaxy is striking and so its frequent explanation using Coriolis force

---

[45] In 1835, French physicist Gaspard-Gustave de Coriolis (1792–1843) suggested a force which emerges due to the differences in the rate of spinning of the objects moving in a rotating system as described by the observer in that non-inertial system.

[46] e.g., Vinogradov (2000, p. 79).

[47] Sommerfeld (1952, pp. 59–60).

Summary of the Theoretical Progress

Our excursus to the history of inertia and the force of inertia has come to its end. A complex and impressive conceptual story has unfolded, a continuous discourse of the ideas and conceptions produced by the bright minds of Kepler, Galileo, Descartes, Euler, d'Alembert, Newton, and Huygens, to name several well-known names (Fig. 4.11).

Kepler was the first to introduce the idea of inertia while remaining close to the conceptual world of Aristotle and considered it to be a passive resistance of matter to motion (a *feature* of natural *sluggishness*).

Galileo introduced inertial motion as the separate kind, neither natural nor violent (Aristotelian categories). He coined physical indifference of bodies to the inertial uniform motion.

Descartes denied the idea introduced by Kepler. He wrote[48]:

I do not recognize any inertia or natural sluggishness in bodies...

His worldview and deep belief was that there is nothing in the world but a conservation of [quantity of] motion, and that principle alone governs all changes of the motion of bodies. Descartes had no use of forces in his qualitative claims.

Newton conceived inertia as an active internal *force* with which the body preserves its state of uniform rectilinear motion or rest and continuously resists its change as long as external force is exerted on the body. Newton distinguished between inertial mass and weight (which he considered to be identical to gravitational force). The equivalence of inertial and gravitational masses was established and empirically verified. Centrifugal force was considered as the counterpart of a centripetal one. The unique frame of reference included the whole universe.

Euler refused to consider inertia to be a force and claimed it to be a *faculty* of the body by which it resists to the change of its state of motion or rest. He introduced the new form of the second Newton's law (in practice from then on) $F = ma$, in which inertia is represented by the *inertial mass* m.

Huygens was the first who introduced inertial (centrifugal) force into the account of reality by the rotating (non-inertial) observer. Unlike Newton, he considered it as a real force on the rotating body itself. Huygens determined the functional dependence of the centrifugal force to be expressed by the famous formula:

$$F_{c.f.} = \frac{mv^2}{r} \qquad (4.9)$$

Huygens further demonstrated the essential similarity between the nature of centrifugal force and gravitational force. Centrifugal force as understood by Huygens was the first inertial force introduced into physics. Much later, in the nineteenth century, another inertial force was introduced—the Coriolis force, the force exerted on the body moving in the rotating frame of reference.

---

[48] Descartes (1976).

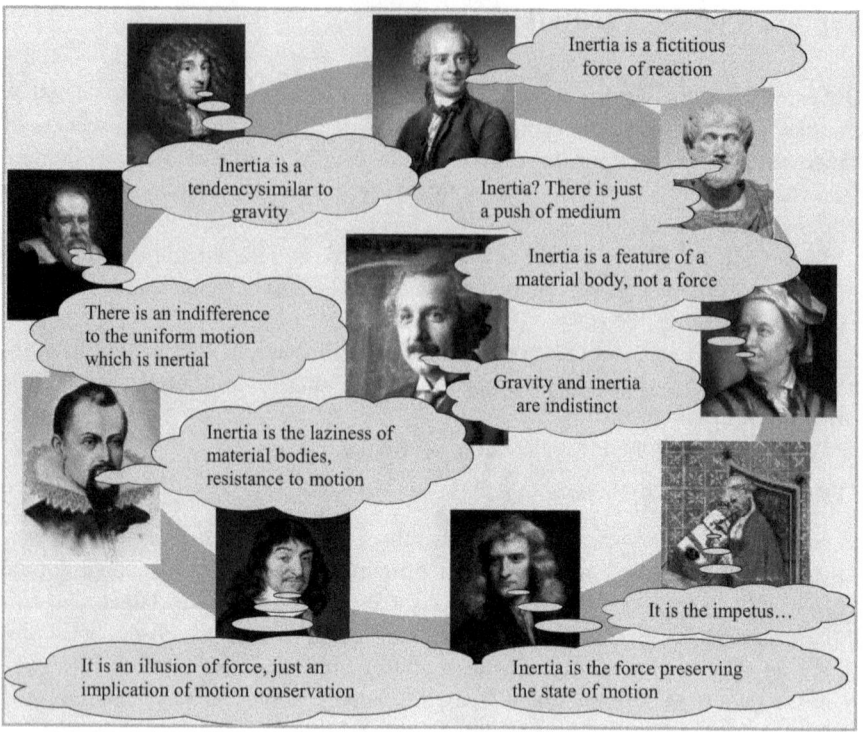

Fig. 4.11 The diachronic dialogue of the ideas regarding inertia

D'Alembert obtained the general form of inertial force, $\mathbf{F}_I = -m\mathbf{a}$, but this step was considered to be a trick by which any dynamic problem could be reduced to a case of equilibrium of forces—the d'Alembert principle. Only in modern physics did this same manipulation become a legitimate description of reality by non-inertial observers moving in accelerating frames of reference.

Einstein changed the framework of physics knowledge. In his version, any observer became equally valid for describing reality. Modern physics legitimizes operational definition of forces. Within this framework, inertial forces became real by their effect interpreted by a non-inertial observer. Within this new understanding, the label "fictitious" lost its dismissing meaning. The equivalence principle (the identical nature of gravitational and inertial forces) empirically established by Newton[49] as the equivalence of inertial and gravitational masses received physical understanding by Einstein within the new theory—the general theory of relativity (the new theory of gravitation).

---

[49]Confirmed with very high accuracy by Roland von Eötvös a century ago, in 1909. Later in the twentieth century, the accuracy was even further increased.

Galileo and Centrifugal Force

During the most of his scientific carrier, Galileo argued for the heliocentric system of the world and paid a heavy price for that since this was not the official posture of the church which was of major influence on people at that time. The case should not be oversimplified as the whole intellectual life then, including science and philosophy, was maintained and led by the church. Thus, Galileo was involved in the debate on the worldview issues (and Copernicanism was such an issue) with numerous clerical authorities, including the highest level—the Pope Urban VIII and Cardinal Bellarmine of the Inquisition. In his arguing for the correctness of the heliocentric system, Galileo, as a scientist, produced several pieces of fundamental evidence.[50] Among them, and perhaps the central for Galileo himself, was his explanation of tides as the direct and incontrovertible evidence of the Earth's motion.[51] Though appeared to be incorrect, it presents interest to the learner, especially in the context considering the concept of centrifugal force.

Galileo explained tidal motion of seas by the difference in summarizing of the diurnal and annual motion of water which is more flexible than rocks of the Earth's ground. Galileo's reasoning was intuitive and stated creation of water wave due to the diurnal change of water from higher velocity at night to the lower velocity during the daytime (Fig. 4.12).

In reality, however, during the 24 h, the annual motion of the Earth is practically rectilinear and uniform in comparison with the diurnal rotation which actually removes any effect. This is regardless other inconsistences widely elaborated (wrong periodicity and spring tides).

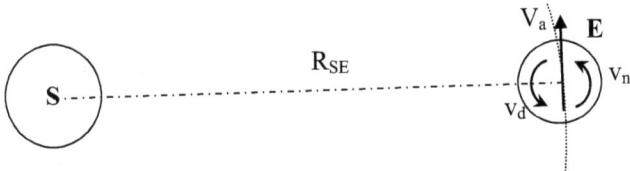

Fig. 4.12 The model by Galileo for explanation of the ocean tides. During the night, the velocity of water due the diurnal rotation v_n adds to the annual velocity of the Earth, V_a, and during the daytime, the velocity of water v_d subtracts from the V_a. The difference causes water turbulence

---

[50] The more famous are—the nature Moon surface, the phases of Venus, the satellites of Jupiter, sunspots, and seasons explanation.

[51] The original title of his *Dialogo* was *Dialogo del flusso e reflusso del mare* (*Dialogue on the Ebb and Flow of the Sea*) (Galilei, 1632/1953, Day Four). Under the pressure of the Pope Urban VIII (and seemingly fortunately for Galileo),the title was modified to the *Dialogue Concerning the Two Chief World Systems* (Drake, 1999, pp. 23–36).

However, Galileo's effect is not an absurd. Indeed, the two motions, rotation and revolving, imply a change in the centrifugal force on the ocean water (velocity changes at practically the same radius for the observer on the Sun, Eq. 4.9), causing a single pair of water ebb and flow diurnally, instead of two in reality. Lacking dynamics of *vis centrifuge* introduced only later by Huygens, Galileo could not develop this kind of explanation using his model, even if it would remain not the true explanation of the real tides. That was provided by Newton within his dynamical theory.

Historical and Epistemological Considerations

Natural in Physics

A close look at the theoretical development of the inertia concept reveals important aspects which are necessary for its proper understanding. One may pay attention to the fact that the concepts of inertia and inertial motion emerged much after the first theories of motion. They came together with the scientific revolution of the seventeenth century. To understand inertia, we need to understand this delay, the reason for it, in order to see the conceptual predecessors of this concept.

Inertia deals with motion. As defined by Aristotle, science looked for the courses of phenomena. Motion was the central subject for physics, so why was there no inertia concept until the seventeenth century, while scholars observed the same phenomena constantly?

In Aristotelian physics, there was no place for inertia given that inertial motion is motion without a cause. There was no such case. In any violent motion, there was always a mover that pushed or pulled being in contact with a body. Medieval impetus did not change the situation but transferred the mover—impetus—inside the moving object.

But there was also *natural* motion—motion with no cause observed. This status of being *natural* is important because it remained in science regardless of the theory. It was such motion that was ascribed to the celestial objects (supralunar world) circularly uniform natural motion. It took 2000 years before the boundary between supralunar and sublunar worlds was dismissed in the seventeenth century. Galileo introduced the motion of the third type, besides natural falling and violent arbitrary motion—indifferent (neutral) motion—and that was the circular motion around the center of the world and/or the Earth. That motion, still natural, was a clear predecessor of the inertial motion. Kepler ascribed inertia to the celestial objects in the meaning of the weight for the sublunary objects in Aristotelian physics, as the resistant faculty of objects to their motion. Descartes dismissed all of that and stated rectilinear motion as being a natural state. It was a cardinal change toward the classical mechanics of Newton which stated the same status of *natural* to the rectilinear uniform motion. Importantly, Newton split inertia from inertial motion. Inertia was postulated as the force of resistance of any body to the change of its natural state—*vis insita*. Looking for clarification of the status of *natural* is thus what unites all the scholars throughout history.[52]

Newton united two phenomena, which looked different to people—to resist increasing the motion (including its start) and to resist decreasing (maintain) the motion. Newton provided a single cause for both—inertia, the resistance to any *change of motion*. To unite phenomena under the same cause suited one of his *Rules for the Study of Natural Philosophy*, Rule 1[53]:

> *No more causes of natural things should be admitted than are both true and sufficient to explain their phenomena.*

Moreover, it also fits well with the second rule[54]:

> *Therefore, the causes assigned to natural effects of the same kind must be, so far as possible, the same.*

Indeed, this kinship appeared in the relativity of motion. We observe a body accelerating from the rest ($v_1 = 0$ at $t_1 = 0$) to the motion at 1 m/sec ($v_2 = 1$ at $t_2 = 1$). The observer moving with the velocity $V = 1$ m/s in the same direction would observe the same body as stopping: from $v'_1 = -1$ (at $t_1 = 0$) to $v'_2 = 0$ (at $t_2 = 1$). The first observer saw the body *accelerating* from the rest and the other—*decelerating* to the rest! The relativity of this kind could lead Newton to infer that the physics theory should not distinguish between "resistance to motion" and "motion preservation." This universal faculty is *inertia*, and it is considered as *natural* faculty.

Newton's unification regarding inertia is reminiscent of that in astronomy where the scholars unified the bright evening star at sunset with the bright morning star at dawn. They were the same object, the planet of Venus.

Observer

The concept of the principle of relativity brings the learner to the concept of observer, which had never appeared in physics before Galileo and Huygens. Newton remained silent regarding this. He placed the entire universe in an all-inclusive "box" of absolute time and space which implied one unique observer depicting the reality. For this observer, his theory was valid. The whole universe was considered from the unique perspective of the observer being at rest in absolute space and time.

However, Newton could not find any physical evidence of the straight homogeneous motion relative to absolute space, the principle of Galileo. In accordance, if a frame of reference was found in which Newton's laws are valid, they remain equally valid for any observer moving at a constant relative velocity—the inertial observer.

---

[52] Time will pass and free falling will return to be natural motion within another theory—the motion without force along the geodesic—in the general theory of relativity.

[53] Newton (1687/1999, p. 440).

[54] Newton (1687/1999, p. 441).

Huygens' exploring the perspective of the rotating observer preceded the *Principia* and could not change Newton's general perspective of the unique observer. The situation changed much later, after the status of the *inertial forces* was clarified as being of a noninteractive nature in comparison with interactive forces, such as gravitational force. In accordance, observers (together with frames of reference) split to *inertial* and *non-inertial* (required inertial forces). Inclusion of inertial forces allows any observer to apply the same framework of laws of motion (the first and the second). Such an expansion of the area of validity represents a fundamental feature of science: seeking the most inclusive framework of a universal and objective account of the natural phenomena. Therefore, Einstein tried to remove the limitation of classical mechanics to a particular type of observer, the inertial observer.

To summarize the changing role of the observer in physical theory, we may summarize the three phases of its understanding:

I. In the physics of Aristotle, the universe, or cosmos, was considered to be spherical, finite, static, and geocentric. The observer, by contemplation, presuming logical analysis, was able to reveal nature, the law-like dependence (the causes) that governs reality.
II. In the classical physics of the seventeenth to nineteenth centuries, this picture changed. Newton understood space to be the absolute, immobile, and *infinite* reservoir of the world: material bodies moving in absolute time. However, the observer's inquiry was now taken to include empirical investigation. Furthermore, the physical account was understood to be unaffected by whether the observer was at rest or moving with constant velocity relative to this space, Galileo's principle of relativity. Such observers were defined as *inertial*. Only these observers were legitimate.
III. In the relativistic physics of the twentieth century, any observer has become legitimate for providing a physical description of reality. However, unlike inertial observers, the non-inertial observers must use inertial forces in their dynamical account of reality. Physics has reached a universal picture of the world by means of laws valid for any observer.[55]

Furthermore, the principle of equivalence, introduced by Einstein, states the identity of the force of inertia to gravitational force. It allowed a new understanding of the concept of weight, as independent of the kind of observer, or as simply the result of weighing. Weight can be explained by gravitation and/or by inertial forces.[56] Making all observers equally legitimate for the physical account of reality has become a philosophical principle of fundamental importance (Fig. 4.13).

---

[55] Physicists call such form of physical laws—*covariant* form. It is used in the relativity theory of Einstein.

[56] Chapter 5.

Fig. 4.13 The allegoric representation of multiple observers for illustrating the new approach of validity of multiple observers in the new theory of mechanics by Einstein. The illustration uses the same fresco by Michelangelo mentioned above, but this time expands the fragment from the unique observer to the multiple observers—the other figures. Their perspectives are represented by the sight directions (the arrows)

Is it a Force?

The history of inertial force—the question: Is the inertial force an illusion?—raises the issue of the meaning of force in general. Newton created an all-inclusive paradigm of classical mechanics with the concept of force at its center: the parameter characterizing the interaction of material bodies. The universe was considered as an empty space containing masses which interact in pairs by means of central forces (acting along the line connecting between the interacting bodies). However, Newton's definition of *impressed* (external) force in the *Principia* ignored interaction[57]:

> *An impressed force is an action exerted upon a body, in order to change its state, either of rest, or of moving uniformly forward in a right line*

This definition does not restrict force to being interactive or central (as the gravitational force is) but being the cause of changes of state of motion. Perhaps it is unfair to expect Newton to consider anything other than the central force. The interactions which seemed to him fundamental (gravitational, static electricity, and natural magnets) acted as central. It was only later with the discovery of electromagnetic action (Oersted and Ampere in the nineteenth century) that a noncentral, Lorentz force, was introduced. Though non-Newtonian, this force could change the state of motion.

---

[57] Newton (1729/1952, Definition 4, p. 6).

Mach, Kirchhoff, and Hertz,[58] in the second part of the nineteenth century, introduced a new type of *operational* definition for physical quantities within positivist philosophy of science.[59] This was an epistemological revolution. Einstein applied an operational definition to *simultaneity* in establishing a new account of reality—the theory of special relativity, in 1905.

The requirement for an operational definition for concepts is at the heart of the philosophical trend of *operationalism*.[60] Later it was upgraded[61] to the necessity for both *nominal* (theoretical) and *operational* (epistemic) definitions of concepts. The inertial forces, defined by their effects on a force-meter, became legitimate and real. Inertial force, as a measurable quantity, is a part in a non-inertial observer's account of a situation. However, lacking an interactive partner, the inertial force remains enigmatic.[62] In any case, the introduction of the operationally defined inertial forces as *real* forces presents a striking change in the mechanics of modern physics.

The "problem" (the feature) of inertial forces, in their modern form, is that they are not interactive in nature (no action-reaction pair can be shown) contradicting the third Newton's law. Einstein revealed that inertial forces could be interpreted by an observer as a gravitational field which suggests looking for the gravitational source of inertia. This idea led Mach to hypothesize (Mach principle)[63] the source of inertial forces to be the interaction with distant galaxies, the measure of the interaction depending on the relative acceleration between the masses. This consideration, however, presents an open problem in physics due to the complexity of evaluating interaction of this kind.[64] Meanwhile, one cannot be very certain when negating the noninteractive nature of inertial forces.

Inertia and the Cumulative Nature of Scientific Knowledge

The development of scientific knowledge of inertia sheds light on the features of scientific knowledge construction which started from an idea of theoretical revision. This evolved into formal and precise accounts of natural motion and eventually arrived at mathematically arranged principles within the old theory and the new more advanced one. Qualitative and quantitative accounts developed synergistically and were interwoven. Let us return to Figure 4.11 and organize and rephrase its content (the diachronic dialogue of the ideas) in the way that would represent the sequential steps in the genesis of the concept of inertial force in physics (Fig. 4.14).

---

[58] Mach (1883/1919), Hertz (1899/1956).

[59] Mach (1883/1919, pp. 216–218).

[60] Bridgman (1927/1952, 1927/1952a).

[61] e.g., Margenau (1950, pp. 220–244).

[62] Physicists continue to discuss the option that the distant stars cause the inertial forces. This hypothesis is known under the name "Mach principle." The fact that this discourse remains speculative does not imply any practical restriction to the use of inertial forces.

[63] This idea may be traced back to Berkeley (1710/1952, §§111–117, pp. 434–436).

[64] e.g., Sciama (1971, pp. 197–199).

Observing this diachronic debate of ideas accompanied by the complementary progress in problem-solving and growing mathematization of the account, we may learn about the way scientific knowledge is constructed. Scientific knowledge clearly does not present a collection of products arranged in groups of different periods—archeological layers of different reigning paradigms exchanged—an image that might emerge from Kuhn's description of scientific revolutions. The discourse on inertia presented one tree of thought on the specific subject, the debate through centuries within different worldviews, theoretical frameworks (paradigms), and various approaches, while sharing the subject and considering the previous results, adopting empirical and theoretical information.

Kepler launched the idea of *inertia* in a specific sense in the celestial context – a resistance to motion.

Galileo introduced a new type of motion corresponding to the *inertial motion* while dismissing its alternative account by means of impetus.

Descartes dismissed the idea of *inertia* by introducing the state of motion and the principle of conservation of quantity of motion.

Newton returned to inertia as an *internal force* of resistance to the change of motion state (the first law). In parallel, (*inertial*) *mass* was introduced in the second law of motion within the quantity of motion.

Euler modified the second law of motion to its differential form. Inertial force was dismissed and replaced by *inertial mass* as a faculty of resistance to the external force causing acceleration (F=ma).

Huygens introduces *centrifugal force* for the rotating observer and shows its nature as similar to gravitation. Newton redefines the centrifugal force to be a "regular" force in the account by the unique observer at rest.

d'Alembert introduces a *fictitious inertial force* (F=-ma) as a formal trick of reduction of the dynamic account of material system to the static one. Centrifugal force becomes a kind of fictitious force.

Einstein introduces observer dependent mechanical accounts of material systems. Centrifugal force becomes a legitimate operationally defined force – *inertial force* – in the account of reality by a non-inertial observer. The equivalence of the nature of inertial and gravitational forces was understood/interpreted.

Fig. 4.14 Sequential steps in the development of the concept of inertial force in physics

The debate about inertia (Fig. 4.14) converged to Einstein's elucidating *interpretation* of the empirical reality in his theory of general relativity. The use of non-inertial reference frames implied inertial, noninteractive forces. Einstein interpreted the equivalence of inertial and gravitational masses as the evidence of their reciprocal dependence. The *gravitational* mass of a body curves space-time, which influences its motion, thus manifested itself in its *inertial* mass. In its turn, motion of the body changes the distribution of gravitational masses and that causes new profile of space-time curvature and so on.[65] This way both masses cannot escape but being firmly related. Despite the differences, the conceptual content the scientists worked on is preserved in its essence and evolved in the iterative process to the collective product, not an independent discovery of the latest maverick thinker. It is in this sense that scientific knowledge is *cumulative*.

This feature of scientific knowledge is important because the opposite claim is frequently made. As mentioned, Thomas Kuhn depicted the history of science as a succession of revolutionary changes of the reigning theories (paradigms) and the abandoning of the old knowledge. As seen in the history of inertia, such perspective is inaccurate. Importantly, in the social sciences, the situation is rather different. Michel Foucault depicted the situation where the perspectives (*epistemai*) regarding the same social phenomenon were actually incommensurable and entirely replaced each other. Foucault termed his approach *Archeology of Knowledge*, hinting at the independent layers in excavation and the difference between everyday knowledge and scientific knowledge. Moreover, he pointed to the difference in the type of knowledge produced in social and natural sciences.[66] The latter drew on the scientific method committed to *objectivity* and *conceptual continuity* even in the stream of changing theories. Such perspective creates the holistic image of inertia in the form represented in Fig. 4.14 instead of extraction from it its last line or any other.

Educational Aspects

On the Nature of Scientific Concepts

Consider the inertial force shed light on the nature of the scientific concepts. The inertial force "exists" for non-inertial observers and "does not exist" for inertial observers. This special situation invites clarification of the meaning of *existence* with regard to concepts—the important ontological aspect of knowledge rarely addressed and often used. Students often consider interactive forces as if objects

---

[65] This mechanism was epitomized by Wheeler: "…spacetime grips mass, telling it how to move. In this way, mass grips spacetime how to curve" (Wheeler, 1990, p. 5).

[66] Foucault (1972, p. 202) illustrated by psychiatric medicine in which the subjectivity prevailed due to the social reality which essentially intruded into the discipline (theory). He stated the need of "careful distinguishing between scientific and archeological territories," different in "their principles of organization," alien to unity.

existed materially—the naïve reification of an abstract idea.[67] In what sense do forces exist?

Teachers often silently presume as obvious, that force (as any other physical concept) presents an artifact, a mental tool to describe a certain objective reality. It seems not sufficient.

The concept of force belongs to the conceptual nucleus of classical mechanics, whereas fundamental physics theories—quantum and general relativity—reconsidered its role.

In the quantum theory, the notion of force became problematic, clashing with the new principles of the theory. Forces are excluded from formalism and solving problems. They are avoided in the quantum picture of the world. The energy-momentum account became fundamental. Yet, even so, force is preserved to connect the quantum picture with the classical mechanics and macroscopic account as a residual concept.[68]

The theory of general relativity, in its turn, does not employ gravitational force but manipulates with the curvature of space-time as related to the energy-momentum content. Yet, the resultant apparent "attraction" between material objects can be interpreted as the force of gravitation, allowing the obtaining of the Newtonian law of gravitation in the limit of small masses, close enough to neglect the time delay in their interaction.

Thus, both great theories, quantum and relativistic, do not break with forces entirely. They keep them because they depict the world as commonly perceived by us out of the perspective of physics, neither micro, nor macro—but an everyday experience. The strength of different types of interaction, gravitational, electromagnetic, strong, and weak is usually compared in terms of forces. Addressing wider perspective can shake the naïve belief in "existence" of force—a common misconception of reification of abstract theoretical concepts.

However, one does not need to expand beyond classical mechanics. The artifact nature of inertial forces comes to the fore in the freedom to use or avoid inertial force. It is beyond mere simplicity or intuition. The virtual existence of force becomes clear in the juxtaposing of the two frameworks, with and without inertia force. Consider, for instance, a container orbiting the Earth within a rocket. The container holds water. Accounting for the shape that this volume of water takes allows two approaches, that of the inertial observer (the reference frame related to the Earth) and that of the non-inertial observer (in the reference frame of the rocket itself) (Fig. 4.15).[69]

---

[67] Reiner et al. (2000); Galili & Hazan (2000a, b).

[68] For instance, Van der Waals *force* is used to depict macroscopic interaction between atoms and molecules, yet on the microscopic level, the contribution Pauli's Exclusion Principle which is underneath does not rely on any force.

[69] The split to two accounts is essential and cannot be avoided. Attempts to "combine" them, avoiding addressing the frame of reference, leads to confusion (e.g., Harrison, 2000, pp. 57–58).

Inertial observer(Newton)	*Non-inertial observer* (Huygens-Einstein)
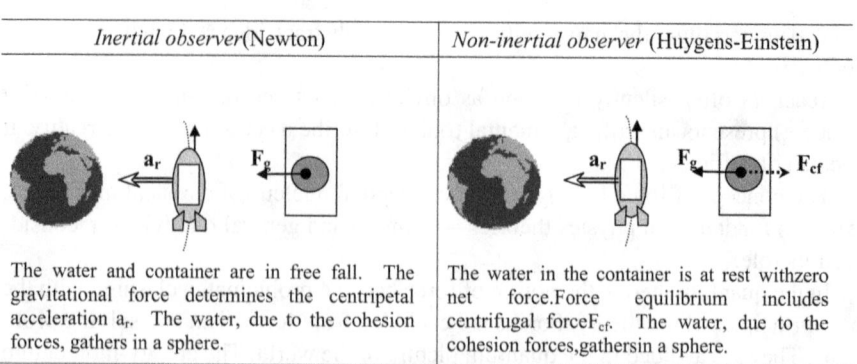	
The water and container are in free fall. The gravitational force determines the centripetal acceleration a_r. The water, due to the cohesion forces, gathers in a sphere.	The water in the container is at rest withzero net force.Force equilibrium includes centrifugal forceF_{cf}. The water, due to the cohesion forces,gathersin a sphere.

Fig. 4.15 Juxtaposition of the two accounts of physical situation by inertial and non-inertial observers

Considering the status of the force concept can be traced to medieval philosophy. There, the debate started between the *realists* stating the existence of certain entity manifested by the concept and the *nominalists* dismissing such existence beyond using the concept in descriptions of objects. Inertial force may illustrate the contrast of the ontological status of the real world against its nonunique description in physical theories. Einstein explained the nonunique product of knowledge construction by lacking direct access to the subject[70]:

> *Physical concepts are free creations of the human mind, and are not, however it may seem, uniquely determined by the external world. In our endeavor to understand reality we are somewhat like a man trying to understand the mechanism of a closed watch. He sees the face and the moving hands, even hears its ticking, but he has no way of opening the case.*

In essence, this process presents the modeling of reality. In doing this, one should refrain from delegalizing other interpretations as happened historically with the forces of inertia. They, these interpretations, became particularly confusing in education due to their depiction in terms of "fictitious" and "pseudo" forces introduced in the nineteenth century. These strong terms clearly drew on the Newtonian unique observer and his framework of solely interactive forces.[71] This is not the way physics considers inertial forces today. The cardinal change took place with the legitimizing of multiple frames of reference for a dynamic accounting of reality and the operational definition of physics concepts, forces, in our case.

---

[70] Einstein & Infeld (1938, p. 33).

[71] The formal designation of $F_{in} = - ma$ as a force of inertia was made more than 100 years after d'Alembert by French mathematician and astronomer Charles Delaunay who presented the results of d'Alembert. We, however, may not change the tradition of ascribing this step to d'Alembert himself.

Conceptual Change in Learning

We have followed the concepts of inertia and the forces of inertia through their rich history. Only after the adoption of non-inertial observers were inertial forces considered legitimate. Therefore, considering inertial forces in physics class obliges using modern epistemology of physics. School curricula are very slow in matching this development. The gap in teaching classical dynamics is already over 100 years wide, despite no formal complexity. Inertia and inertial forces are commonly ignored even in advanced placement physics courses.

Our approach displays the historical growth of understanding. However, one cannot ignore the charges of disappointment regarding historical approach[72]:

> ... *even materials produced for teachers, for example, those produced in the UK ... are not used. Attempts to produce restructured courses that put history at the center of the enterprise ... have enjoyed only marginal success, as have those that have sought to introduce a more rigorous and current view of the philosophy of science ...*

The explanation is straightforward. Firstly, teachers often lack any background in HPS that is not included in either the disciplinary or the pedagogical training program of preservice studies. Clearly, lacking background, people cannot be effective in teaching such content.

Secondly, historical materials might seem to be obsolete in conceptions that are confusing to the novice. Yet, as shown by empirical studies, students who were never instructed in inertial forces overwhelmingly included them in their account of accelerated systems (on an inclined plane, in a car, on a rotating disc, in orbiting satellite, etc.) and frequently showed the following conceptions[73]:

- Inertial force pushes forward the driver performing a quick stop (similar to the impetus conception and incorrect within the Newtonian framework)
- Centrifugal forces act on the rotating bodies (Huygens' account, erroneous in Newton's framework)
- Equilibrium of forces acting on the objects in an orbiting satellite, including active centrifugal force (Borelli's erroneous conception, erroneous in Newton's framework)
- The force of resistance to motion (Kepler's erroneous understanding of inertia)

This reality removes the speculation regarding misleading and confusing impact of learning historical conceptions: very often they are already there, spontaneously produced by the students. On the contrary, the remedial impact is expected by addressing such conceptions and their analysis—the strategy of cognitive resonance.[74] Intuitive interpretation of the sense experience in accelerating motion causes spontaneous schemes of knowledge to be deeply entrenched in students'

---

[72] Monk and Osborne (1997).

[73] Galili & Bar (1992); Galili & Kaplan (2002).

[74] Galili & Hazan (2000a).

cognition, regardless of class teaching. This requires a radical conceptual change causing the exchange of knowledge elements in the triadic structure of knowledge (nucleus-body-periphery) (Fig. 4.16).[75]

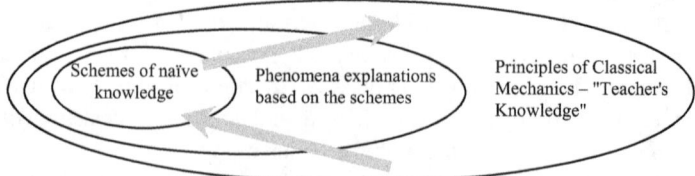

Fig. 4.16 Students' conceptual change in the course of learning can be represented as an exchange of contents (shown by arrows) in the knowledge structure

Operational Definition of Concepts in Teaching Physics

Being rigorous in defining concepts is a characteristic feature of scientific knowledge. This is unlike liberal arts, where variation in meaning serves as a creative resource. Scientists try to avoid dependence on interpretation by using operational definitions. This approach is especially important regarding fundamental concepts. To clarify the need, one may consider Newton's definition of mass as the *quantity of matter*. It might look intuitively clear. However, it is not difficult to realize that this definition is problematic: how to count that quantity? This way the learner may appreciate as more appropriate defining mass in terms of a *body's inertia*—a subject of direct measurement in the first place—and only then infer regarding the *quantity of matter* in the body.

Another example is the concept of weight.[76] Since the gravitational force is identical in effect to the inertial force, as stated by Einstein's principle of equivalence, the theoretical definition of weight is not effective, whereas the operational definition ("weight is the result of standard weighing") is. It allows both gravitational and inertial forces to equally contribute to the weight of an object. One may illustrate this by teaching weight as being due to the centrifugal force in an imaginary rotational space station (Fig. 4.18c). Huygens foresaw this nature of weight as early as in 1659.

The Knowledge Progression in Science

"Culture of science" implies a dialogue of ideas in the scientific progress. Scientific knowledge is constructed in such dialogue. The diachronic debate regarding inertia and inertial forces included the ideas by Kepler, Descartes, Newton, Huygens, Euler, d'Alembert, and Einstein (Fig. 4.13). This approach to teaching creates a

---

[75] Chapter 6.
[76] Chapter 5.

space of learning in which the target concept emerges from its variations.[77] The product of such learning is a conceptual range in which a certain idea prevails in the contest preserved as an environment—cultural content knowledge.[78]

Thus, scientific progress should be distinguished from the linear, climbing toward a truth reached by sequential steps of replacement of false ideas. The image to serve as a contrast is climbing a ladder as the pro- gression in faith through trials and temptations (Fig. 4.17). This image should be challenged with the alternative progress in a conceptual dialogue (Figs. 4.11 and 4.13 versus Fig. 4.17).

In a way, this comparison also implies the impor- tance of conceptual understanding while pointing to insufficient teaching, limited to a mere problem-solving.

Fig. 4.17 The image of evolution of religious knowledge using the metaphor of climbing a ladder motivated by angels (on the right side) and impeded by the temptations of dev- ils—on the opposite side

Media for Learning

Technology provides a supportive environment for teaching inertial forces. Numerous mechanisms are normally explained by inertial force.[79] The range expands from the steam machines of the eighteenth century to the rotating space stations and spaceships of the future (Fig. 4.18). Every driving instructor teacher warns the trainee not to drive too fast on curves because of the too large centrifugal push. Luna park instructions warn of the extreme centrifugal forces in roller coast- ers. Laboratory manuals refer to centrifugal force in explaining the principle of centrifugal separators and centrifugal analysis. Texts of technology, medicine, geo- physics, and astrophysics widely use oversimplified explanations, ignoring the observer-dependent nature of inertial force.[80] Numerous appliances provide a huge stage for teaching inertia and inertial forces, both formally and informally. Seemingly, inertial forces are used everywhere except in the school physics curricu- lum. Ironically, confusion and ignorance in such accounts characterize just those who took physics classes and are warned not use fictitious forces as "unreal."

---

[77] Marton & Tsui (2004).

[78] Chapter 6.

[79] e.g., Vinogradov (2000).

[80] e.g., Harrison (2000).

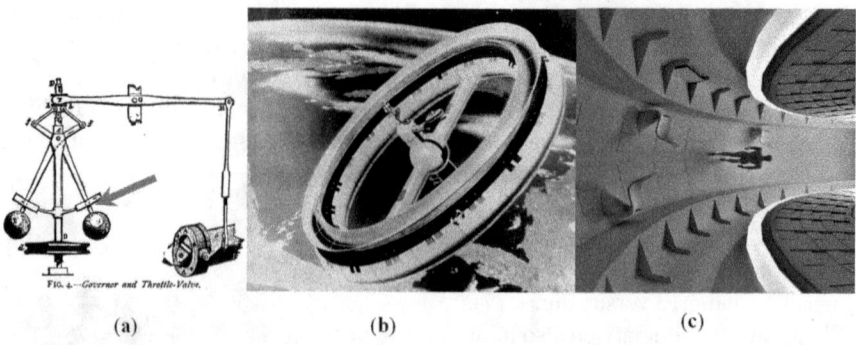

(a)   **(b)**   **(c)**

Fig. 4.18 (**a**) Centrifugal force regulator by James Watt in his steam machines (eighteenth century). It contains two masses rotating around a shaft. Masses displace proportionally to the speed of rotation and produce regulating impact. (**b**) Rotational space station suggested by von Braun in 1952. (**c**) A sketch of the rotating cabin in an imaginary spaceship. The weight provided to the astronauts is often explained by the centrifugal force (In the following chapter (Chapter 5), we will expand more regarding the weight in a rotating space station (Fig. 5.17), and in a rotating cabin of a spaceship (sketch by Guy Galili) (Fig. 5.23))

Questions for Reflection

- Kepler introduced the notion of inertia in the seventeenth century. How could scholars manage to describe the same reality of motion before Kepler?
- Descartes did not recognize inertia and sufficed with the account by means of conservation of (quantity of) motion in collisions. Was this approach preferable?
- Is inertia a force? Illustrate your answer.
- When we set in motion a heavy revolving door, what do we struggle with: (a) the gravitation, (b) friction, (c) inertia?—Explain your choice.
- What is the force of inertia? Give a list of examples of the action of inertial forces.
- Describe physical phenomena, devices, and machines known to you which use inertial force. Try to explain them without inertial force.
- In the past, physicists labeled inertial forces *fictitious*. Why? What seemed to them wrong with those forces?
- Descartes and Huygens tried their best to explain gravitation by means of the centrifugal effect. How was it possible to obtain a force outward while the attraction they wanted to explain pulled bodies inward?
- Explain the movement of the body of a driver in a suddenly stopping or accelerating car.
- A small helium balloon is floating inside a car. Show the inertial force which is exerted on it and explain the movement of the balloon while the car is stopping/ accelerating. Relate your answer to the explanation of gravitational attraction by Descartes and Huygens.
- Imagine a debate between Newton and Einstein regarding the inertial forces. Newton would reject them and Einstein would adopt them. Who is right in this debate? Explain.

- Huygens did not consider his centrifugal force fictitious, but real. Feynman in his course of lectures called inertial forces "effective" (acting as a force but still not really a force—pseudo-force). Hertz thought that all forces were fictions. Who was right? Explain their rationale.
- In Newton's famous thought experiment with a bucket, water surface receives a parabolic profile in rotation.[81] Newton believed that through considering four situations (Fig. 4.19), he demonstrated that the concave profile of the water is not due to the relative motion of bucket and water but due to the absolute motion of the water relative to the absolute space. Explain the phenomenon in two ways, as Newton and Huygens would do. Which description is easier? Which is correct?

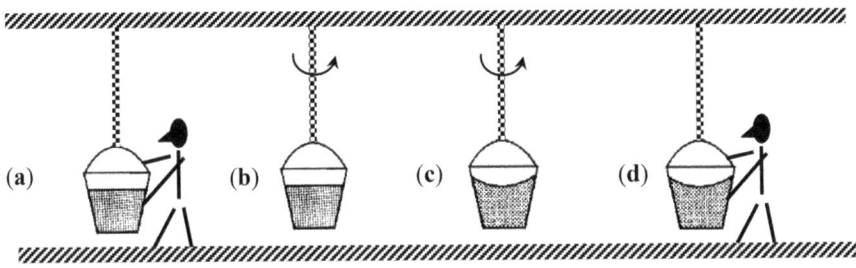

Fig. 4.19 The thought experiment with a bucket of water suspended by a twisted cord. Four sequential moments are depicted. (**a**) The bucket and water are at rest. Water level is horizontal. (**b**) The bucket is released and begins to rotate, while the water remains at rest and only slowly begins to move. (**c**) Some time has passed; the bucket and the water share the rotation; water level is concave. (**d**) After a while, the bucket was stopped while the water continued to rotate. Water level is concave

- Figure 4.20 presents a model that demonstrates Descartes' idea of the origin of gravitation and, at the same time, the mechanism of the device termed separator. While rotating, a separation according to the masses of the particles immersed in a fluid takes place. Mass M is of greater density than the fluid and mass m—of lesser one. In the state of rotation, M will rise to the right from the axis, and m will go down toward the axis. Explain this phenomenon with inertial forces and without.

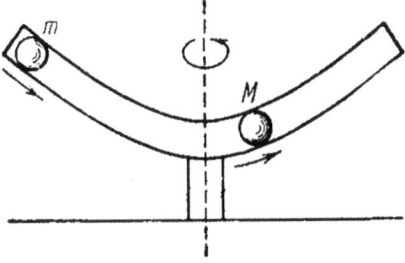

Fig. 4.20 Separator of particles in accordance with their density

- Figure 4.21 shows a river in Portugal which both banks are high. What can be inferred about its direction? Explain.

---

[81] Newton (1687/1999, Definitions, Scholium, p. 407).

Fig. 4.21   Douro River in Porto (Photo by the author)

Concluding Remarks

The concepts of inertia and the force of inertia might look problematic as extending teaching beyond the Newtonian perspective—considering non-inertial observers. In fact, however, teaching classical mechanics continuously changes in light of the progress in our physical worldview, understanding of the scientific method and its appropriate representation for learning. We cannot afford to continue teaching of classical mechanics from the obsolete perspective held before the twentieth century.

One cannot ignore the epistemological dictum of the operational definition of concepts (inertia, force, weight), addressing the dependence of the physics account on the observer (frame of reference); the need to match, rather than ignore, the more advanced theories which shed light on the meaning of the classical mechanics. The latter is not to be replaced. It remains fundamental, objective, and correct in the area of its validity. Introducing inertial forces is not a mere replacement of the old vision but a part of the dialogical nature of scientific knowledge. It is in this sense that the historical approach becomes pivotal. It creates in the learner genuine understanding of classical mechanics and cultural content knowledge. The story of inertial force should become a pedagogical tool in a new physics curriculum which will include inertial forces. This step is unavoidable in contemporary schools considering the benefits and interests of the wider population of science students in modern society.

The new horizon

Picture Credits

- The Pinwheel Galaxy. Image Credit: ESA/Hubble & NASA. Flickr user Det58. CC Attribution 3.0 Unported license. https://www.nasa.gov/content/hubble/grand-swirls-from-nasas-hubble
- Portrait of Kepler by unidentified painter. Public Domain. https://commons.wikimedia.org/w/index.php?curid=470711
- Portrait of Galileo Galilei by J. Sustermans, 1636. Public Domain. https://commons.wikimedia.org/w/index.php?curid=230543
- Portrait of Rene Descartes after Frans Hals,1640. Crédit communal de Belgique. Public Domain. https://commons.wikimedia.org/w/index.php?curid=2774313
- Portrait of Christiaan Huygens by C. Netscher, Kunstmuseum Den Haag. Public Domain. https://commons.wikimedia.org/w/index.php?curid=44047
- Portrait of Newton by Godfrey Kneller, 1689. Public Domain. https://commons.wikimedia.org/w/index.php?curid=146431
- Photo of Einstein by F. Schmutzer, 1921. Public Domain. https://commons.wikimedia.org/w/index.php?curid=34239518
- Hammer thrower. Illustration picture from C. Ray (1935, p. 260), The Popular Science Educator. London: The Free Way House
- Jump of a seal. Photo by the author
- A souvenir. Photo by the author.
- Title page of the Principia, 1787. Public Domain. https://commons.wikimedia.org/w/index.php?curid=2681838
- Portrait of Johannes Kepler by unidentified painter. Public Domain. https://commons.wikimedia.org/wiki/File:Johannes_Kepler.jpg
- Portrait of Euler by J. E. Handmann, 1756, Public Domain. https://commons.wikimedia.org/w/index.php?curid=1001511
- The title page of *Mechanics* by Euler, 1736. Public Domain. https://archive.org/details/mechanicasivemot02eule

- Portrait of d'Alembert by M. Q. de La Tour, 1753. Louvre Museum. Public Domain. https://commons.wikimedia.org/w/index.php?curid=657700
- The title page of *Dynamics* d'Alembert. 1743 Public Domain. https://commons.wikimedia.org/w/index.php?curid=96278204
- Fig. 4.2, Statue Discobolos by Myron, 2nd c. CE. CC Attribution-Share Alike 3.0 Unported license. https://commons.wikimedia.org/w/index.php?curid=547351
- Fig. 4.3 Illustration from Descartes, 1644, p. 56
- Fig. 4.5a, b Huygens' drawings in *On Centrifugal Force*, 1659/1703, pp. 256, 259. https://www.princeton.edu/~hos/mike/texts/huygens/centriforce/huyforce.htm
- Figs. 4.6 and 4.13 Michelangelo, fresco, 1511, Cappella Sistine, the Vatican. Public Domain. https://commons.wikimedia.org/wiki/File:Michelangelo,_Separation_of_the_Earth_from_the_Waters_00.jpg
- Fig. 4.9a Illustration from magazine *Le_petit_Parisien*, 2nov 1902: Public domain. https://upload.wikimedia.org/wikipedia/commons/2/23/Le_petit_Parisien_illustre_2nov1902.jpg
- Fig. 4.10a Cyclone Chapala. Photo Credit: NASA. Public Domain. https://commons.wikimedia.org/w/index.php?curid=44628566
- Fig. 4.10b Spiral Galaxy, December 16, 2011. Photo Credit: NASA, ESA, and the Hubble Heritage (STScI/AURA)-ESA/Hubble Collaboration. https://www.nasa.gov/multimedia/imagegallery/image_feature_2132.html
- Fig. 4.11 Bust of Aristotle after Lysippos - Jastrow (2006), Public Domain, https://commons.wikimedia.org/w/index.php?curid=1359807
- Fig. 4.11 Portrait of Kepler. Illustration from the book from *Mechanics* of Mach, 1883/1919.
- Fig. 4.11 Portrait of Oresme. Bibliothèque Nationale, fonds français 565, fol. 1r. Paris, France. Public Domain. https://commons.wikimedia.org/w/index.php?curid=436922
- Fig. 4.11 Portrait of Euler by J. Handmann, 1849, Kunstmuseum Basel. Public Domain. https://commons.wikimedia.org/w/index.php?curid=893656
- Fig. 4.17 *The Ladder of Divine Ascent* by John Climacus. Monastery of St Catherine, Mount Sinai. Public Domain. https://en.wikipedia.org/wiki/The_Ladder_of_Divine_Ascent#/media/File:The_Ladder_of_Divine_Ascent_Monastery_of_St_Catherine_Sinai_12th_century.jpg
- Fig. 4.18a Centrifugal governor. Image from *Discoveries, Inventions of the Nineteenth Century*, Routledge, 1900. Public Domain. https://commons.wikimedia.org/w/index.php?curid=231047
- Fig. 4.18b Werner von Braun's project of a rotating station. Picture credit: NASA. Public Domain. https://commons.wikimedia.org/w/index.php?curid=14867654
- Fig. 4.20 Illustration from *Mechanics,* Mach, 1883/1919, p. 162, Fig. 106.
- End of the chapter photo: Flying carousel. Photo by the author

References

Berkeley, G. (1710/1952). *Principles of human knowledge*. Encyclopedia Britannica.

Bridgman, P. (1927/1952). *The logic of modern physics*. : Macmillan.

Bridgman, P.. (1927/1952a). *The nature of some of physical concepts*. New York: Philosophical Library.

Cohen, I. B. (1971). *Introduction to Newton's Principia* (pp. XVI, 28). Cambridge: University Press.

D'Alembert, J. (1758). *Traité de dynamique*. David Libraire.

Descartes, R. (1976). Letter to Mersen. In C. Adams & P. Tannery (Eds.), *Oeuvres de Descartes* (Vol. 2, p. 466). J. Vrin.

Drabkin, I., & Drake, S. (1960). *Galileo on motion and on mechanics*. University of Wisconsin.

Drake, S. (1964). Galileo and the law of inertia. *American Journal of Physics, 32*, 601–608.

Drake, S. (1999). Galileo: Scientific method and philosophy of science. In *Essays on Galileo and the history and philosophy of science* (Vol. II). University of Toronto Press.

Einstein, A., & Infeld, L. (1938). *Evolution of physics*. Cambridge University Press.

Euler, L. (1736/2008). *Mechanics or the science of motion analytically demonstrated* (Vol. 1).

Euler, L. (1765/2009). *Theory of the motion of solid or rigid bodies* (Vol. 1. Ch. 4). http://www.17centurymaths.com/contents/euler/mechvol3/genmotch4.pdf

Foucault, M. (1972). *Archaeology of knowledge*. Routledge.

Galilei, G. (1613/1957). Second letter on sunspots. In S. Drake (Ed.), *Discoveries and opinions of Galileo*. Doubleday.

Galilei, G. (1632/1953). *Dialogue concerning the two chief world systems – Ptolemaic and copernican*. University of California Press.

Galilei, G. (1638/1914). *Dialogue concerning two new sciences*. Dover.

Galili, I., & Bar, V. (1992). Motion implies force. Where to expect vestiges of the misconception? *International Journal of Science Education, 14*(1), 63–81.

Galili, I., & Hazan, A. (2000a). Learners' knowledge in optics: Interpretation, structure, and analysis. *International Journal in Science Education, 22*(1), 57–88.

Galili, I., & Hazan, A. (2000b). The influence of historically oriented course on students' content knowledge in optics evaluated by means of facets-schemes analysis. *Physics Education Research, American Journal of Physics, 68*(7), S3–S15.

Galili, I., & Kaplan, D. (2002). Students' interpretation of water surface orientation and inertial forces in physics curriculum. *Praxis der Naturwissenschaften Physik in der Schule, 51*(7), 2–11.

Galili, I., & Tseitlin, M. (2003). Newton's first law: Text, translations, interpretations, and physics education. *Science & Education, 12*(1), 45–73.

Giancoli, D. C. (1988). *Physics for scientists and engineers*. Prentice Hall.

Harrison, E. (2000). *Cosmology. The science of the universe*. Cambridge University Press.

Hertz, H. (1899/1956). *The principals of mechanics*. Dover.

Huygens, Ch. (1659/1703). *On centrifugal force*. From *De vi Centrifuga*. In *Oeuvres Complètes* (Vol. XVI, pp. 255–301). Translated by M.S. Mahoney. Huygens - On Centrifugal Force (princeton.edu)

Jammer, M. (1961). *Concepts of mass in classical and modern physics*. Harvard University Press.

Kepler, J. (1621/1952). *Epitome of Copernican astronomy* (p. 845). Britannica.

Lagrange, J. L. (1788/1938). *Mechanique Analitique*. Paris: Desaint. Translated into English (1997). *Boston studies in the philosophy of science* (Vol. 191). New York: Springer.

Lanczos, C. (1949/1970). *Variational principles of mechanics*. Toronto: University of Toronto.

Mach, E. (1883/1919). *The science of mechanics, a critical and historical account of its development*. The Open Court.

Marton, F., & Tsui, A. B. M. (2004). *Classroom discourse and the space of learning*. Lawrence Erlbaum.

Monk, M., & Osborne, J. (1997). Placing the history and philosophy of science on the curriculum: A model for the development of pedagogy. *Science Education, 81*(4), 405–424.

Newton, I. (1687/1999). *The Principia. Mathematical Principles of Natural Philosophy*. Translated by B. Cohen & A. Whitman. Berkeley, CA: University of California Press.

Newton, I. (1729/1952). *Mathematical principles of natural philosophy*. Revised edition of Motte's translation by F. Cajori. Chicago: Britannica Great Books.

Reiner, M., Slotta, J. D., Chi, M. T. H., & Resnick, L. B. (2000). Naive physics reasoning: A commitment to substance-based conceptions. *Cognition and Instruction, 18*(1), 1–34.

Sciama, D. W. (1971). *Modern cosmology*. Cambridge University Press.

Sommerfeld, A. (1952). *Mechanics. Lectures on theoretical physics* (Vol. 1). Academic Press.

Vinogradov, O. (2000). *Fundamentals of kinematics and dynamics of machines and mechanisms*. CRC Press.

Wheeler, J. A. (1990). *A journey into gravity and spacetime*. New York: Scientific American Library; and (1999). *Journey into gravity and spacetime*. New York: Freeman.

Chapter 5
Weight Concept: From Aristotle to Newton and Then to Einstein

© Springer Nature Switzerland AG 2021, corrected publication 2022
I. Galili, *Scientific Knowledge as a Culture*, Science: Philosophy, History and
Education, https://doi.org/10.1007/978-3-030-80201-1_5

A deceitful balance is an abomination before the Lord: and a just weight is his will.

The Book of Proverbs, 11:1

The Lord demands exact weighing. That obliged us to know what is weight and how should weight be determined? The Bible does not help us in that, only physics does.

Abstract The concept of weight emerged in physics through the periods of Ancient Greece, the medieval world, the scientific revolution of the seventeenth century, classical mechanics, and modern physics. Along the way, the knowledge and the method of knowledge construction changed; they are connected to each other. Weight was always among the fundamental concepts of mechanical theories, and its definition changed until it was finalized in the twentieth century. However, school curricula in many countries did not copy this progress and remained with the Newtonian definition of weight. In contrast, some textbooks define weight in the modern way, matching the equivalence principle of Einstein and modern epistemology. Following the conceptual growth along the history and philosophy of science brings familiarity with the scientific discourse and promotes construction of the cultural content knowledge of weight and its relation to gravitation. The implications for the teaching of weight drawing on the results of physics education research are considered, being illustrated with a brief review of weight impact on society, situations of everyday life on our planet and beyond.

Fig. 5.1 Weighing on the scale. The scene from the Egyptian Papyrus (c. 1275 BCE)

Understanding of Weight before Newton

The evolution of the weight concept in *science* started with the two notions of heaviness (*gravity*) and lightness (*levity*). Greek natural philosophy employed only qualitative characteristics, which came in pairs of opposites: hard-soft, wet-dry, and heavy-light. Both extremes were considered equally fundamental and intrinsic to the elements and the objects that they composed.

The concept of levity was dismissed when modern science (seventeenth c.) tried

to manipulate levity quantitatively. Galileo asked: do two light objects create a lighter one when combined?[1] The ensuing controversy dismissed the concept of levity and left a single quantifiable characteristic of heaviness—the weight of material objects.

Regarding weight, there were two explanatory conceptions discussed in Greek science. The first was attributed to Plato (428–348 BC). Plato introduced a very peculiar understanding of weight as a manifestation of attraction of bodies of similar kind:[2]

Plato

...the tendency of each towards its kindred element makes the body which is moved heavy, and the place towards which the motion tends below, but things which have an opposite tendency we call by an opposite name.

Though it looks to us rather more naïve than scientific, it was in fact a sort of anticipation of the reciprocal gravitational attraction between material bodies to be introduced by Newton 2000 years later.

Aristotle (384–322 BC) had a different perspective.[3] For him, weight was a part of cosmology —the theory of the organized world, cosmos. Weight manifested the *tendency* of objects to restore the violated order of elements with respect to the center of the universe. In the ordered state, the fundamental elements (earth, water, air, and fire) would be organized in radially arranged layers, from the center of the Universe outward, to the Heavens. Aristotle stated that material bodies permanently seek their respective states of rest and reach it to the extent they can. Their seeking this arrangement constituted the *teleological* cause of the natural motion of objects, while their weight designated the *efficient* cause of such motion.

Aristotle

Aristotle[4] ascribed absolute *weight* to earth (the element) and absolute *levity* to fire (the element), while the other two elements possessed a relative weight and levity. The weight of a compound object was in accordance with the ratio of its heavy and light components.

In the natural motion of objects, weight and levity served as the cause of motion— falling or ascending. Falling took place in accordance with the principle: the greater the weight—the quicker the motion. In contrast, in violent, unnatural motion, weight

---

[1] Galilei (1638/1914, p. 78).

[2] Plato (1952, *Timaeus*, 63, p. 463).

[3] Aristotle (1952, *On the Heavens*, Book II, Ch. 13, 295a, b, 296a).

[4] Aristotle (384–322 BC)—one of the greatest Greek philosophers—actually founded science as a discursive activity dealing with the organization of nature. He systemized the rules of logical reasoning and suggested the first scientific holistic picture of the world, which has been preserved and later modified throughout the whole history of science.

resisted the mover: the greater the weight, the less quickly the object would move. We express these regularities symbolically as follows:

$$v \propto \frac{F}{W} \ (\text{violent motion}) v \propto \frac{W}{R} \ (\text{natural motion}) \qquad (5.1)$$

Here v, the speed of motion; F, the intensity of the mover; W, weight of the body; R, resistance of the medium.

Two manifestations of weight were recognized: weight causes the natural falling of nonsupported objects, and weight causes the heaviness of the object, the downward pressure exerted by it on its support (a holder, when available). Celestial bodies in the Heavens were obviously not supported, yet they did not fall (Fig. 5.2a).

(a)   (b)

Fig. 5.2 (**a**) Ancient Egyptians did not share the vision of the Greeks: in their view, gods supported celestial bodies, preventing their falling. (**b**) The balance scale and weights used for weighing in Roman times. (Photos by the author)

Therefore, inferred Aristotle, celestial bodies were weightless, seemingly being comprised of a fifth weightless element—ether.

An alternative approach to weight appeared soon after Aristotle, in Hellenistic science. Euclid, who lived in the third century BC, understood a body's weight, practically, as the pressure of the body upon its support as measured by balance. This was the first operational definition of weight:[5]

Euclid

> *Weight is a measure of the heaviness and lightness of one thing, compared to another by means of a balance.*

The balance scale had served as an instrument for measuring weight long before the calibrated spring scale. It was also in use before any theoretical account of weight in physics was established; indeed, the balance had been used since the very dawn of human civilization. The two terms "gravity" and "weight" are often used without any differentiation. Yet, *gravity* stood for the qualitative idea of heaviness, while *weight* stood for the result of the procedure of measurement comparing

---

[5] Euclid (1959, p. 24).

between the heaviness of a certain object with that of the object chosen as a standard—the weight (Figs. 5.1 and 5.2b).

Archimedes (287–212 BC) related weight to other factors—forces. As a practitioner, he saw weight as the quality that competes with the buoyant force pushing upward the objects immersed in water and thus determining if the body floats or sinks. For example, he stated:[6]

Proposition 6. If a solid lighter than a fluid be forcibly immersed in it, the solid will he driven upwards by a force equal to the difference between its weight and the weight of the fluid displaced.

Archimedes

Or, consider a loaded lever, the applied force on one side of a lever competes with the weight on the other side. Archimedes established laws on both subjects under his name—the law of buoyancy and the law of the lever. Both provided the way for the quantitative account of the situation.

Medieval science preserved the Aristotelian interpretation of weight as an intrinsic *inclination* of a body, not as a force. Thomas Aquinas, a devoted follower of Aristotle during that period, rather originally elaborated on the distinction of weight from force:[7]

A thing moved by another is <u>forced</u> if moved against its own inclination; but if it is moved by another giving to it its <u>own inclination</u>, it is <u>not forced.</u> For example, when a heavy body is made to move downwards by that which produced it, it is not forced. In like manner God, while moving the will, does not force it, because He gives the will its own inclination (emphasis added).

When medieval scholars discovered that objects accelerate while falling, the original concept of weight was upgraded. Nicole of Oresme split weight into two components, the natural (habitual) *still-weight* (or *pondus*) which always remained unchanged, and the *actual gravity, accidental weight* (gravitas), reflecting the apparent rise in speed during falling. Indeed, within the Aristotelian framework, the increase in speed (the effect) was itself a testimony to the increasing weight (the cause of falling).[8] From the perspective of Aristotle, one may say that the dual medieval concepts of weight represented *potential* and *actual* gravity.[9]

Nicole of Oresme

Another interesting derivative of weight at that time was *positional gravity*. Historians inform us about the view belonged to Jordanus Nemorarius of the thirteenth century, the Dominican Master General, introduced the new concept:[10]

---

[6] Archimedes (1952, Book I, p. 540).

[7] Aquinas (1267/1952, Part I, Q. 105, Art. 4; Part I, Q. 2, Art. 3) This elaboration strongly connotes with the human context and hints to the historical origin of being natural in physics of the animated nature.

[8] Brown (1978, p. 180).

[9] Albert of Saxony (1959, in Clagett, 1959, pp. 139, 566).

[10] Moody and Clagett (1952, pp. 123–124), Chapter 8.

*This is the concept of "positional gravity" (gravitas secundum situm), which is the compo-
nent of the force of a body's natural gravity, directed along whatever path of movement the
body can take as constrained by its connections with other bodies in a single system.*

The explanation followed:

*Positional gravity is measured by that fraction of the body's natural weight determined by
the deviation from the vertical of its path of displacement; this "obliquity" in turn is measured by the ratio of the vertical projection of a small descent along the oblique path, to the
length of the corresponding oblique descent. The application of this concept of positional
gravity to problems of equilibrium involves the general principle that the force by which one
weight can balance another, on a lever or pulley system, varies according to the amount of
vertical descent which it can accomplish through a movement compatible with the constraint system. Similarly, the resistance offered by the other weight, on the opposite arm of
the balance or at the other end of the pulley arrangement, depends on the amount of vertical
ascent which would be involved in a movement compatible with the constraints.*

Years will pass and Galileo will use this positional weight as causing resistance
to motion and apply exactly this knowledge in his derivation of the new kind of
motion—inertial motion, which does not require either force or weight to support it
(Fig. 8.1).[11]

In medieval science, another concept also arose—*impetus*. It was defined as
weight multiplied by speed. Within the new theory of motion, Buridan and Oresme,
from Paris, used impetus to replace the concept of actual gravity in their descriptions of the acceleration of falling objects.[12]

The concept of weight changed considerably when the Earth lost its central position in the Aristotelian cosmos. In the new Copernican picture, the universe was left
without the geocentric symmetry essential for the Aristotelian definition of weight.
An urgent need arose in science to explain why all free-falling objects fall to the
ground. The old Platonic idea of attraction between "things alike" was revived, and
it justified the falling of objects toward the Earth. It was imagined that a similar
attraction may exist in the vicinity of other planets. This vision
replaced the Aristotelian tendency to seek the center of the universe.

Galileo, in the seventeenth century, followed the same path in
his learning. He began with the adoption of the medieval conception of motion and weight. In 1608, he suggested a way to measure
the difference between "dead weight," weight at rest (pondus), and
weight in motion (gravitas).[13] Yet, Galileo preserved the central
idea of Aristotle regarding weight and refined it by specifying the
nature of falling:[14]

Galileo

*I begin by saying that a heavy body has an inherent tendency to move with a constantly and
uniformly accelerated motion toward the common center of gravity, that is, toward the
center of our earth...*

---

[11] Chapter 8.

[12] Clagett (1959, pp. 541–556); Grant (1977, pp. 45, 53).

[13] Drake (1978, pp. 126–127).

[14] Galilei (1638/1914, p. 74).

However, the idea of weight as heaviness of a body was used in a manner closer to Archimedes. Galileo considered specific gravity and regarded weight as being proportional to the amount of matter in the object (akin to Newton's mass). The commitment of Archimedes to weighing experience brought Galileo to stating that:[15]

> ... as has been often remarked, the medium _diminishes_ the weight of any substance immersed in it...

How could weight diminish in water if it stands for the amount of matter which remains the same? Without question, this is an obscure comment. In fact, weight "reduction" indicates an operational connotation—the experience of weight. It is indeed easier to support a body immersed in water. Summarizing, Galileo's "mass," "gravity," and "weight" were used as very close synonyms.[16] In his descriptions, they all conveyed the same idea of load, burden, and heaviness measured through weighing while clearly refusing to proceed to a deeper theoretical analysis of the weight concept as "not really worthwhile" for the moment.[17]

Descartes

The picture of the weight concept prior to classical mechanics is incomplete without the original idea of Descartes,[18] who tried to explain weight in an essentially different manner. Descartes considered weight to be the residual centripetal push exerted upon a body by a vortex of surrounding matter of the medium ["matiere subtile"].

Descartes believed fine particles pervade the pores of all objects and the rest of the universe. Being in a permanent, very fast whirl, they experience _centrifugal_ tendency. Their radial tension _outward_, however, creates the effective _centripetal_ (inward) push on the bodies, making them "heavy" and compelling them to move (fall) toward the ground, which is toward the axis of rotation. Descartes illustrated that idea by a thought experiment: A large bowl of gun pellets with a few pieces of light cork among the metal balls. During the rotation of the bowl, the pieces of cork move to the center of rotation because the metal balls move outward.

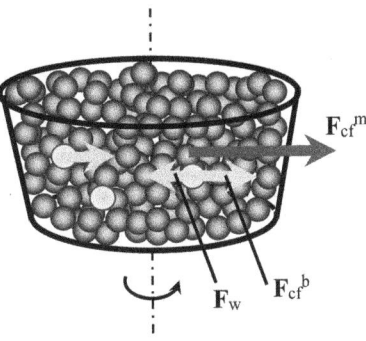

Fig. 5.3 Descartes' model to illustrate creation of the centripetal attraction on objects downward

Descartes' model for the creation of gravitational pull on objects toward the axis of rotation (Fig. 5.3) involved the particles of the medium in a rotating container. The particles experience the radial outward tendency (conatus, or the centrifugal force, $\mathbf{F}_{cf}^{m}$). The particles of immersed bodies (light balls in the figure) also experience radial outward tendency—$\mathbf{F}_{cf}^{b}$. This latter outward tendency is, however, smaller than $\mathbf{F}_{cf}^{m}$. The resultant influence on the immersed body is effectively in the

---

[15] Galilei (1638/1914, pp. 75, 77–78, 81).

[16] Jammer (1957, p. 100, 126).

[17] Galilei (1638/1914, p. 166).

[18] Descartes (1644/1982, pp. 191–192).

inward direction, creating weight force, $\mathbf{F}_w$ toward the axis of rotation. This was the mechanism of weight creation in the view of Descartes.

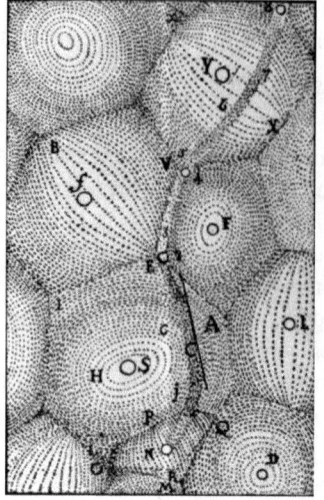

Needless to say, the situation represented in this thought experiment of Descartes does not even remotely resemble the reality around us, since real bodies are surrounded by air, a much less dense material. Yet, Descartes was not discouraged. Above all, he sought for the mechanism which could somehow explain the observed phenomenon "in principle" and found such an idea in the residual centripetal push, which he held and developed (Fig. 5.4) regardless of reality.[19] That is the salient feature of his rational method to take into account when one considers the physical claims of Descartes.

Fig. 5.4 Descartes' drawing from his treatise *The World* showing his vision of the space of the solar system filled with vortexes of the fine matter in which all the regular bodies (Sun and planets) are immersed

Questions for Reflection

1. Why can we talk about the heaviness of the bodies as their physical characteristic and cannot do the same with levity?
2. Aristotle did not consider weight to be a force but a tendency of a body. What was the difference?
3. Scholars of medieval science were not satisfied with a single concept of weight (gravity) and distinguished between *still-weight* always unchanged, and *actual gravity* or *accidental weight*. What was the rationale of this division?

Fig. 5.5 Distribution of matter in a spiral galaxy

4. What was the mechanism by which Descartes tried to explain weight? Was it reasonable in our view? Explain.
5. Figure 5.5 depicts a matter whirlpool in a huge galaxy as seen through the space telescope Hubble (NASA). It might look just like a vortex that Descartes talked about in his theory. Could it be evidence supporting Descartes? Could this vortex explain weight?

Weight in Newtonian Mechanics

Following Galileo and Kepler, the search for the cause of gravity shifted from the terrestrial to the astronomical context. The solar system, in a sense, provided a simpler physical system for an initial understanding of gravity as it was free of the

---

[19] Aiton (1959).

major distractor—friction and contact forces. Huge distances separated big objects (planets and Sun) which still created one system. The solar system was scrutinized by Newton in England as it had never been before.

Firstly, Newton introduced a special framework of the universe's organization: all processes were related to specific forces.[20] The great conjecture of Kepler that all bodies in the universe apply mutual attraction—gravitation—was adopted as an idea.[21] Newton's approach was, however, more fundamental, mathematically and conceptually advanced, and correct. He established an inclusive theory based on a new paradigm of the world organization. Its nucleus included three basic laws of motion and the basic interaction between bodies—their gravitation.

The prevailing motion in the solar system was apparently the circular motion.[22] Newton treated it in the new way. Within the new paradigm of mechanics, a special force—a centripetal force, acting on the rotating body toward the center of rotation—stipulated the circular motion. The introduced centripetal force (F_{cp}), he elicited, was "decreasing always in the duplicate proportion of the distances" (the inverse proportion to the square of the distance to the center—r_{12}):[23]

$$F_{cp} \propto \frac{1}{r_{12}^2} \qquad (5.2)$$

It was a central force (that is, directed along the straight line connecting the bodies) acting on each planet. This kind of force was generalized into the Law of Universal Gravitation—the force between each pair of material bodies. It was this force that Newton related to the familiar feature of gravity acting on each body around us, placed on the ground. It was far from obvious to relate these two forces of different manifestations: the centripetal force on planets and the feature of the gravity of regular objects. This step was elaborated in an imaginary scenario in which he considered the Moon gradually descending to become an object next to the Earth's surface:[24]

> If now the moon is imagined to be deprived of all its motion and to be let fall so that it will descend to the earth with all that force urging it by which (by Cor. Prop. III) it is [normally] kept in its orb... that force by which the moon is kept in its orbit in descending from the moon's orbit to the surface of the earth comes out equal to the force of gravity here on earth, and so (by rules I and II) is that very force which we generally call gravity.

How could Newton make any inference regarding the force acting on the Moon or any other celestial body? In doing that, he firstly calculated the centripetal acceleration for the Moon's observed motion, its observable "falling" under the assumed action by the force in according with Eq. 5.2 for the distance to the Earth's center. Then, Newton compared the obtained acceleration of the Moon with the

---

[20] Chapter 2.

[21] Kepler (1992, p. 55), Koyre (1943), Hecht (2019).

[22] In our line of conceptual presentation, we do not need here the more accurate Kepler's law of elliptical orbits.

[23] Newton (1687/1999, Book III, Proposition I, II, III, pp. 802–803). Newton's treatment was in contrast to the Kepler's inverse dependence of gravitation on the distance, nonsymmetrical in magnitude and considered as a magnetic faculty.

[24] Newton (1687/1999, Book III, Proposition IV, p. 804).

acceleration reported by Galileo by which a body next to the ground falls under the Earth's attraction at the distance of its radius. The ratio of the two accelerations matched the square of the inversed ratio of the distances. This fact proved the inversed squared dependence of the attraction force on the distance from the center of the Earth (Eq. 5.2).[25] In this analysis, Newton, applied the principle from his *Rules of Reasoning*,[26] when he concluded that the force exerted on the Moon and the force acting on the bodies next to the Earth are the same force:[27]

> And therefore (by Rule I and II) the force by which the moon is retained in its orbit is that very same force which we commonly call gravity;

These rules were:[28]

> Rule I: We are to admit no more causes of natural things than such as are both true and sufficient to explain their appearances.
>   Rule II: Therefore, to the same natural effects we must, as far as possible, assign the same causes.

He then reasoned by the fallacy of the contrary assumption:[29]

> For if gravity were different from this force, then bodies making for the earth by both forces acting together would descend twice as fast.

Indeed, a different celestial attraction force would add to the action by the Earth. The final claim appeared in the Scholium:[30]

> The force which retains the celestial bodies in their orbits has been hitherto called centripetal force; but it being now made plain that it can be no other than a gravitating force…

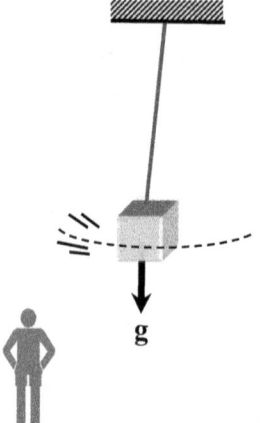

$$F_{grav} \equiv F_{cp} \qquad\qquad (5.3)$$

This was the moment of a great discovery. The hitherto inherent faculty of objects—gravity, or weight—from now on became *the force of universal gravitation* or as we call it now—*the gravitational force*. Weight, or gravity, was now equated to the gravitational force.

Newton proceeded with the important feature of weight:[31]

> …That all bodies gravitate towards every planet; and that the weights of bodies towards the same planet, at equal distances

g

---

[25] Though it looks as a straightforward proof, its realization was, in fact, not immediate for the problems with accuracy (e.g., Brooks, 2010, p. 36).

[26] Newton (1687/1999, Book III, Rules of Reasoning in Philosophy, pp. 794–796).

[27] Newton (1687/1999, Book III, Scholium, p. 805).

[28] In effect, these both rules—rules I and II—described the principle of Parsimony—not to multiply concepts, to manage with a minimum number in accounting for natural phenomena. The Principle of Parsimony (simplicity) is known as Ockham's Razor (the rule of economy, or parsimony) from the fourteenth century.

[29] Newton (1687/1999, Book III, Proposition IV, p. 804).

[30] Newton (1687/1999, Book III, Scholium, p. 806).

[31] Newton (1687/1999, Book III, Proposition VI, p. 806).

from the centre of the planet, are proportional to the quantities of matter which they severally contain.

We see here, firstly, the relational nature of weight: "the weight of bodies towards..." and, secondly, the proportionality of gravitational force to the quantity of matter (inertia). The latter was demonstrated by Newton through experimenting with pendulums of equal geometry and shape but differing in material. He wrote:[32]

> *I tried the thing in gold, silver, lead, glass, sand, common salt, wood, water, and wheat. I provided two wooden boxes, round and equal: I filled the one with wood, and suspended an equal weight of gold (as exactly as I could) in the centre of oscillation of the other. The boxes hanging by equal threads of 11 feet made a couple of pendulums perfectly equal in weight and figure, and equally receiving the resistance of the air. And, placing the one by the other, I observed them to play together forward and backward, for a long time, with equal vibrations.*

Thus, by showing that gravitating force does not depend on the kind of matter (exactly the same period of oscillation), Newton concluded the gravitational force to be proportional to what he called the quantity of matter, and we, the inertial mass:

$$F_g \propto m \qquad\qquad (5.4)$$

Finally, for the symmetry of force interaction (Law III of mechanics), one should add another mass into the dependence of the attraction force of gravitation:

$$F_g \propto m_1 \cdot m_2 \qquad\qquad (5.5)$$

If one combines these results, the famous Law of the Universal Gravitation emerges:

$$F_g \propto \frac{m_1 \cdot m_2}{r_{12}^2} \qquad\qquad (5.6)$$

Newton made another important modification. Until Newton, mass and gravity were considered to be a single concept. The medieval concept of *impetus* was defined as the product of *weight* and *speed*, whereas momentum—the quantity of motion for Newton—was a product of *mass* and *velocity*. It was the revolutionary split of gravity (weight) and mass.[33] The gravitational force was proportional to the mass, not the "weight of the body."

Newton could not accomplish the law of gravitation (Eq. 5.6) beyond the proportionality for he could not measure the gravitational force between two known masses in the laboratory. He could not infer the unknown mass of the Earth from the celestial context.

Only 100 years later, Cavendish, who like Newton worked in Cambridge University, measured the gravitational forces between two lead balls with known

---

[32] Newton (1687/1999, p. 807).

[33] Jammer (1957, p. 118).

mass. The experiment was named, "the weighing of the Earth" (Fig. 5.6) since it allowed calculating the mass of the Earth given the acceleration of a free fall and the radius of the Earth.

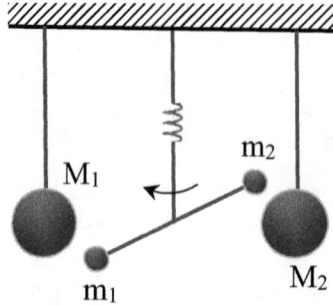

Fig. 5.6 The schematic idea of the experiment by Cavendish in 1798. In the torsion balance, the attraction forces between lead balls in two pairs M_1m_1 and M_2m_2 ($m = 730$ g, $M = 158$ kg) created torque on the balance shaft causing the twist in the suspending cord. The very small turn angle of the beam was measured. The experiment allowed the calculation of the constant of gravitational attraction G in the Newton law of gravitation

After Cavendish, physicists could write (Eq. 5.6) as:

$$F_g = G \frac{m_1 \cdot m_2}{r_{12}^2} \tag{5.7}$$

The *identity* between celestial attraction and the weight (heaviness) of objects on the Earth seemed natural to Gilbert, Descartes, Huygens, and of course, to Newton. It took over two centuries before the equating of the cause (the gravitational force) to its effect (the weight of the object) became peculiar enough to merit further inquiry.

Following Newton's understanding of the nature of weight as caused by the interactive force of gravitation, the Newtonian weight was always relational:[34]

>...the weights of the planets <u>towards the sun</u> must be as their quantities of matter....

The weight of the Earth toward the Sun was equal to the weight of the Sun toward the Earth, and the weight of the Earth toward the Moon was equal to the weight of the Moon toward the Earth. The weight of the Earth toward the Sun was different from the weight of the Earth toward the Moon. Newtonian weight was not a characteristic of one body, but of a pair of bodies. Weight ceased to be a characteristic of the object itself, while mass remained as such.

The Newtonian definition of weight could not be useful in practice where the only meaning was the weight of things toward the Earth. Newton did not forget to define weight operationally in a procedure of weighing:[35]

>*Thus the weight is greater in a greater body, less in a less body; and, in the same body, it is greater near to the earth, and less at remoter distances. This sort of quantity is the centripetency, or propension of the whole body towards the centre, or, as I may say, its <u>weight</u>; and*

---

[34] Newton (1687/1999, Book III, Proposition 6, pp. 806–810).

[35] Newton (1687/1999, Definition VIII, pp. 407–408).

it is always known by the quantity of an equal and contrary force just sufficient to hinder, the descent of the body (emphasis added).

Here, however, problems began. What about the relation of weight to the weighing actually mentioned here? On the one hand, weight was defined by Newton as the gravitational force on the object. On the other, weight was measured by weighing, sensitive to the supporting force preventing the falling of that object. Are both quantities always the same?

What we observe is that despite the great elucidating impact, the gravitational identity of weight was never perfect. It was known that the same body weighs differently at different latitudes on the surface of the Earth. How could one explain that? Does it indicate a different attraction to the ground? The correctness of the equating weighing result to the gravitational force was questioned.

The discrepancy was resolved in trade and commerce by careful indication of the place where the weights and spring-based weighing tools were calibrated. The necessary corrections at various geographical locations were tabulated. Yet, the Newtonian definition of weight as the gravitational force on the object preserved while waiting for a better account.

In fact, this problem with weighing indicated the much more general limitation of validity of Newton's laws which hold only for certain observers—inertial frames of reference. The rotating Earth was not such a frame. To apply Newtonian theory correctly, one needs to imagine oneself outside the Earth, being at rest relative to the stars. In that sense, it was known that weighing does not reliably indicate gravitational force.

The comprehensive understanding of weight was reached in the twentieth century with a new approach incorporating other observers.[36] Newton's concern was rather different. He professed one "true" picture of the world—the one reached in terms of absolute space and time.

What really bothered Newton was the unknown origin of the gravitational force. In his *Principia*, he refrained from any speculation regarding its origin. He summarized his vision in the last lines of this seminal treatise:[37]

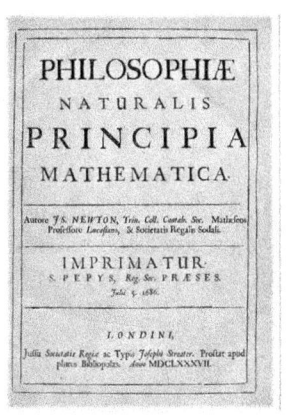

Hitherto we have explained the phenomena of the heavens and of our sea by the power of gravity, but have not yet assigned the cause of this power. This is certain, that it must proceed from a cause that penetrates to the very centres of the sun and planets, without suffering the least diminution of its force; that operates not according to the quantity of the surfaces of the particles upon which it acts (as mechanical causes used to do), but according to the quantity, of the solid matter which they contain, and propagates its virtue on all sides to immense distances, decreasing always in the duplicate proportion of the distances. ...

But hitherto I have not been able to discover the cause of those properties of gravity from phenomena, and I frame no hypotheses; for whatever is not deduced from the phenomena

---

[36] Chapter 4.

[37] Newton (1687/1999, General Scholium, p. 943).

is to be called an hypothesis; and hypotheses, whether metaphysical or physical, whether of occult qualities or mechanical, have no place in experimental philosophy. In this philosophy particular propositions are inferred from the phenomena, and afterwards rendered general by induction. Thus it was that the impenetrability, the mobility, and the impulsive force of bodies, and the laws of motion and of gravitation, were discovered. And to us it is enough that gravity does really exist, and act according to the laws which we have explained, and abundantly serves to account for all the motions of the celestial bodies, and of our sea.

Newton left the problem of understanding gravitation for further exploration which has continued ever since.

Euler

Leonard Euler[38] lived in the eighteenth century just after the scientific revolution, at the time when Newtonian mechanics became the most developed fundamental theory in physics. In his treatise on mechanics, Euler presented Newton's ideas about weight as an already adopted claim of physical science, although in reality the scientific community still debated the validity of Newtonian gravity (action at a distance) versus the Cartesian idea of vortexes which caused weight by the pressure from the surrounding ether. Euler wrote:[39]

Euler

Definition 16
179. Gravity is the force, by which all bodies near the surface of the earth are forced downwards; and the force, by which any body is acted on by gravity, is called the weight of this body.
Corollary 1
180. Gravity is the external cause, which forces terrestrial bodies downwards; and therefore, it <u>cannot be a property assigned to certain bodies themselves.</u>
Corollary 2
181. Thus a body sent off near the surface of the earth, even if it should be at rest, is urged on in a downwards motion and meanwhile it sinks until it comes upon obstacles preventing the fall.
Corollary 3
182. Moreover as long as the fall is impeded, either the body being held immobile pressing on an object or it suspended, the <u>weight of this body</u> exerts itself by pressing down.
(emphases added).

In accordance with weight as defined by Newton, Euler mentioned that gravity "cannot be a property assigned to certain bodies themselves" (Cor. 1), meaning that

---

[38] Leonhard Paul Euler (1707–1783) was an outstanding Swiss mathematician and physicist who worked in Germany and Russia. Euler made important discoveries in mathematical analysis (he introduced the notion of function—central in mathematics and physics). He is renowned for his contribution to mechanics, fluid dynamics, optics, and astronomy.

[39] Euler (1765/2009, p. 127).

weight is due to a pair of objects, not one. Almost like Newton, Euler ascribes to weight the cause of the downward pressure (Cor. 3).

Euler defined the force of gravity, or weight, similarly to Newton. No cause is mentioned. His confusion is expressed in the Scholium. Euler speculates and mentions that in the absence of material evidence, the Cartesian idea of gravitation by "the action of some more subtle matter that escapes the notice of our senses" might be somewhat better than the Newtonian lack of any account but the supernatural:[40]

> ***184***. *Therefore the force of gravity is not a material force acting on the body, truly thus con-nected with the earth, in order that with this removed, the force likewise would vanish; and likewise it is therefore as if a certain spirit should move rapidly to force bodies downwards; for how otherwise the force itself is able to propagate through great distances without the support of any kind of intermediate material, cannot in any manner be considered to be understood. …What is perhaps more likely to be true is that <u>the force of gravity arises from the action of some more subtle matter that escapes the notice of our senses;</u> …when the admirers of attraction say that the attractive force has been put in place by a God of the earth, they say nothing else, except that bodies are to be impelled immediately by this god himself* (emphases added).

Euler went further in establishing a working framework for physics. He demon-strated the possibility of using weight for measuring forces and masses:[41]

> ***191***. *We may express the forces acting consistently through the equal weights from these.*
> ***192***. *This expression of the forces by the weights gives no difficulty; for since the weight of each body is a force, by which that is acted on downwards, the forces acting and the weights are quantities homogeneous between each other; and whatever body may be acted on by some force, … so that just the weight of this body will show the measure of that force.*

It is within this tradition that we calibrate a force-meter in the school laboratory by suspending different weights on a spring and use such a calibrated spring to measure other forces.

Huygens

The impact of Newton on physics was enormous. However, there was another brilliant mind at work at the same time, Christian Huygens,[42] a prominent Dutch, natural philosopher who worked almost in parallel with Newton and produced alternative ideas

Huygens

---

[40] Euler (1765/2009, p. 128).

[41] Euler (1765/2009, p. 131).

[42] Christian Huygens (1629–1695) was a renowned Dutch physicist who invented the first pendu-lum clock, which greatly increased the accuracy of time measurement. He was a pioneer researcher in mechanics, optics, astronomy, and mathematics.

regarding the fundamental issues in mechanics and optics.[43] In the previous chapter, we already elaborated on his fundamental idea of centrifugal force and its similarity in nature with the force of gravity.

Huygens' view was highly original, and his *centrifugal force*[44] was not interactive. Huygens was the first to describe a physical situation from the point of view of the observer in a rotating system. In Chap. 4, we saw how Newton removed Huygens' approach by reducing centrifugal force to the interactive pair centripetal-centrifugal. It was not the vision of Huygens who, unlike Newton, considered centrifugal force as exerted on the rotating body, not as on the agent causing the rotation. Today, physicists distinguish between the descriptions of the situation by observer N, at rest (Fig. 5.9), and that made by observer H, on the rotating wheel. Observer N, the only one considered later by Newton, presents an inertial observer. Observer H, introduced by Huygens, needs an additional force −ma to nullify the net force on the ball, at rest for H. This additional force (−ma) does not exist for N(ewton), but it is required by H(uygens)—the *inertial force*. Observer H is a *non-inertial observer* and the rotating wheel presents a *non-inertial frame of reference*.

Furthermore, lacking Newton's theory of gravitation and laws of motion, Huygens' treatment was necessarily limited. Nevertheless, he made an ingenious conjecture, by which he was able to demonstrate that gravitational force (gravity) was inherently similar to centrifugal force. Were he to know the formulas for both forces, as we do today, the claim would be obvious since he would immediately see that both are proportional to the mass of the object, and therefore they are essentially similar. This implies that observer H might be misled regarding the gravitational force while interpreting the result of weighing, that is, the tension in the thread, the heaviness of the suspended body at the state of rest (mg* in Fig. 5.7). Yet, Huygens could draw only on Galileo and Descartes.

Unlike Newton, Huygens defined gravitation by its manifestation in the tension of the thread tied to the ball: *Heaviness is a tendency to fall.* Two forces contribute to this tension: gravity and centrifugal. Huygens showed that they were identical in nature in a special way. He considered what would happen if one cuts the rope and the ball flies along the tangent (Descartes' second law of motion). Huygens showed that the radial distance, from the ball to

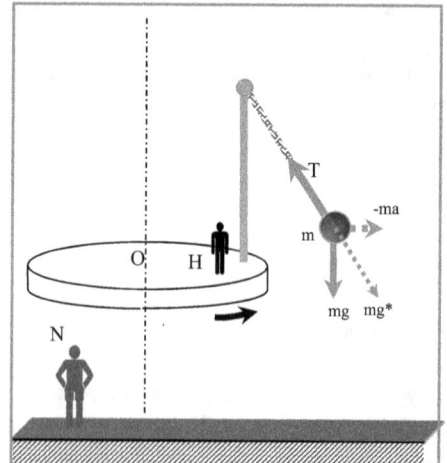

Fig. 5.7 Observers N on ground (at rest) and H on the rotating disc account for the conical pendulum m on the disc

---

[43] In optics, Huygens confronted the idea of particle (light rays) paradigm with his wave theory of light (elastic distortions in the ether medium).

[44] Huygens (1659/1703).

the center of the wheel, would increase the rate with which a body increases its distance in falling.[45] That sequence—1, 4, 9, 16, etc.—was obtained by Galileo in his study. Huygens concluded regarding the identity of the two forces: centrifugal force and gravity.

Huygens' treatment of the centrifugal force was truly remarkable in the mechanics of the seventeenth century. It was highly appreciated by his contemporaries, including Newton, who rarely praised his colleagues. He wrote[46]:

> And by such propositions, Mr. Huygens, in his excellent book De Horologio Oscillatorio, has compared the force of gravity with the centrifugal forces of revolving bodies.

Yet, Newton and all other scholars of that time viewed the world as a theater with the action on a unique stage—a universe of absolute space and time. The legitimacy of other observers was dismissed, and so, the centrifugal force and its contribution to weight was ignored, no need of it. Although praised, Huygens' work was never truly evaluated, not even by Huygens himself. The time was not yet ripe for another perspective which would change it all.

What Is Gravity? The Answer According to Classical Mechanics

Following the ambitious idea of Descartes, Huygens tried to explain gravitational force by means of centrifugal force created due to the medium in which our world is immersed. In a way, Descartes, in his model (Fig. 5.3), revived the Aristotelian idea of antiperistasis[47] and the Archimedean mechanism of buoyancy:[48]

> The mechanism envisaged by Huygens involved a fluid vortex rotating at such a high speed that all bodies on the earth are pushed toward its center because their centrifugal force is smaller than that of an equivalent volume of the vortex. Thus gravity results from a difference of centrifugal forces, and in this sense centrifugal force does produce motion, i.e., every time a heavy body falls.

This attempt to see the gravity of bodies as caused by a residual centrifugal force was, however, deficient. Huygens' attempt failed, just as it had failed for Descartes and would fail later for Leibnitz. Newton removed this scenario. He converted the centrifugal force (from the force on the body) to the force fitting his paradigm of force interaction (the force on the agent). For him, the centrifugal force was the companion of the centripetal force acting on the constraint (the sling). Discussing the body rotating inside a hollow cylinder he wrote:[49]

---

[45] Chapter 4.

[46] Newton (1687/1999, Book 1, Section II, Prop. IV, Theorem IV, Scholium, p. 452).

[47] Chapter 2.

[48] Meli (2006).

[49] Newton (1687/1999, p. 453).

This is the centrifugal force with which the body urges the circle; and the opposite force, with which the circle continually repels the body toward the center, is equal to this centrifugal force.

In practice, however, it was Huygens' meaning of centrifugal force that survived. This is because the Earth is, in effect, a gigantic revolving wheel as considered by Huygens. By virtue of the Earth's revolution, the results of weighing change with latitude. Obviously, the consumers in trade and technology could not ignore this, on the one hand (as it created significant loss), and could not know anything about observers, on the other hand, to explain the discrepancy in the weighing results (assuming of course, fair commerce).

The free fall acceleration at the equator is less than elsewhere and so the weight, mg, is less there. The period of a pendulum's oscillations: $T = 2\pi\sqrt{L/g}$ is longest at the equator. Huygens' framework provided an easy explanation. This fact, however, did not change the Newtonian framework. Even so, Newton himself used centrifugal force in Huygens' sense in explaining the Earth's being flattened at the poles, it was simply easier. Newton's vision on weight, identified as gravitational force, remained undisputable especially in the general picture of the world in public understanding. The difference in meaning between gravity (heaviness) and gravitation (interaction) did not exist. *Force of gravity* meant *force of gravitation*.

Weight became even more important after Lavoisier's discovery of matter conservation in chemical reactions. It was stated as *weight conservation* (instead of *mass conservation*), matching the practical evidence of such constraint. Atomic weights were used in the sense of mass. Weight became central in the new organization of elements in the Mendeleev periodic table.[50]

Not before the nineteenth century did questions arise, along with the dawn of modern physics. Still, the general vision of the world by a unique observer was preserved although physicists began to question the foundation of the Newtonian framework. Ernst Mach demanded an empirical (operational) definition for physical concepts (he started with inertial mass).[51] Yet, the big perspective he held did not change, and he never asked about weight-gravitation identity.[52] As long as the old framework was preserved, all deviations in weighing, including falling (zero weighing result!), were explained within the Newtonian framework and adjusted to all practical needs. That was done by the conceptual split of notions to *true* weight (gravitational force) and *apparent* weight (the result of weighing). The strong connotation of the former to be "true" implied the latter to be a sort of masked truth. That changed in the conceptual reconstruction within the new scientific revolution.

Questions to Reflect on and Discuss

1. How did Newton show that gravitational force is proportional to the quantity of matter (inertial mass)?

---

[50] Merz (1904/1965, p. 395).
[51] Mach (1883/1919, pp. 216–218).
[52] Galili (2019).

2. Consider in what way Newton changed the medieval concept of weight.
3. What mechanism for gravity did Descartes suggest? Was it plausible? Explain.
4. Characterize Newton's concept of weight. What features of the Newton's weight were abandoned?
5. Why was Newton not satisfied with his theory of gravitation? Why did he publish that theory?
6. What was, in Euler's view, the condition needed for weight to manifest itself?
7. What mechanism for gravity did Euler imagine to himself?
8. Huygens demonstrated that gravity force was similar to the centrifugal force. What strategy did he use?
9. What was the interpretation of the centrifugal force by Newton? How did this differ from Huygens' interpretation?
10. Huygens was the first to ask about the account by a rotating observer. Why was this approach abandoned?

Weight in Modern Physics

Einstein: The Principle of Equivalence

A great change took place in physics at the beginning of the twentieth century. Physicists revealed the special role of *observer*. It was Albert Einstein[53] who, in 1905, put an observer into the center of the physical account of the world with his demand for the laws of physics to be indistinguishable for inertial observers (Galileo's principle). This demand, together with the universality of the speed of light, brought to relativistic physics its first step—the *special theory of relativity*. It changed the foundations of mechanics—its concepts of time, space, mass, and, hence, of many other concepts (energy, force, etc.). Yet, in a sense, the program was not accomplished. The special status of inertial observers still looked unnatural to Einstein. His ambition was to obtain a physical picture valid for *any* observer. This led him to the *general theory of relativity* in 1916. This unification was reached due to the new idea—to build a physics account produced by an observer in a "closed" laboratory, that is, drawing on the tools in the laboratory. It was the divorce from metaphysics—the unfounded assumptions of the theory which do not correspond to reality.

Einstein

As the first step toward the new approach, we may consider the thought experiment of Huygens and his *vis centrifuga*. Einstein completed the program which Huygens introduced in considering a conical pendulum on the rotating wheel (Fig. 5.7) in the account by the observer H on the wheel. If observer H had no knowledge about the world outside, he could infer only the effective gravity through its local measurement. This was the application of operational definition. As

---

[53] Albert Einstein (1879–1955), the outstanding physicist and physics philosopher who shaped modern physics in the twentieth century.

Huygens showed, the inertial (centrifugal) force was identical in effect to gravitational force.

The description of nature by a non-inertial (accelerating) observer can be reduced to the account made by an inertial observer given that the latter introduces inertial forces and considers them as gravitation.[54] This idea was the *principle of equivalence*:[55]

> There is no experiment observers can perform to distinguish whether acceleration arises because of a gravitational force or because their reference frame is accelerating.

With regard to our subject, weight, it became clear that Einstein's principle of equivalence implied a fundamental change to this concept. Here is how Einstein presented this principle. He performed a thought experiment, known as *the experiment of the accelerated elevator* (Fig. 5.8). This experiment should guide our reconstruction of the weight concept. Reichenbach[56] wrote:[57]

> *Imagine a box of a room size, in which a physicist suspends a spring with the attached weight. The box has no windows and sits on the ground. Suppose that the box is being pulled up by a rope, like an elevator, in the direction of arrow **a**. Would the physicist inside notice it? – Yes, he would. Indeed, due to its inertia **m** [the weight] would remain slightly behind the motion; the length of the spring would increase a little, accompanied by an increase in its tension. The accelerated movement would thus result in lengthening of the spring. Note, that if the motion were at a constant velocity, no expansion of the spring would take place. This follows Galileo's principle of relativity.*
>
> *Now, says Einstein, was it necessary that the physicist inferred that the box moved? Certainly, not. This is because there is another scenario that could explain the extension of the spring. The other cause that could produce the same effect is gravitation. Indeed the same extension of the spring would happen if we assume that for some reason the mass of the planet suddenly increased. Its attraction would act on the weight in the direction of the arrow **g** and pull it down. Therefore, from the observed lengthening of the spring the physicist could not decide what exactly happed. The point is, as Einstein stated, that there is no other way of distinguishing between these the two possibilities. No experiment within the box could differentiate between gravitational attraction and an accelerated motion... Two entirely different phenomena, inertia and gravitation, are placed here parallel to each other and either of them leads to the same effect, namely, to the increased tension of the spring...*

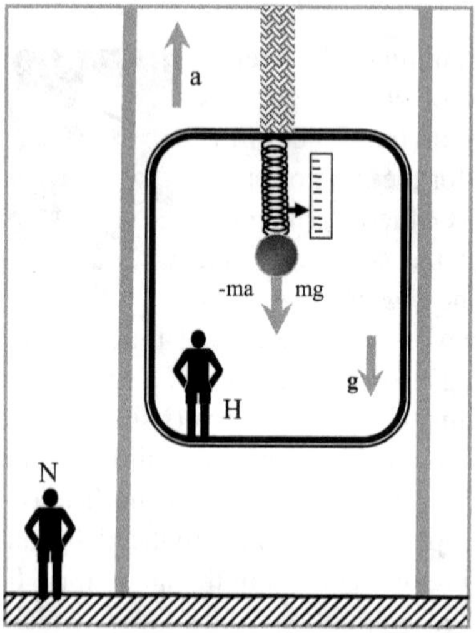

Fig. 5.8 The thought experiment by Einstein, weighing in an accelerating elevator

---

[54] Einstein (1916/1923), Born (1962), Wheeler (1990).

[55] Giancoli (1988, p. 155).

[56] Hans Reichenbach (1891–1953) was a prominent philosopher of science associated with the Vienna Circle group, who developed the new philosophy of science—logical positivism.

[57] Reichenbach (1927/1942, pp. 86–89).

There is another way to demonstrate the meaning of Einstein's claim. As we learned from Newton, any two bodies gravitate toward each other with forces proportional to their inertial masses m_i. In principle, one could expect that bodies attract each other in proportion to their gravitational masses m_g: after all, the phenomenon of gravitation is different from the context of accelerated motion in general. Accordingly, we may infer regarding the acceleration of one of the bodies while writing its equation of motion:

$$m_i a = G \frac{m_g \cdot M_g}{r^2} \text{ and } a = \frac{m_g}{m_i} G \frac{M_g}{r^2} \tag{5.8}$$

This outcome fits the famous empirical result by Galileo with regard to the free falling of objects. The fact that different bodies fell with the same acceleration indicated that their gravitational and inertial masses were proportional (or equal, if the units are selected appropriately).[58] In fact, this is the meaning of the constant G. We thus obtain the acceleration of free fall (labeled as g):

$$a = G \frac{M_g}{r^2} \equiv g \tag{5.9}$$

The practical equivalence of gravitational and inertial masses $m_i = m_g$ was discovered by Newton in real experiments. Why should these masses be equal and bodies fall with the same acceleration? The explanation was no more than: it is so. Newton could do no better than that and remained merely surprised. The same principle was behind the thought experiment of Huygens. Reichenbach wrote:[59]

Reichenbach

Although the equality of the inert mass and the heavy [gravitational] mass was long known, nevertheless Einstein was the first man to recognize the basic significance of this fact. He realized that here lies the reason why the distinction between accelerated motion and gravitation cannot be made and why the physicist in the box cannot, therefore, determine whether he is moving upward in an accelerated motion or whether the gravitational field has acted upon it. Hence, Einstein calls both conceptions equivalent, and maintains that it is meaningless to look for a truth-distinction between them.

Yet, Einstein surpassed this enigma. In his theory, he explained the enigma by establishing a very special connection of matter with the nature of space-time which appeared to be in mutual determinative influence:

---

[58] This is in approximation of the infinitely large Earth mass, which is a very good one in the case of regular objects next to the ground. For the significantly different masses, the situation changes (Lehavi & Galili, 2009).

[59] Reichenbach (1927/1942, p. 93).

The same entity – mass – determines the nature of the space-time and the motion of objects in it.

Multiple experiments performed by physicists, and the applications in technology made since then proved this theory to be correct.

New (Old) Definition of Weight

As we see, the identity of inertial and gravitational masses, rediscovered in modern physics, implies uncertainty in the interpretation of the procedure of weighing in the sense that that the results of weighing cannot provide certain inferences regarding the actions of gravitational force. In this situation, there is no choice but to change the definition of the concept.

In 1928, soon after the introduction of the principle of equivalence by Einstein, Reichenbach wrote:[60]

> *What is the basis of this indistinguishability? According to Einstein, its empirical basis is the equality of gravitational and inertial mass. This new distinction must be added to the usual distinction between mass and weight. There are therefore three concepts: inertial mass, gravitational mass and weight.*

Newton's distinction between mass and gravitational force became insufficient. Now there was a need for further refinement—to distinguish between gravitational force and weight force. After more than 200 years of partnership, gravitational force was conceptually divorced from weight (Fig. 5.9). Weight of a body was now defined using the theoretical definition:

> Weight of the object is the force that it exerts on its support at the state of rest as claimed by an observer in the correspondent system of reference.

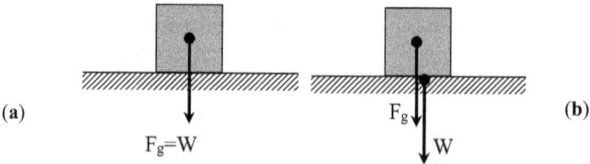

$$F_g = W$$
$$F_g \quad W$$

(a)   (b)

Fig. 5.9 (a) Old definition of weight: weight force, W, is the gravitational force on the object, F_{gr}. (b) New definition of weight: weight force, W, is the force on the support

In accordance with this definition, weight is the force that is measured by the calibrated spring, exactly as the observer in Einstein's thought experiment of elevator measured it. Next to this theoretical definition, we may provide the complementary operational definition:

> Weight is the result of standard weighing.

---

[60] Reichenbach (1928/1958, p. 223).

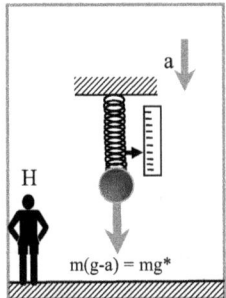

Fig. 5.10 Weighing in an accelerating elevator

This definition is coherent with the inability of the observer to know whether the origin of the weight he measures is due to gravitation or accelerated motion of the laboratory. For him:

$$W = F_{\text{elastic}} = mg* \tag{5.10}$$

where $g*$ signifies the effective acceleration of falling as measured by the *internal* observer H (Fig. 5.10). The *external* observer N (Fig. 5.8) may interpret weighing differently, such as:

$$W = F_{\text{elastic}} = m(g - a) \tag{5.11}$$

The two observers interpret the same reality differently (unless $a = 0$). Yet, they obtain the same magnitude, the same weight, sustained by the support.

Thus, we say here: "Weight is the force sustained by the support." Compare this with similar words used by Newton "...the weight sustained by the bottom of the vessel..."[61] and see the difference. For Newton, the weight is sustained which means that another force is acting on the support, not the gravitational force, but due to gravitational force. The operational definition states that force on the support is the weight of the object.

The situation is especially interesting in the case of free-falling: $a = g$. This is the state of weightlessness inside the laboratory: $W = 0$. The external observer may explain it by the cancellation of inertial and gravitational forces. The non-internal observer H may remain ignorant regarding the origin of weight or its absence, reflecting the inability to distinguish between inertia and gravitation.

This approach moves physicists to define weight in its original sense of heaviness. The origin of weight may be unknown. Importantly, the force of weight is spread across the body and increases in the downward direction (Fig. 5.11). In fact, it is this force which defines the downward direction. Weight causes the pressure which each layer inside the object exerts on the layer supporting it. The gradient of tension within the body accompanies weight. This is why magnetic boots, which might be able to attach a body to a surface, cannot replace weight in the state of weightlessness when the whole body remains weightless.

Fig. 5.11 Schematic representation of weight decrease across a body from the full weight on the feet. (Photo by the author)

---

[61] Newton (1687/1999, Book 2, Proposition 36, Corollary 4, p. 739).

Weight in Rotating Systems

Weight in a continuously revolving or spinning system is an especially important case. Examples include weight changes on the spinning Earth. The particularly important case is a satellite in the form of a revolving circular tube connected to the axis of rotation. Huygens was the first to consider weight phenomena in rotating systems. He revealed that the object on a rotating disc is subject to two similar forces: regular gravitation and centrifugal force. Both forces contribute to the weight of the object (its "gravity"). The dependence of the centrifugal force on the rate of rotation (and its radius) allows monitoring of the magnitude of the weight in a rotating system. This is especially important for the need to create weight during the free gravitational motion such as in a satellite otherwise being at the state of weightlessness. Humans cannot survive weightlessness for a long time. Irreversible biochemical changes eventually cause serious damage to the living organisms and to a deterioration of health. As living organisms, we essentially need weight. Rotation of the space station is the only way to avoid weightlessness. This idea, thus, is crucial for future space travel, which will take a long time.

Questions to Reflect on

1. What was the idea of splitting the weight concept used after Newton? What was the rationale of such a split?
2. What is more essential for the functioning of the human organism, gravitation, or weight? Why do you think so?
3. Could one solve the problem of the weightlessness of the human body by using magnetic boots? Explain.

Historical and Philosophical Summary

In summary, the concept of weight underwent essential changes. This concept was introduced in Hellenic science as an *inherent feature* of the *heaviness* of material objects paired with the opposite levity. From the beginning, the idea of weight corresponded to the sense perception related to the quantity of matter. For Aristotle, it manifested in the intention of a body to move to its natural place in accordance with the element composition. But Hellenistic science quantified and operationally defined weight as a measure of heaviness established by weight balance. Weight was directly related to the quantity of matter. Medieval science proceeded along the same path but further modified the weight concept as it also related it to motion: *natural* and *accident weight*. Galileo proceeded with this approach of heaviness. It was inherent, but it was inferred operationally; the weight of the body immersed in water was reduced.[62] Descartes rejected the weight as inherent to the bodies and related to the residual effected as being caused by centrifugal forces of ether vortices. This merely qualitative model was essentially deficient (axial instead of radial symmetry, contradictive features that were never observed) and was abandoned very soon.

Newton defined gravitational attraction between any two material objects to be the cause for their weights. The gravitational attraction seemed strange to Newton (it acted at a distance without any agent and delay, but it worked to account for numerous phenomena). Newton suggested that the support preventing falling can be sensed and its strength evaluates the weight of the object supported. Euler repeated this understanding. Newton's concept of weight was characteristic of a particular pair of objects. The amount of matter was characterized by Newton with a different quantity—inertial mass. Newton discovered that inertial mass determines the gravitational attraction. For this very reason, in everyday practice, one may often ignore the difference between mass and weight.

The concept of weight was used mainly in the terrestrial environment where the only important gravitational interaction is between the Earth and the object considered, not between the objects. In this context, weight lost the Newtonian meaning as a characteristic of a *pair* and became the force of gravitational attraction exerted on the body toward the Earth. As such, weight often returned to be a characteristic of the object itself, much like in its pre-Newtonian use.

In twentieth-century physics, the accounts of all observers (inertial and non-inertial) became legitimate. Weight lost its univocal correspondence to the gravitational force. Inertial forces were considered legitimate contributors to weight. Weight was defined as equal to the results of standard weighing or, equally, as the force that the object exerts on its support being in the rest frame for a certain observer. The new definition of weight does not require transition to the full-scale relativity theory. Classical mechanics allows the change of weight definition without any change in the problem-solving using the traditional apparatus of mechanics, making it logically consistent and closer to a sense experience. The major historical steps of the concept of weight can be seen in the flowchart of Fig. 5.12.

The important point here is that within the operational definition, weight appears to be *observer independent*. The results of weighing remain the same regardless of the forces by which the particular observer explains it. The answer to the question "What causes weight?" is, however, observer dependent. For a non-inertial observer, the deformation of a spring may indicate the activity of radial gravitation, whereas an inertial observer would identify only the elastic force of the spring functioning as centripetal force.

---

[62] This understanding of Galileo was corrected by Newton (1687/1999, Corollary 6, p. 691).

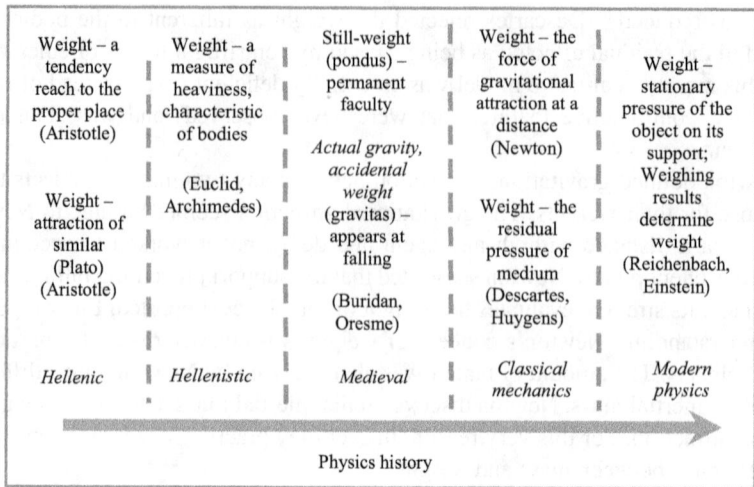

Fig. 5.12 Understanding of weight—historical summary

Epistemological Development

If the historical picture is clarified, one may be able to make sense of the perspective of approach to the specific phenomenon—of material objects. One may identify three major approaches which physics demonstrated during the 2500 years of development. It seems possible to identify them in accordance with the refinement of understanding natural when regarding this phenomenon and designate them under the names of mechanical theories adopted at the corresponding periods: ancient, classical, and modern (Fig. 5.13).

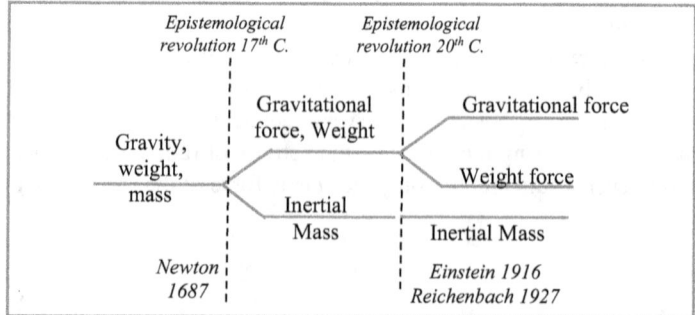

Fig. 5.13 Emphasizing the major changes in the approach to weight concept through the history of physics

1. *The ancient physics approach*. During this period, the feature of heaviness was treated as ***natural*** without further explanation ("natural does not require force"[63]). In Hellenic natural philosophy, heaviness was considered only qualitatively, heavy as opposite to light (Fig. 5.14a, b). In the following period of Hellenistic science, weight was quantified and defined in a weighing procedure by means of a calibrated balance (Fig. 5.14c). It was the period of the natural weight of objects determined by the features ascribed to the elements of matter. The addition of actual and accidental weights in medieval science did not change the fundamental approach to weight based on the sense perception of heaviness and the effort required in stopping and preventing falling.

(a)   (b)   (c)

Fig. 5.14 (a) The "true" weighing of the Earth by Atlas, the Hellenic idea of weight on a street in Sydney. (b) The modern version of Atlas weighing the universe in New York. Is it physically correct? (c) Hellenistic *calibrated* balance in a modern museum demonstration. Its arms are not equal, and its explanation is based on the law of levels. (Photos by the author)

2. *The classical mechanics approach*. In the seventeenth century, Descartes tried to explain weight with the residual pressure of the surrounding medium of ether, but failed. Newton postulated the gravitational interaction between material bodies without any mechanism. The gravitational interaction was taken as *natural*, that is, as a given without explanation. The universe was perceived as an object, from a unique perspective, as if looking from the side, or flying above it. It was a natural container in which the world was placed in absolute space and time; it was the basis, taken as self-evident, without further reduction,[64] "known to all." Within this framework, the gravitational force could be always identified theoretically, regardless of measurement. Therefore, there was no difficulty to identify weight with the gravitational force on the object, regardless of weighing: indeed, all other bodies which gravitate to the considered body are observed; what is left is to calculate the gravitational force exerted on the object. Weighing is, thus, superfluous and may even mislead. Yet, in practice, this definition was not used. In everyday activities, weight is taken as a result of weighing and never

---

[63] There is a strong connotation with the theological understanding of *natural* as a type of common-sense idea (Aquinas, 1267/1952, pp. 541–542).

[64] Newton (1687/1999, Definitions, Scholium, pp. 408–415).

as a result of calculating gravitational force from the Earth, Moon, Sun, and so on. For this, distinguishing between gravitation and weight concepts was needed.

Thus, the notion of *true-weight* was reserved by Newton for the gravitational force on the body. As to *apparent-weight* (relative, common), he affiliated it to the results of weighing (perception) in the presence of impeding factors, which might deceive a practitioner, but not a philosopher. Newton illustrated this by referring to buoyant force as a misleading factor. Addressing an object immersed in liquid he wrote:[65]

> But those things (immersed in water) which neither by preponderating descend, nor, by yielding to the preponderating fluid, ascend, although by their <u>true weight</u> they do increase the weight of the whole, yet comparatively, and as <u>commonly understood</u>, they do not gravitate in water.
>
> The gravity of bodies in fluids is therefore two-fold: the one true and absolute: the other apparent, common, and relative. <u>Absolute gravity</u> is the whole force with which a body tends downwards; <u>relative or common gravity</u> is the excess of gravity with which the body tends downward more than the surrounding fluid. … Those things which are in air, and do not preponderate are commonly looked upon as not heavy. Those which do preponderate are commonly considered to be heavy, inasmuch as they are not sustained by the weight of the air. The <u>common weight</u> is nothing but the excess of the <u>true weight</u> above the weight of the air (emphasis added).

This approach addressed equally the fundamental concepts of motion, space, and time. Newton distinguished between *true* and *apparent rest*, as well as *true* and *apparent motion*. He understood *true* rest and *true* motion relative to absolute space in contrast to apparent rest or motion, as perceived by the common observer who fails to detect absolute quantities. He proceeded:

> … I must observe that the common people conceive those quantities under no other notions but from the relation they bear to sensible objects. And from these arise certain prejudices, for the removing of which it will be convenient to distinguish them into absolute and relative, true and apparent, mathematical and common. But we may distinguish rest and motion, absolute and relative, one from the other by their properties, causes and effects. … it follows that absolute rest cannot be determined from the position of bodies in our regions… Absolute time, in astronomy is distinguished from relative, by equation or correction of the apparent time.

In the case of rotation, Newton believed in the ability to distinguish between relative and absolute movements by the appearance of the "true" motion.[66] Newton never considered *frames of reference*. Koyré summarized:[67]

> In the Newtonian world and in Newtonian science, it is not man, but God, who is the measurer of things.

3. *The modern mechanics approach.* The relationship between theoretical and operational definitions of weight never arose in classical mechanics. The operational definition existed in the form of refinement and illustration, subdued to

---

[65] Newton (1687/1999, Book II, Theorem 15, Corollary 6, p. 691).

[66] Newton illustrated this by the thought experiment of a rotating bucket of water (ibid.: pp. 412–413).

[67] Koyré (1956, p. 183).

theoretical. In the twentieth century, within the new trend in the philosophy of science—positivism—the situation changed, even reversed with regard to weight. The Einstein theory of special relativity drew on the operational definition of simultaneity which implied a whole new theoretical framework: a revolution regarding the fundamental concepts—time and distance—previously considered *metaphysical*, that is, external to physics, a priori given by intuition. Physicists always tried to avoid claims addressing such concepts. In contrast, positivism demanded direct measurement as a procedure of determining the objective meaning of any concept in use.[68]

The definition of a concept drawing upon theoretical knowledge was termed—nominal (*nomos* is law in Greek). With regard to weight, Newton was the pioneer of the *nominal gravitational* definition of weight:

The weight of an object is the gravitational force exerted on that object.

As was mentioned, this definition matched the newly developed by Newton account of gravitational attraction between any two material bodies. Weight was defined in pairs of forces, hence, the weight of an object was different with relation to different bodies, and the weight of an apple was equal to the weight of the Earth. This is not what physics textbooks state, but this was how Newton introduced weight.

In the twentieth century, Bridgman[69] further developed this epistemological trend regarding scientific concepts termed as operationalism. In particular, after the introduction of the equivalence principle between gravitational and inertial forces, the *nominal gravitational* definition of weight became problematic. It had to be replaced. The new definition addressed the force that allowed direct measurement:

Bridgman

Weight is the force that a body exerts on its support at the state of rest as viewed by a certain observer in the corresponding system of reference.

The related operational definition was:

The weight of a body is defined as the result of its weighing.

Since the *operation* itself (weighing) is not defined here, a more accurate definition is:

The weight of a body is defined as the result of standard weighing.

Or, to clarify the *standard* one may write:

The weight of a body is determined by weighing at the state of rest by means of a calibrated spring (spring scale).

Specifying the *spring* scale was necessary because a *balance* scale, using horizontal lever, compared the forces applied on its arms (torques, in general). The

---

[68] Mach exemplified his by the operational definition of inertial mass (Mach, 1883/1919, pp. 216–218).

[69] Percy Williams Bridgman (1882–1961)—the Nobel Prize winner in Physics (1946). He explored the scientific method and established the trend of operationalism in the philosophy of science (Bridgman, 1927/1952).

balance scale, therefore, rather informs about the mass ratio of the objects and is insensitive to the changes due to geographical latitude. In contrast, spring deformation indicates the action of the weight force in linear proportion (Hooke's law). A calibrated spring device became a legitimate tool for weight determination.

It is an important requirement of the modern philosophy of science that a pair of coherent definitions—operational and nominal—should represent each physical concept.[70] They are complementary in establishing the meaning of physical concepts and neither of them is sufficient by itself. Not to forget, physics is a theory of nature. Indeed, though weighing was practiced at each of the three periods of weight understanding, its meaning and status changed together with the theory that reigned at the time.

Altogether, the historical debate of ideas about weight, adopted and refuted, makes the knowledge of weight an especially elucidating stage for demonstrating the role of epistemology of science. We will further specify this perspective in considering the topics, questions, and materials which are potentially appropriate for the effective teaching of weight.

Questions to Reflect on

1. What did Newton mean by *true* and *apparent* (absolute and relative) with regard to physical concepts? In what sense were these notions different from their modern use?
2. What were true and apparent weights? How were they defined?
3. Provide an operational definition of true weight. Discuss the difficulty in the observer dependence perspective.
4. Why do you think it is important to provide operational definitions to physical concepts?
5. Why is it important to mention *standard* operation for the weight measurement?
6. Why does standard weighing use a spring scale and not a balance?
7. What are the advantages of a balance scale?
8. Why do you think there is a need to provide a pair of definitions for each physical concept? Is an operational definition sufficient to define a physical concept? Explain your answer with regard to weight.
9. The apparatus on the sketch of Fig. 5.15 is comprised of two weighing devices. It measures the weight of a stone immersed in water.

Relate this situation to the quoted above text from the *Principia* in 1687:

   ...*bodies placed in fluids have a two-fold gravity: the one true and absolute, the other apparent, common, and comparative... The common weight is nothing but the excess of the true weight above the weight of the air.*

Fig. 5.15 Weighing a stone immersed in water

---

[70] Margenau (1950, Ch. 12, pp. 220–244).

Discuss weighing results with and without water. What weight would Newton consider as the true weight of the stone? What weight would Newton consider as the apparent weight of the stone? What would be the reading of the lower scale? What would be the reading of the upper scale?

Educational Aspects

Three levels of schooling, elementary, middle, and high, are usually distinguished. They all differ in the nature of content and its account.

- The *first level* of the *elementary* school curriculum (6–12 years of age) usually includes an *operational* definition: weight of a body is its heaviness and is measured by weighing. At this level, weight is associated with the permanent features of the known objects, conceived as their nature. Weight understanding creates an intuitive scheme by which children ascribe heaviness to familiar objects and classify them as "light" and "heavy."[71] Weight, within this scheme, draws on the tactile perception of pressure which in nature is not different from the weighing by scale.
- The curriculum of the *middle* school (12–15 years of age) split in two optional coverage of weight and gravitation. In a more frequent approach, weight is defined as gravitational force. The inconsistency with the previous instruction stems from the discrepancy between the weighing results and gravitational force. It is especially apparent in considering the state of falling when the gravitational force is acting but weighing results are zero. The teacher has to disregard weighing results as non-indicative of gravitation. He/she may explain that "the issue is not that simple" or postpone the explanation for the high school studies.[72]

Alternatively, the teacher may define the weight of the object as the force it exerts on the support (or suspending cord) which correspond to the results of weighing. This definition relates weight to the perceived pressure on the support or tension of the cord. The operational and nominal definitions appear consistent. The teacher may discuss the meaning of standard weighing and illustrate its violations such as by additional support and media. Gravitational force (the invention of Newton) is distinguished as the cause of falling, gravitational motion of planets and satellites. The changes in weight with geographical location (latitude) are explained qualitatively by the Earth's rotation. Gravitational force is taught as possibly contributing to weight but not identified with it and possibly different.

- At the *third level*, the curricula of *high* schools (15–18 years of age) continue to be twofold. The trend of the gravitational definition of weight introduces a split

---

[71] e.g., Piaget (1972); Galili and Bar (1997).

[72] For the majority of students in my country, it means forever, (or never), as learning physics in high school is elective.

of weight into two concepts, the *true weight* F_g (gravitational force exerted on the body) and the *apparent weight* F_p (weighing results).[73] The only way to determine the true weight is to calculate the gravitational force in accord with the universal law of gravitation. The apparent weight receives operational definition and is related to the pressing force, equal to the normal force N (Fig. 5.16). The gravitational force F_g on the object is not necessarily equal to the pressing force F_p.

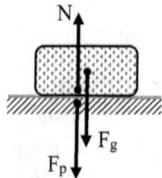

Fig. 5.16 The forces in the case of an object lying on the support. The gravitational force F_g on the object. The pressing force F_p is the force which the body exerts on its support. F_p is equal to the normal force N (its partner in action-reaction pair). F_p is the force measured by the spring scale or perceived by a supporting hand

Alternatively, the teacher may proceed with the instruction according to the operational definition of weight ($W = F_p$) as different from gravitational force F_g. This way of instruction may optionally involve inertial forces (non-inertial observers in accelerated systems). The important feature of the operationally defined weight is that its magnitude being related to a local measurement (weighing) is independent of force identification, that is, observer independent.

A special didactical impact could be reached if the teacher discusses reality on board of an imaginary rotational space station (Fig. 5.17). Explained by centrifugal force (the simplest way) or without it, this example strengthens the conceptual distinguishing between weight and gravitation which are saliently independent in this case both in direction and magnitude.[74]

Inertial forces and multiple observers are usually outside the scope of the traditional school curricula. In reality, however, while considering weight, students very often identify themselves with the observer inside the system under acceleration.[75] Thus, the introduction of non-inertial observers makes this common intuition legitimate and removes unnecessary confrontation and misconceptions. As was shown experimentally, such teaching is feasible and beneficial already in the middle school.[76]

---

[73] e.g., Young and Freedman (2004, pp. 120, 441, 459–460).

[74] For the rotating space-station and space-ships see Stanly Kubrick's celebrated movie *2001: A Space Odyssey.*

[75] Galili and Kaplan (2002).

[76] Stein and Galili (2018).

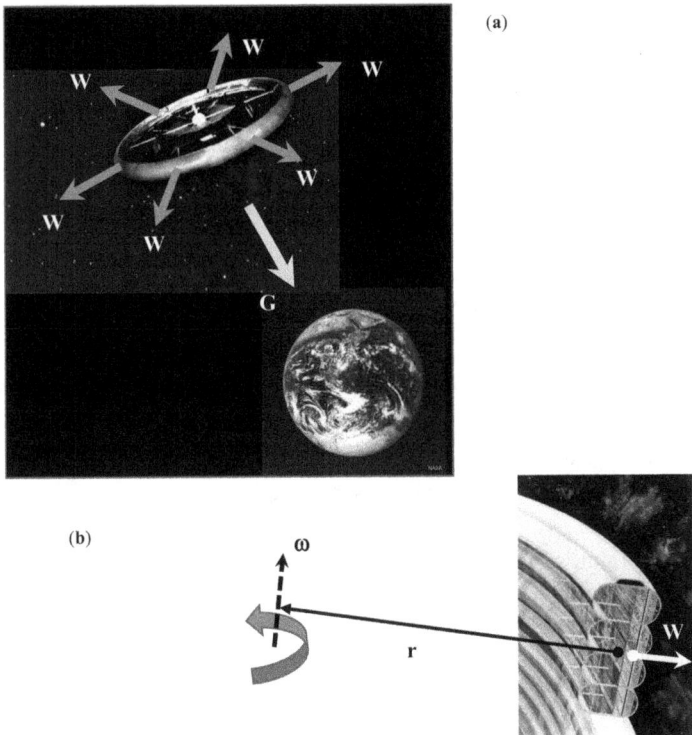

Fig. 5.17 (**a**) Schematic depiction of a rotational space station—a physical system in which gravitational force **G** and weight **W** are independent in direction and magnitude. (**b**) Radial weight **W** of the objects placed on the "wall" is determined by a simple formula of the centrifugal force: $W = m\omega^2 r$ (m, mass of the object; ω, the angular velocity of the station; r, radius of the circular path of the object). It provides the weight W of the object created in the station. The formula is used to design the dimensions of the station for providing regular intensity of artificial gravitation

Impact on Students' Knowledge

Several studies reported the low success of the teaching drawing on the gravitational definition of weight. Strong confusion was reported in students' interpretations of weighing results and its relationship with gravitational force.[77] Students' conceptions can be reduced to several schemes of knowledge which are held in considering and practicing the concept of weight:[78]

- Manipulating with a unique weight concept (no use of true and apparent weights as provided in the instruction).

---

[77] e.g., Ruggiero et al. (1985), Galili and Kaplan (1996), Galili and Lehavi (2003), Galili et al. (2017).

[78] Galili and Kaplan (1996), Galili and Lehavi (2003).

- Weight, or weight force, is unconditionally related to the results of weighing.
- Weighing informs about the gravitational force on the object.
- The weight of a body decreases with the increase of the distance from the sources of gravitation.
- Weight/gravitation can be changed/neutralized by other forces or agents (air, water, soil).
- Weight is created by the medium (air, water, soil) and transferred by it.
- Weight can change in motion.
- Vacuum causes weightlessness.
- Weight is an inherent and invariant quality of a body.

These schemes of knowledge "produce" even more abundant conceptions—facets of knowledge—which are context dependent. Among them, one may find:

- Satellites (and any objects inside) are (almost) weightless.
- Objects "in space" are weightless.
- Weight increases in going below the Earth's surface and becomes infinite at its center.
- Objects are (almost) weightless on the Moon, in spaceships, under a glass dome with evacuated air.
- Swimmers in water have lower weight.
- Floating in water indicates weightlessness.
- Weighing on the surface of the Earth should take into account attraction to the Moon (or the Sun).

These student conceptions testify as to the failure of the instruction which mingles weight and gravitation, which presents them as one. It goes against a strong natural intuition which connects weight with the heaviness of objects as causing their behavior in specific environments and motion. The mismatch of the curricula of elementary and middle schools apparently contributes to the abundant and various misconceptions in students' knowledge and teachers' confusion.[79] In contrast, teaching adopting the operational definition of weight as well as addressing the history of weight-gravitation causes reduction of these misconceptions and of conceptual confusion.[80]

Media for Learning

Weight in Physics Textbooks

Physics textbooks are divided in their choice of weight-gravitation account. The majority of textbooks still follow the gravitational approach to weight definition.[81]

---

[79] Teachers of physics seek and praise consistency of the concept definitions (Galili & Lehavi, 2006).
[80] Galili (2017); Stein and Galili (2014); Stein et al. (2015).
[81] Young and Freedman (2012, p. 108).

The gravitational force that the earth exerts on your body is called your weight.

To illustrate the salient feature of this definition, one may quote its generalized version as it appears for many years in this and other popular textbooks:[82]

The weight of a body is the total gravitational force exerted on the body by all other bodies in the universe.

In this form, it may represent the holistic vision of Newton. Understandable at that time, it has become rather meaningless in our days. It presents weight as a quantity impossible to calculate and measure, lacking any practical use, even if it were known. The improvement by splitting weight to its/a true and apparent weights merely transfers the problem to the true weight.

As to the *operational definition*, it appears in textbooks in three forms which are epistemologically close and numerically equivalent (Fig. 5.18). Worth mentioning is the option **c** that corresponds to the international conventions but is rather rare in textbooks.[83]

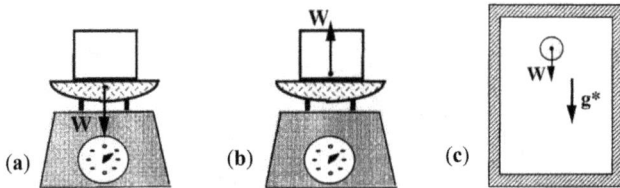

Fig. 5.18 Operational definitions of weight: (**a**) the prevailing version: weight is the force exerted on the support (e.g., Baruch and Vizansky (1937), Chaikin (1947/1963), Orear (1967, p. 82), Marion and Hornyak (1982, p. 129), Lerner (1996), Halliday, Resnick and Waker, J. (2001, p. 80), Hewitt (2002), Walker (2008, p. 95), Knight (2013, p. 146)); (**b**) weight is the force exerted on the object by the support (French (1971, pp. 129–130), Levin (2016)); and (**c**) weight is the force exerted on the object causing its spontaneous falling with acceleration g^*, as measured in the laboratory (Keller et al. (1993, p. 99))

In light of the historical and epistemological background addressed in this excursus, the educational goal is not about how to save the gravitational definition of weight, which is both theoretically deficient, and less than effective, but how to promote its replacement with that which matches modern epistemology. There is much work to do on this issue, however.

---

[82] This definition of weight was inherited from Sears and Zemansky in the 50s and continuously preserved by Young and Freedman (e.g., 2012, p. 406). Gravitational definitions appeared in the past in Resnick and Halliday (1966, p. 286), Ohanian (1989, p. 126), Hewitt (1992), Knight (2004, p. 129), Jewett and Serway (2008, p. 106), to name a few.

[83] Taylor and Thompson (2008); Thompson and Taylor (2008).

Weightlessness

Besides being the paradigmatic case representing the epistemological revolution in the physics of the twentieth century, the state of weightlessness presents the most effective probe to reveal the stance of particular authors with respect to weight. Textbooks following the old gravitational definition of weight mention that objects in a free fall preserve their (true) weight but experience a merely *apparent weightlessness*, an illusion (Fig. 5.19). Textbooks that adopt the operationally defined weight consider weightlessness as true, the state of *lacking of weight*.[84] However, the most unsatisfactory feature of the introduction to (!) physics textbooks is the highly frequent ignoring of the topic of weightlessness. Even the ambitious educational project *Science for all Americans. Project 2061*[85] lacks any attempt to deal with the topic, which already now, much before 2061, presents the salient feature of space exploration activities, globally lead by the celebrated National Aeronautics and Space Administration, NASA.

Fig. 5.19 Two accounts of the state in which an object is in a free gravitational motion—illusive weightlessness by Newton and true weightlessness by Einstein

Quite interesting is that NASA materials usually avoid the label—*weightlessness*, replacing it with *microgravity*. Seemingly this label reflects the obscure meaning of weight in educational literature. Besides being apparently confusing, microgravity has some advantages. Firstly, it can be distinguished from "microgravitation," which may lead the reader to the elucidating discovery of *gravity* as not being the same as *gravitation*! Secondly, it may hint to the existence of other manifestations of gravitation—*tidal forces*, a residual product of gravitation which, unlike gravitational force, is a subject of direct measurement, clearly observed in the environment of weightlessness. However, it might be perceived as surpassing the limits of the school curriculum, as it can be interpreted as providing the learner

---

[84] French (1971, p. 229); Levin (2016); Balukovic and Slisko (2018).
[85] Rutherford (1990).

with an elucidating contrast to the gravitational force and as such, presents a power-ful tool purely for the sake of general education.[86]

Addressing weightlessness in numerous texts on technical and applied sciences often presents a real challenge of understanding by school students and educators. The texts frequently avoid mentioning the observer-dependent nature of the account in which inertial forces appear, the observer dependence as a physics perspective. As a result, dubious conceptual constructs may appear, deficient in the view of physics theory. Here is an example:

In a serious treatise presenting biomedical impacts of weightlessness on the human body, the authors introduced the force of inertia F_i = ma (resembling the historical d'Alembert force) which "balances" the gravitational force causing weightlessness. Frame of reference is not mentioned. A physics teacher, trying to relate to the regular curriculum, has to figure out that despite the "balance" the body continues to fall with acceleration. What kind of balance is it, then? The full account should include the force of reaction from the support of the object, the force which is measured in weighing. The introduced "force of inertia" which always exists with any acceleration, simply misleads a novice. It brought the authors to the claim:[87]

> Orbital weightlessness. A vehicle in a circular orbit around the Earth is constantly attracted toward the Earth's center of mass CM_e by the gravitational force F_{ge}. This produces an equal but opposite force of acceleration of F_{iv}. These forces are exactly equal and this produces weightlessness of the orbiter.

It is stated that the orbiting spacecraft is weightless because the gravitational force on it is balanced. In physics, it is a bluntly wrong statement, because the orbiting object is under the unbalanced force of gravitation and due to this fact, it is orbiting. This is a good place to remind students about the idea of the pre-Newtonian Italian scholar, Giovanni Alfonso Borelli, who in the seventeenth century stated exactly the same balance of the forces, toward and away from the attracting Sun, exerted on the orbiting planet.[88] In contrast, Newton stated that the forces on the planet are not balanced and therefore its trajectory is not a straight line. This is what we teach in physics classes.

In the account equating weight with the gravitational force, exact weightlessness never happens because of the presence of the acting gravitational attraction as seen by any observer. Instead, the authors consider the "effective weightlessness" and apply the concept of *apparent weight*—the result of weighing (pressure on the support). If one wants to use inertial force in its standard form, it appears only in the account of the observer in accelerated frame of reference, such as in a spacecraft. A non-inertial observer in an orbiting satellite may use the inertial force which cancels the gravitational force. So, the explanation of weight measurement in the spacecraft depends on the observer, though its result (or simply weight, in its operational definition) remains zero, regardless of the observer.

---

[86] Galili and Lehavi (2003).

[87] Thornton and Bonato (2017, p. 10).

[88] Chapter 4.

Gravitational "Weightlessness"

To appreciate the misleading potential of the gravitational definition of weight, one may read the episode depicted by Jules Verne (1828–1905)—the famous science fiction novelist of the nineteenth century—in his novel *From the Earth to the Moon* (1874) (Fig. 5.20). There, he explained in length, as if being a teacher, his understanding of weightlessness. He drew upon the straightforward application of Newton's tenet: weight is the gravitational force.

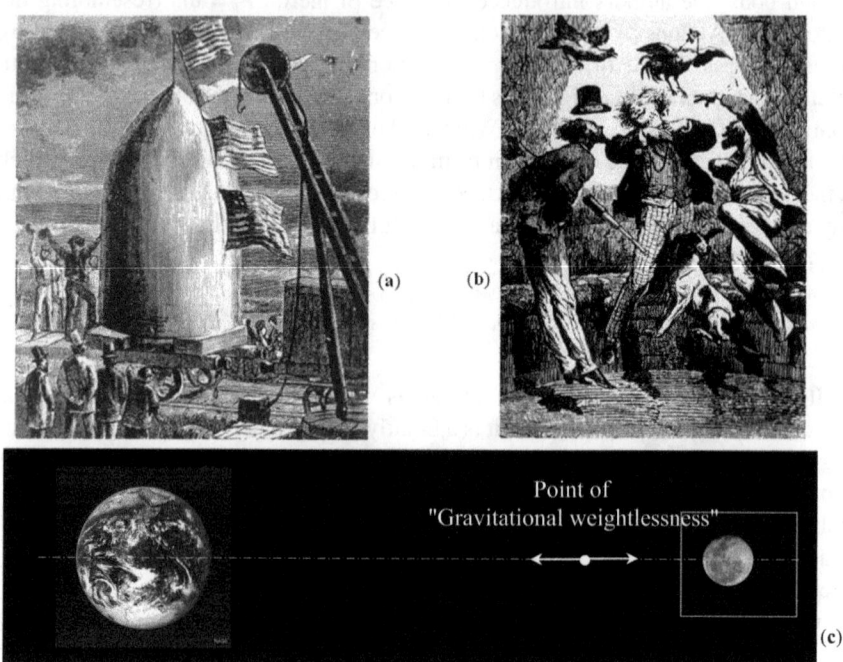

Fig. 5.20 (**a**) Jules Verne sent his astronauts, looking for adventures, in a gigantic cannon shell fired from the Earth toward the Moon. (**b**) The moment of weightlessness inside the shell. Is the depiction of weightlessness in these old drawings equally correct in both pictures? (**c**) The point of weightlessness according to Jules Verne (1874)

The well-educated novelist, who of course had never observed real space flights, explained the state of weightlessness as that which is reached, in his view, when the spaceship arrives at the point of force equilibrium, that is, where the attraction to the Earth is balanced with the equal force of the Moon's attraction (Fig. 5.22c).

The salient feature of the state of weightlessness is that the concept up-down direction loses its regular meaning. For Jules Verne, it happens only in one point (also, only if neglecting other attractions such as from the Sun, which is not small, etc.). On their way to the Moon, the spaceship was strictly oriented, and his travelers stood inside the cabin with their feet toward the Earth. For the unexpected influence by a bypassing asteroid, their cannon shell deviated from the original trajectory. They (fortunately) lost the chance to fall on the Moon. In the attempt to correct the trajectory of their inertial motion, they fired a rocket cartridge attached to the shell. They did not succeed and actually performed the same trajectory around the Moon and back to the Earth as the astronauts of Apollo-13 had to make in the not less unfortunate circumstances about 150 years later. Here is the quotation from the book which illustrates the Newtonian understanding of gravity as equal to gravitation by Jules Verner:[89]

> "Are we falling?" asked Michel Ardan, at length. "No," said Nicholl, "since the bottom of the projectile is not turning to the lunar disc!"

Students may discuss this erroneous description which implies continuous up-down reality in the free falling toward the Moon and the Earth, as well as the state of weightlessness for an instant of "gravitational equilibrium" during the space flight, comparing it with the real appearance of Apollo astronauts who were weightless all the way to the Moon. Discussing and analyzing this understanding could serve an effective remedy for several misconceptions documented in physics education research.[90]

Weightlessness/Overweight: The Impact on Human Organism

Textbooks point to the very real and dangerous impact of continuous being in a state of weightlessness on humans.[91] This topic may provide an interesting interdisciplinary discussion regarding the role of weight in our life as organisms essentially dependent on weight,[92] the fact that weight determines the way we eat and drink, move, and function. Weight is essential for the functioning of our organism, weight and not gravitation.

Life, as we know it, is stipulated by weight. For a human organism, weightlessness implies radical changes in a range of aspects: muscle degeneracy, calcium loss in bones, deterioration of hearing and vision, and impact on cardiovascular system, to name a few. A different dynamic of chemical reactions influences cell metabolism. Numerous aspects of this subject are under intensive biomedical investigation.[93] It is clear that the influence of weightlessness is predominantly destructive for people: researchers carefully explore the impact of weightlessness for longer and longer periods of time, currently not extending for more than a year. The problem is clear, and its ultimate solution is to liberate people from this constraint of

---

[89] Verne (1874, Ch. XIX).

[90] e.g., Galili and Kaplan (1996), Stein and Galili (2014), Stein et al. (2015).

[91] e.g., Crowell (2006, pp. 109–110); Thornton and Bonato (2017).

[92] This claim draws on various arguments of medical, physiological, biochemical, and other origin deeply rooted in biological evolution.

[93] Thornton and Bonato (2017).

long-term space exploration by creating weight due to the inertial effect rather than gravitation. This means constructing space stations, spaceships, and eventually space colonies based on rotation.

The opposite case—overweight—emerges in the much rarer context of short-duration accelerations in supersonic jet planes, pilot training in centrifuges, preparing them for the overweight experience during flight maneuvers associated with extreme acceleration. Yet, unlike weightlessness, in an overweight experience, the damage to the human organism could be very fast. The resistance of the organism in this case only lasts for up to a few g of acceleration. The failure of the human heart as a blood pump in providing blood to the brain takes place at about 7–8 g lasting for about 10 seconds.[94] At lower accelerations, a range of damage is caused to the heart and blood vessels. Importantly, no special clothing, no special space suits, can help the heart or internal blood vessels to withstand high acceleration.

Rotational Space Station

The use of suitably designed rotating space vehicles is an obvious way of establishing weight in a free gravitational motion (falling, in the case of objects near the ground). Considering rotating space stations is a special topic of space engineering, or serves as a short conceptual question/problem to solve.

Fig. 5.21 Hermann Potocnik (Noordung) and one of his first projects, the rotating space station— *the residential wheel* (1929)

The history of the rotating space station perhaps starts with Hermann Potocnik[95] [Noordung] (1892–1929), the Slovene rocket engineer. He was the first to describe the idea of a rotating space station in his book *The Problem of Travel in the World Space*.[96] His *residential wheel* had the form of a very big torus, in which regular weight was reached by rotation of the station at a particular rate (Fig. 5.21).

---

[94] No need to provide here the exact numbers regarding the tolerance of the human body to overweight as obtained in actual experience during flights and centrifuge training of pilots and astronauts. Importantly, no precaution helps the human heart in facing inertial overweight.

[95] Noordung (1929, pp. 187–188).

[96] Noordung (1929).

Yet, the apex of this history is related to the leading rocket engineer and developer, Wernher von Braun. In March 1952, he published an article *Through the Last Border*. There, he described a huge rotating space station and a full range of space techniques, including reusable launch vehicles (the space shuttle) to provide transportation between the space station and Earth (Fig. 5.22). He described a colony in space—serving the needs of astronauts and scientists involved in space exploration outside the Earth. Incidentally, this discussion may expand on a topic not directly related to weight. Wernher von Braun was the person who led the greatest technical project in the history of human civilization—the Apollo project that took people to the Moon. Nevertheless, the personality of the man behind this outstanding professional legacy is controversial—a Nazi criminal, a senior SS-officer in Nazi Germany, directly responsible for the bombardment of London by V-2 rockets (the first long-distance, ballistic military rocket). He was heavily involved in the development of the rocket industry during WW II using slave labor, and the deaths of thousands of prisoners who worked there. Only 10 years after the war, he was already leading the US space program.

Fig. 5.22 Wernher von Braun—the head of the US Apollo Program of reaching the Moon in 1970 and his earlier project of a rotating satellite serving as an inhabited space station proposed in 1952

Various questions may be raised in this connotation:

- May a scientist collaborate with antidemocratic and evil regimes for the sake of his scientific dreams or just to solve a difficult problem?
- May a scientist afford himself the coverage of "I am not a philosopher," as Hans Bethe stated in his period of working on the Manhattan project? (He later changed his stance essentially…).
- Is there a special moral responsibility that scientists should take on themselves in contemporary society?
- Should the names of Nazi scientists, such as Stark and Lenard,[97] even be mentioned in our classes?[98]
- While addressing the greatest technological achievements of humankind, should we also mention the role of von Broun?

---

[97] Both were Nobel Prize winners in physics.

[98] In 2014, I was stunned to discover the name of Lenard in the calendar of the physics Department of the University of Sydney among the dates to celebrate. Some local academics, with whom I shared my feelings, as a guest from far away, just shrugged their shoulders and remained silent…

This side discussion, to the weight in space subject, may expand to Heisenberg and the German atomic project during World War II. The well-known play *Copenhagen* revealing the debate between Niels Bohr and Werner Heisenberg can be used as a culturally rich event in school life. Connection can be made to the activities of Andrey Sakharov, who was among the creators of the thermonuclear bomb in the USSR, and the moral split in the Los Alamos command working on the American atomic project when the time came to decide on proceeding to the second stage of thermonuclear weapon (Julius Robert Oppenheimer, Hans Bethe versus Edward Teller, and Stanislav Ulam regarding continuing the development). Finally, in this regard, the form of "balancing position" practiced by Albert Einstein and

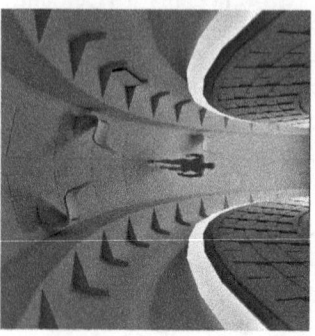

Niels Bohr can be mentioned as an interesting, though not at all convincing resolution of this dilemma.

The essential aspects of the weight concept were artistically represented in science fiction movies and novels depicting long-distance space flights. The problem of weightlessness was resolved there

Fig. 5.23 Schematic representation of an astronaut in a long space voyage walking on the wall of the rotating cabin of the spaceship (sketch by Guy Galili)

by using rotational structures. Such was the impressive representation of this idea in the movie *2001: A Space Odyssey* by Stanley Kubrick in 1968, drawing on the novel by Arthur Clarke. In the movie, the rotating station, *Ferris wheel*, provided weight to the space explorers in the station orbiting the Earth. The movie also showed a spaceship on its way to Jupiter, an especially long-term mission. The spaceship provided weight to the astronauts of a regular magnitude in the cabin spinning around its axis. Astronauts walked on the internal wall with their heads toward the center of rotation (Fig. 5.23).

Today, the 30-ton wheel model ordered by Kubrick for $750,000 is an exhibit in a museum, continuing to ignite people's imagination of the future. In the much more distant future, however, the reality will surpass it. People will construct huge space colonies in the form of gigantic wheels. They will shelter the numerous inhabitants who will leave the Earth forever while enjoying regular weight conditions (Fig. 5.24).

The pertinent discussion in a physics class may include the calculation of the radius of the space station, the rate of its spinning, and tangential and angular velocities of the rotational station taking into account that the created gravity intensity (g) in the rotating torus should not change more than 1% along the height of a person ($h = 2$ m).[99] One may examine the dimensions of the spaceship going to Jupiter as shown in the Kubrick movie "2001: Space Odyssey" and ask whether the dimensions shown in the movie (radius about 20 m) were realistic.

---

[99] See http://visions2200.com/SpaceHabitat.html for possible help.

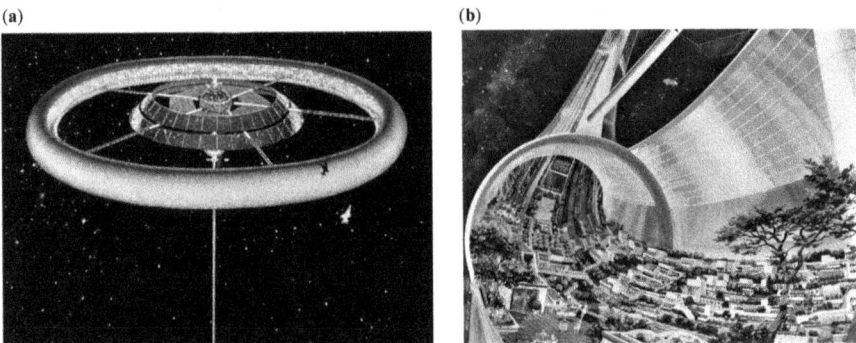

Fig. 5.24 A permanent colony in space, as imagined by scientists today, where many people will live their regular lives in space possessing regular weight without planetary gravitation. Image Credit to NASA Ames Research Center (**a**) ID AC76-0525; (**b**) ID AC75-1086-1

Weightlessness versus Buoyancy

A conceptually deep comparison between weight and buoyant force is extremely rare in literature, if it exists at all. At the same time, it is highly beneficial for understanding weight concept.

The buoyant force and weight force present action-reaction pair of forces. They act in opposite directions: up and down. They are collinear and actually define the direction of up and down.

Consider a container with water (Fig. 5.25a). Within the water, imagine a spherical drop (dashed line). At the state of rest, the gravitational force on the drop F_g (a volume force) is balanced with the net force surrounding this drop—F_b (a contact force). In fact, buoyant force is parallel to the force N in Fig. 5.16. The buoyant force is the force of "support" summarizing multiple forces F_1, F_2, and so on from all around the drop. The force F_b is the reaction force to *the weight force* of the drop applied on the surrounding medium (parallel to the force F_p in Fig. 5.16). *At rest*, there is a gradient of pressure from the bottom to the surface of the water in the container, including through the drop that we are considering.

In contrast, when the whole container is *in a free fall*, each drop of the water is accelerated in the same way, and there is no pressure inside the water. This presents the essential difference between the state of floating in a water basin on the ground and the state of a body in falling (Fig. 5.26). This implies the difference in the nature of the forces: the buoyant force, the weight force, and the force of gravitation.

In reality, the difference between flowing in water and "floating" in falling (Fig. 5.26) can be appreciated by the fact that a person with closed eyes in a free-falling situation (in a vacuum, or with a diminishingly small air resistance) cannot differentiate between up and down, whereas a diver knows very well the up-down direction regardless of whether he sees anything. Being upside down in water is as equally dangerous as being upside down outside of water. Human organisms cannot stand it for more than a few minutes. It is for this very reason that the "anti-g" suits

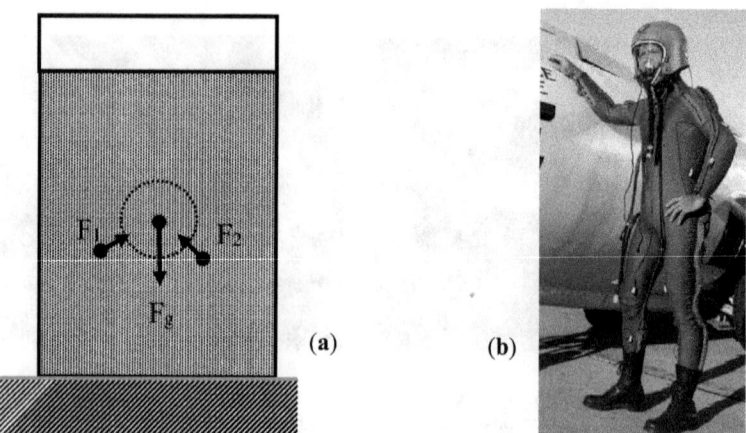

Fig. 5.25 (a) Water container at rest. Multiple forces F_1, F_2, and so on from the water surrounding the drop summarize to the buoyant force on the drop balancing the gravitation force F_g on the imaginary drop inside the water. (b) Supersonic jet pilot Joseph Walker in a pressure suit in 1958. (Picture credit: NASA)

(Fig. 5.25b),[100] although reducing the harmful impact on the close to skin blood vessels by redistributing the pressure of the increased weight during acceleration, cannot help protect the heart of the pilot. The limitation to the acceleration for human surviving remains exactly the same.

Fig. 5.26 (a) Floating in free falling from a plane. (b) Floating in diving

We have explained this difference without using inertial force. The explanation using inertial force, that is, in the frame of reference where the container is always at rest, regardless its possible acceleration, is even simpler, providing the same result: buoyant force is determined by weight (operationally defined of course) not by the gravitational force.

---

[100] Also called "pressure suits," which they actually wrap the body of the pilot with a layer of compressed air.

Questions to Address and Discuss

Learning to distinguish between weight/gravity and gravitation may involve conceptual questions of the kind to be illustrated by the following questions:

1. Consider two satellites orbiting the Earth at heights of 100 and 200 km (Fig. 5.27a). Astronauts in each satellite weigh a mass of 1 kg. What are the results of weighing obtained, and what are the inferences regarding the weight in each case? Explain your considerations.

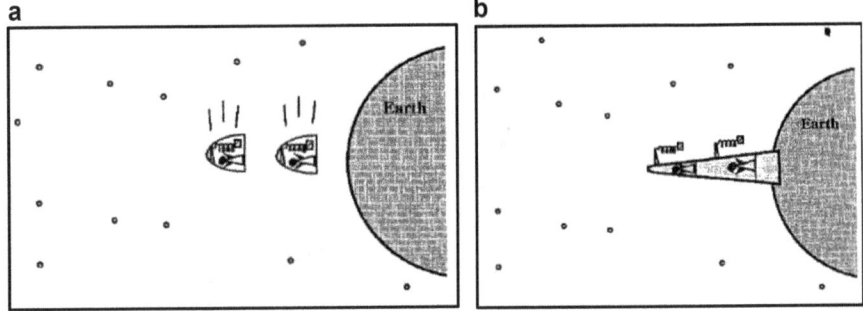

Fig. 5.27 Weighing in satellites (**a**) and weighing on an imaginary high tower (**b**)

2. Consider an imaginary tower, 200 km high (Fig. 5.27b). Two researchers weigh a mass of 1 kg at the heights of 100 km and 200 km from the ground. What are the results of weighing obtained and what are the inferences regarding the weight in each case? Explain your considerations

3. Suppose a person weighs a box by means of a spring scale of extremely high sensitivity. During the measurement, the Moon is passing over the person (Fig. 5.28a). Will this event influence the results of the weighing? Did the weight of the box change? Explain your considerations.

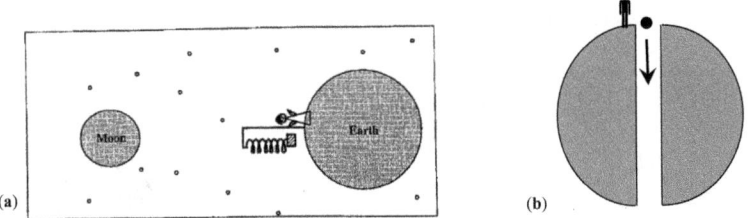

Fig. 5.28 (**a**) An object is weighed on the ground, while the Moon is passing over. (**b**) An object is dropped into the imaginary shaft through the globe of the Earth

4. In the fourteenth century, medieval scholars at the University of Paris debated the following imaginary situation: A body is dropped into a tunnel through the Earth globe (Fig. 5.28b). Consider the motion of the falling object. What happens to the weight of the body? Explain your considerations.

5. Consider four cases: jumping from a plane, floating under water surface, descending with a parachute, and jumping from a cliff. Compare these cases in terms of the weight changes of the people.

6. The following photograph was taken, while the astronaut left a satellite for a free "walk" in space (Fig. 5.29a) and the sketch of a person in a free-falling elevator cabin, next to the ground (Fig. 5.29b). Compare and characterize their situations in terms of weight and gravitational force. Are the situations different or similar in physical account?

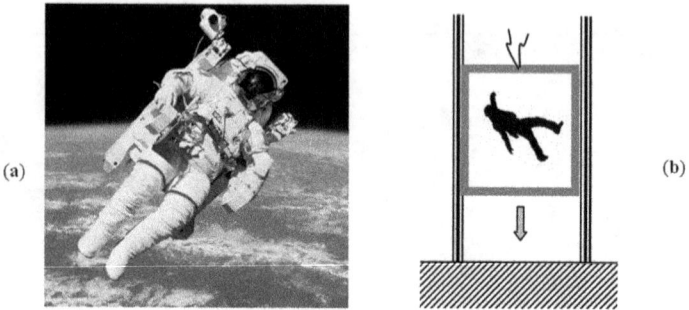

Fig. 5.29 (**a**) Astronaut Bruce McCandless II, STS-41-B mission in 1984, in space near the orbiting satellite. Image credit: NASA. (**b**) An image of a passenger in a free-falling elevator

7. Falling objects are in a state of weightlessness. Consider a person on the ground throwing a ball at an angle to the horizon. At what part of its trajectory may one say that the ball is weightless, if at all? Ignore air resistance.

8. What is the weight of the Moon? What would be the answer given by Newton? Are the two answers (yours and Newton's) different? Explain.

9. NASA train people to function in a state of weightlessness in a plane especially prepared for that mission (Fig. 5.30).[101] Explain how it is possible to reach weightlessness flying inside the atmosphere. At what trajectory is the state of weightlessness reached?

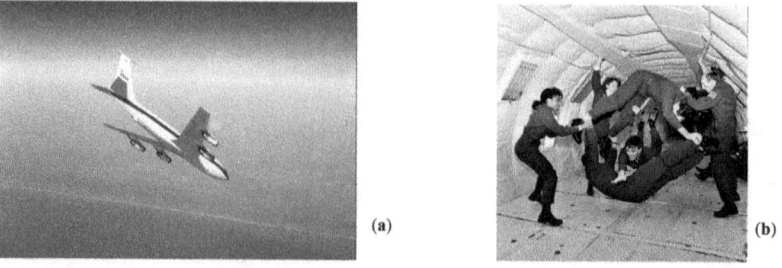

Fig. 5.30 (**a**) The NASA plane used to train weightlessness. (**b**) Inside the plane, at the state of weightlessness. Picture credit: NASA

---

[101] Levin (2016).

10. In the movie "2001: Space Odyssey," Kubrick also showed an additional way to cope with weightlessness. In a short-distance flights from the Earth to the satellite station, a waitress used magnetic boots which allowed her walking on the walls of the spacecraft (Fig. 5.31a). Consider whether the waitress indeed possessed weight? Explain your answer in comparison with:

   – The way astronauts will experience weight in the rotating cabin (Fig. 5.23).
   – The way used in physical training of the astronauts on board of the real space station currently orbiting the Earth. There, springs pull the astronaut in three directions (Fig. 5.31b).
   – In the regular on ground environment where people experience weight stress distribution across their bodies (Fig. 5.31c).

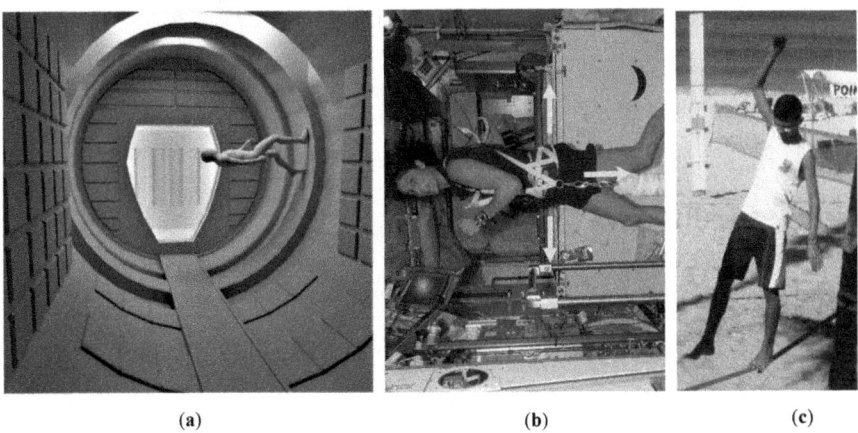

(a)   (b)   (c)

Fig. 5.31 (**a**) A sketch imitating the situation in the movie *2001: Space Odyssey*. It was suggested that magnetic boots created her weight and ability to walk on the walls (sketch by Guy Galili). (**b**) The weight created by three springs pulling the body of the astronaut in the international space station allowing her physical exercises (arrows show elastic forces on the body of the runner). (**c**) Weight as experienced in a regular environment. (Photo by the author)

11. Compare a diver with an astronaut in a coasting satellite space station. Both close eyes and consider where is up and down. What is the difference between their experiences?

12. The National Aeronautics and Space Administration—NASA—uses pools of water for training astronauts (Fig. 5.32). What could be the rationale of using water pools in astronauts' training? Does floating in water put the divers into the state of weightlessness as in a satellite? Discuss the differences and similarities with respect to weight.

(a)   (b)

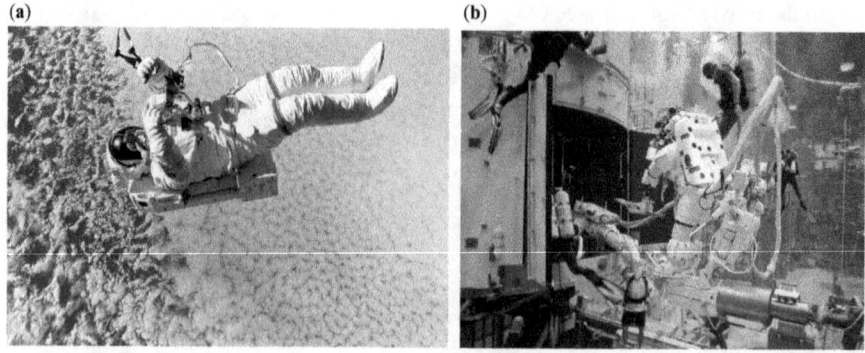

Fig. 5.32 (a) Astronaut floating in space. (b) Astronauts floating in the water pool in the training center. Pictures credit: NASA. (Pictures credit: NASA)

*** *

Picture Credits

- Stanford torus, 1975 NASA concept by Don Davis, NASA Ames Research Center (ID AC76-0525), Public Domain. https://commons.wikimedia.org/w/index.php?curid=1423294
- Bust of Aristotle after Lysippos – Jastrow (2006), Public Domain. https://commons.wikimedia.org/w/index.php?curid=1359807
- Portrait of Archimedes. Public Domain. https://commons.wikimedia.org/w/index.php?curid=653804
- Portrait of Rene Descartes, after Frans, Hals1640. De eeuw van Rembrandt, Bruxelles: Crédit communal de Belgique. Public Domain. https://commons.wikimedia.org/w/index.php?curid=2774313
- Portrait of Christiaan Huygens by Caspar Netscher, 1671, Kunstmuseum Den Haag. Public Domain. https://commons.wikimedia.org/w/index.php?curid=44047
- Portrait of Newton by Godfrey Kneller, 1689. https://commons.wikimedia.org/w/index.php?curid=146431
- Portrait of Einstein by Ferdinand Schmutzer. Public Domain. https://commons.wikimedia.org/w/index.php?curid=34239518
- Fig. 5.1 Weighing of the Heart. Ancient image. British museum. Public Domain. https://commons.wikimedia.org/wiki/File:BD_Weighing_of_the_Heart.jpg
- Bust of Plato. Copy of Silanion. The CC Attribution 2.5 Generic license. https://commons.wikimedia.org/w/index.php?curid=7831217
- Euclid statue in Oxford University. Photo by Mark A. Wilson. The CC Attribution-Share Alike 4.0 International license. https://commons.wikimedia.org/w/index.php?curid=50414450
- Portrait of Galileo Galilei by Justus Sustermans, 1640, National Maritime Museum. Public Domain. https://commons.wikimedia.org/w/index.php?curid=230543
- Portrait of Nicole Oresme in "Traité de l'espère", 1377. Public Domain. https://en.wikipedia.org/wiki/Nicole_Oresme#/media/File:Oresme.jpg
- Fig. 5.4 Illustration from Descartes, *The Principles of Philosophy*, 1644, *Plate VI*

- Fig. 5.4 Spiral galaxy. Photo by *ESA/Hubble & NASA*. https://commons.wikimedia.org/w/index.php?curid=13586042
- Title page of Newton's Principia, 1687. Public Domain. https://commons.wikimedia.org/w/index.php?curid=2681838
- Portrait of Euler by Jakob Emanuel Handmann, 1756. Public Domain. https://commons.wikimedia.org/w/index.php?curid=1001511
- Portrait of Reichenbach Hans, 1921, Vienna museum photo. Public Domain. https://commons.wikimedia.org/w/index.php?curid=39811611
- Portrait of Percy Williams Bridgman, 1946, Nobel prize cite. Public Domain. https://commons.wikimedia.org/wiki/File:Bridgman.jpg
- Fig. 5.17a Rotational space station image NASA ID #AC76-0492.1 Picture credit: NASA. https://space.nss.org/settlement/nasa/70sArtHiRes/70sArt/art.html
- Fig. 5.17b Inside rotational space station image NASA ID Number #AC78-0330-4. https://space.nss.org/settlement/nasa/70sArtHiRes/70sArt/art.html
- Fig. 5.19, 29a Space walking by Bruce McCandless in 1984. Picture credit: NASA. https://commons.wikimedia.org/w/index.php?curid=48209
- Fig. 5.20a An illustration from Jules Verne, 1872 by Henri de Montaut. Public domain. https://en.wikipedia.org/wiki/From_the_Earth_to_the_Moon#/media/File:'From_the_Earth_to_the_Moon'_by_Henri_de_Montaut_39.jpg
- Fig. 5.20b An illustration from Jules Verne, 1872. Public domain. https://er.jsc.nasa.gov/seh/vernewei.gif
- Fig. 5.21 Portrait of Herman Potocnik, 1929. Public Domain. https://commons.wikimedia.org/wiki/File:Herman_Potocnik_Noordung.jpg
- Fig. 5.21 Illustration picture in Noordung 1929. Public Domain. https://commons.wikimedia.org/wiki/File:Noordung%27s_Space_Station_Habitat_Wheel_-_GPN-2003-00101.jpg
- Fig. 5.22 Werner von Braun at NASA headquarters, 1970. Picture credit: NASA. Public Domain. https://en.wikipedia.org/wiki/Wernher_von_Braun#/media/File:19700202-wernher-von-braun-nasa.jpg
- Fig. 5.22 Braun's project, 1952. Picture credit: NASA. Public Domain. https://commons.wikimedia.org/w/index.php?curid=14867654
- Fig. 5.24a, b. A permanent colony in space. Images credits: NASA ID #AC76-0525 and NASA ID # AC75-1086-1. https://space.nss.org/settlement/nasa/70sArtHiRes/70sArt/art.html
- Fig. 5.25b Supersonic jet pilot, 1958. Picture credit: NASA. https://upload.wikimedia.org/wikipedia/commons/7/71/Joe_Walker_X-1E.jpg
- Fig. 5.26a Jumping from the plane. Photo credit: <u>NASA, Center Snapshot: Steve Riddick.</u> https://www.nasa.gov/centers/langley/news/researchernews/snapshot_sriddick.html
- Fig. 5.26b Photo of a diver by Summitandbeach. The CC Attribution-Share Alike 4.0 International license. https://commons.wikimedia.org/w/index.php?curid=42295587
- Fig. 5.30a The plane for weightlessness training. Picture credit: NASA – Zero-Gravity Plane. https://www.nasa.gov/vision/space/preparingtravel/kc135onfinal.html
- Fig. 5.30b Weightlessness inside plane. Picture credit: NASA. https://www.nasa.gov/sites/default/files/images/601194main_kc135-microgravity-training.jpg
- Fig. 5.31b Walking in the International Space Station. Picture credit: NASA. https://www.nasa.gov/mission_pages/station/main/2001_anniversary.html
- Fig. 5.32a Astronaut in space. Picture Credit: NASA. https://cnet2.cbsistatic.com/img/Kko0wAwXQKOD2O2_S1mLHKv_xXU=/980x0/2018/10/10/64aa5535-142f-4988-a291-931cab655b58/marklee.jpg
- Fig. 5.32b NASA Astronaut training. Picture Credit: NASA. https://upload.wikimedia.org/wikipedia/en/thumb/7/78/Nasa_astronaut_training_at_NBL.jpg/1280px-Nasa_astronaut_training_at_NBL.jpg?1607700400717

References

Aiton, E. J. (1959). The Cartesian theories of gravity. *Annals of Science, 15*(1), 27–49.

Albert of Saxony. (1959). Questions on the four books of the heavens and the world of Aristotle. In M. Clagett (Ed.), *The science of mechanics in the Middle Ages.* Madison, WI: The University of Wisconsin Press.

Aquinas, T. (1267/1952). *Summa Theologica.* Chicago: Encyclopaedia Britannica.

Archimedes. (1952). *On floating bodies.* Chicago: Encyclopedia Britannica.

Aristotle. (1952). *On the heavens.* Chicago: Encyclopedia Britannica.

Balukovic, J., & Slisko, J. (2018). Teaching and learning the concept of weightlessness: An additional look at physics textbooks. *European Journal of Physics Education, 9*(1), 1309–7202.

Baruch, A., & Vizansky, A. (1937). *Physics.* Tel-Aviv: Dvir.

Born, M. (1962). *Einstein's theory of relativity.* New York: Dover.

Bridgman, P. (1927/1952). *The logic of modern physics.* New York: Macmillan.

Brooks, R. A. (2010). *Fields of color: The theory that escaped Einstein.* Silver Spring, Maryland: Universal Printing, LLC.

Brown, J. E. (1978). The science of weight. In D. C. Lindberg (Ed.), *Science in the Middle Ages.* Chicago: The University of Chicago Press.

Chaikin, S. E. (1947/1963). *Mechanics. General physics course* (Vol. 1). Moscow: Gosizdaetlstvo Physmath Literaturi.

Clagett, M. (1959). *The science of mechanics in the Middle Ages.* Madison, WI: The University of Wisconsin Press.

Crowell, B. (2006). *Newtonian physics.* Fullerton, CA: Fullerton.

Descartes, R. (1644/1982). *Principles of philosophy.* Dordrecht: D. Reidel.

Drake, S. (1978). *Galileo at work.* New York: Dover.

Einstein, A. (1916/1923). *The foundation of the general theory of relativity in the principle of relativity* (pp. 111–164). New York: Dover.

Euclid. (1959). The book of balance. In M. Clagett (Ed.), *The science of mechanics in the Middle Ages.* Oxford: Oxford University Press. Def. 1.

Euler, L. (1765/2009). *Theory of the motion of solid or rigid bodies* (Vol. 1). Ch. 4. http://www.17centurymaths.com/contents/euler/mechvol3/genmotch4.pdf

French, A. (1971). *Newtonian mechanics.* New York: Norton.

Galilei, G. (1638/1914). *Dialogue concerning two new sciences.* New York: Dover.

Galili, I. (2017). Chapter 8: Scientific knowledge as a culture – A paradigm of knowledge representation for meaningful teaching and learning science. In M. R. Matthews (Ed.), *History, philosophy and science teaching research. New perspectives* (pp. 203–233). Dordrecht: Springer.

Galili, I. (2019). On the influence of Ernst Mach on contemporary physics curriculum at schools: The concept of weight. In *Ernst Mach – Life, work, influence.* Springer.

Galili, I., & Bar. (1997). Children's operational knowledge about weight. *International Journal of Science Education, 19*(3), 317–340.

Galili, I., Bar, V., & Brosh, Y. (2017). Teaching weight-gravity and gravitation in middle school – Testing a new instructional approach. *Science & Education, 26*(3), 977–1010.

Galili, I., & Kaplan, D. (1996). Students operation with the concept of weight. *Science Education, 80*(4), 457–487.

Galili, I., & Kaplan, D. (2002). Students' interpretation of water surface orientation and inertial forces in physics curriculum. *Praxis der Naturwissenschaften Physik in der Schule, 51*(7), 2–11.

Galili, I., & Lehavi, Y. (2003). The importance of weightlessness and tides in teaching gravitation. *American Journal of Physics, 71*(11), 1127–1135.

Galili, I., & Lehavi, Y. (2006). Definitions of physical concepts: A study of physics teachers' knowledge and views. *International Journal of Science Education, 28*(5), 521–541.

Giancoli, D. C. (1988). *Physics for scientists and engineers.* Englewood Cliffs, NJ: Prentice Hall.

Grant, E. (1977). *Physical science in the Middle Ages.* Cambridge, MA: Cambridge University Press.

Halliday, D., Resnick, R., & Waker, J. (2001). *Fundamentals of physics.* New York: Wiley.

Hecht, E. (2019). Kepler and the origins of the theory of gravity. *American Journal of Physics, 87*(3), 176–185.

Hewitt, P. G. (1992). *Conceptual physics*. Upper Saddle River, NJ: Pearson Prentice Hall.

Hewitt, P. G. (2002). *Conceptual physics*. Upper Saddle River, NJ: Pearson Prentice Hall.

Huygens, CH. (1659/1703). *On centrifugal force*. From *De vi Centrifuga*, Oeuvres Complètes. XVI, pp. 255–301, (M. S. Mahoney, Trans.)

Jammer, M. (1957). *Concepts of force*. Cambridge, MA: Harvard University Press.

Jewett, J. W., & Serway, R. A. (2008). *Physics for scientists and engineers*. Belmont, CA: Thomson Higher Education.

Keller, F. J., Gettys, W. E., & Skove, M. J. (1993). *Physics*. New York: McGraw Hill.

Kepler, J. (1992). *New astronomy*. Cambridge: Cambridge University Press.

Knight, R. D. (2004). *Physics for scientists and engineers*. Reading, MA: Pearson.

Knight, R. D. (2013). *Physics for scientists and engineers* (3rd ed.). Reading, MA: Pearson.

Koyré, A. (1943). Galileo and the scientific revolution of the seventeenth century. The philosophical review. *Journal of the History of Ideas, 52*(4), 333–348.

Koyré, A. (1956). Influence of philosophic trends on the formulation of scientific theories. In P. G. Frank (Ed.), *The validation of scientific theories*. New York: Collier.

Lehavi, Y., & Galili, I. (2009). The status of Galileo's law of free-fall and its implications for physics education. *The American Journal of Physics, 77*(5), 417–423.

Lerner, S. L. (1996). *Physics*. Sudbury, MA: Jones & Barlett.

Levin, W. (2016). *MIT physics lectures* (Lecture 7: Weight, Perceived Gravity, and Weightlessnes; 8.01 Classical Mechanics). https://www.youtube.com/watch?v=M0mxyPOMcw0

Mach, E. (1883/1919/1989). *The science of mechanics, a critical and historical account of its development*. Open Court.

Margenau, H. (1950). The role of definitions in science. In *The nature of physical reality*. New York: McGraw-Hill.

Marion, J. B., & Hornyack, W. F. (1982). *Physics for science and engineering*. New York: Saunders.

Meli, D. B. (2006). Inherent and centrifugal forces in Newton. *Archives of History of Exact Science, 60*, 319–335.

Merz, J. T. (1904/1965). *A history of European scientific thought in the nineteen century*. New York: Dover.

Moody, E. A., & Clagett, M. (Eds.). (1952). *The Medieval science of weights*. Madison, WI: The University of Wisconsin Press.

Newton, I. (1687/1999). *The Principia. Mathematical principles of natural philosophy* (B. Cohen & A. Whitman, Trans.). University of California Press.

Noordung, H. (1929). *Das problem der Befahrung des Weltraums*. Berlin: Schmidt & Co. (*The problem of space travel*. NASA History Office, 1995).

Ohanian, H. C. (1989). *Physics*. New York: Norton.

Orear, J. (1967). *Fundamental physics*. New York: Wiley.

Piaget, J. (1972). *The child's conception of physical causality*. Totowa, NJ: Totowa Adams Littlefield.

Plato. (1952). *The dialogues of Plato. Timaeus*. Chicago: Encyclopedia Britannica.

Reichenbach, H. (1927/1942). *From Copernicus to Einstein*. New York: Philosophical Library.

Reichenbach, H. (1928/1958). *The philosophy of space and time*. New York: Dover.

Resnick, R., & Halliday, D. (1966). *Physics*. New York: Wiley.

Ruggiero, S., Cartelli, A., Dupre, F., & Vinncentini, M. (1985). Weight, gravity and air pressure: Mental representations by Italian middle school pupils. *European Journal of Science Education, 7*(2), 181–194.

Rutherford, F. J. (1990). *Science for all Americans. Project 2061*. New York: Oxford University Press.

Stein, B., & Galili, I. (2018). Introduction of observer dependent concepts into physics teaching of middle school. In *Proceedings of the 11th conference of the European Science Education Research Association (ESERA), Book of oral papers*, Dublin, Ireland.

Stein, H., & Galili, I. (2014). The impact of operational definition of weight concept on students' understanding of physical situations. *International Journal of Research in Science and Mathematical Education, 13*(6), 1487–1515.

Stein, H., Galili, I., & Schur, Y. (2015). Teaching new conceptual framework of weight and gravitation in the middle school. *Journal of Research in Science Teaching, 52*(9), 1234–1268.

Taylor, B. N., & Thompson, A. (2008). *The international system of units (SI)* (NIST Special Publication 330) (p. 52). Commerce Department, National Institute of Standards and Technology (NIST).

Thompson, A., & Taylor, B. N. (2008). *Guide for the use of the international system of units (SI)* (Special Publication 811). Gaithersburg: National Institute of Standards and Technology (NIST).

Thornton, W., & Bonato, F. (2017). *The human body and weightlessness. Operational effects, problems and countermeasures*. Springer.

Verne, J. (1874). *From the earth to the moon*. New York: Schribner, Armstrong & Co.

Walker, J. (2008). *Halliday-Resnick fundamentals of physics* (p. 95). New York: Wiley.

Wheeler, J. A. (1990). *A journey into gravity and spacetime*. New York: Scientific American Library. (1999). *Journey into gravity and spacetime*. Freeman.

Young, H. D., & Freedman, R. A. (2004). University physics (pp. 120, 441, 459–460). Pearson, Addison Wesley.

Young, H. D., & Freedman, R. A. (2012). *University physics* (pp. 406, 421–422). Pearson, Addison Wesley.

Part II
Perspectives

Chapter 6
Scientific Knowledge as a Culture: A Paradigm of Knowledge Representation for the Meaningful Teaching and Learning of Science

Old understanding –
Simplicio - Aristotle

Adoption of the new
knowledge - Sagredo

New understanding –
Salviati - Galileo

I. Galili, *Scientific Knowledge as a Culture*, Science: Philosophy, History and Education, https://doi.org/10.1007/978-3-030-80201-1_6

Abstract This chapter reviews a cultural approach to the science/physics curriculum. Scientific knowledge is considered as a culture related to a discipline, which we term as discipline–culture (DC); accordingly, disciplinary content knowledge is upgraded to cultural content knowledge (CCK). Physics knowledge comprises fundamental theories that are hierarchically structured in a triadic pattern: nucleus–body–periphery to represent the discipline–culture. This structure effectively displays the meaning of each theory within the discipline and their relationship to each other. The meaning of fundamental theories (nucleus) is emphasized by their contrast to alternative theories (periphery). From an epistemological perspective, the cultural approach suggests that different approaches—rationalist and empiricist—are complementary threads interwoven into an integrated method of science. The DC-oriented curriculum incorporates content of history and philosophy of science and clarifies their role in science education. A metaknowledge (big picture) of science is established, which may be attractive to a wide variety of learners having different interests and cognitive preferences. Three ways to implement a DC-oriented approach were empirically explored and are briefly described: a new curriculum, conceptual excursus, and summative lecture.

> *Besides the future engineers are other less numerous pupils, destined in their turn to become teachers, and so they must go to the very root of the matter; a profound and exact knowledge of first principles is above all indispensable for them.* Poincaré (1903)

Ontology in the Cultural Approach to Knowledge

One of the central missions of science education is to explore the ways in which scientific knowledge is represented in teaching and learning. It is therefore not only necessary to identify what are the most representative and essential contents of scientific knowledge but also necessary to understand the structural network within which these contents interrelate. As a result, we may need to revise the common curriculum, which presents disciplinary contents as an unfolding sequence of concepts, building from the simpler to the more complex. This unfolding presentation ignores the holistic image of scientific knowledge, its structure, and hierarchy. However, it is the holistic image, which is required by the science teacher who seeks to present a big picture of scientific knowledge and does not merely wish to practice science in the context of everyday needs or professional problem solving. In this sense, the interests of the science teacher are closer to those of the science philosopher than to those of the science practitioner. The science teacher requires a deeper knowledge of science than other consumers of science (the epigraph). Science teachers therefore need precise knowledge of a specific kind. It is argued here that cultural content knowledge (CCK) is the appropriate form of required knowledge. It implies a new structure of curriculum, based on a discipline–culture (DC) rather than on a discipline as traditionally taught.[1]

---

[1] Galili (2012, 2017)

We consider knowledge about science as a sort of metaknowledge, which involves HPS as we specify in this chapter. *Scientific knowledge* presents a *cultural system*, which implies the knowledge of a discourse of ideas rather than univocal claims. Such orientation not only makes the curriculum closer to the nature of science but also makes it easier for mediation with the learners who construct their understanding of science.

We will describe this new perspective and its applications to the science curriculum, the way it presents the scientific knowledge, the principle of choosing which HPS contents to address, and its impact on the conceptual change in students and on the typology of their preferences. We will specify the implementations of the new perspective in the form of three curricular approaches: the DC-based curriculum, a conceptual excursus, and a summative lecture. We will then expand the cultural perspective on the epistemological issue—the special role of philosophy of science in the science curriculum—and argue for its importance in learning science.

Scientific Knowledge as a Culture

The notion of culture is extremely inclusive. It was introduced in anthropology to designate the entirety of human products.[2] Hofstede reduced this vast scope to a subset of material and spiritual products and the activities, which distinguish one group of products from another. Thus, science is clearly distinct from history, religion, philosophy, and arts, even if it addresses them. Science is of a specific nature and employs specific methodology, goals, and values. Bruner defined: "a culture – an established, almost irreversibly stabilized way of thinking, believing, acting, judging"[3] and following Vygotsky, he considered the essence of education as enculturation.[4] These authors apparently distinguished science *and* culture and consider them in interaction.[5] Another vision elaborates the features of activities, behavior, and the relationships of scientists as a social group.[6] Still another meaning of culture presumes traditional knowledge of different ethnic groups and populations in their account of nature, social behavior, and spiritual heritage.[7] We address here a still different perspective on science knowledge as rational knowledge about nature, produced in a specific discourse of an objective sense. This knowledge is sometimes termed "Western science" because it was invented in the Greek civilization about 2500 years ago and obtained its form as modern science in Western Europe in the seventeenth century. However, the "Western" label is now obsolete, as science presents a specific kind of knowledge universally accepted and practiced by the whole of humankind.

---

[2] Tylor (1871/1920); Hofstede (1991).

[3] Brunner (1996, p. 103).

[4] Also Russell (1932/2010, p. 15).

[5] For example, Fehl (1965), Bevilacqua et al. (2001), Jaroszyński (2007).

[6] For example, Latour (1987).

[7] For example, Aikenhead (1997); Ma (2012); Liu (2015).

The Greek innovation was to construct an objective *theory* of nature, which, unlike mythology, was independent of any individual will. The objective theory of nature became the subject of natural science. Over time, natural science split into several domains, addressing different aspects of reality. Physics knowledge, the most fundamental in natural science, as we know it, is composed of a small number of fundamental theories. Each fundamental theory of physics presents a very inclusive, organized, and coherent cluster of interrelated concepts, principles, models, laws, and their derivatives. Together, these theories describe and interpret the features of world organization in a unified structure.[8] Each such theory presents a specific "picture of the world,"[9] broad, but valid within a certain domain of experience and within a certain degree of accuracy/reliability.

There is a perspective in which knowledge itself is considered as a culture. Lotman distinguished two types of such constructs: the *culture of rules* and the *culture of texts*.[10] Within the first type, well-defined rules regulate relationships among elements in a certain area of theoretical knowledge, making clear their correct–incorrect status (e.g., jurisprudence). By contrast, the culture of texts allows knowledge elements to be grouped around canonical exemplars (e.g., art). Thus, in art, the meaning of *correct* is very different; it is vague and not particularly/truly/altogether representative of a cultural system.[11] In scientific knowledge, the status of correct (and of wrong, of course) comes to the fore. It is considered as a *culture of rules*, which is composed of several disciplines, each drawing on a fundamental theory. An example would be physics as composed of classical mechanics, electromagnetism, thermodynamics, etc. A disciplinary theory presents an inclusive group of knowledge elements, which can be structured in terms of nucleus–body–periphery, together termed discipline–culture (DC) (Fig. 6.1).[12] Such representation not only is hierarchical but also reflects scientific knowledge as inherently discursive, explicit in validation in terms of correct–incorrect with regard to the epistemological norms included in the nucleus. Considering scientific knowledge as a *culture* creates a more complete expression of scientific knowledge. Expanding *discipline* to *discipline–culture* refers to the meaning of culture in the sense of multiplicity, including even those additional elements that contradict the norms of the nucleus. However, unlike in art, the conflicting content is clearly distinguished from the fundamentals by being associated with the periphery. Accordingly, the hierarchy of correct–incorrect is explicit as is the dialogical nature of scientific knowledge, which excludes a mere ignoring of the alternative.[13]

---

[8] Though we address the fundamental theories of physics (Heisenberg, 1959/1971; Weizsacker, 1985/2006; Weinberg, 1992), the stated, regarding theory representation, holds also regarding theories in biology (e.g., theory of evolution) and chemistry (e.g., theory of elements, substances, and their transformation in interaction).

[9] Einstein (1918/2002).

[10] Lotman (2010).

[11] Other examples of similar cultural systems could be history, traditional medicine, etc.

[12] Tseitlin and Galili (2005). To represent a discipline, one may start from the dual taxonomy, nucleus–body, and expand it to the triadic by adding other elements at odds with the nucleus–periphery.

[13] Appendix 1.

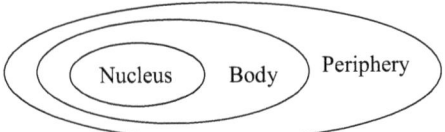

Fig. 6.1 Schematic representation of the discipline–culture structure of a scientific theory. Elements of knowledge are affiliated to three different areas

Discipline–culture codifies scientific theory in a tripartite structure. Its first region, nucleus, includes fundamentals—the principles of an ontological and epistemological nature, the paradigmatic model of the theory and its basic concepts used as principles, and assumptions to be justified by further success in application. The second region, body knowledge, includes the elements subdued to the nucleus. They could both be derived from and reduced to the fundamentals or be empirically based statements, associated with the theory. Body knowledge incorporates specific laws, concepts, models, explanations of natural phenomena, experiments (actual and theoretical), technological applications, etc.[14] These two regions of knowledge elements represent a structure of scientific disciplines as they are traditionally taught in classes.[15]

However, to better represent the nature of scientific knowledge as a culture, the theory representation should be expanded to include elements of a third type—*periphery*. The elements of the periphery are at odds with the pertinent nucleus. In addition, the periphery includes challenging open problems and alternative accounts of problems and phenomena. By challenging certain elements within the nucleus, the periphery of that theory refines the meaning of the nucleus and also helps to establish the boundaries of the theory, which are often left implicit and not known to the novice. In this sense, the periphery relates to the tradition of apophatic and comparative approaches in philosophy.[16] Scientific knowledge, codified and structured in terms of discipline–cultures, establishes the cultural content knowledge and suggests a framework for a new type of science curriculum.

---

[14] It is actually with regard to the *body of knowledge* that we can adopt Kuhn's statement about theories ("disciplinary matrix") as being taught in classes through providing exemplars of the concrete problem–solutions in representative settings of inclined plane, circular motion, the conical pendulum that students encounter in their education, in laboratories, on examinations, or exercises (Kuhn, 1962/1970, p. 187). They comprise a toolkit for a science practitioner.

[15] This definition, to a great extent, escapes the identification of a theory as either *syntactic* (based on axioms and derived theorems) or *semantic* (composed of models) type (van Fraassen, 1980, p. 44; Giere, 1988, p. 48). DC incorporates organization of both types. It is hierarchically structured and conceptually elaborated, and it also provides models. Its syntax includes axioms (principles, ontological, and epistemological) and theorems (derived and affiliated laws). Its semantics includes models in the broad sense, including concepts, working and representative models, laws, experiments, and methodological rules. In fact, the dual organization, nucleus–body, depicts a scientific theory as it is commonly presented in disciplinary courses: mechanics, electromagnetism, thermodynamics, and so on.

[16] For example, Libbrecht (2009).

The concept of discipline–culture fuses two concepts that were seen by Kant in opposition. He wrote:[17]

> *The restraint which is employed to repress, and finally to extirpate the constant inclination to depart from certain rules, is termed <u>discipline</u>. It is distinguished from <u>culture</u>, which aims at the formation of a certain degree of skill, without attempting to repress or to destroy any other mental power, already existing.*

Indeed, knowledge elements within a scientific discipline presume to share a commitment to its fundamentals, incorporated in the nucleus of its theory. The knowledge elements of the periphery in the correspondent discipline–culture may represent other ways to account for the same subject matter, sharing other commitments.[18] Different *disciplines*, however, belong to the same *culture* of science. It may be surprising that this perspective expressed by Kant at the time when only one scientific discipline practically existed—the Newtonian mechanics. In fact, however, fusing the ideas of *discipline* (thesis) and *culture* (antithesis) in *discipline–culture* (synthesis) was started much earlier and represented the nature of science. Aristotle, unlike many other scholars of the time, normally considered a variety of other views in presenting his own. We may mention Galileo's *Dialogues* (1632) and *Discourses* (1638), among the first treatises of the Modern Science during the scientific revolution. The chosen form of content presentation by Galileo well matched the structure of a discipline–culture (see the front piece of this chapter).

The framework of discipline–culture may serve as guidance in selection of relevant HPS materials, clarifying their role and possible involvement in teaching strategy. This approach was exemplified in developing the historico-philosophical perspective on teaching optics.[19]

Furthermore, since learning scientific concepts could be considered somewhat similar to the process of learning a foreign language[20]—the learning "from outside"—identifying scientific knowledge as a culture cannot be left for students to discover alone. Instruction should introduce the triple codification of scientific knowledge and its meaning explicitly.

With regard to the perspectives accumulated in HPS, the triadic structure of a fundamental theory allows the central claims regarding science knowledge to be visualized, including their functioning, developments, and changes. Thus, this structure may provide a platform upon which to explain various philosophical

---

[17] Kant (1781/1952, p. 210). Kant reminds us the original meaning of the term *discipline*, which signifies the obligation to follow certain rules of behavior whether or not we understand them. In science, it is common to use this notion in a different sense, signifying a certain domain of knowledge. For example, we call mechanics to be a discipline, which together with other disciplines comprises physics curriculum.

[18] Thus, the periphery of the classical mechanics includes the elements of other mechanics prior and after classical. We will expand on this in the following.

[19] Galili (2014), Chap. 7

[20] Vygotsky (1934/1986).

concepts such as Popper's three worlds,[21] Lakatos's research programs,[22] Khun's scientific revolution,[23] Feyerbend's alternative knowledge,[24] and Schwab's stable and fluid enquiries,[25] and the debate regarding the cumulative nature of scientific knowledge within research tradition[26] can all be displayed and discussed in terms of the triadic DC structure. This will become clear in the following section.

CCK Implications

Discipline–culture is beyond a working model; it can serve as a broad approach in the representation of important issues in science, science history, ontology, and epistemology.

Scientific Revolution and History of Physics

We can apply the introduced structure to represent conceptual change taking place in science during scientific revolutions, when one fundamental theory replaces another. Instead of merely stating the new concepts, DC suggests an inclusive and specific depiction of the conceptual change. The change is never abrupt and often creates a complex picture. Giere stated that:[27]

> *At almost any period in history, one can find a vast range of ideas existing simultaneously. The important question is which of the variety of ideas available at an earlier period got adopted and transmitted to later periods and thus shaped later interpretations.*

Consider, for example, the account of the cosmic organization starting from the geocentric theory. Its *nucleus* included principles of Aristotelian physics of motion—the state of rest as the natural state of matter in the under-lunar world and the homogeneous concentric circular motion in the supra-lunar world. These were the basic principles of a spherical universe. In accordance, the *body knowledge* of that theory included the celestial model of Eudoxus that is coherent with the nucleus, but not only. The body knowledge also included the Ptolemaic model, its auxiliary concepts (deferent, epicycle, and equant), the successful account of seasons, equinoxes, eclipses, and other phenomena.[28] Though the latter did not match the

---

[21] Popper (1978), e.g., Chap. 7

[22] Lakatos (1978), Appendix 2.

[23] Kuhn (1962/1970), Appendix 3.

[24] Feyerabend (1993) prescribes to the periphery the central role in the scientific progress.

[25] Schwab (1964) and Elkana (1970).

[26] Laudan (1996, pp. 119–120, 122).

[27] Giere (1999, p. 88).

[28] Kuhn (1957).

Aristotelian principles, it was used as an affiliated working model, empirically justified and much simpler than that of Eudoxus. It presented a pragmatic compromise for the under-lunar world.[29] The body knowledge implied so called *stable enquiry*,[30] allowing systematic data collection, the production of instruments (quadrant and Jacob's staff), and accounts for the observed phenomena (parallax and shadow patterns).

Importantly, however, the rival accounts of reality by Pythagoras, Heraclites, and Aristarchus (heliocentric setting and Earth spinning) were present from the very beginning, establishing a permanent debate with the nucleus.[31] In the course of time, various scholars also perform *fluid* enquiry,[32] introducing new elements of the periphery, challenging the nucleus. The accumulation of such elements increases tension with the nucleus ultimately causing the breakthrough.[33] That happened during the Copernican revolution. The heliocentric/heliostatic model occupied the nucleus, and the geocentric model moved to the periphery. The body elements transformed to adjust to the new nucleus. Besides the appearance of new elements of knowledge, the scientific revolution manifests itself in the relocation of numerous other elements accumulated in science. The DC structure of the fundamental theory makes the dynamics of the conceptual change observable. The elements of the new and old knowledge appear interrelated in one inclusive conceptual system. One may perceive this spirit of conceptual unity and internal dynamics from Einstein's words addressing the introduction of relativity to physics. He wrote:[34]

> *With respect to the theory of relativity, it is not at all a question of a revolutionary act, but of a natural development of a line which can be pursued through centuries.*

DC-based description visualizes this pursuit in the space of scientific discourse, displaying the crisis and the following revolutionary change as described by Kuhn. Importantly, in the *cultural* perspective, the elements of the old nucleus do not disappear. After being refuted, they are shifted to the periphery where they remain within the horizon of the theory as its conceptual background.

---

[29] Within the Lakatos perspective, the epicycle-based working model of Ptolemy is considered as a *protection belt* of the theory *hard core* (Lakatos, 1978).

[30] Schwab (1964), Elkana (1970).

[31] Ptolemy (1952, pp. 6–14); Heath (1966).

[32] Schwab (1964) and Elkana (1970).

[33] Kuhn (1957, 1962/1970).

[34] Miller (1986).

Theories Relationship

The controversy of fundamental physics theories is often addressed by claiming their *incommensurability* – an essential mismatch of their scientific paradigms.[35] The tripartite code of discipline culture allows this relationship to be presented. In particular, the nucleus of one fundamental theory ought to be located in the periphery of the other (Fig. 6.2). Yet, at the same time, the body areas of these theories may overlap, representing the cases where the same phenomenon or problem is treated by the two theories producing *commensurable* results.

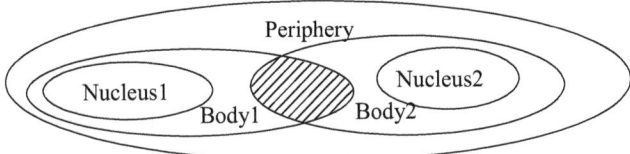

Fig. 6.2 Two fundamental theories in physics structured in DC form. Each nucleus is placed in the periphery of the other theory representing conceptual *incommensurability* of the nuclei, and the overlap of the bodies (the shaded area) represents their practical *commensurability*

For example, in the limit of low velocities (v < <c), both classical and relativistic mechanics provide physically equivalent results despite the essential contradiction between the nuclei of these two theories. This way the tripartite structure illustrates the meaning of the principle of correspondence in physics knowledge. It is in this sense that continuity of scientific knowledge is maintained, showing its cumulative nature as one of its major characteristics. Importantly, being cumulative does not imply that all the claims and problems included in the body knowledge of classical theory must be included in the body of the relativistic theory although it is more advanced. Alongside the results of relativistic theory, which cannot be reached by classical mechanics, classical theory may provide results that are valid for, say, simplicity and possessing sufficient accuracy. This is expressed in the only partial overlap of the bodies of knowledge in the diagram.[36]

Physics knowledge cannot be reduced to a single theory; it includes different theories, complementing each other in a special manner. They establish relationships known as family resemblance, sharing some features and differing in others. It allows more advanced theories to draw on the old accomplishments and problem solutions, rather than replacing altogether the previous knowledge and ignoring it. The DC structure envisions this type of relationship clarifying the nature of

---

[35] Kuhn (1962/1970).

[36] This comment could refer to the claim of Laudan (1996, pp. 21–23) regarding the limited value of cumulativity in scientific progress. Though it deserves refinement, the cumulative character of scientific knowledge is not dismissed but stipulates any progress in science. Positive and negative experience is equally valuable as guidance in research and exploration.

scientific holism and refining the metaphor of "patchwork" for the scientific knowledge.[37] Structuring the multiplicity of knowledge elements in a conceptually related web leads the learner to establishing *metaknowledge* of science, defined as an organism in terms of its essential functioning.[38] CCK does this in terms of the tripartite structure of theories.

Students' Learning

The DC tripartite structure suggests that we should consider learning as a dynamic process, often termed "conceptual change," within the CCK. Posner and associates[39] saw a similarity between the learning in the individual and conceptual change in science. Posner stated the conditions under which conceptual change may occur—dissatisfaction with old knowledge and intelligibility, plausibility, and fruitfulness of a new one. Since the same factors are considered with regard to the change of collective knowledge, the individual conceptual change can be represented in a similar way—as an exchange of content between the nucleus (the naïve "schemes of knowledge"[40])—and the initial peripheral content created by instruction, the disciplinary knowledge. The derivatives of the schemes, the "facets of knowledge,"[41] create the body knowledge of the particular subject matter. In effect, they serve as a protection belt of the nucleus, a barrier against change—blocking the motion of the periphery items into the nucleus. During knowledge transformation in the course of learning, the new knowledge migrates from the periphery obtaining new affiliation to the nucleus or body knowledge. At the same time, the patterns of naive knowledge do not disappear but shift from the nucleus to the periphery creating a new triadic structure from the naïve knowledge of a rather fragmentary organization.[42] Nevertheless, faced with novel situations or nonstandard problems, students may retrieve their naïve conceptions and apply them again.[43] In a way, they preserve those conceptions within their knowledge horizon, as potential tools for possible future application. The metaphor of breaking through the barrier of body knowledge may explain the difficulty of the required conceptual change (a higher "potential difference" between nucleus and periphery may be needed for a more experienced,

---

[37] Cartwright (2005).

[38] Novak and Gowin (1984).

[39] Posner et al. (1982).

[40] Schemes of knowledge present general (context independent) cognitive constructs, which serve as patterns of reasoning by students making sense of reality (Piaget, 1972).

[41] Facets of knowledge present cognitive constructs of a more concrete (context dependent) content also used in reasoning and problem solving strategy (Minstrell, 1992).

[42] "Knowledge in pieces" (diSessa, 1993).

[43] Galili and Bar (1992).

confident learner[44]) and how it is essentially different from upgrading old software in a computer.

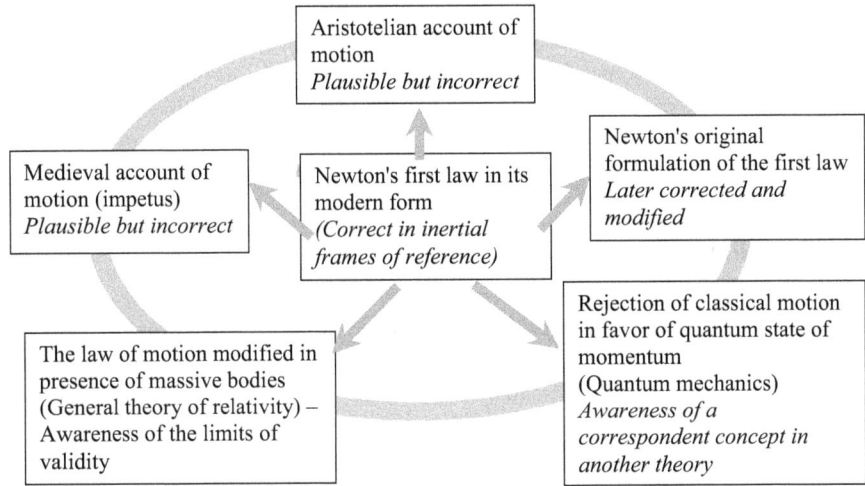

Fig. 6.3 The CCK with regard to the first Newton law requires enlarged space of conceptual variation

Elements from the periphery may facilitate meaningful learning by providing comparison stimulating understanding. The options for comparison create and extend a "space for learning"[45]—a very important setting for a learner to grasp the essence of the considered concept. The periphery provides such conceptual variation. For instance, seeking genuine understanding in teaching mechanics, the teacher should expand upon the usual univocal presentation. For example, Newton's first law—the cornerstone of classical mechanics—may be presented against the background of alternative accounts of motion, those by the Aristotelian and medieval impetus theories (Fig. 6.3).[46] A deeper understanding comes from recognizing the difference between the modern (addressing an inertial observer) and original (Newtonian) formulations of the law.[47] Further understanding comes from mentioning that the first law of classical mechanics does not hold in quantum theory and is

---

[44] This understanding may hold regarding scientists as well, explaining the famous comment of Max Planck (1936/1963, p. 97) that "An important scientific innovation rarely makes its way by gradually winning over and converting its opponents: it rarely happens that Saul becomes Paul. What does happen is that its opponents gradually die out and that the growing generation is familiarized with the idea from the beginning: another instance of the fact that the future lies with youth."

[45] Marton et al. (2004).

[46] In contrast to what is often taught in class, the first law is not a special case of the second law, but the central claim of classical mechanics that is refined in the second law (e.g., Galili and Tseitlin, 2003).

[47] Galili and Tseitlin (2003).

essentially modified in the theory of general relativity. This way, the rich meaning of the law is constructed by the students together with another important aspect of knowledge—the domain of validity of the unavoidable in any physics law.

The creation of CCK naturally incorporates the historical background of the subject matter of a special kind. It supports the comparison of conceptual accounts. It does not significantly expand the time of teaching but does so regarding students' thinking about subject matter. Making such metaknowledge valuable presents meta-learning; they are interconnected.[48]

Physics Curriculum

The DC paradigm implies essential changes to the physics curriculum. Typically, during teaching, the theory-based nature of physics knowledge is scarcely emphasized. Instead, an unfolding sequence of concepts, laws, models, instruments, experiments, problems to solve, and phenomena to explain flood the learners with a vast amount of knowledge elements without their hierarchy and structure. Teaching proceeds from simpler to complex in building a unique homogeneous construction. The learner often has an impression of a simple accretion and infers the more one learns, the more one knows – more models, solved problems, explained phenomena, concepts, and laws. Principles are barely distinguished from laws emerging in endless multiplicity. In this scenario, some laws originally learned might later appear as incorrect and replaced. So laws might appear "...neither universal nor necessary – nor even true."[49]

The knowledge items organized within the DC organization reveal a different picture. Indeed, the laws of physics are not universal since they all hold specific areas of validity. Physics knowledge is composed of fundamental theories, and the theories are structured hierarchically. They are more than rules for model construction. As van Fraassen put it:[50]

> A well-constructed scientific theory will tell a story, a narrative in which the why is as clearly explained as the what, and we come to understand not only 'what happens' but 'what is really going on'.

The theories are different but are related, epistemologically and ontologically. The diminished status of a theory in the science curriculum may impede adequate appreciation of scientific knowledge as a culture. If not explicitly taught as theory-organized, the scientific knowledge is structured in other ways, being guided by

---

[48] Novak and Gowin (1984, p. 9) argued that learning about the nature and structure of knowledge helped students to understand how they learn (meta-learning) and how humans construct knowledge.

[49] Giere (1999, p. 90).

[50] Van Fraassen (2008, p. 266).

different pragmatic and cognitive factors because *structuring* presents a natural and unavoidable cognitive activity of knowledge accretion. Yet, such spontaneous structures are not organized in the way science suggests.[51]

Instead of a sequential presentation of apparently unrelated elements of knowledge, the DC-based curriculum assigns each item to a position within the triadic structure. Important aspects of the content may emerge as a result of this approach. Consider Newton's account of specular reflection and refraction in two separate theories of motion and light. Within the theory of light, *Opticks*, Newton positions specular reflection and refraction laws within the nucleus as empirically supported principles. However, in Newton's theory of mechanics, the *Principia*, the same laws appeared as elements of body knowledge being derived from the general principles of mechanics. Similarly, presenting thermodynamics, one may place the state equation of gases (Mendeleev–Clapeyron) in the nucleus as an empirically based fundamental law, whereas in statistical physics, the same law would be affiliated to the body knowledge backed by its microscopic underpinning. Such manipulations in teaching cause mature knowledge.

The DC approach emphasizes the nucleus of each theory, often missed in physics class. If the concepts of space and time, inertia, relativity, and type of interaction are ignored, the essential difference between mechanics and electromagnetism becomes masked. The contrast between the nuclei of classical mechanics and classical electromagnetism are rarely emphasized. Whereas mechanics includes *action at a distance* for gravitation, electromagnetism is essentially a relativistic theory in which there is only *contact* interaction between a body and a field that combines central (electric) and noncentral (magnetic) components.

In the case of a short introductory course, as modern physics (quantum mechanics) is commonly taught in high school, the DC structure has become almost the only way to meaningfully represent this area of scientific knowledge. The typical historical introduction of a new topic is doomed to stop during the early steps of its understanding, because there is not enough time, nor are there the necessary mathematical tools. By contrast, the DC-based teaching of quantum mechanics is structured in triads—a basic principle, an illustrative experiment, and a contrasting parallel from classical mechanics. This structure has a chance to provide an initial understanding of the quantum theory. Such a curriculum emphasizes *how* modern physics is different from the classical (Fig. 6.4). The validity of each theory is explicitly related to the certain span of parameters (space, time, and mass). Notably, this teaching approach to modern physics provides a better understanding of the meaning of the classical theory of motion in retrospective.[52]

---

[51] The studies of Koponen and his colleagues shed light on such organization in terms of conceptual nets (Koponen & Pehkonen, 2010; Koponen & Nousiainen, 2013).

[52] Weisman et al. (2019a, b).

Nucleus	Body	Periphery
Quantum particle (quanton) and quantum state Superposition principle Particle-wave nature (duality)	Two slit experiment for electrons and photons	Particles and particle state, the results of two slit experiments for particles and water waves in classical mechanics
	"Schrodinger cat" thought experiment with a quantum object	"Schrodinger cat" thought experiment with a real cat
	Mach–Zehnder interferometer for electrons	Mach–Zehnder interferometer with detection of the path
	Photoelectric phenomenon	The predicted results of photoelectric phenomenon according to classical theory

Fig. 6.4 Fragment of the curriculum content of the introductory quantum theory course in high school

An important feature of the CCK-based curriculum is, therefore, the inclusion of knowledge items considered to be wrong in the theory taught. Such are the concepts of Aristotelian mechanics and medieval impetus in teaching classical mechanics (Fig. 6.5). The involving of alternative ideas guides the choosing of the relevant HPS content, which surpasses mere enrichment and scientific literacy. It reveals the creative role of the diachronic conceptual unity and dialogic nature of science, thus promoting its meaningful learning.

"CORRECT" KNOWLEDGE	"INCORRECT" KNOWLEDGE
Eratosthenes measurement of Earth radius	Aristotle's laws of motion
Archimedes laws of levers	Geocentric theory of solar system (explanations of season, eclipses)
Archimedes laws of buoyancy	Hellenic and Hellenistic theories of vision
Aristarchus measurement of astronomical distances	Impetus theory of motion
Specular reflection and refraction of light	Arab theory of vision
Telescope and microscope inventions	Theories of heat
	Theory of ether
Easy to include in historical setting but *not essential*	*More difficult* to include but *essential* for understanding the disciplinary content

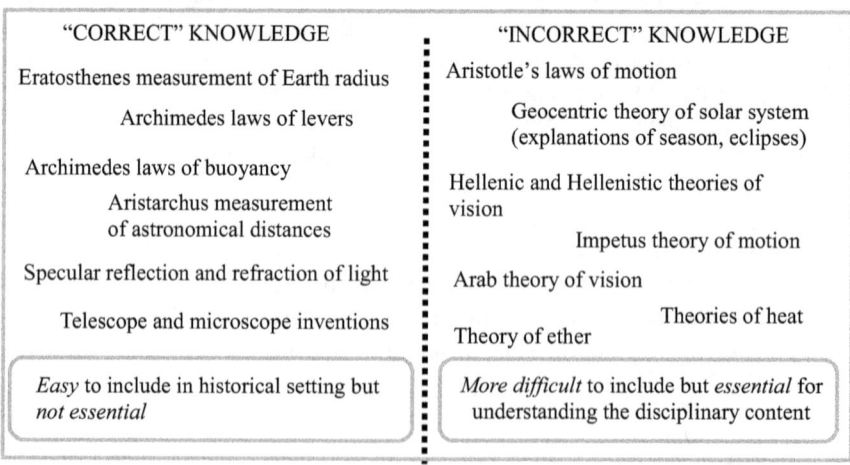

Fig. 6.5 Examples of historical materials of two types included and to be included in physics curriculum at school

Students' Typology

The tripartite structure of the discipline–culture suggests a new typology of students with regard to their interests, intentions, and cognitive preferences. Rather than dividing students into those who are "good" and "bad" at science, or "realists" and "humanists" ("physicists" and "poets"), representing the natural division of "two cultures,"[53] students may instead be distinguished by their preferences toward the three facets of scientific knowledge corresponding to the three types of knowledge elements in discipline–culture (Fig. 6.6).

Some students may show interest in the overall world organization and regularity (the nuclei), but remain reluctant to solve problems by applying laws and principles, developing appliances, and preferring to be informed about science and its derivatives and applications in technology. Other students may prefer solving practical problems (body), taking general principles as given. Such individuals are focused on certain problems that attract them either by challenging their ambition of mastering scientific knowledge or being encouraged by their needs of a different nature. Becoming a competent practitioner implies seeking proficiency in modeling and problem solving. The third type of student is interested in a very different aspect of scientific knowledge. The type of student who challenges the authoritative claims of the teacher shows the type of curiosity that Einstein expressed by the question he allegedly said he would ask God if a chance were granted: why these laws and not others? The content of the periphery facilitates consideration of precisely this aspect of interest.

b. "Aha, I got it... I know now exactly how it works, wonderful... I can do it myself, great..."

c. "Aha... now I see why I was wrong, it is not the way I thought it was, it is different because..."

a. "Wow, what a beautiful idea... it is so nice..."

Fig. 6.6 Students at a lecture. Fragment from bas-relief on the tomb of a university professor in Bologna (1383). Different interests of the students are represented by the artist. Our speculative identification of the students showing major interests to (**a**) nucleus, (**b**) body, and (**c**) periphery

---

[53] Snow (1959).

Teachers often accuse students of the first type—"spectators"—of being superficial, not practical, and not serious enough. They are advised to practice problem solving, to master mathematical formalism, and to stop dreaming. Such students often leave the science class in favor of the humanities. This is a real loss for the society, which badly needs a scientifically literate population in order to make educated decisions in the modern environment. It is worth noting the words of Steven Weinberg, one of the leading physicists of our time:[54]

> *What is important in science (I leave philosophy to others) is not the solution of some popular scientific problems of one's own day, but understanding the world.*

The students attracted to the body knowledge—"engineers"— do not need to be pressured. They often please their teachers and are usually supported by the administration. Such students are often represented in the media as they are easy to present and appreciate. Indeed, they provide a great reserve for science practitioners, normal scientists, and people of technology.

The third, and often less addressed type of student, is attracted to the periphery knowledge – the debate of ideas. Such students—"investigators"—often challenge teachers with "philosophical" questions. Correct and incorrect subject matter equally interest and attract them. Since such students may impede the flow of instruction, they are considered as annoying and do not meet with a favorable attitude. Yet, before calling these students to order, one may recollect that such questions in fact inspire progress.

The founders of quantum mechanics were inspired by their interest in archaic philosophical ideas such as *potentiality* and *actuality* from Aristotelian physics, which were in no way presented in the physics curriculum but in another discipline taught in the classical gymnasium. The antique idea of plenum, adopted by Descartes and dismissed by the Newtonian paradigm, led physicists to the introduction of the *field* concept to account for interaction in electromagnetism. The particle–wave historical debate regarding the nature of light in seventeenth to nineteenth centuries was surprisingly reopened with the introduction of photon particle and resolved within the quantum field theory.[55]

Recycling of ideas is a norm in physics research and a virtue of intelligence. Therefore, students of the third type who raise doubts and compare accounts will nourish the new generation of researchers. They deserve support and encouragement, as they act as science pioneers in producing new knowledge through the doubt of the taught—an extremely important skill.

This variation implies an important feature of the science curriculum, which should speak in three voices corresponding to the three aspects of CCK, matching the interests important for society from a wider perspective. It is plausible to expect that people may combine all three aspects in different proportions and change them through the course of maturation and learning. Cognitive preferences are subject to development, and one should be exposed to all three aspects of scientific knowledge

---

[54] Weinberg (2015, p. 24).

[55] For example, Brooks (2010).

to recognize the possible variations in dealing with science and personal preferences for future occupations. All the aspects together possess a special value of
holistic account that pleases and promotes a multifaceted intellectuality common in
the student population. Initial empirical evidence of the relevance of the tripartite
typology of preferences was received in studies that introduced discipline–culture.

The Ways to Provide CCK

After a certain innovation is theoretically considered, the ways to implement it
become a subject of experimentation. Feasibility and the nature of impact of any
innovation should be checked through real teaching. We have explored a number of
ways to facilitate construction of CCK in students. A brief account follows.

DC-Based Curriculum

The *first* and the most comprehensive way to encourage CCK in students is the
production of a DC-structured curriculum in a certain area of knowledge. We produced such for teaching optics—a theory of light and vision—in high school. A
yearlong teaching experiment was performed, and a special textbook was developed.[56] The new curriculum unfolded the discourse with regard to the nature of
vision and light, the debate of competitive accounts throughout the history of
science.[57]

Drawing on our research in optics knowledge, we found evidence of a certain
recapitulation in the knowledge of students. This implies similarity of ontogeny of
individual development with the correspondent phylogeny and the development of
the pertinent collective knowledge. In the domain of optics and vision, such parallelism is presented in Table 6.1.

Given this similarity, one may expect a productive cognitive resonance between
the historical conceptions and students' ideas and beliefs in the course of learning.
The critique of historical conceptions provides a remedy for the misconceptions and
a form of immunization preventing similar misunderstanding in the course of learning, even if the student had not developed a particular alternative conception.
Students' naïve ideas are often not well reasoned. Therefore, the pertinent historical
debates enrich students with appealing augments regarding parallel conceptions.
Vague, intuitive ideas become distinct, allowing their analysis and refutation. The
process may be compared with increasing the potential difference between the

---

[56] Galili and Hazan (2004, 2009). Only the first part of the course was experimentally tested (Galili
& Hazan, 2000, 2001).

[57] Lindberg (1976, 1978); Galili (2014).

Table 6.1 Conceptual parallelism in optics knowledge

Historical conceptions	Students' conceptions
Pythagorean active vision (fifth c. B.C.)	Active vision scheme
Euclidean dichotomy of vision and light rays (Ptolemy—Second c., Alkindi, ninth c.)	Rays of sight and rays of light
Atomists' conception of eidola (moving replica, simulacrum, from the observed object) (fifth c. B.C.)	Conception (scheme) of holistic image moving in space
Biblical dichotomy of light entity and light perception (lumen–lux in Latin) and in photometry	Conception of static light located in/around light sources (halo, bright sky) and illuminated surfaces and moving light
Ibn Al-Haytham's theory of light and vision (eleventh c.) in its account for pin-hole image(s) and image creation in vision	Light composed of rays and image projection scheme to account for optical image and its transfer by means of light rays
"Pure" (white) light and color as a pigment color mixing	"Pure" (white) light and color as a pigment/color mixing

nucleus and periphery promoting a breakthrough—the required conceptual change. Our experiment confirmed this remedial impact on students' misconceptions.[58]

The positive impact of the DC-structured curriculum sustained in the case of a short course, such as is usual in high school in addressing quantum mechanics. In this case, the teaching does not follow the historical line but preserves the content structured in triads of complementing nature: nucleus, body, and periphery.[59]

Conceptual Excursus

The *second* way to facilitate construction of CCK does not address the regular curriculum. It facilitates the cultural upgrading of conceptual knowledge by means of adding complementary learning of conceptual excursus to the historical consolidation of particular concepts, identifying the major steps in such a process, thus providing the space of variation causing genuine understanding. This constitutes historical *excursus* that steps outside regular class teaching to elaborate on a certain subject in its HPS background and pertinent pedagogical suggestions.[60] This genre is common in historical studies.[61] It, however, is marginal in regular disciplinary teaching–learning. The first chapters of this book illustrate excursus to several concepts both interesting and relevant for an introductory physics course (collision, motion, image, weight, and inertial force). Their complex nature calls for the attention of the learners and suggests required reinforcement of teaching.

---

[58] Galili and Hazan (2000).

[59] Weissman et al. (2019a, b).

[60] Chapters 1, 2, 3, 4 and 5.

[61] For example, Jammer (1957); Lindberg (1976); Stinner and Williams (1993).

Summative Lecture

The third way, the *summative lecture*, though lacking the inclusiveness of the novel curriculum and the depth of historical excursus, might be, however, the most affordable and easily implemented.

In the past, David Ausubel suggested a special tool—*Advance Organizer*—to fortify learning.[62] Before teaching a new topic, students are instructed regarding the framework of the knowledge to be considered. The instruction may include the tools to be used in the following teaching, such as the required aspects of mathematical formalism. Yet, the major goal in such an approach is displaying the overall idea, the concept that may unite the course content in certain aspects. For example, in a biology course, introducing the concept of natural selection may serve as an advance organizer. In physics, the big picture of the course would require addressing the nucleus of the particular fundamental theory. However, providing such metaknowledge as the advance organizer might be problematic for addressing too many unknown specific concepts. Therefore, we tried the reversed order. We designed a summative, reviewing lecture, which might be considered as a posteriori or *delay organizer*. The lecture addresses the students after they learned the traditional course.

The summative lecture is designed to rearrange the already learned fragments in a theory-based structure. The lecture identifies the elements of knowledge as affiliated to three types: nucleus, body, and periphery of a certain theory. Though the lecture addresses the already known contents, to reach the CCK of the subject, the teacher should provide, even if only qualitatively, certain elements for the periphery of the considered theory utilizing the pertinent historical context. Such a lecture creates a holistic view on the subject matter and rearranges the mosaic of its elements into a unifying structure.

During the summary, the tripartite structure of DC is introduced. The major confusion is frequently the clash between the idea of theory and model, since the latter is often known to the learner and sometimes used in the meaning of the former. The students often ask for clarification of the theory–model relationship when they need to affiliate different models to the DC structure of a theory. In responding to this question, the teacher can depict different models used in the considered course (say, optics), affiliating them to the three areas of the corresponding theory (Fig. 6.7). Thus, the models of ray, wave, and photon are attributed to the nuclei of the three learned optical theories as their *paradigmatic* models. The models of thin lenses, paraxial rays, and point sources, for example, are identified as *working* models and located in the body of the ray theory. Finally, the models from the periphery could be exemplified by photons in the wave theory or wave—in the ray theory, and so on. They are used as *heuristic* models.

---

[62] Ausubel (1968).

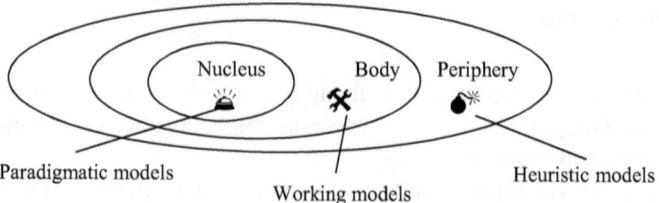

Fig. 6.7 Models may appear in all areas of theory structure, being of different natures

When interviewed, teachers of the classes involved in the experiment expressed appreciation of DC as a framework for the subject matter, unifying various knowledge elements—models, laws, experiments, etc.—in a related web. They mentioned that such inclusion of HPS infused new meaning into the regular teaching. In particular, they pointed to the ability to display the progression of physical theories, the conceptual change in science, and transitions from one theory to another, salient in school optics. Using representative diagrams of knowledge structure and of supportive artistic images in the short summative lecture was seen as effective, apparently for the holistic appeal imbedded in the nature of these pedagogical tools.[63]

Epistemology in the Cultural Approach to Knowledge

The impact of the DC approach to the science curriculum with respect to epistemological aspects is different to the impact on the content knowledge. This is because philosophy as knowledge corresponds to a different type of culture. While the content knowledge of science—a group of fundamental theories—can be considered as a *culture of rules* (a clear hierarchy of the nucleus and periphery within each theory), the epistemological knowledge of science is rather a *culture of texts*, implying a more flexible perception of correct–incorrect opposition.[64] Different epistemological approaches complement each other rather than exclude. Together they create a complex construct of scientific methodology.

In particular, there is a continuous contest of rationalist and empiricist methods of knowledge construction throughout the history of science. Aristotle combined them in the inductive–deductive repetitive ("circular") procedure of scientific investigation drawing on contemplation and logical analysis. Platonic–Pythagorean rational analysis preceded Aristotelian. It drew on the idea of an a priori transcendent order and mathematical logic to uncover the hidden forms projected onto perceived reality. Theory was introduced to account for and represent reality, representing it in human knowledge.

---

[63] Chapters 7 and 10, Galili (2013).
[64] Lotman (2010).

Throughout history, the scientific method was developed and modified. Initially focused on theory-based speculations, it was reinforced by experimentation in Hellenistic and Muslim sciences. Science became more specific, concrete, and numerical. In this shift of the scientific method, medieval science introduced "prerogatives of experimental science," that distanced it from theoretical speculations, which, nevertheless, developed the rationalist apparatus of logical analysis termed the resolution–composition method.[65] The medieval nominalist–realist philosophi-

cal opposition regarding scientific concepts nourished the following opposition of Baconian empiricism and Cartesian rationalism in early modern science. Science kept in use both approaches on the way to their synthesis by Newton in modern science (Fig. 6.8).

Fig. 6.8 Plato and Aristotle in the center of the Raphael fresco *The School of Athens* in the Vatican (c. 1511). This image became emblematic of Western science as having two images in its focus. The gesturing of the two philosophers (the added arrows) is interpreted as representing of rationalism ("theory first") and empiricism ("experience first"). Their integration is expressed in their sharing of the converging point of the perspective

Epistemology is different in the relationship of its perspectives from the relationship of different components of the content knowledge of ontology. For instance, classical mechanics, in its account for motion either refutes the Aristotelian account or just distances itself, as with respect to the relativistic theory. Both alternatives, Aristotelian and Einsteinian, belong to the periphery of classical mechanics and are considered as contradictory to the nucleus of classical mechanics. Though they remain within the horizon of the theory, they are not complementary, as epistemological items can be.

The cultural perspective in epistemology recognizes the accounts of *discovery* in the Baconian "interrogation of nature" (extreme empiricism) and *invention* or construction of knowledge (such as by applying a research program[66]). The whole range between discovery and construction presents scientific practice. One may use artistic images to represent the poles of this veracity (Fig. 6.9).[67] In reality, however, discovery and construction of knowledge were never purely isolated.[68] In each discovery, one can find both types of component. Moreover, they stipulate each other: there was no discovery without much theoretical construction, which directed and provided meaning to it, and also no theorizing without the guidance of the newly discovered aspects, features, and phenomena.

---

[65] For example, Losee (1993), Pedersen and Pihl (1974).

[66] For example, Lakatos (1978).

[67] Chapter 10.

[68] It reflects the iterative nature of the scientific method. We will return to the nature of science in Chap. 9.

Fig. 6.9 The images representing the two extremes of scientific methodology. (**a**) Discovery was chosen as emblematic for classical physics as depicted on the Nobel Prize medal (names emphasized). (**b**) Empirical construction of the new knowledge could be represented using the image of Pygmalion (the mythological sculptor) who produced the sculpture of Galatea (his ideal) to animate and fall in love with her

The pluralistic approach to the scientific method might look like inconsistency. Indeed, in their activities, scientists employ methods identified with different epistemologies. Einstein wrote:[69]

> He [scientist] therefore must appear to the systematic epistemologist as a type of unscrupulous opportunist: he appears as realist insofar as he seeks to describe a world independent of the acts of perception; as idealist insofar as he looks upon the concepts and theories as free inventions of the human spirit (not logically derivable from what is empirically given); as positivist insofar as he considers his concepts and theories justified only to the extent to which they furnish a logical representation of relations among sensory experiences. He may even appear as Platonist or Pythagorean insofar as he considers the viewpoint of logical simplicity as an indispensable and effective tool of his research."

Yet, scientists remain committed to standards including empirical verification, drawing on theories, objectivity, and open disciplinary discourse.[70] We may know about the standards from cases when their violations caused the failure to produce valid scientific knowledge, clearly demonstrating that not anything goes. The teaching within the DC approach displays this plurality while specifying the range of variation, pointing to the identified types of perspectives and claiming their complementarity.

The role of epistemology can be depicted using another important tool—the semiotic triangle. Numerous cultural entities can be addressed by the triad, object—its sign and its meaning—depicted as a triangle.[71] The scale can be greatly expanded

---

[69] Einstein (1949/1979, pp. 683–684).

[70] For the specific scope of this account, we do not address the ethical standards, which are, however, no less important.

[71] For example, Löbner (2013, p. 24).

to include the whole cultural domain as scientific knowledge.[72] Thus, if the disciplinary content of a theory constitutes the object vertex, then, its sign vertex could be identified with the science curriculum. What is especially important is that in such a framework, the vertex of meaning of the object should be reserved for the adopted epistemology (Fig. 6.10). This representation makes explicit the role of epistemology in science education. The choice of epistemology determines the nature of representation of science content and the relationship and preferences given to the empirical or theoretical aspects of the subject matter.[73]

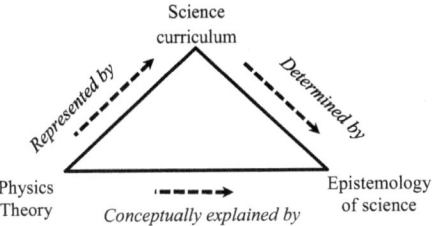

Fig. 6.10 Semiotic triangle of science education

For example, pragmatic, instrumental philosophy[74] would demand a problem-based curriculum, learning by doing, personal experience, and conceptions of the learner (educational constructivism) as the way to mastering science. This perspective, however, may easily miss the overall view, the epistemological status of knowledge in the theory-based structure, identification of fundamentals, concepts, principles, and the interrelation of the constituents. That content comes to the fore in the curriculum based on rationalistic epistemology.[75] The latter provides knowledge that does not emerge from practicing standard problems, but should be explicitly taught, illustrated, and discussed.

Different perspectives on modeling can be indicative. Within the semantic view, one may consider theory as merely a set of models constructed according to certain rules. In classical mechanics, such an approach is pointed to the set of basic models.[76] It included the uniform motion as such, next to circular and oscillatory motion.[77] This approach neglects the fact that uniform rectilinear motion possesses a special status, different from all other types of motion as stated in Newton's *Principia*.[78] Ignoring the nucleus of classical mechanics, the relativity principle,

---

[72] Tseitlin and Galili (2006).

[73] Thus, one may compare the introductory physics textbooks of the Nuffield project in the United Kingdom, which strongly emphasize laboratory work and less the elaboration of theoretical aspects with other physics textbooks of the same level.

[74] For example, Dewey (1938).

[75] For example, Frank (1957).

[76] For example, Giere (1999, pp. 110–111).

[77] For example, Halloun (2006, pp. 140–141).

[78] Galili and Tseitlin (2003).

concepts of time and space, state of motion, the concept of force, central interaction, etc. lead classical mechanics to losing its status as a fundamental theory. Instead, within the DC perspective, theoretical models present only partially the content of the theory contributing to all three areas of theory structure, where they play different roles (Fig. 6.7).

Apologists of the pragmatic curriculum sometimes quote Einstein:[79]

> *If you want to find out anything from the theoretical physicists about the methods they use, I advise you to stick closely to one principle: don't listen to their words, fix your attention on their deeds.*

This advice, however, addresses the methodology of practitioners in the context of inquiry, not the context of science education. For that, one needs a wider scope of the cultural vision of epistemology. It is especially explicit in considering concept definition. Mach[80] was the first who stated the requirement of the operational definition of physical concepts. He introduced the new definition of inertial mass through the measured accelerations of colliding bodies. Einstein followed him with regard to simultaneity in his theory of relativity.[81] The initial claim of operationalism that any "concept is synonymous with the corresponding set of operations"[82] was, however, transformed to the more adequate philosophical account requiring a pair of complementary nominal and operational definitions for each concept.[83] This requirement integrates empirical and rational approaches matching the cultural curriculum. It can be illustrated with respect to teaching the concept of weight.[84]

Concluding Remark

Altogether, considering scientific knowledge as a culture reveals the dialogic nature of the scientific pursuit, revealing its conceptual discourse − ontological and epistemological—which is often missing from strictly disciplinary teaching. For the disciplinary contents, a DC approach suggests that the curriculum be organized in a tripartite hierarchical structure of fundamental theories: nucleus–body–periphery. Such a curriculum reveals the big picture of scientific knowledge and promotes critical thinking in science education, awareness of epistemological status, and the hierarchy of knowledge structure.

The discipline–culture–based curriculum promotes metaknowledge appealing to the broad population of learners possessing different interests and cognitive

---

[79] Einstein (1934/2011).

[80] Mach (1883/1989, p. 218).

[81] Reichenbach (1928/1958) − in philosophy of science, and Karplus (1981) and Arons (1965, 1990) − in physics education.

[82] Bridgman (1927).

[83] Margenau (1950).

[84] Chapter 5.

preferences beyond the simple consuming and pragmatic orientation. The DC paradigm provides a framework for the inclusion of historical content and specifies its involvement as a curricular necessity. Introducing understanding of science as a culture in school curricula matches the tradition of dissemination of scientific literacy and enlightenment.

Appendices

Appendix 1

It can be helpful to relate DC knowledge organization to Foucault's very inclusive cultural vision that he called the *archeological* approach.[85] Through it, he addressed systems of ideas producing knowledge elements, which shared a somewhat similar trend of thought, fundamental assumptions, and the worldview prevailing in the realm of thought in a certain cultural period. He called them *epistemes*. They present layers of culture, and that is where the term "archeology" comes in. Episteme represented the common factor in a huge cluster of features of cognitive account practiced in different areas of intellectual activities. Epistemes are present across scientific and nonscientific knowledge, and in practice and in the theoretical products of a specific time period. For instance, one may consider medieval culture as an episteme of Faith, which penetrated throughout different cultural products creating a certain wholeness. We may exemplify different epistemes by addressing the two frescoes of Rafael in the Vatican: *The School of Athens* and *the Disputa*.[86] One may identify their epistemes—*Cosmos* (the organized universe to be observed revealing its order and underlying logic through contemplation) in classical Greece and *Faith* (unconditional adoption of certain religious assumptions, a priori to any cultural activity) in the medieval–Renaissance time period. The DC paradigm of knowledge organization differs from episteme by addressing solely scientific knowledge, being hierarchical and empirically verified, coherent and possessing the well-defined meaning of being correct. It represents not only synchronic but also diachronic discourse while adopting the epistemological commitment of seeking objectivity.

Appendix 2

To allow comparison, we mention here the perspective of the Scientific Research Program (SRP).[87] The two knowledge organizations, SRP and DC, have different structures, components, and status of elements and seek different goals.

---

[85] Foucault (1972, pp. 200–212).

[86] Chapter 10.

[87] Lakatos (1978, pp. 47–51).

The methodology of SRP considers a theory as a machine for problem solving and does not address the full scientific structure of a scientific theory, its components, and status. SPR represents a working model of science practice, a scientific theory that establishes itself through solving problems. The model is minimalistic.

Fig. 6.11 Schematic depiction of the Scientific Research Program

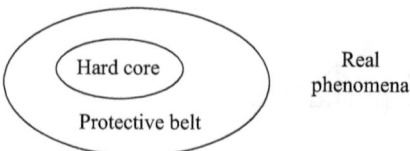

SRP possesses a dual structure of hard core and a protection belt (Fig. 6.11). The *hard core* incorporates the basic ontological statements of the theory, not a subject of any critique in its application ("negative heuristic"). The second area—*the protective belt*—includes working models, constructs drawing on the principles of the core ("positive heuristic"), and provides a scientific account of real phenomena (solved problems and explaining phenomena). It is these working models, inspired by the core, that are judged for their success in matching reality. If successful, a model remains in the theory as adequate and useful. If it fails, it is removed, and new models are suggested. The core remains untouched. The models of the protection belt may deviate from the core dogma in particular aspects (such was the Ptolemaic model of motion of celestial bodies), thus presenting certain compromises made for the sake of empirical success.

This pattern of understanding was suggested by Lakatos within his improving of the principle of falsification by Popper. Poppers' principle of falsification stated the fallibility of any scientific theory, which implied its removal based, in principle, on a single test of a crucial experiment. Lakatos, in his scenario, provided a much more realistic picture of working science preserving the demand of empirical verification for the applied models but making the whole process of theory exchange more complex, gradual, and not abrupt. In fact, a partially successful theory could coexist with other theories in a continuous competition, if not a dialogue.

Appendix 3

Kuhn's account of the organization of scientific knowledge is close to the DC approach but still differs. Instead of a fundamental theory, Kuhn considered "disciplinary matrix," a constellation of ordered elements of various sorts.[88] They shared commitments to certain *paradigm*, not comprehensively discussed (Fig. 6.12). Kuhn ascribed to the paradigmatic model a kind of metaphysical status. However,

---

[88] Kuhn (1962/1970, pp. 182, 184, 187).

Kuhn's main attention was directed to the body of knowledge—the extensive activity of application of the paradigm to problem solving ("puzzle resolving"). This activity was labeled "normal science." Disciplinary teaching traditionally includes exemplars of the concrete problem–solution type, providing a toolkit for science practitioners, tested for being mastered by numerous examinations and exercises in textbooks. However, no part of the disciplinary matrix included questioning regarding legitimacy of the paradigm or intent to test and change it.

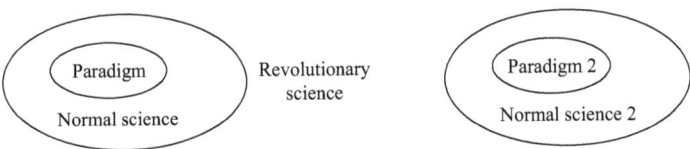

Fig. 6.12 Schematic depiction of Kuhn's approach to the reality of science practice

Yet, Kuhn recognized the appearance of anomalies, which break the routine of normal science with unexplained phenomena, unexpected empirical data, and failures of apparatus to adequately facilitate research in such cases. Such problems require new procedure, heuristic principles, and a new framework of thought. This is the context of revolutionary science violating the old paradigm. It causes tension in science, reveals inadequacies, and casts doubt on the adopted paradigm. They motivate revolutionary studies, which ultimately cause the major shift—scientific revolution. Its scenario is rather similar to the nucleus–periphery exchange within the DC paradigm. In Kuhn's picture, however, different theories, for example, mechanics and electrodynamics, ignore each other (they are also being taught in different courses). There is no stage for their comparison beyond the claim of the incommensurability of paradigms of different theories.

Although Kuhn's approach uses triadic code in a practical way, the nucleus–body–periphery in terms of DC, their content is different in the DC paradigm. In particular, the nucleus in DC expands more on the metaphysical premises tacitly implied in the body knowledge but usually not articulated in everyday teaching (e.g., state, absolute time and space in classical mechanics, central and instantaneous interaction in void, and action at a distance). The nucleus of DC also includes fundamental concepts (e.g., mass, force, and acceleration) and epistemological principles (e.g., the demand of operational definitions, and the explicit area of validity in terms of the range of basic parameters).[89] Body knowledge in the DC paradigm includes derived laws and experiments and experimental tools. They all are beyond problem solving. Finally, periphery, which is premature in Kuhn's model, presents an important component in the DC model. It includes not only anomalies and unsolved problems but also alternative conceptions from the past (obsolete and incorrect), as well as correct but alternative ontological statements from the nuclei of competitive theories. For instance, the periphery of classical mechanics includes

---

[89] Nucleus is the place to include declaratively the metaphysical features of the nature of science (Chap. 9) and subsequently exemplify them in treating the elements of body knowledge.

the basic claims of relativity theory regarding space-time, interaction through a field, relativistic velocity addition, and noncentral interaction. Periphery introduces conceptual polyphony, and this way establishes the *cultural content knowledge* of science, which provides a more inclusive and representative picture of science and scientific knowledge.

***

Picture Credits

- First page image. The title page of the *Dialogue* by Galileo, 1632. Public Domain. https://en.wikipedia.org/wiki/Dialogue_Concerning_the_Two_Chief_World_Systems#/media/File:Galileos_Dialogue_Title_Page.png
- Fig. 6.6 Students in the relief by P.P. Masegne and J. da Bologna, 1383. Photo by De I. Sailko. CC Attribution-Share Alike 3.0 Unported license. https://commons.wikimedia.org/w/index.php?curid=4942241
- Fig. 6.8 Detail from the fresco by Raphael *The School of Athens*, 1509. Public Domain. https://commons.wikimedia.org/w/index.php?curid=75881
- Fig. 6–9(a) The Nobel Prize medal belonged to Albert Einstein. With permission from: Albert Einstein Archives, the Hebrew University of Jerusalem
- Fig. 6–9(b) Schematic drawing of Pygmalion casting Galatea. Sketch by unknown author. Public Domain. https://lh4.googleusercontent.com/proxy/m4QqhD7iWfKTUtv2otWbtDDrM NljS2q-9HZ_dFeuh_1IJ0VA_nM-4s7OI7XgVl7qMhvtp1FYD25RGXiZWBDCIMTAI8L5c WFgELcXVHoz=w1200-h630-p-k-no-nu

References

Aikenhead, G. S. (1997). Towards a first nations cross-cultural science and technology curriculum. *Science Education, 81*(2), 217–238.

Arons, A. (1965). *Development of concepts of physics*. Reading, Mass: Addison-Wesley.

Arons, A. (1990). *A guide to introductory physics teaching*. New York, NY: Wiley.

Ausubel, D. P. (1968). *Educational psychology: A cognitive view*. New York: Holt, Rinehart & Winston.

Bevilacqua, F., Giannetto, E., & Matthews, M. (Eds.). (2001). *Science education and culture. The contribution of history and philosophy of science*. Dordrecht, the Netherlands: Kluwer.

Bridgman, P. (1927). *The logic of modern physics*. New York: Macmillan.

Brooks, R. A. (2010). *Fields of color: The theory that escaped Einstein*. Silver Spring, Maryland: Universal Printing, LLC.

Bruner, J. (1996). *The culture of education*. Cambridge, Mass: Harvard University Press.

Cartwright, N. (2005). *The dappled world. A study of the boundaries of science*. Cambridge: Cambridge University Press.

Dewey, J. (1938). *Experience & education*. New York: Kappa Delta Pi.

diSessa, A. (1993). Toward an epistemology of physics. *Cognition & Instruction, 10*, 105–225.

Einstein, A. (1949/1979). Autobiographical notes. In P. A. Schilpp (Ed.), *Albert Einstein: Philosopher-scientist*. Harper.

Einstein, A. (1918/2002). *Principles of research. The collected papers of Albert Einstein: The Berlin years, 1918–1921* (pp. 42–45). Princeton University Press.

Einstein, A. (1934/2011). On the method of theoretical physics. In *Essays in science*. Open Road.

Elkana, Y. (1970). Science, philosophy of science, and science teaching. *Educational Philosophy and Theory, 2*, 15–35.

Fehl, N. E. (1965). *Science and culture*. Hong Kong: Cheng Chi, The Chinese University of Hong Kong.

Feyerabend, P. (1993). *Against method*. London: Verso.

Foucault, M. (1972). *Archaeology of knowledge*. London: Routledge.

Frank, M. (1957). *Philosophy of science. The link between science and philosophy*. Englewood Cliffs, NJ: Prentice-Hall.

Galilei, G. (1632/1953). *Dialogue concerning the two chief world systems – Ptolemaic and Copernican*. University of California Press.

Galilei, G. (1638/1914). *Dialogue concerning two new sciences*. Dover.

Galili, I. (2012). Promotion of content cultural knowledge through the use of the history and philosophy of science. *Science & Education, 21*(9), 1283–1316.

Galili, I. (2013). On the power of fine arts pictorial imagery in science education in science education. *Science & Education, 22*, 1911–1938.

Galili, I. (2014). Teaching optics: A historico-philosophical perspective. In M. R. Matthews (Ed.), *International handbook of research in history and philosophy for science and mathematics education* (pp. 97–128). Dordrecht, the Netherlands: Springer.

Galili, I. (2017). Chapter 8: Scientific knowledge as a culture – A paradigm of knowledge representation for meaningful teaching and learning science. In M. R. Matthews (Ed.), *History, philosophy and science teaching research. New perspectives* (pp. 203–233). Dordrecht: Springer.

Galili, I., & Bar, V. (1992). Motion implies force. Where to expect vestiges of the misconception? *International Journal of Science Education, 14*(1), 63–81.

Galili, I., & Hazan, A. (2000). The influence of historically oriented course on students' content knowledge in optics evaluated by means of facets-schemes analysis. *Physics Education Research, American Journal of Physics, 68*(7), S3–S15.

Galili, I., & Hazan, A. (2001). The effect of a history-based course in optics on students' views about science. *Science & Education, 10*(1–2), 7–32.

Galili, I., & Hazan, A. (2004). *Optics – The theory of light and vision in the broad cultural approach*. Jerusalem, Israel: Science Teaching Center, The Hebrew University of Jerusalem.

Galili, I., & Hazan, A. (2009). *Physical Optics – The theory of light in the broad cultural approach, Parts II and III* (Optics of Waves and the Modern Theory). Jerusalem, Israel: Science Teaching Center, The Hebrew University of Jerusalem.

Galili, I., & Tseitlin, M. (2003). Newton's first law: Text, translations, interpretations, and physics education. *Science & Education, 12*(1), 45–73.

Giere, R. N. (1988). *Explaining science. A cognitive approach*. Chicago: University of Chicago Press.

Giere, R. N. (1999). *Science without Laws*. Chicago: The University of Chicago Press.

Halloun, I. A. (2006). *Modeling theory in science education*. Dordrecht, the Netherlands: Springer.

Heath, T. (1966). *Aristarchus of Samos*. Oxford: Clarendon Press.

Heisenberg, W. (1959/1971). *Physics and philosophy. The revolution in modern science*. Harper.

Hofstede, G. (1991). *Cultures and organizations: Software of the mind*. London: McGraw-Hill.

Jammer, M. (1957). *Concepts of force*. Cambridge, MA: Harvard University Press.

Jaroszyński, P. (2007). *Science in culture*. New York: Rodomi.

Kant, E. (1781/1952). The critique of pure reason. In *Britannica great books* (Vol. 42). Chicago: The University of Chicago Press.

Karplus, R. (1981). Educational aspects of the structure of physics. *American Journal of Physics, 49*(3), 238–241.

Koponen, I. T., & Nousiainen, M. (2013). Pre-service physics teachers' understanding of the relational structure of physics concepts: Organizing subject contents for purposes of teaching. *International Journal of Science and Mathematics Education, 11*, 325–357.

Koponen, I. T., & Pehkonen, M. (2010). Coherent knowledge structures of physics represented as concept networks in teacher education. *Science & Education, 19*, 259–282.

Kuhn, T. S. (1957). *The Copernican revolution. Planetary astronomy in the development of Western thought*. Cambridge, Massachusetts: Harvard University Press.

Kuhn, T. S. (1962/1970). *The structure of the scientific revolution*. Chicago: The University of Chicago Press.

Lakatos, I. (1978). Falsification and the methodology of scientific research programmes. In J. Worrall & G. Currie (Eds.), *Imre Lakatos philosophical papers: Vol. 1. The methodology of scientific research programs* (pp. 8–101). Cambridge: Cambridge University Press.

Latour, B. (1987). *Science in action: How to follow scientists and engineers through society*. Cambridge, MA: Harvard University Press.

Laudan, L. (1996). *Beyond positivism and relativism. Theory, method, and evidence*. Boulder, Colorado: Westview Press, Harper Collins Publishers.

Libbrecht, U. (2009). Comparative philosophy: A methodological approach. In N. Note, R. Fornet-Betancout, J. Estermann, & D. Aerts (Eds.), *Worldviews and cultures*. Dordrecht, the Netherland: Springer.

Lindberg, D. C. (1976). *Theories of vision from Al-Kindi to Kepler*. Chicago, IL: The University of Chicago Press.

Lindberg, D. C. (1978). The science of optics. In D. C. Lindberg (Ed.), *Science in the Middle Ages* (pp. 338–368). Chicago: The University of Chicago Press.

Liu, Y. (2015). Overview of ancient Chinese science and technology. In Lu (Ed.), *A history of Chinese science and technology* (Vol. 1, pp. 19–40). Heidelberg, Germany: Springer.

Löbner, S. (2013). *Understanding semantics*. London, UK: Routledge.

Losee, J. (1993). *A historical introduction to the philosophy of science*. New York: Oxford University Press.

Lotman, Y. (2010). The problem of learning culture as a typological characteristic. In *What people learn. Collection of papers and notes* (pp. 18–32). Moscow: Rudomino.

Ma, H. (2012). *The images of science through cultural lenses. A Chinese study on the nature of science*. Rotterdam: Sense Publishers.

Mach, E. (1883/1919/1989). *The science of mechanics, a critical and historical account of its development*. Open Court.

Margenau, H. (1950). The role of definitions in science. In *The nature of physical reality*. New York: McGraw-Hill.

Marton, F., Runesson, U., & Tsui, A. B. M. (2004). The space of learning. In F. Marton & A. B. M. Tsui (Eds.), *Classroom discourse and the space of learning* (pp. 3–40). Mahwah, NJ: Lawrence Erlbaum.

Miller, A. I. (Ed.). (1986). *Frontiers of Physics: 1900–1911. Selected essays*. Boston, MA: Birkhauser.

Minstrell, J. (1992). Facets of students' knowledge and relevant instruction. In R. Duit, F. Goldberg, & H. Niedderer (Eds.), *Research in physics learning: Theoretical issues and empirical studies* (pp. 110–128). Kiel: IPN.

Novak, J. D., & Gowin, D. B. (1984). *Learning how to learn*. Cambridge, UK: Cambridge University Press.

Pedersen, O., & Pihl, M. (1974). *Early physics and astronomy*. London: McDonald & Janes.

Piaget, J. (1972). *The child's conception of physical causality*. New Jersey: Totowa Adams Littlefield.

Planck, M. (1936/1963). *The philosophy of physics*. Norton.

Popper, K. R. (1978). *Three worlds. The Tanner Lecture on Human Values*. The University of Michigan. http://www.tannerlectures.utah.edu/lectures/documents/popper80.pdf. Accessed 24 Sept 2015.

Posner, G. J., Strike, K. A., Hewson, P. W., & Gertzog, W. A. (1982). Accommodation of a scientific conception: Toward a theory of conceptual change. *Science Education, 66*, 211–227.

Ptolemy, C. (1952). *Almagest*. Chicago: Encyclopedia Britannica.

Reichenbach, H. (1928/1958). *The philosophy of space and time*. Dover.

Russell, B. (1932/2010). *Education and the social order.* Routledge.

Schwab, J. J. (1964). Problems, topics, and issues. In S. Elam (Ed.), *Education and the structure of knowledge* (pp. 4–47). Chicago: Rand McNally.

Snow, C. P. (1959). *The two cultures and scientific revolution*. Cambridge, UK: Cambridge University Press.

Stinner, A., & Williams, H. (1993). Conceptual change, history, and science stories. *Interchange, 24*(1 & 2), 87–103.

Tseitlin, M., & Galili, I. (2005). Teaching physics in looking for its self: From a physics-discipline to a physics-culture. *Science & Education, 14*(3–5), 235–261.

Tseitlin, M., & Galili, I. (2006). Science teaching: What does it mean? – A simple semiotic perspective. *Science & Education, 15*(5), 393–417.

Tylor, E. (1871/1920). *Primitive culture* (Vol. 1). John Murray.

van Fraassen, B. C. (1980). *The scientific image*. Oxford: Clarendon Press.

van Fraassen, B. C. (2008). *Scientific representation: Paradoxes of perspective*. Oxford: Clarendon Press.

Vygotsky, L. (1934/1986). *Thought and language*. The MIT Press.

Weinberg, S. (1992). *Dreams of a final theory*. New York: Pantheon Books.

Weinberg, S. (2015). *To explain the world: The discovery of modern science*. New York: Harper Collins Publishes.

Weissman, E. Y., Merzel, A., Katz, N., & Galili, I. (2019a). Teaching quantum mechanics in high-school — Discipline-Culture approach. In *Proceedings of GIREP-2018 conference*, San Sebastian, Spain. *Journal of Physics Conference Series, 1287*, 012003. https://doi.org/10.1088/1742-6596/1287/1/012003.

Weissman, E. Y., Merzel, A., Katz, N., & Galili, I. (2019b). Teaching quantum physics in high school. In *Proceedings of ESERA-2019 conference*. Italy: University of Bologna.

Weizsacker, C. F. (1985/2006). *The structure of physics*. Springer.

Chapter 7
Teaching Optics: A Historico-Philosophical Perspective

There are very few, if at all, new things in this world. Therefore, the agenda of a person is to find a new, fresh interpretation of those familiar.

Giorgio Morandi

Italian artist of the twentieth century

© Springer Nature Switzerland AG 2021, corrected publication 2022
I. Galili, *Scientific Knowledge as a Culture*, Science: Philosophy, History and Education, https://doi.org/10.1007/978-3-030-80201-1_7

Abstract Optics—the theory of light—is presented from the perspective of the theory-based science. Four separate theories have sequentially dominated in the domain of optics: Geometrical optics (theory of rays), Newtonian theory (theory of particles), physical optics (theory of waves), and modern optics (theory of photons). Three of these theories are still taught in high school today. Here we argue that when teaching about light, these theories should be presented in their discipline–culture structure, making explicit the contents of the nucleus of each theory and the corresponding elements of the body of knowledge. Presenting the historical sequence of theory exchange, rather than mere "replacement of the wrong theory," can clarify the area of validity of each theory. The conceptual change between theories of light exemplifies the features of the nature of science and is presented drawing on the triadic structure of a scientific theory.

Introduction

This chapter depicts how history and philosophy of science (HPS) can be included in teaching about light and vision within the paradigm of discipline–culture (DC). Light is conceived in science as one of the two components of physical entities— light and matter. This dichotomy stems from the difference between photons of zero mass and the rest of the elementary particles, possessing mass. Our scientific knowledge of light is organized in the form of optical *theories*. The history of science includes an astonishing saga of knowledge construction, evolving understanding through different periods of increasing complexity, leading to the modern theory of light unified with matter. This emerged in a diachronic dialogue of theories in which, at each period, one theory dominated, essentially governing the views on the nature of light and vision.

Within the liberal education tradition, the major curricular question is how to represent a vast amount of knowledge about light in order to create in the learners an inclusive and essentially representative big picture of this knowledge. Accordingly, it is common to identify the main periods of knowledge transformation in the domain and thus create a roadmap to follow and be scrutinized in the course of learning.

Observing the very long history of light and vision from this perspective, one finds that physics teaching at schools usually skips over the long history up to the seventeenth century and deals with three optical theories consolidated since then (Fig. 7.1). These are usually presented in separate sequential courses: (i) geometrical optics (the theory of rays), (ii) physical optics (the theory of waves), and (iii) modern optics (the theory of photons). The latter is not addressed as a new theory of light but as fragments of modern physics.

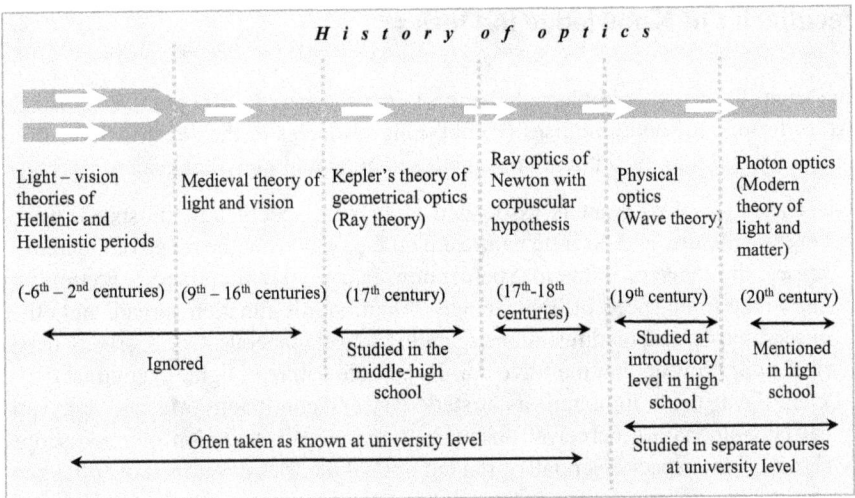

Fig. 7.1 The flowchart of historical development of optical theories and their coverage in high school and university curricula

The earlier history, however, addresses the conceptions of light and vision in the Hellenic period, the theory of rays of the Hellenistic, medieval European, and Muslim periods. School teaching skips over much of the discourse establishing the knowledge of light. Most students will never know about Newton's theory of color rays and his original explanation of interference within the ray theory, nor will they know anything of Huygens' theory of light as pressure waves from the ray–particle–wave debate of the seventeenth to eighteenth centuries. Although Newton's conception of light particles dominated until nineteenth century, in the end it too was supplanted.

From the perspective discussed in the previous chapter, knowledge of theories of light (optics) can be considered as discipline-culture of the subject. We argue that it stipulates meaningful learning of the school curriculum, justifying and explaining it.

DC approach guides selection of the relevant pieces gleaned from the centuries of optics history to facilitate modern curriculum. To appreciate the relevance of such enrichment, we consider it against the background of students' knowledge as elicited by education studies. While doing that, one reveals their conceptual proximity. Following this trend, we will depict the new way of optics teaching at schools.

Peculiarity of Knowledge in Optics

The scientific account of light and vision was produced over 2500 years. It is instructive to formulate some premises (P) outlining obstacles to the construction of our knowledge about optical phenomena both scientific and individual.

P-1. The physical parameters associated with light, specifically, its speed, wavelength, pressure, and discrete nature are all far away from the range of the human senses, the range of individual perception. In everyday life, there is no time lag due to the finite speed of light. Image formation, illumination spread, and other observable optical manifestations seem to be instantaneous. Our everyday experience confirms neither the wave nor the particle nature of light. In contrast to the scientific account, light appears as stationary and continuous. Macroscopic optical phenomena are perceived through unconscious integration of microscopic signals. This aspect essentially affected optical theories developed over the generations giving rise to speculative interpretations alongside verifiable claims.

P-2. Optical phenomena are commonly observed within a medium (air and water), which greatly modify their appearance. This intrusion is so ubiquitous that it is often difficult to account for. Optical phenomena, which are modified by environment (such as light scattering, creation of halos, and light glows) impede interpretation in terms of fundamental qualities.[1] Everyday exposure to these

compound phenomena trigger their naive interpretation corresponding to direct perception. Artists and writers express the sense perception of the sky as luminous, full of light. Spherical glows surround light sources.[2] Light rays may appear as observed reality, as if informing us that light is composed of a bundle of rays (Fig. 7.2).

Ernst Mach (1926, p.10) recorded his impression:

Having spent my youth on a great plain, I often saw how, when the setting sun cast its rays through rifts in the clouds, these rays, forming great circles in the heavens, coalesce again on the horizon at the opposite point from the sun as a perspective image of a parallel bundle of rays on the spherical dome of the sky.

Fig. 7.2 The entering skylight as a *parallel bundle of rays* from the spherical dome of the Holy Sepulcher church in Jerusalem

---

[1] Minnaert (1940) depicts numerous atmospheric optical phenomena, often highly familiar and yet, still requiring nontrivial account.

[2] For example, Van Gogh's typical glow portrayed round a candle in *Le Sedia di Gauguin*, and halos of stars illuminated the starry sky out of the coffee bar filled with bright light in his *Esterno di Caffe di Notte*.

P-3. The observer in optics is an integral part of the optical system. Normally, in other domains of classical physics, significant effort is invested to exclude intrusion by the observer and to account for the reality as it is without being observed. It is often not so in optics, where an account of the observed phenomenon to be explained presumes the active participation of the human eye as an inherent part of the physical system. Indeed, it is easy to miss the role of the observer since vision and light delivery to the eyes are not accompanied by any perceptible muscular effort somehow controlled. The process of seeing (i.e., recognizing and interpretation of visual images), like breathing, operates subconsciously.

P-4. Language brings problems of a psychological nature. Historically, language was developed under the influence of visual perception, and well before our scientific understanding of vision was reached. Thus, many linguistic constructs do not conform to scientific knowledge. Phrases such as "her eyes shine," "his face radiates light," "she casts a glance," "light fills the room," "mirror reflects images," and "trees cast shadows" are usually interpreted regardless any optics. This mismatch presents a serious problem for physics instructors.

P-5. Accounts of light phenomena are essentially interdisciplinary. Physics (the nature of light), physiology (the functioning of the eye), and psychology (the interpretation of visual and color perception) all mix in their contribution to the account provided for optical phenomena.[3] This is certainly so at the introductory level of teaching optics at school.

P-6. Optics knowledge representation heavily involves graphic symbolism, which is subject to interpretation. A number of commonly adopted conventions of graphical coding are often tacitly presumed. As a result, the consumers of optics often misinterpret some common auxiliary graphical tools, such as light rays.[4] It is especially true at the introductory level of instruction to students with culturally various backgrounds.

In that respect, students of optics are somewhat similar to researchers of the past. Naive reasoning may greatly mislead the learner.

In view of these objective premises, it is not difficult to anticipate specific barriers to the consolidation of optics knowledge both collectively, during the history of science, and privately, in the individual learners. Moreover, since the same barriers impede the process, the similarity of the conceptual growth is unsurprising. In psychology, the idea of similarity emerged first in Kofka's conception of recapitulation in 1925.[5] It was adopted and developed to the broad conception of Piaget's genetic epistemology.[6] This vision faced strong criticism. Vygotsky, in Russia, pointed to the aspect, which undermines the proclaimed similarity. Even being an individual researcher, a child, from a Marxist perspective, essentially remains a product of society, and his/her knowledge is nothing but a product of enculturation: "No

---

[3] For example, Feynman et al. (1964); Gregory (1979).

[4] For example, Beaty (1987); Galili et al. (1991).

[5] Kofka (1925, p. 44).

[6] Piaget (1970, p. 13); Piaget and Garcia (1989).

Mowgli (a feral child) is possible."[7] No progress is made in science without the influence of socially accumulated scientific knowledge, transferred to the individual researcher or child equally. Thus, Piaget's child—an independent researcher—is a fantasy and a scientist, a hermit in his laboratory.[8] As is often the case, both these perspectives were correct in the sense that they are complementarity and symbiotic in nature. Therefore their similarity and difference.

Students' Knowledge of Optics

Stating complementarity is not much use in education if it does not define the teaching strategy and content. A more comprehensive picture is therefore required.

Exploring students' knowledge about light and vision became a special domain of physics education research under the paradigm of educational constructivism central in modern education.[9] Within it, fundamental importance is ascribed to the process of rational construction of knowledge, emphasizing making sense of knowledge claims, discovery, and invention of meaning on behalf of an individual. Educational constructivism considers learning as a conscious conceptual change from naïve knowledge to scientific one in the manner somewhat similar to that taking place in science.[10] This understanding emerged from criticism of an oversimplified image of learning as encouraged accretion of knowledge (behaviorism in education[11]). Behaviorism and constructivism and adoption and construction (conceptual change) do not exclude but rather refine and complement each other. Is this similar to the continuous discourse that produces scientific knowledge? Does the progression in individual conceptions in the course of learning have something in common with the change of theories of science?[12]

Numerous alternative conceptions in optical knowledge were elicited, investigated, and documented.[13] Their abundance testifies the counterintuitiveness of optics theories of the curriculum. Naïve "commonsense" simply often fails in revealing the true account of processes nothing of which can be "seen." The universality of misconceptions across different countries indicated their objective origin

---

[7] Mowgli is a fictional character of Rudyard Kipling's *The Jungle Book* stories. Grown up by wolfs the wild boy could later return to human society.

[8] Flavell (1963).

[9] For example, Duit et al. (2005), Duit and Treagust (1998).

[10] Driver and Bell (1986); Posner et al. (1982).

[11] Skinner (1968/2003).

[12] This similarity in optics is striking. The similarity in vision mechanism was addressed by Dedes (2005). There is a clear parallel between the theory of optical image by Alhazen (Ch. 3) and the Image Projection Scheme in students' knowledge (Galili, 1996).

[13] For example, Guesne (1985); Beaty (1987); Feher and Rice (1988, 1992); Schnepps and Sadler (1989); Ramadas & Driver (1989); Bendall et al. (1993); Galili et al. (1991); Osborne et al. (1993); Atwood and Atwood (1996); Selley (1996); Langley et al. (1997); Colin and Viennot (2001).

undermining the variety of psychological, social, ethnical, and educational background. Such conceptions were registered from kindergarten to PhD graduates, teachers, and textbooks. The amount of reports continuously grew, but interest in them decreased. A mere listing of misconceptions[14] lacking suggestive analysis and generalization has limited impact as a pedagogical tool.

The large amount of accumulated data invited reduction to a structure, establishing a theory. Among the first suggestions for such structures was considering students' conception of motion as a theory reminiscent of the medieval theory of impetus.[15] diSessa[16] developed a more fragmented and comprehensive depiction. He identified intuitive explanatory constructs—*phenomenological primitives* (p-prims)—that governed understanding in accordance with "naïve sense of mechanism." For example, the p-prim "force as a continuous mover" suggests each motion to be affiliated with an agent (force). This is reminiscent of the account for "violent" motion in Aristotelian theory.[17] Minstrell formulated students' conceptions in terms of *facet of knowledge*[18]—a cognitive construct describing students' reasoning or strategy in a specific context. Facets represented conceptual, operational, and representative ideas and beliefs by which they account for certain situations. For example, students may state that in a collision, a bigger car applies bigger force. That facet would correspond to the inclusive p-prim: the more ... the more (regardless the context).

In view of such structures in other domains, students' knowledge of optics—vision and light—was considered. It was suggested that naïve knowledge be organized into a two-level structure of scheme—facets of knowledge. *Scheme of knowledge* represented a general concept, a certain mechanism, and cause–effect relationship between physical factors. Examples of such *schemes* include: "Light is comprised of light rays" or "An image is transferred as a whole entity."[19] *Schemes* often manifest themselves in context specific *facets*—concrete realizations of the correspondent schemes. For example, the Image Holistic Scheme is related to a cluster of facets each applying the same idea to the various contexts of vision, mirrors, lenses, pinhole, or prisms. Some of them are presented in Table 7.1. In all of these facets, an image moves and stays as a whole.

---

[14] As a general anthropic intention of comprising of lists was impressively depicted by Ecco (2009).

[15] McCloskey (1983).

[16] diSessa (1993).

[17] Chapter 2.

[18] Minstrell (1992).

[19] Galili and Hazan (2000a, b). Unlike p-prims, schemes of knowledge present explanatory cognitive constructs, which were produced not only prior to instruction but possibly under the influence of the instruction, for its misinterpretation or any other reason.

Table 7.1 Examples of scheme-facets structure of students' knowledge[23]

Scheme	Facet	Context
A corporeal replication of an object—image—might move, remain stationary, or turn as a whole	The image stays in the mirror whether or not it is observed. (Even if you do not see it, others can) 	Plane mirror
	The image moves from the object toward the mirror, where it stays (the observer is not involved in image existence) 	
	A half lens produces a half image. The rest of the image (rays) is blocked 	Convex lens
	When the screen moves toward or away from the lens, the image will become bigger or smaller but remain sharp.	
	Triangular prism splits the image of an object behind it 	Prism

This reduction to a structure does not reduce the number of conceptions but organizes them into a meaningful picture refining the general claim of "Knowledge in pieces" by diSessa. Our study provided a list of eight schemes (Table 7.2), which reduces a much longer list of facets and empowers teachers to recognize numerous manifestations of the same schemes.

Table 7.2 Schemes of knowledge, which structure students' knowledge in optics

	Scheme	Content
1	Spontaneous vision scheme	*Seeing happens naturally by eyes*
2	Corporal light scheme	*Light is reified to a material object, continuous (filling space) or discrete (rays)*
3	Flashlight scheme	*Individual rays are emitted from the points of a light source*
4	Image holistic scheme	*A corporeal replication of an object—image—might move, remain stationary, or turn as a whole*
5	Image projection scheme	*Each image point is related to its correspondent object point by a single light ray, which transfers its appearance*
6	Shadow image scheme	*A shadow is a kind of image separated from the object*
7	Shadow associative scheme	*Shadow depends on the environment*
8	Color-pigment scheme	*Color is different from light*

The scheme-facets structure undergoes modification under the influence of learning and eventually produces knowledge corresponding to the triadic DC structure we elaborated for scientific theory (Fig. 7.3). Importantly, this structure suggests that students, even after learning and upgrading their perspective, preserve their previous conceptions (the periphery) from where they can be retrieved when the student faces a novel situation to tackle new problems and explain unfamiliar phenomena.[20]

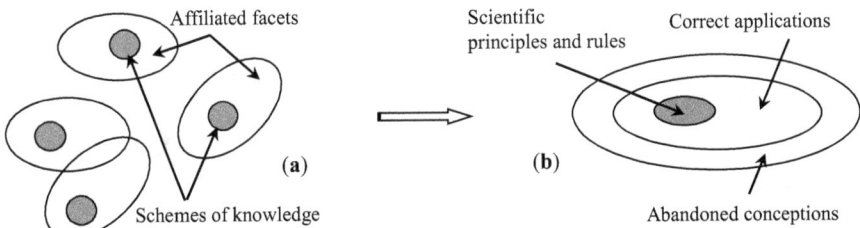

Fig. 7.3 (**a**) Schematic representation of the knowledge in pieces by means of scheme-facets structure of students' knowledge. Small circles designate schemes, while the oval areas—the clusters of facets affiliated to a certain scheme. Since the same facet may match more than one scheme, the areas of the clusters may overlap. (**b**) The new structure of students' knowledge of specific curricular domain. The previously held conceptions, considered "abandoned," are kept in the periphery

[20] An empirical evidence of this scenario regarding the knowledge of mechanics is in Galili and Bar (1992).

The Cultural Perspective on Teaching Optics

Multiple Theories and Their Relationship

The central feature of teaching from the DC perspective is the emphasis on theory as a major component of optics (scientific) knowledge. Of the four fundamental theories constructed in optics, three are normally learned in classes to a greater or lesser extent. These are the theory of rays (or geometrical optics), the theory of waves (physical optics), and the theory of photons (modern physics). All three are valid theories, each within the specific range of parameters. The fourth theory is Newton's theory of light particles. Though never developed to a comprehensive theory, it reigned supreme in the scientific thinking of the whole of the eighteenth century and was refuted only in the nineteenth century. All four theories should be mentioned to the students when each of the three valid theories is included in the curriculum. This strategy aims for a highly important goal, recognizing the status of a physics theory. Each scientific theory reflects reality but is not identical to it. It captures certain aspects and creates certain picture valid in a certain area of parameters. Optics, with its several fundamental theories, all from the school curriculum, presents a unique opportunity for learning about the nature of science.

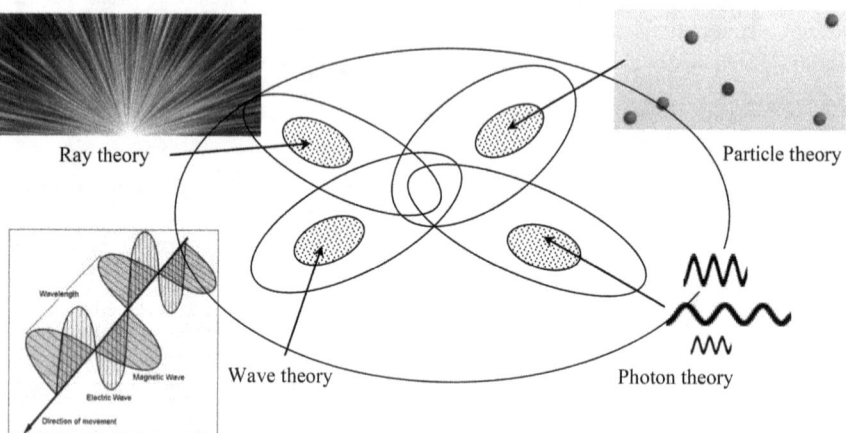

Fig. 7.4 Schematic relationship of the four theories of light (area has no indicative meaning). Each theory draws on a paradigmatic model of its nucleus, claiming certain nature of light, which determines the way of account for optical phenomena in the body. Overlapping body areas include the phenomena/problems treated by several theories

Representing the big picture of optics within the DC framework displays an intricate structure (Fig. 7.4) revealing relationship between the fundamental theories of light. The principles (nucleus) of one theory are incoherent and contradict the principles of the other. Therefore, each nucleus appears being affiliated to the periphery of other theories.

Table 7.3 Validity areas of the four fundamental theories of light in terms of physics parameters

Ray theory	$\Rightarrow I_{tot}/I_1$ decreases $\Rightarrow$	**Particle theory**
$\Downarrow \lambda/L$ increases $\Downarrow$		$\Downarrow \lambda/L$ increases $\Downarrow$
Wave theory	$\Rightarrow I_{tot}/I_1$ decreases $\Rightarrow$	**Photon theory**

In high school, there are all the necessary tools to specify the areas of validity for each theory in terms of physical parameters (Table 7.3). The required parameters are the ratio of the wavelength λ to the typical size of the objects L in the setting $-\lambda/L$ and the intensity of illumination I_{tot} to the energy of a single photon $I_1 - I_{tot}/I_1$. Teaching can include only the claims of sufficient increasing and decreasing of the ratios.

Nature of Light and Vision

The cultural teaching of optics does not start from geometrical optics (as usually presumed) but from the pre-theory conceptions. The first ideas of light and vision within Hellenic science furnished the conceptual discourse and determined its topics and notions for many years to come. It started with the questions how we see and what is the role of light in this process. They remained relevant for the novice student. In Hellenic science, scholars debated on whether vision presents an intromission or an extra-mission process, whether a whole image (eidolon) enters the eye (as the atomists thought) or the eye emanates something (visual fire) that touches the environment causing visual perception (as the Pythagoreans believed). Or, perhaps, the process goes both ways (Plato and Empedocles). Or the observed object, by its color, put the medium around it into tension, which transmits the "idea" of the object into the observer eye (Aristotle).[21]

This discourse continuously expanded on the following periods of science. To frame the problem and to set the questions were no less important in science than to provide answers. The diachronic enterprise developed conceptual setting to deal with the questions about light and vision. For example, Aristotle discovered the observation with a pinhole—camera obscura—that looked to him as a paradox. Why do we observe an image of sun on the screen after a small whole (Fig. 7.5a), whereas a big lit area of the shape of the opening (window) is seen behind the large opening (Fig. 7.5b)?

Aristotle was stunned to observe a continuous transition from one case to the other with growing size of the opening. This paradox was resolved only by Al-Haytham about 14 hundred years later using the ray theory. He explained how images emerged in the camera obscura (Fig. 7.5e), a primitive but fundamental apparatus of geometrical optics. His account drew on the principle of light diffusion from each point of any object in *all* possible directions (Fig. 7.5d). Importantly, the

---

[21] Chapter 3.

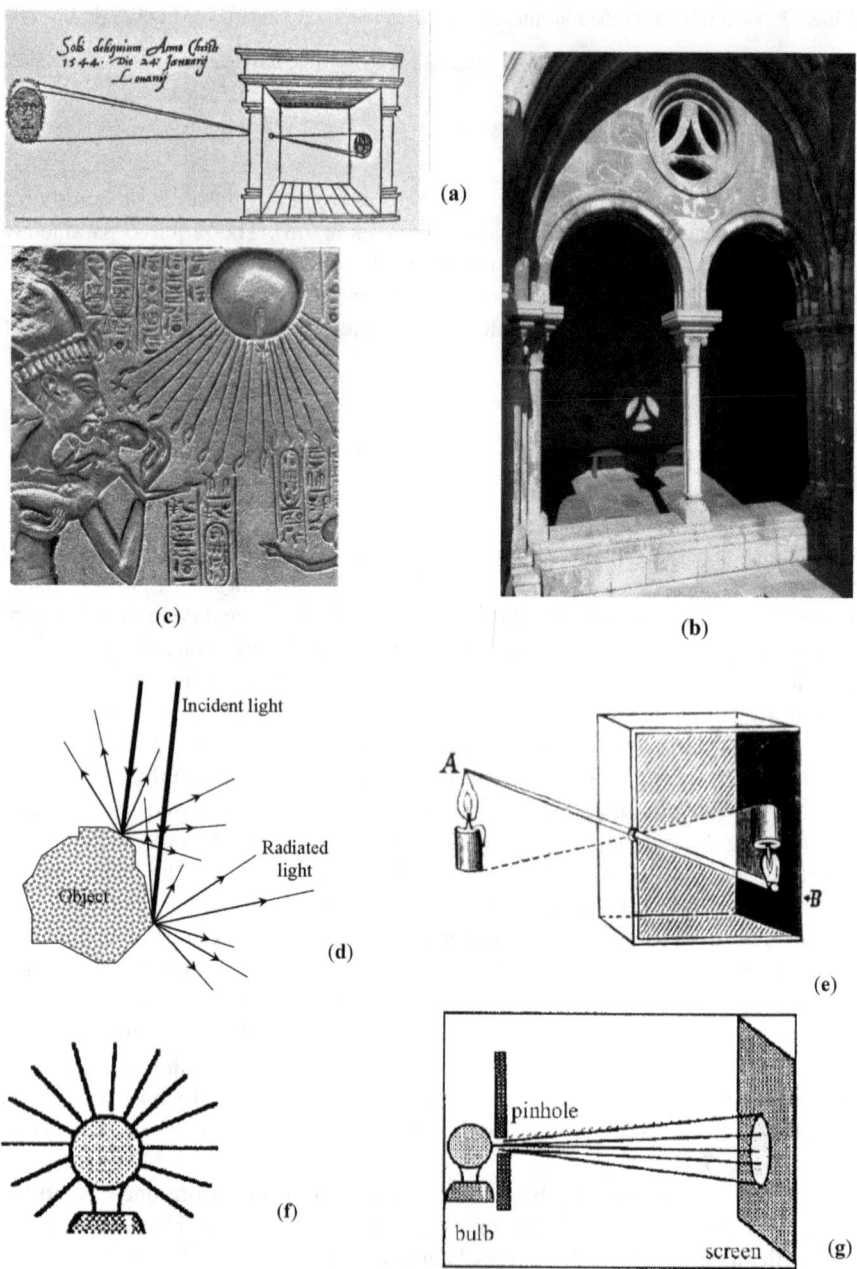

Fig. 7.5 (**a**) The image of Sun observed by Aristotle behind a small hole. Illustration from Fricius (1545). (**b**) The illumination pattern after a big opening lit by Sun. (Picture by the author) (**c**) The understanding of solely radial light expansion from the Sun in the ancient Egyptian art. (**d**) The principle of Al-Haytham of light emission. (**e**) The explanation of the pinhole image revealed by Al-Haytham in the eleventh century. (**f**) The popular student conception of light expansion from a source, and (**g**) its manifestation in the prediction of a clear circular spot of light on the screen against the opening by the light bulb attached to the opening

deficient understanding of light expansion prior this discovery by Al-Haytham (Fig. 7.5c) was very similar to the erroneous understanding of modern students (Fig. 7.5f, g). Both the problem and its solution are essential for understanding optics and both disappeared from modern school textbooks that often skip over pinhole image on their way to the camera with a lens. The story of this understanding is diachronic, and it should be presented as such.[22] The students are led to appreciate the critical aspects of light expansion, which allow to account for observation in the range of settings. The affective impact of the pinhole image is significant: students are normally more excited and surprised to observe an image after an *empty* hole than after a lens—an instrument.

Consolidation of the Ray Theory

The first theory of light—geometrical optics—is a standard component of the physics curriculum. Yet, it is scarcely called a physical theory in classes. By contrast, Einstein surprised his colleagues in academy when he called geometry to be the first *physical* theory. He had his own reason, as he wanted to emphasize the *empirical* basis (not a priori) for this apparently abstract mathematical theory on the way to his own theory of space-time—the theory of relativity as drawing on the *empirically* based metric. Indeed, geometry is known as the first mathematical theory, structured in a deductive hierarchical way. It was comprised in the third century BC by Euclid from numerous results obtained by several scholars. The same Euclid suggested a theory of light, organized in a similar manner. It was not the first knowledge about light, but it was the *first theory*. It contrasted the previous conceptions of Hellenic philosophers—we may place those in the periphery of the ray theory.[23] The nucleus of Euclid's theory incorporated the concepts of light and visual rays but was silent about their nature. Their basic features were postulated.[24] The body knowledge of the theory included the basic ideas of perspective and the procedure of shadow pattern construction. It implied numerous problems to be solved. Among them are Thales' solution of the height measurement of distant objects, Aristarchus' determination of distances and sizes of the Moon and Sun, and Eratosthenes' experimental evaluation of the Earth radius. Ptolemy included the first version of the law of refraction, Heron—the proof for the law of specular reflection and Archimedes—the principle of reversibility of light path in the context of light reflection. Geometrical optics continued to develop into medieval times. Al-Haytham introduced his version of vision mechanism, Grosseteste suggested account for the

---

[22] Teaching may include camera obscure at use by the artists of the seventeenth century (Steadman, 2002).

[23] Chapter 3.

[24] Darrigol (2012, p. 9). In parallel, in his geometry, while introducing straight line as a basic concept, Euclid illustrated it by the sunray penetrating a small opening (Fig. 7.2).

focusing of light passing through a glass sphere, and Theodoric tried to explain the rainbow under new principle of multiple refraction and reflection in water drops.[25]

Even if not all accounts were correct or precise (such as those of vision mechanism and refraction law), the major challenge to the theory of rays was perceived by scholars in the seventeenth century when two new phenomena were discovered, double refraction and diffraction. They joined the old controversial issue of color, which was not treated within the ray theory at all. All the three open problems were left by the theory of rays without even an attempt at an explanation. Thus, the DC structure of the ray theory could be presented as in Table 7.4.

Table 7.4 Representative elements of content knowledge of the theory of rays structured as a discipline–culture. The situation corresponds to the seventeenth century

Nucleus	Body	Periphery
Light ray (Newton) Specular reflection (Heron) Refraction law (Snell) Principle of reversibility of light path (Archimedes) Principle of extremal path (Fermat)[a]	Shadow phenomenon (shadow construction and the principle of Al-Kindi) Construction of illumination patterns (umbra–penumbra shadow patterns) Measurements using shadow (Thales, the height of distant objects; Aristarchus, the size and distance to the Moon and Sun; Eratosthenes, the radius of the Earth globe) Al-Haytham's explanation of imagery in camera obscura Kepler's account of vision Imagery in lenses and mirrors, burning mirrors (Archimedes) Optical instruments (spectacles, microscope, telescope, and prisms)	*Obsolete conceptions* Hellenic conceptions of light and vision (intromission and extra-mission) Visual rays (removed by Al-Haytham) Principle of minimal path of Heron Refraction law of Ptolemy Color conception of Aristotle Visual image as constructed by single rays (Al-Haytham) *Open problems* Colors, color fringes Double refraction (Bartholinus) Diffraction (Grimaldi)

[a]Ross (2008, p. vi)

The astronomical endeavors of Aristarchus and Eratosthenes mentioned here as the body knowledge elements of the ray theory are often affiliated to the history of astronomy.[26] Yet, they in fact present striking examples of using rectilinear propagation of light. As such, they belong to optical theory and can be used in the class following DC-based curriculum.

---

[25] Ptolemy (1940/1948), Boyer (1987), Sabra (2003).
[26] Berry (1898/1961, pp. 34–35. 39–40).

The image transferred point by point by a single ray

Fig. 7.6 (**a**) Annunciation (1437) by Fra Angelico. (**b**) Depiction of a transfer through space, each point by a single ray (fragment from the Annunciation). (**c**) Depiction of an image transfer in accordance with students' misconception: Image Projection Scheme

The important contribution to the ray theory is due to the paramount Arabic scholar Al-Haytham in the eleventh century. It led scholars along 600 years, up to the seventeenth century.[27] His understanding of optical image production, point to point from the object to its image inside the observer eye, by a single ray for each point, seemed plausible to people (Fig. 7.6) as it remains to our students who spontaneously produce this naïve conception in a wider context of optical settings beyond vision.

The Transition from the Ray Theory

The transition from the ray theory is of special interest as it did not come about through a direct contest between the new and old theory. Rather, a competition developed between two new conceptions, that of waves by Huygens and that of particles by Newton. Even more strange is that neither new theory adequately solved

[27] Lindberg (1976).

the problems raised by the old theory of light. Both of the new theories managed to solve some problems but failed in others.

Huygens followed Descartes and suggested that the world could be considered to be filled with specific medium—luminiferous ether. The pressure waves of this medium present light and the directions of pressure expansion were represented by light rays. The expansion of light was considered through appearance of secondary sources (principle of Huygens). Each point of ether reached by light wave becomes such a new source. This way the world is filled with waves, which all interfere, amplifying and destroying each other. Huygens could not make exact accounts of waves except very simple cases of spherical and plane waves of light, which form propagating wave fronts due to the summing of secondary waves from the line of secondary sources (Fig. 7.7a). A perpendicular to such frontier was the light ray. Based upon such a model, Huygens succeeded in explaining specular reflection of light. The other success of Huygens was the explanation of refraction (Fig. 7.7b). The model was simple to understand. Consider an axis with two wheels moving at an angle to the surface between two areas with different speed of possible motion. Meeting the surface first with one wheel naturally causes turning of the direction of motion of the axis—refraction.

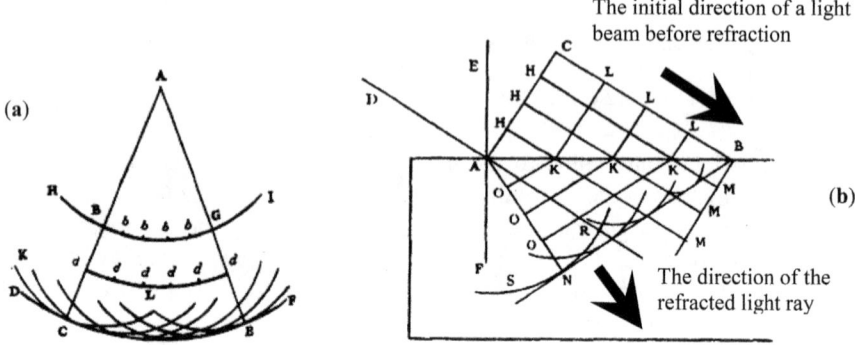

Fig. 7.7 (**a**) The drawings of Huygens explaining creation of wave of light. (**b**) The diagram explaining refraction of light in a crystal. Arrows and captions are added

The greatest success of Huygens was his explanation (though premature) of the double refraction. He understood that in a crystal, the expansion of light is not equally isotropic; there is a dependence of the direction of light expansion due to the structure of the crystal. Therefore, in the crystal of calcite (Iceland spar), a light beam split into two of which one—known as the extraordinary ray—violates the law of refraction. Huygens showed how nonspherical secondary waves of light in the crystal create a wave front that expands non-perpendicularly to the front (Fig. 7.8).

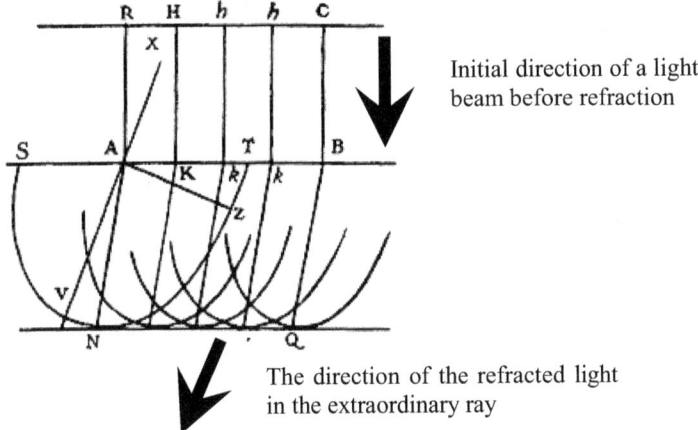

Fig. 7.8 The drawing of Huygens explaining the extraordinary ray of light in a crystal after refraction. Arrows and captions are added

Yet, if two such crystals are place sequentially in the certain identical orientation, the experiment showed that the two beams caused by the first crystal do not split for the second time when reach the second crystal. The secondary beams could preserve their identity in the refraction behavior (Fig. 7.9a) or exchange it if the second crystal was turned at right angle (Fig. 7.9b).

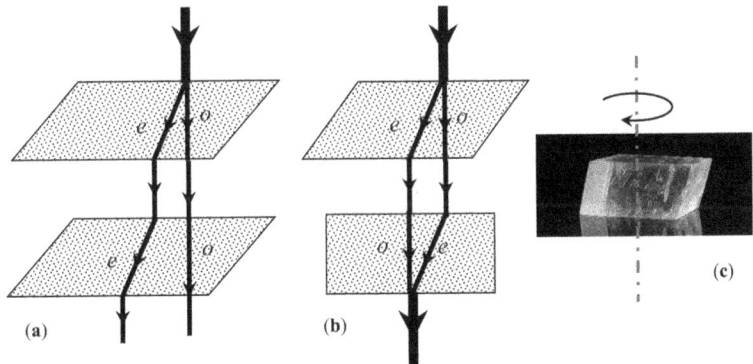

Fig. 7.9 The idea of the experiment that puzzled Huygens. At certain orientation of the crystals, the light beams do not split again entering the second identical crystal. (**a**) Yet, when the two crystals are placed at the same orientation, the two beams preserved their identity in the second refraction. (**b**) When the second crystal is turned at right angle, 90°, the two produced beams exchanged their identity in the second refraction. *o* stands for ordinary, and *e* stands for extraordinary beams. (**c**) Island spar crystal of the kind considered by Huygens. The direction of turn of the second crystal is shown

That behavior of light Huygens could not explain and the splitting of the light beam into two. He described the experiment and stopped saying the following:[28]

---

[28] Huygens (1690/1912, p. 92).

Before finishing the treatise on this Crystal, I will add one more marvelous phenomenon which I discovered after having written all the foregoing. For though I have not been able till now to find its cause, I do not for that reason wish to desist from describing it, in order to give opportunity to others to investigate it. It seems that it will be necessary to make still further suppositions besides those which I have made…

No question, it was a dramatic moment for him. Huygens succumbed to the complexity of light behavior in crystal. Nevertheless, his *Treatise on Light* founded wave optics and developed the approach of treating light in crystals—nonisotropic medium (the modern crystal optics). Also, Huygens did not touch on the other two problems of the ray theory waiting for solution—color and diffraction. All that was watched by Newton who understood the origin of Huygens' failure in his account for light. Newton saw that failure as evidence of a more complex nature of light: light rays may have sides,[29] and light is not a pressure wave (which is scalar wave), as Huygens supposed, but what is it?

One could not find a better example illustrating the dialogue of the two basic philosophical approaches to physics—rationalism and empiricism. Huygens—a devoted follower of Descartes—wrote his *Treatise of Optics* in a pure rationalist style. He stated light to be a wave of ether and implemented this idea in the account for the reality of optical phenomena. Newton—facing the great difficulty of addressing the hidden underpinning of optics phenomena—chose to go a pure empiricist way. He did not claim what light was but preferred to develop the old theory of rays in hope of arriving at the necessary inference of the particle nature of light. He intended to "deduce" the theory "from the phenomena" while avoiding "hypothesis."[30] That is why he started with the operational definition of his central tool—the light ray in the first lines of his *Opticks*:[31]

The least Light or part of Light, which may be stopp'd alone without the rest of the Light, or propagated alone, or do or suffer any thing alone, which the rest of the Light doth not or suffers not, I call a Ray of Light.

This strategy may deceive one into thinking that Newton's *Opticks* was "neutral" as regarded the particle–wave controversy. It is indicative, in this regard, to compare Newton's accounts for the light in his *Opticks* and his *Principia*. In the latter, using the mechanistic theory of particles Newton *demonstrated* the law of (specular) reflection and the Snell law of refraction through considering particles momentum, items of body knowledge,[32] while in *Opticks*, he *postulated* the same laws as empirical for rays, thus refraining from interpretation and placing them in the nucleus of the ray theory.[33]

---

[29] Newton (1704/1952), in his *Opticks*, Book III, Query 26, wrote: "Every Ray of Light has therefore two opposite Sides, originally endued with a Property on which the unusual Refraction depends, and the other two opposite Sides not endued with that Property." It was an ingenious conjecture anticipating much later understood polarization of light waves involved in the triumph of the wave theory in the nineteenth century (Lipson, 1968; Kipnis, 1991).

[30] Newton (1687/1999, p. 943).

[31] Newton (1704/1952).

[32] Newton (1687/1999, pp. 623–625).

[33] Newton (1704/1952, p. 5).

Then, Newton treated the problem of color. Addressing color dispersion, Newton introduced light rays varying in refrangibility—color rays as an empirical reality. Applying the classical method of *resolution* and *composition*[34] (Fig. 7.10a, b), he decomposed sunlight into color spectrum and then, to remove speculations about the role of prism, he synthesized the color rays back to the white light beam.

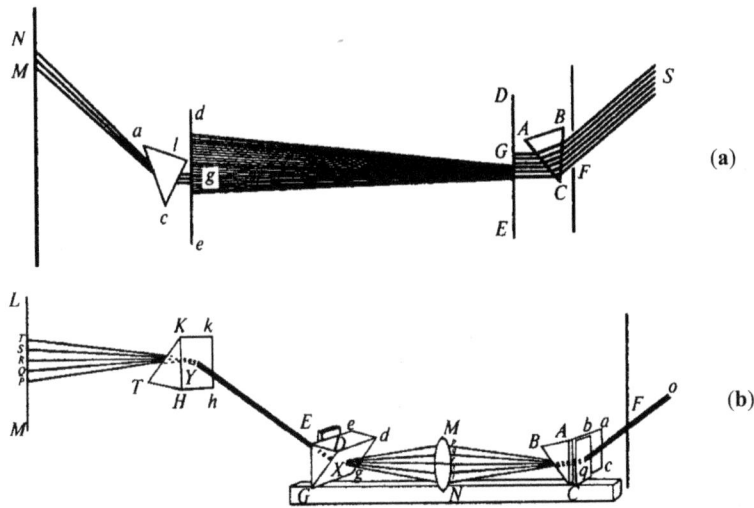

Fig. 7.10 (**a**) Newton's diagram from his *Opticks* (1704, Book I, Plate 2) explaining his experiment in which showing no further split happens to a single-color ray. (**b**) Newton's diagram from his *Opticks* (1704, Book I, Plate 4) explaining his experiment, which showed that white light beam can split into color rays and can be reproduced back by their composition

Newton proceeded with rays to explain the pattern of color rings due to a thin layer of air of varying thickness between a lens and a plate (Newton's rings)—which we now consider an interference phenomenon. Not for Newton, however, he explained it within the ray theory and without interference. Instead, he ascribed to each ray periodicity of "fits," predispositions to reflection from or penetration into the transparent medium by the light ray (Fig. 7.11).[35] Mach mentioned:[36]

> *Newton's experiments were the first to give a clear demonstration of the relevance of the periodic nature of light...*

Newton, then, turned to light diffraction and meticulously reproduced and refined the experiments of Grimaldi.[37] Newton rejected Grimaldi's split into regular and extraordinary light to explain fringes (light strips) next to the edge of geometrical

---

[34] The principle is known from Aristotle and medieval scholars (Losee, 1993, p. 28).

[35] Tyndal (1877); Westfall (1962, 1989); Kipnis (1991); Shapiro (1993). Newton's numerical results of ray periodicity were of unprecedented to his time accuracy: for yellow-orange ray, it was 1/89000 inch (Newton, 1704/1952, p. 285) well conforming to the half wavelength known today.

[36] Mach (1913/1926, p. 141).

[37] Grimaldi (1665).

shadow[38] and replaced Grimaldi's *diffraction* with *deflection* of rays. Yet, he failed to establish any general theoretical and mathematical account of light diffraction. After a detailed description of the phenomena in several settings, he abruptly stopped for "being interrupted," in his words. It was a dramatic recognition for Newton of the unsurmountable complexity of the diffraction phenomenon within the ray theory, in a clear parallel to Huygens' difficulty with respect to double refraction, Newton wrote:[39]

> *When I made the foregoing Observations, I design'd to repeat most of them with more care and exactness, and to make some new ones for determining the manner how the Rays of Light are bent in their passage by Bodies, for making the Fringes of Colours with the dark lines between them. But I was then interrupted, and cannot now think of taking these things into farther Consideration.*

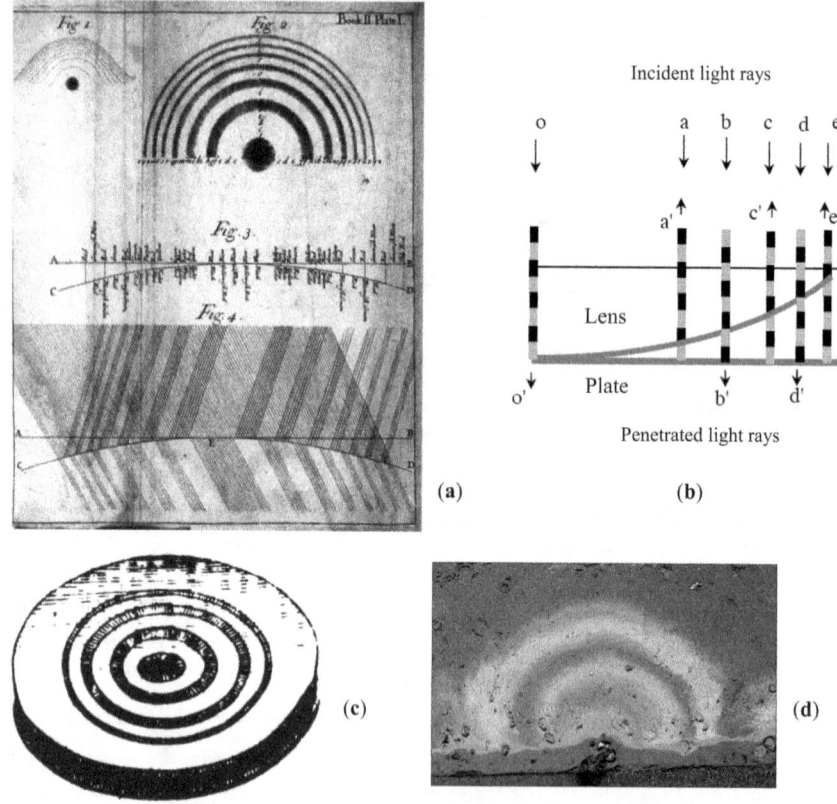

Fig. 7.11 (**a**) A page from *Opticks* (1704). Newton showed creation of light rings (his Fig. 2), mentioned position of rings of different colors (his Fig. 3), and showed separation of light rays to penetrated and reflected fluxes (his Fig. 4). (**b**) The mechanism of fits introduced by Newton explained schematically. Periods of fits of rays are marked. The incident light rays o, a, b, c, d, e... split at the inner surface of the lens to reflected rays a′, c′, e′...and o′, b′, d′...—penetrated rays. (**c**) Illustration of Newton's rings from Britannica (1771, Vol.3, Plate CXVI, N. 39). (**d**) Newton' rings as observed in a spot of oil. The diminishing thickness of the oil film from the center of the spot radially reproduces the geometry of Newton's experimental setting

---

[38] Gliozzi (1965, pp. 121–122); Taylor (1941, p. 516).

[39] Newton (1704/1952, pp. 338–339).

Yet, after the main text of *Opticks*, Newton could not resist his own need to expose his thoughts. He added a special section—*Queries*—where he described his considerations and hypothesis regarding the nature of light and apparently addressed the future researchers (again, similar to Huygens…). Newton wrote:[40]

> And since I have not finish'd this part of my Design, I shall conclude with proposing only some Queries, in order to a farther search to be made by others.

Only there, in *Queries*, did Newton allow himself to speculate: "Are not rays of light small particles emitted by shining substances…?" and argued for the advantages of the *corpuscular* nature of light over the *wave* theory suggested by Huygens. Addressing the double refraction (birefringence), Newton further stretched the ray theory and introduced *sides* to the light rays—a primitive version of light polarization.[41] By this, he suggested a qualitative explanation to the situation of two consecutive crystals, which puzzled Huygens.[42] In the nineteenth century, Malus provided the quantitative account for polarization of the reflected light (the Malus law). He introduced polarization of light "molecules" instead of Newton's sides of rays and also rejected Huygens' understanding since he treated the reflected light.[43]

However, what served as a major support for the particle nature of light was the rectilinear expansion of light (shadow experience), which could be considered for the rectilinear motion of bodies in harmony with the major state of matter—a rectilinear motion of a free particle; a beam of light preserved itself. Newton quit his struggle with the enigma of light, not before that he expressed his preference for the *particle* nature of light.

Thus, in the contest of two theories in the seventeenth century—particles and rays—each presented its successes in the account of light. Both scholars ultimately fall short in their attempts to produce an accomplished theory. Yet, the theory of light particles although treated not as Newton would want[44] (as he did in the *Principia*) was preferred over its rival—the wave theory of Huygens. From about 1700, the *Treatise on Light* by Huygens was almost completely ignored even in research reports from within the medium tradition.[45] We may summarize the accomplishment of the two great scientists (Table 7.5).

Neither of these theories was accomplished nor offered satisfactory mathematical ability, and neither had a full breadth of coverage. We can only speculate as to why it was Newton's theory that was overwhelmingly preferred by the

---

[40] Newton (1704/1952, pp. 339–406).

[41] For example, Mach (1913/1926, p. 189). See also footnote 30.

[42] Newton (1704/1952, Query 26, pp. 358–361).

[43] A more mature investigation was provided later by Brewster (Darrigol, 2012, pp.191–193).

[44] One may add to the list of Newton's successes the dynamical account of light behavior in *Principia* and his polemics there with Descartes' paradigm of plenum.

[45] Hakfoort (1995, p. 53).

majority of scientists.[46] Seemingly, a central role was played by Newton's explanation of rectilinear light expansion (shadow phenomena and stability of light beams) and the alleged light pressure on comets' tails. Newton's inability to explain diffraction or light appearing in shadow was ignored as if being a minor effect arising from the interaction of light particles with the edges of the screen and their extreme smallness. Such "details" could wait further investigation for an exact accounting.

Table 7.5 The accomplishment of Newton and Huygens

Newton	Huygens
Ray of light operationally defined / Light beam stability / Specular reflection (corresponding to the momentum conservation parallel to the surface)	Wave nature of light / The principle of light expansion—Huygens' principle / Wave-based explanation of specular reflection of light wavelet
Refraction law (corresponding to the momentum increase due to the mistaken claim of higher speed of light in a denser medium)	Wave-based explanation of light wavelet refraction
Natural light as a composition of color rays of different refrangibility (color spectrum)	The inference regarding lower speed of light in a dense medium
Explanation (mistaken) of rings pattern in thin films by ascribing periodicity within a single ray	Explanation possibility of crossing light beams moving from different sources
Correct estimation of the period corresponding to the light wavelength ("length of fits") / Correct guessing regarding polarization of light	Explanation (partial) of the double refraction
Failure to explain diffraction, interference and double refraction	Failure to explain color / Failure to explain diffraction and interference / Failure to explain stability of straight beam of light

[46]Yet, with significant exceptions like Euler who wrote (Darrigol, 2012. p. 154): "Thus, above all, I lay down that light is propagated by means of pulses through a certain elastic medium in a manner similar to sound; and just as sound usually spreads through air, I assume that light propagates through another elastic medium of some kind, which fills not only our atmosphere but all the universal space of the world that separates us from the most distant fixed stars."

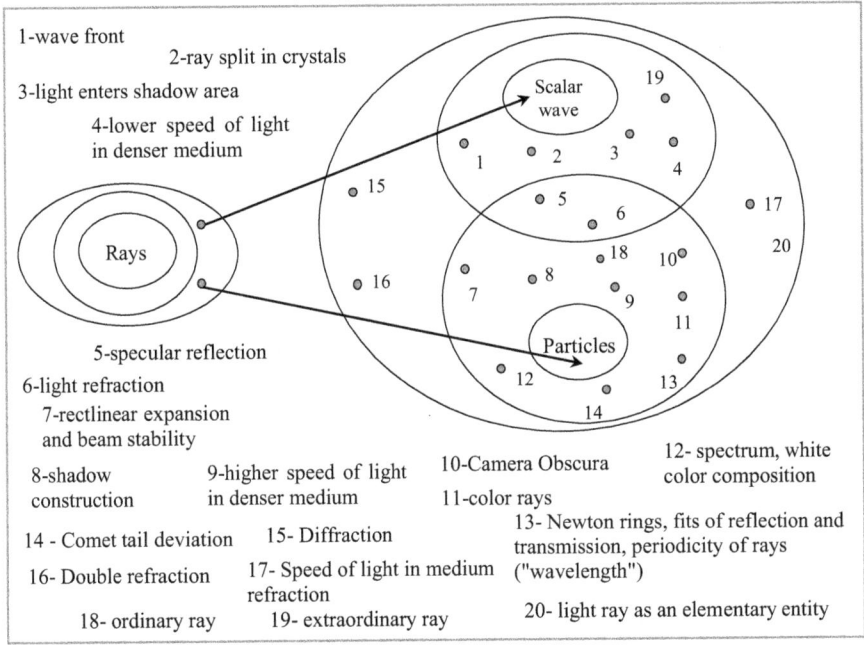

1-wave front

2-ray split in crystals

3-light enters shadow area

4-lower speed of light
in denser medium

5-specular reflection

6-light refraction

7-rectlinear expansion
and beam stability

8-shadow
construction

9-higher speed of light
in denser medium

10-Camera Obscura

11-color rays

12- spectrum, white
color composition

14 - Comet tail deviation

15- Diffraction

13- Newton rings, fits of reflection and
transmission, periodicity of rays
("wavelength")

16- Double refraction

17- Speed of light in medium
refraction

18- ordinary ray

19- extraordinary ray

20- light ray as an elementary entity

Fig. 7.12 The symbolic representation of the optical knowledge development in the seventeenth to eighteenth centuries and the contest of the two optical theories using DC structure for a physical theory

A holistic view of the transition from the ray theory of light and the contest of the empiricist theory of Newton with the rationalist theory of Huygens can be displayed using the DC diagrams of Fig. 7.12. This representation may suggest that Newton's theory of particles was more prolific and showed higher potential than Huygens' wave theory. In addition, Newton's light particles theory seemed a natural extension of the ray theory. In particular, considering light rays as trajectories of particles of light naturally relates optics to the already accepted Newton's laws of motion in mechanics (so well verified). It was declared:[47]

Light consists of an inconceivably great number of particles flowing from a luminous body in all manner of directions.

The aforementioned operational definition of ray, which Newton introduced in his Opticks, was replaced with the theoretical definition:[48]

A ray of light is a continued stream of these ~ articles, flowing from any visible body in a straight line.

---

[47] Encyclopeadia Britannica (1771/1979, p. 416).

[48] Encyclopeadia Britannica (1771/1979, p. 417).

The particle theory of light perfectly matched Newtonian mechanics in a unified mechanical picture of the world, together revealing a homogeneous divine design of the universe, as he imagined to himself.

Fig. 7.13 The old graving by Riccioli (1651) modified for lecturing the contest of Newton's and Huygens' theories of light (see the undisturbed image in Chap. 9). The icons of the authors Newton (the preferred theory), Huygens (rejected theory) on the balance, and Euclid-Ptolemy-Alhazen theory of light rays (abandoned theory) on the ground were added to facilitate the ascribed new meaning of the picture. The original picture illustrated the preference of the Tychonic world system over Copernican and of the Ptolemy system abandoned on the ground

The dramatic contest between the two theories illustrates the bigger picture of scientific knowledge as organized into theories (Fig. 7.13). Such a picture of optics knowledge could be presented in a summative ("vista point") lecture after a regular optics course[49] or as a structuring framework of the curriculum explicitly presented in class.[50]

The Transition to the Wave Theory

Newton's conception of light particles remained dominant throughout the whole of the eighteenth century[51] until Thomas Young in England, Augustin Jean Fresnel, and François Arago in France revived the undulatory (wave) theory and raised it to a new level. They introduced the principle of interference of waves to account for optical phenomena and more mature mathematical tools by which they succeeded in demonstrating a clear priority of the wave nature of light.[52] Their accounts of several experiments were of key importance and should be mentioned as a scenario of development toward a new theory of light.

There could be an elucidating discussion around the question why this experiment was not performed by Galileo, Kepler, Descartes, even by Al-Haytham, and especially by Newton, many years before. The answer might be that even such a straightforward experiment requires specific theoretical ideas and the knowledge of

---

[49] Levrini et al. (2014).

[50] Galili and Hazan (2000b).

[51] *Britannica Encyclopedia* (1771/1979).

[52] Fresnel (1866–1870), Lipson (1968), Shapiro (1973), and Kipnis (1991).

the setting parameters. In particular, one needs to imagine the light *interference* before looking for its evidence. In a way, one needs to know what to look for in order to see it. The person who could do that before was Huygens. Still, even he did not.

The central barrier to observe light interference is the requirement of *coherence* between the sources. We rarely see light interference as a salient pattern, despite the abundance of light sources around...[53] This is because the requirement of coherence is hard to reach in the case of natural light.[54] We can admire the intuitive recognition of this condition by the pioneers of physical optics.[55] They succeeded to solve this problem in the very original way. In all experiments with natural light interference, coherence is reached by using a *single* light source and splitting it before meeting again (Fig. 7.14).

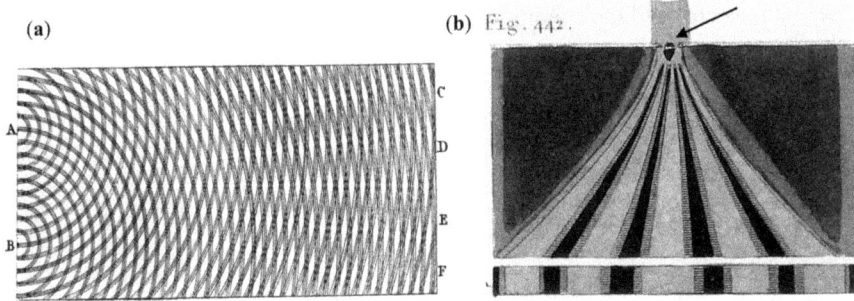

Fig. 7.14 (**a**) The drawing of Young in 1802 for the interference of water waves from two sources A and B. Young mentioned the maxima created in points C, D, E, and F. (**b**). The drawing by Young from the same lecture in which he explained his experiment. The obstacle in the opening (pointed at by an arrow) creates two light sources at a very close distance. Below the major drawing, Young schematically reproduced the interference pattern, which in reality was much less distinct but still presented a very clear evidence of the light interference

The experiment of Young did not close the questions regarding the nature of light. Poisson suggested a challenging observation that could decisively probe the truth of the wave theory. If that were correct, one would observe illumination, a spot of light on the axis just behind a disc screen. This was, however, not simple to

---

[53] Today it is a simple observation of a CD disc on sunlight. At that time, the settings though available were far from trivial to recognize (such was the peacock colors produced by many layers of his feathers). The introduction of laser also makes observation of interferences much easier, but it is a rather artificial situation.

[54] The idea of interference is very simple in the case of two simple sine waves, but myriads fragments of simple wave with arbitrary phase and amplitude comprise the natural light. The only way to overcome this chaos is to split it into two halves and meet between them—also with a short difference of optical paths, if at all.

[55] Mach (1913/1926, pp. 156–157).

provide coherence, that is, a sufficiently good "point" source of light. Nevertheless, Arago succeeded in producing such demonstration, and the so-called "Poisson spot" (or more appropriate, Poisson–Arago–Fresnel) in the middle of the disc shadow was indeed observed. It became a part of physics ethos and regular curricula.

So, light is a wave, but why did Huygens fail to understand his observation with double refraction in crystals (Fig. 7.9)? Fresnel and Arago in Paris answered. Huygens erred in the type of wave light. Light presents a *transverse* (oscillating perpendicular to the direction of propagation) wave, neither scalar nor longitudinal (i.e., oscillating in the direction of propagation). That explains why the two beams after the first crystal did not split at the second crystal one—they became polarized. Huygens thought that regular light left the first crystal, but in fact, it was not. The first crystal split regular light into two perpendicular polarizations, the ordinary and extraordinary rays. The polarized beams meet the second crystal and did not split further. Moreover, the turn of 90° placed each ray in the condition of the opposite type and thus changed the condition of the ray with respect to the second crystal.

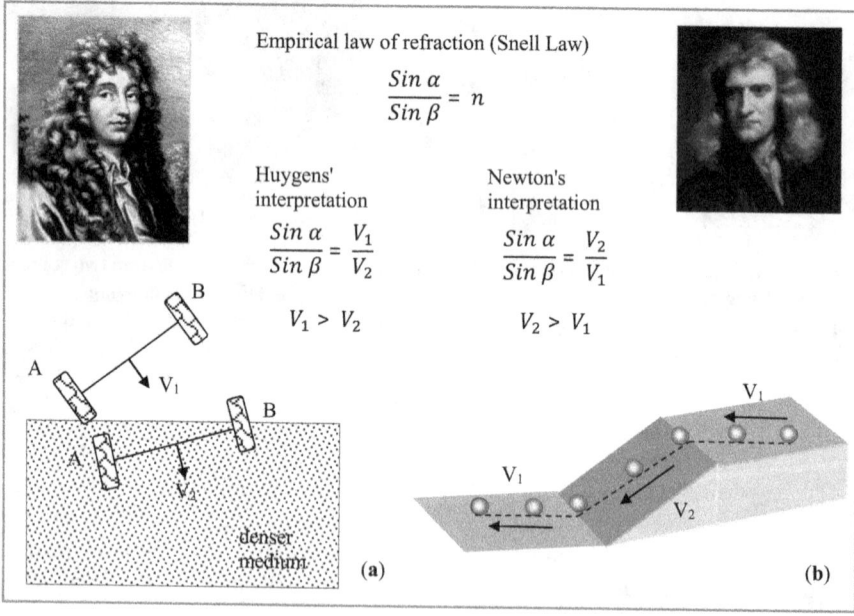

Empirical law of refraction (Snell Law)

$$\frac{Sin\ \alpha}{Sin\ \beta} = n$$

Huygens' interpretation

$$\frac{Sin\ \alpha}{Sin\ \beta} = \frac{V_1}{V_2}$$

$$V_1 > V_2$$

Newton's interpretation

$$\frac{Sin\ \alpha}{Sin\ \beta} = \frac{V_2}{V_1}$$

$$V_2 > V_1$$

(a)

(b)

Fig. 7.15 (**a**) The model illustrating Huygens' interrelation. An axis of two wheels representing wave front moves from less to denser medium. Since wheel A enters the denser area first and moves slower than wheel B ($V_1 > V_2$), the axis turn toward the denser medium. This represents the turn of the wave front of the incident light. (**b**) A ball rolls toward the slope with velocity V_1. While rolling down, the ball accelerates ($V_2 > V_1$), and its trajectory approaches the perpendicular to the border between the slope and plane. Given the particle nature of light, the denser medium stronger attracts the particles of the light beam (sucks them in). The acceleration of the ball on the slope imitates the increased attraction, which cause the angle change on the trajectory of light

In 1816, to demonstrate the transverse nature of light waves, Fresnel and Arago further developed the setting of Young. Two slits light passed through the calcite crystal, which produced the pair of ordinary (o) and extraordinary (e) rays. That provided an opportunity to investigate interference of three combinations of rays: *o-o, e-e,* and *e-o.* The result was overwhelming: the two sets *o-o* and *e-e* produced interference pattern; *e-o* did not! This was a clear evidence of the transverse nature of the waves involved. Indeed, interference may take place between the waves in the same plane, and *o* and *e* rays apparently were not such. To avoid additional doubts because of different path length of the *o* and *e* rays, they broke the crystal into halves that could be separately turned. The final triumph was to turn one of the crystals on 45° (instead of 90°) and observe all three cases of interference together. The verdict regarding the transverse wave was sealed.[56]

Perhaps to finish with the argument between the theories of light, the teacher may point to the experiment made in 1850 by Hippolyte Fizeau and Leon Foucault in Paris. Using rotating mirror devices, they could measure the speed of light in air versus water. The problem was that both theories of light, that of waves by Huygens and that of particles by Newton, successfully explained the empirical law of light refraction (Fig. 7.12) but interpreted in opposite ways (Fig. 7.15). According to Newton's particle theory, light speed through water was thought to be higher than in air (a less dense material). Both theories held with models, which looked plausible, and there was no way to falsify either of them.

The experimental result, however, was clear: the light speed in water was lower and Huygens' model was correct and Newton's false. That resolved the opposition of the two theories regarding the speed of light in medium (Table 7.5) in favor of the wave theory.

Finally, the oldest argument of the straight beam of light, we so often observe, came to the fore. Isn't this rectilinear beam a clear proof of particle nature? Why does the beam not disperse according to the Huygens' principle? The wave theory succeeded in this case too. Lord Rayleigh suggested constructing a special screen having concentric openings (Fig. 7.16). This device is known as Fresnel zone plate.

Consider two points A and B. In accordance with particle theory, light comes from A to B directly. Yet, within the wave theory, light expands in any direction and can be considered as going in various ways between the points. For instance, it could be the paths AL_1B, AL_2B, and AL_3B. They are such paths that the difference in the neighboring ones will be a half wavelength (Fig. 7.16a):

$$AL_1B - AL_0B = AL_2B - AL_1B = AL_3B - AL_2B = \lambda/2$$

We can, then, design a pattern of circular openings in a screen and cover the areas through which the difference of light paths are any odd number of $\lambda/2$ on average,

[56] Physicists love to tell that after this demonstration, the members of the French academy stood up and greeted the researchers in a standing ovation—a touching moment of the history of science. It could be true, at least, the authors definitely deserved full credit.

while the other circular areas being left open.[57] In this way, we will receive the zone plate of Fig. 7.16b. Alternatively, we may cover the zones through which the differences of light paths are on average an even number product of $\lambda/2$ including central area. In this case, we receive the screen structure of Fig. 7.16c.

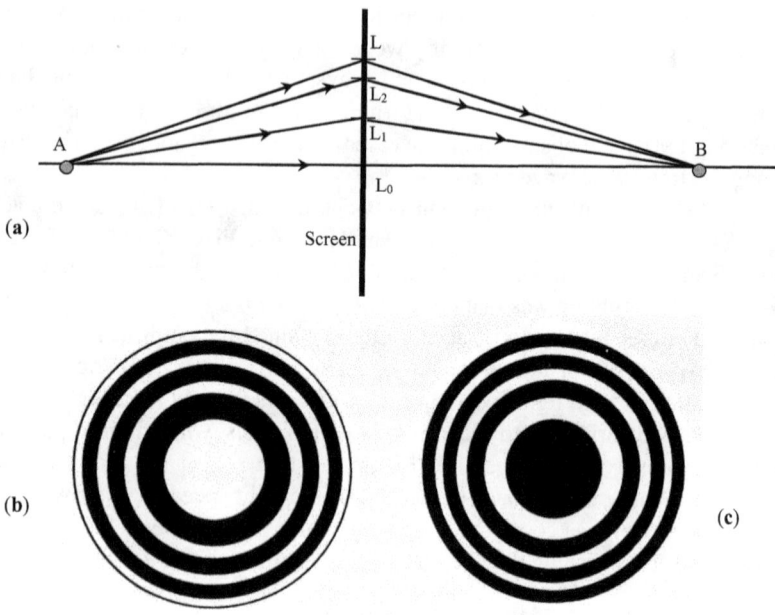

Fig. 7.16 (**a**) The principle of construction of the Fresnel zones. Two types of Fresnel zone plates are shown. (**b**) The plate that covered even zones. (**c**) The plate that covered odd zones. The plates are complementary for the same wavelength

We, then, place one screen plate or the other in the middle point L_0. Surprisingly, as a result of blocking selected paths by which the light may travel, amazingly we achieve a significant *increase* of illumination in point B! There can be only one meaning to that: it is not a hypothesis, light does go all other ways besides the straight path but due to the interference between the various beams, the light coming the other ways, not directly, cancel each other in destructive interference. By the way, we realize now that even the "visual" light does not exactly run on in a straight line but is rather spread around that line in a decreasing gradient. What a great

---

[57] For the full mathematical account for zone plates, see Wood (1911) and Hecht (1998).

discovery![58] It not only resolved the thousands years old enigma of light expansion but also inspired Richard Feynman, our contemporary great physicist, to expand the Huygens principle to matter within the advanced quantum theory.

The Transition from the Wave Theory to Photons

By the end of the nineteenth century, it looked that the wave theory provided a perfect account of light and reliably described its behavior. Yet, this feeling of all-inclusiveness was somewhat spoiled by a few "clouds" in the sky of numerous solved problems and explained phenomena. The further progress brought identification of light as the electromagnetic waves achieved by Maxwell (for the closeness of expansion speed of both kinds of waves). Visual light became a part of much wider spectrum of electromagnetic waves—EM waves. From then on, the theory of light became a part of electromagnetism, sharing its advantages and problems.

The specific problems started with the discovery of discrete linear spectrum of atomic radiation. It appeared that each atom radiates electromagnetic waves, which might be visual light too. Yet, its spectrum is not like that of the Sun, but rather a set of discrete lines associated with characteristic, for each specific kind of atom, frequencies. Why was that? The empirically found numerical rules described the spectra as mysterious combinations of terms (Ritz's Combination Principle[59]).

The next problem appeared with the attempt to measure the difference of the light speed along the direction of the Earth's movement and perpendicular to it. Despite great efforts and various attempts (the famous Michelson–Morley experiment was one of them[60]), only a zero result was ever obtained. The speed of light appeared to be the same in both orientations.

Another problem involved the "black body radiation." Every piece of matter radiates electromagnetic radiation dependent on its temperature. The frequency distribution of this radiation at the state of thermal equilibrium with matter is known as "black body radiation" (simulated in laboratory by the isolated camera). It appeared that this distribution cannot be reached by the wave theory of electromagnetism.[61]

---

[58] Robert Wood further developed Fresnel zone plate. Instead of coverage of certain zones, he covered them with additional film, which produces the path difference of $\lambda/2$. This modification increased the illumination behind the screen even further (Wood, 1911).

[59] For example, Born (1962a, p. 87).

[60] Panofsky and Phillips (1962, p. 277, Table 15.1).

[61] For example, Born (1962b, pp. 250–259). The term "black" here presumes not lacking radiation but the radiation by a body, which *fully* absorbs radiation and only then radiates. In the state of thermal equilibrium, such a body radiates very differently from the radiation of individual atoms (see just above). The spectrum becomes continuous in frequency and independent of the kind of atoms comprising the body (Kirchhoff law, 1859). Thus, in a good approximation, the radiation of the Sun presents a black body radiation. Good teaching should address the transition from atomic to matter radiation in heat equilibrium (e.g., Morison, 2008, p. 50).

Although continuous in frequency, it essentially presumes the radiation at each frequency composed in equal portions, quanta, as established by Plank in 1900.

Still another problem arose at around the same time. It dealt with the account of interaction between light (EM radiation) and matter—photoelectric phenomena. The established empirical laws of such interaction did not match any known theory.

We, thus, may represent the status of understanding of optics at that time in a DC structure (Fig. 7.15). It was the eve of the new scientific revolution. No one predicted it. Indeed, the wide areas of success of the electromagnetism and optics were very impressive. The mentioned four problems though appeared in the periphery of the great theory could easily look to the physicist like local difficulties, unsolved problems requiring only extra effort, sophisticated application, or, perhaps, small modifications of the theory (Fig. 7.17).

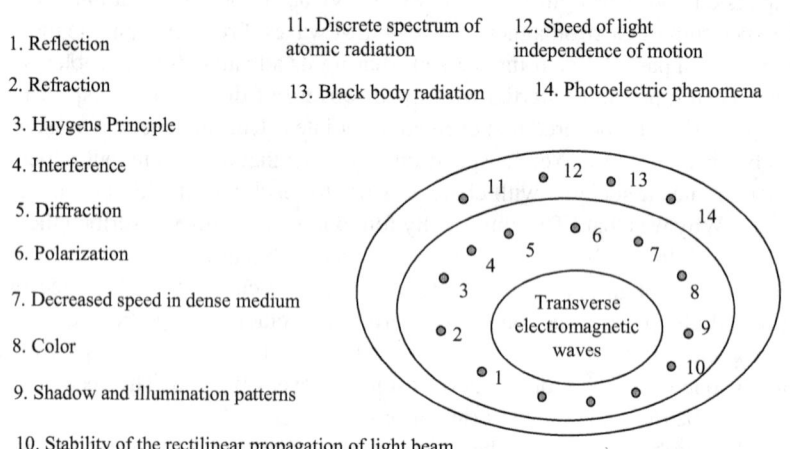

1. Reflection

2. Refraction

3. Huygens Principle

4. Interference

5. Diffraction

6. Polarization

7. Decreased speed in dense medium

8. Color

9. Shadow and illumination patterns

10. Stability of the rectilinear propagation of light beam

11. Discrete spectrum of atomic radiation

12. Speed of light independence of motion

13. Black body radiation

14. Photoelectric phenomena

Transverse electromagnetic waves

Fig. 7.17 The DC structure of the wave theory of electromagnetism/light as perceived at the end of the nineteenth century. The points in the body knowledge area represent a great many achievements of the theory (laws, concepts, rules, models, solved problems, and empirical products). They were, however, in contrast to the four problems of the periphery incompatible with the nucleus of this theory

There were varieties of attempts to solve these problems within old theoretical frameworks. They all failed.[62] Eventually, two new theories consolidated—the theory of the special relativity (by Einstein) and the quantum theory (by Bohr, Einstein, Heisenberg and others). Max Plank in 1900 and Albert Einstein in 1905 produced heuristic models of light quanta, which arrived at the new fundamental light particle—photon. Originally, in the periphery of the wave theory, this model led to the new theory of light—the quantum relativistic theory or the quantum field theory of quantum particles (quantons).

---

[62] For example, Panofsky and Phillips (1962, Table 15.2, p. 282) with regard to the theory of relativity and Wick (1995) with regard to the battle over quantum theory.

The depiction of the photon theory could be very limited in an introductory physics curriculum. Yet, it is possible and actually obligatory in contemporary education. The effective representation can be within the DC approach. In effect, it implies to emphasize the basic features of the photon theory of light in relation and comparison with the three previous light theories. And that implies the triadic content addressing the nucleus, selected elements of the body all contrasted by the pertinent parallels from the previous theories, and the periphery of the photon theory. Altogether, it is the DC-structured curriculum.

The nucleus of the photon theory will list the unique properties of light particles. Photons are quantum and relativistic objects at the same time: they cannot be treated solely by one of these theories. This is because photon has zero rest mass and always "moves" at the same speed (universal constant). As a particle, it has specific energy, momentum, and polarization (that far for the introductory course). Photons can change location with exact speed but that totally excludes establishing its location and trajectory (as place and speed can be possessed simultaneously by quanton). We can thus actually only talk about the event of photon emission and its adsorption in a proper time delay distant determined. Photons are bosons, which implies that unlike classical particles, they appear and disappear causing the change of state of the particles of another type, such as electrons, or their systems, such as atoms.

One highly important feature of photons, however, is that in big numbers they often reproduce the behavior of light waves and rays described by the theories of waves and rays (Fig. 7.4, Table 7.3). To see the specific nature of photons, one should consider the processes of proper energy and setting of other matter around. That could be illustrated with selected experiments affiliated with the body knowledge area of the DC structure. No adherence to the historical sequence of events is needed (even possible) in an appealing and student oriented teaching of this part of the introductory curriculum.

Thus, one may start with the experiment performed by G. I. Taylor in 1909.[63] He used a standard setting for observing light interference after a two-slit screen, but he did it with a very dim light source. In his experiments, the exposure time varied and reached 2000 h, or about three months. It was observed, however, that, at this very low intensity, the known pattern of interference emerges gradually from seemingly random points appearing on the screen behind the slits in an apparently chaotic manner (Fig. 7.18). That showed that light actually is composed of single entities, which create the interference pattern independently of each other. If we talk about interference, each light quantum, photon, interferes "with itself" (as first stated by Dirac[64]). This reality has to be compared with the classical prediction of the two-slit interference predicting a gradual emergence of the illumination pattern without changing its structure. The difference is striking.

---

[63] There is certain controversy regarding the reliability of this particular experiment (Wichmann, 1971, p. 169). Yet, the results were later confirmed in other experiments.

[64] Dirac (1958, p. 9).

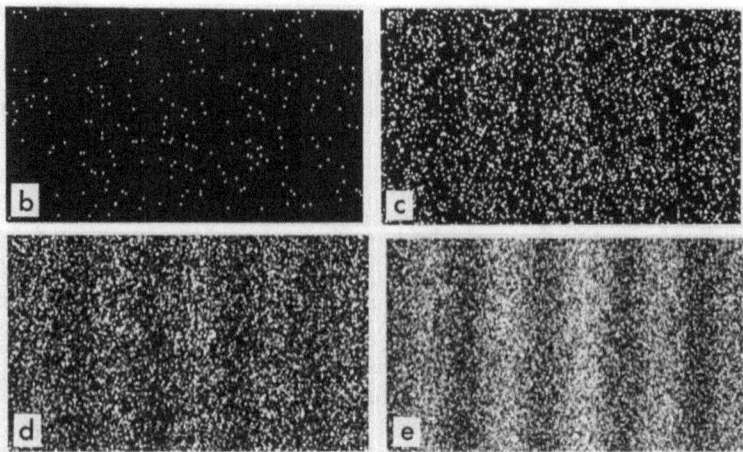

Fig. 7.18 For lacking appropriate original picture of light interference, one may illustrate by several screenshots from the experiment of two-slit interference of electrons in 1989 (Tonomura et al., 1989). The random impact of electrons gradually creates, at increasing numbers of electrons, the pattern very similar for regular two slit interference

The term *photon* for the light particle entered to physics after the Compton experiment in 1923 in which electromagnetic waves (x-rays) were scattered by electrons. This is another illustrative element of the body knowledge of the new theory. In this phenomenon, the EM wave behaved as a stream of particles, and each individually collided with certain electrons. The account keeps with energy and momentum conservation. A shift in the frequency of the scattered radiation (Compton scattering) cannot be explained by the classical wave theory (Thompson scattering), but it exactly corresponds to the account that considers light as particles of radiation—photons—colliding with electrons causing the exchange of their energy and momentum in accordance with the special theory of relativity (Fig. 7.19).

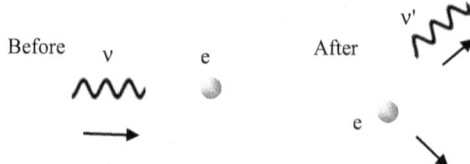

Fig. 7.19 Scenario of Compton scattering. Radiation (photon) with initial frequency ν meets with electron e. After the collision, radiation of the new frequency ν' is recorded. The frequency shift $\Delta\nu = \nu' - \nu$ is accurately explained by considering energy and momentum conservation of photon and electron as relativistic particles

To summarize this, the presentation of the photon theory of light should involve new principles in contrast with the pertinent claims of the previous theories of light. Table 7.6 provides a perspective on the possible curriculum of photon theory at schools. The level of involved formalism may vary from a pure qualitative description to incorporating elementary mathematical tools/symbolism found appropriate for the particular audience.

Table 7.6 Examples of content of photon theory of light at schools

Nucleus	Body	Periphery
Photon (zero rest mass, constant speed, and polarization) Principle of superposition of states for a photon State and principle of superposition Boson nature of photons (Fermion nature of electrons) Spontaneous and induced radiation of photons Probability in photon involved processes The unavoidable impact of measurement Entanglement/nonlocality	Two slit interference for photons. G. I. Taylor experiment (splitting of the interference patterns into spots in the very low intensity of radiation) Compton scattering of radiation with the characteristic shift of frequency of the scattered photon Discrete energy spectrum of atomic radiation Photons birth and death in atoms with discrete energy levels Lack of classical trajectory of photon Photon in a mixed state of different polarizations Probability distribution in the photon transit through a polarizer Tunneling through a potential barrier. Light penetration at the angles more than critical angle of refraction Coherence of great number of photons in the induced radiation (laser) Spontaneous emission of light (photons) in atomic radiation Photon energy proportional to frequency Red frequency limit in the photoelectric emission Thought experiment of two entangled photons with two distant polarizers (nonlocality)	Two slit interference for light waves (Young experiment) Classical scattering of radiation (Thomson scattering) Continuous spectrum of the Sun light Electron decay producing continuous spectrum of radiation–prediction in classical atom Exact trajectory of classical particles The impact of measurement can be reduced or avoided Light polarization/polarizer Malus' law for penetration of polarized light through a polarizer Full internal reflection of light between two mediums (critical angle of refraction) Requirement of coherence in wave interference Lack of spontaneous radiation and of state transitions Absolute potential barrier Energy proportionality to the quadratic amplitude and frequency of wave Classical prediction of the thought experiment with polarized light and two distant polarizers (locality)

Comments Regarding Optics Teaching

The DC-organized knowledge of optics creates its big picture making explicit its *organization* and *syntactic* features.[65] Both these aspects of knowledge determine the *nature of science* (NOS). This subject is addressed in general elsewhere,[66] and here we point to a few important features the optics knowledge culturally organized.

The Role of Theory

Physics curricula often stress modeling as the central feature of science leaving its fundamental structure as theory-based in shade. Theory is often addressed in the opposition theory–experiment that may diminish the centrality of the theory concept in science. In fact, no experiment in physics is conceivable out of a theory. Experiments in science are normally *theory driven* and *theory laden*. Physics knowledge is composed of theories in the meaning of extremely inclusive clusters of coherent knowledge elements on certain subject matter.[67] Optical knowledge demonstrates such organization in terms of theories of rays, particles, waves, and photons (Fig. 7.4). The notion of theory evolved in the course of history, gradually increasing its inclusiveness and sophistication. More than one fundamental theory of the same subject—light—indicates both the multifaceted nature of the subject matter and the polyphonic feature of scientific discourse, making the produced knowledge cultural and dialogical in nature. Indeed, from the beginning, the periphery of geometric optics includes the argument of intro- versus extra-mission conceptions of vision, the competitive ideas regarding the nature of light, and its speed and features. The first optical theory—the theory of rays—consolidated due to the cumulative efforts of numerous scholars: from Euclid and Ptolemy through Al-Kindi and Alhazen to Kepler and Newton. Similarly, the three following theories of light entered the stage facing the challenging problems accumulated in the periphery of the ray theory—colors, light diffraction, and double refraction. While teaching optics, teachers not only enculturate their students but also encourage their creativity illustrating its meaning in terms of resolving challenging problem emerged regarding the nature of vision; creation of optical images in mirrors, lenses, and prisms; explaining natural phenomena; and finding basic astronomical parameters solely by simple laws of shadow and illumination. All these aspects are considered within the conceptually rich space of variation of optical conceptions provided by the DC structure of a theory.

---

[65] Schwab (1964, 1978).

[66] Chapter 9.

[67] Chapter 6.

The Role of Experiment

Experiment is never forgotten in physics class, but its role regarding physical theory is far from intuitively clear. For example, teaching often relates the proof or refutation of a theory with a single experiment. This vision presents oversimplification. In reality, theories consolidate within an environment of various experiments and debate about their interpretation. The DC organized theories of light illustrate invalidity of the naïve view. Each theory drew on multiple experiments affiliated with the body knowledge. At the same time, not less informative and important are those experiments, which provide the negative (apophatic) knowledge regarding the theory. They are the elements of periphery (Figs. 7.9, 7.12, 7.17, Table 7.4). All three radical conceptual transitions between theories of light drew on both types of experiments.

It became clear; one cannot rely on a single, "critical" experiment. Consider the claim of speed of light as observer independent and universal constant. The famous experiment of Michelson and Morley with the moving interferometer is often presented as a decisive proof. In reality however, several competitive valid explanations of the same experiment were provided by other theories (aether drag, emission theory, and length contraction) challenging Einstein's theory. The victory over the rivals emerged from a variety of experiments (13 at that time[68] and much more thereafter). Only such a success can be taken as an *empirical proof* of a theory, which remains under testing in each of its application.

An experiment might be required but not feasible at the time. Thus, the particle and wave theories of the seventeenth century disagreed regarding the light speed through a medium. Though both theories correctly produced the refraction law of Snell, the split regarding the speed of light. Newton's theory implied the speed of light in a dense medium (glass and water) to be higher than in the rare one (air), and the wave theory of Huygens stated the opposite.[69] Accordingly, the refraction index had to be either v_1/v_2 or v_2/v_1. Much later, in 1850, Fizeau and Foucault provided the decisive experiment.

Theories suggest experiments to test their validity. Consider the EPR experiment of entangled photons. It was designed by Einstein in 1935 as a thought experiment, which can pose a question but by itself cannot be decisive in physics.[70] It eventually became a *real* experiment performed by Aspect in 1981 with photons as suggested by Bohm. The experiment intended to demonstrate the incomplete nature of quantum mechanics by pointing to the way to circumvent the limitation of uncertainty principle. It drew on the idea of locality of physical events.[71] The result of the real experiment testified that the quantum mechanics was correct and Einstein was wrong. That implied a paramount discovery: microscopic objects do subvert

---

[68] Panofsky and Philips (1962, p. 240).

[69] For example, Sabra (1981, pp. 217, 302).

[70] Galili (2009).

[71] For example, Cushing (1998, p. 325).

locality.[72] The experiment was confirmed multiple times and had a tremendous impact on our understanding of reality. The EPR experiment migrated from the periphery of quantum theory to its body. The nonlocality is now affiliated to the nucleus of the theory as a new principle of the quantum theory (Table 7.6).

The implication for NOS: theory-experiment is two-way relationship reflecting the iterative development of scientific knowledge.

Cumulative Nature of Science

In his study of scientific revolutions, Kuhn[73] stated that scientific practice comprises periods of conceptually incommensurable paradigms. This thesis is sometimes interpreted as dismissing the cumulative nature of scientific knowledge, breaking its conceptual continuity. Teaching optics allows clarification. The history of optics demonstrates that optical research was highly diachronic. It related different periods through the posed questions, which later obtained answers within different theoretical frameworks while taking into account previous considerations.

For instance, Aristotle[74] suggested his camera obscura image and questioned its explanation (Fig. 7.5). After a small opening, a circular image of the Sun appeared, while behind the big opening, the illuminated area reproduces the form of the opening—a paradox. More than 1300 years later, Al-Haytham resolved this mystery by applying a new principle of diffusive scattering. Another few hundred years later, dela Porta suggested that the eye was similar to the camera obscura with a lens in its opening. In another hundred years, Kepler resolved the enigma of vision with a different understanding of image construction. We thus observe scholars working within different scientific paradigms, using different methods considering the same problems over 2000 years. The account for vision and imagery was continuously accumulated by new methods and ideas, which are all within the horizon of a researcher.

In another example, the principle of rectilinear light propagation was considered over centuries. From the claim by Heron of *minimal* path in specular reflection in the Hellenistic science to Fermat's *extreme* path, it was continuously interpreted. In 1871, by using Fresnel's zone plate, Rayleigh experimentally demonstrated that Huygens principle regarding light expansion was correct and light is not limited to solely straight expansion but moves in all possible paths between any two points. The old argument of Newton against the wave nature of light was removed. Much later, the same idea led Feynman to the generalization of this principle for light and matter (path integrals).

---

[72] For example, Penrose (1997, pp. 64–66); Cushing (1994, pp. 14–16).

[73] Kuhn (1962/1970).

[74] Lindberg (1968).

In this sense, science is cumulative. Its knowledge is not only about what it is but also about what it is not and not only about how to deal with the problem but also about many ways to deal with it. Through centuries, working within different paradigms, scholars maintained a diachronic discourse. As Collingwood expressed:[75]

> The two phases [of science] are related not merely by way of succession, but by way of continuity, and continuity of a peculiar kind. If Einstein makes an advance on Newton, he does it by knowing Newton's thought and retaining it within his own, ... He might have done this, no doubt, without having read Newton in the original for himself; but not without having received Newton's doctrine from someone. ... It is only in so far as Einstein knows that theory, as a fact in the history of science, that he can make an advance upon it. Newton thus lives in Einstein in the way in which any past experience lives in the mind of the historian, ...re-enacted here and now together with a development of itself that is partly constructive or positive and partly critical or negative.

The cultural teaching clarifies the cumulative nature of science. Bernard of Chartres (12th C.) emblemized this feature in a metaphor that holds for NOS: we can understand only drawing on the previously accumulated knowledge, "standing on the shoulders of the giants."[76]

Making Knowledge Scientific

Teaching optics following the historical construction of theories reveals the process of making knowledge less speculative and more scientific. It thus provides an operational definition of what is scientific. Namely, the scientific knowledge emerges as objective, inclusive, empirically verifiable, and reliable. In particular, *only* objective knowledge can be a subject of critical discourse of science, defining its "health."[77] At the same time, as mentioned by Einstein, scientists are *free* to choose particular concepts, a subject of being tested in experiment.[78] The claims of researchers depend on their imagination, beliefs, worldviews, and social constraints.

Consider the principle regulating light path. Heron of Alexandria in his *Catoptrics* demonstrated the specular reflection of light: the path presents the shortest trajectory between any two points including mirror reflection.[79] This is a piece of *objective* knowledge. However, the interpretation of this result as seeking by Nature the most "economical" way to go or Nature does nothing in vain *(natura frustra nihil agit)* looks as *subjective* or metaphysical. Using the method of *Maxima and Minima*, Fermat in the seventeenth century advocated for the *extreme path* of light[80]—the objective truth. Yet, his claim of the "natural intention" presents a subjective

---

[75] Collingwood (1956, p. 127).

[76] Crombie (1959, p. 27).

[77] Popper (1975), Holton (1985), Weinberg (2001).

[78] Einstein (1952/1987); Miller (1986, pp. 44–46).

[79] Cohen and Drabkin (1948, p. 263).

[80] Ross (2008).

interpretation. Indeed, as mentioned by Descartes, how and why can light "decide" about the extreme path? The Snell law of refraction as a ratio of sines of the angles of incidence and penetration presents objective knowledge. Descartes believed that empirical law is not sufficient in science and a mechanism should be revealed, an *explanation*. He suggested an ad hoc mechanism of refraction.[81] He stated analogy with the motion of a ball being hit downward by a tennis racket at the surface of water.[82] Mach called this account "unintelligible and unscientific."[83] However, the question of how it happens remained.

Only in the nineteenth century, by applying interference method to the Huygens principle, the speculations regarding the path of light were removed. The experiment by Raleigh with Fresnel-zones screen demonstrated that light does not chose a unique way to go but expands in all possible directions (Fig. 7.16). Thus, the *subjective* interpretations of Fermat's principle were dismissed (Descartes' critic was correct!). An introductory optics can, thus, introduce students to the advanced scientific picture.

The implication of this story is that scientific knowledge may combine objective and subjective aspects at different stages of its construction. Eventually, in the process of scientific progress, the subjective aspects are abandoned within the better understanding.

The Role of Mathematics

The history of optics illustrates that mathematics is central for physics; it is complex and many-faceted.[84] Euclid and Ptolemy started the mathematical account of optics.[85] In that, they rebelled against the solely qualitative account of nature to be the agenda of physics. Euclid, Ptolemy, and Archimedes employed mathematics in addressing light and depict its behavior precisely. One cannot explain *perspective* (two-dimensional representation of three-dimensional reality) without geometry. At the same time, the mathematical account cannot *explain* light and vision comprehensively. Indeed, many scholars mastered perspective rules and solved important problems but held faulty ideas of vision ("vision rays..." of Euclid, Ptolemy, and Alkindi).[86]

Consider the history of the sine law of refraction. Ptolemy was the first to tackle this problem mathematically.[87] He tried to adjust the collected data to the linear

---

[81] Descartes (1637/1965, p. 79).

[82] Ross (2008, p. *v*).

[83] Sabra (1981, p. 104).

[84] Galili (2018).

[85] For example, Russo (2004); Smith (1982, 1996).

[86] Lindberg (1976), Smith (1996).

[87] Cohen and Drabkin (1948, pp. 271–283), Smith (1982), Mihas (2008).

dependence between the angles of incidence and penetration of *vision* (instead of light) rays. The required ratio of angle sines was difficult to elicit within the views held by Ptolemy. Simplicity, economy, and other general ideas could not help to decipher reality with numerical accuracy. The appearance required invention of a mathematical form.

For centuries, scientists kept trying to get the mathematical form of the refraction law. However, even such a skillful mathematician as Kepler failed.[88] Having nothing better, he used the linear law of refraction by Ptolemy (which presents a good approximation for small angles), and it served him well in designing a telescope.[89] Eventually, several scholars worked out the correct mathematical form of the law.[90] Yet, it did not satisfy physicists looking for the physical mechanism that underpins this rather unusual mathematical form. Descartes failed. Huygens and Newton succeeded, but they drew on the mutually exclusive models of waves and particles providing contradictive mathematical derivations… Which of the two should be adopted as the correct one? The answer came from outside mathematics. The two theories made opposite predictions regarding the light speed in mediums of different density.[91] Only experiment resolved the problem later in the nineteenth century.

We also observed how Newton failed to develop an adequate mathematically supported theory of light, which could treat the phenomena of diffraction and interference. In retrospective, we may see that his mission was impossible to fulfill since an adequate theory of light essentially requires the mathematical tools he lacked, such as functions with more than one variables, equations in partial derivatives, and developed calculus. In mechanics, Newton faced a similar barrier but succeeded to overcome it by invention of calculus, to the extent required there. Yet, he did not make a mathematical revolution for the second time. Later, in the nineteenth century, the required mathematics was available to Fresnel in Paris, and he succeeded in promotion the wave theory, which had to wait for the insufficient level of mathematics.

These cases show that the mathematical account in physics is essential but remains complementary to the qualitative conceptual one. Paraphrasing Einstein's address to the relationship between physics and philosophy, one may conclude that mathematics without physics is blind (i.e., unable to provide qualitative understanding), while physics without mathematics is empty (i.e., destined to speculations of unresolved validity).

---

[88] Herzberger (1966) explained this failure by Kepler trusted the insufficiently accurate old tables of Vitello back to13th c.

[89] The simplified law of refraction (linear approximation) can be used in a qualitative account for a variety of optical phenomena in regular teaching (Galili and Goldberg, 1996).

[90] Ibn Sahl (c. 984), Harriot (c. 1602), Snell (c. 1621), and Descartes (c. 1637) (Rashed, 2002, p. 313; Kwan et al. 2002).

[91] Sabra (1981, pp. 300–302).

Commonsense Controversy

Another important facet of science to be revealed in teaching optics is the appeal to commonsense and the relation to its controversy. Optics history is especially eloquent in this regard. Consider the "upside down" image created on the retina of human eye. For years, many scholars could not accept this fact. It apparently contradicted "commonsense"—the regular perception. The inverted images in the camera obscura and after convex lens were not convincing. The "obviousness" misled Alhazen who erroneously placed a right side up image on the eye lens surface. Later, Leonardo painstakingly searched for *two* successive inversions that could provide a right side up image.[92] Only Kepler, in the seventeenth century, removed this enigma: *the commonsense lies*—the image in the eye is inverted and additionally "inverted" by the mind of the observer, by our cognition.[93]

Another example deals with the nature of light. Due to the fact that we normally receive light without appreciated sense of effort (not perceiving collision of photos with retina), our commonsense does not conceive light as a moving agent but rather as a state of illumination in which light "fills" and "stays" in space. Light as a static entity fits the Biblical description of the world creation and the conception of students.[94] The historical teaching of optics may address this topic revealing to the students the difference between the visual perception of light, on the one hand, and its objective existence, on the other—an epistemological issue.[95] This interesting topic touches on the historical split of the light concept into two: *lux*, corresponding to the perception of illumination and *lumen*, the agent of light itself.[96] *Lux* and *lumen* are Latin terms introduced in the translation of the Bible into Latin (the Hebrew original had no such split in terminology, but in the meaning).

Discussion on the dichotomy of what objectively exists and what is subjectively perceived may lead students to a critique of naïve commonsense and its difference from scientific knowledge[97] and, yet, its necessity in making science. Koyré stated that the role of founders of modern science was "to replace a pretty natural approach, that of our commonsense, by another which is not natural at all."[98] Einstein moderated that claim saying that science continuously refines and corrects commonsense, which we are destined to keep in use. Science continuously upgrades it and transmits the new benefits of the progress to public use and education.

---

[92] Chapter 8.

[93] Kepler (1611/2000), Lindberg (1976), Chap. 3.

[94] Bendall et al. (1973), Galili and Hazan (2000b).

[95] Gregory (1979), Linn et al. (2003).

[96] Middleton (1963), Steneck (1976), Galili and Hazan (2004).

[97] Cromer (1993), Wolpert (1994), Chap. 9.

[98] Koyré (1943).

Summary

We may summarize that teaching optics as a discipline–culture is beneficial in various aspects. It clarifies the *disciplinary content*, the *nature of science*, and scientific knowledge construction. The quality impact of the *cultural content knowledge* established in this way is beyond the remedy of misconceptions. The new curriculum addresses the genesis of optics by means of scientific discourse, diachronic and synchronic dialogue, elucidating the meaning of the knowledge to be learned. This approach matches the claims about similarity, although not identity, of the ideas regarding light and vision in the history of science and those on behalf of learning students. The research-based evidence of beneficial impact of this approach calls for a pertinent restructuring of the optics curricula. The essential role of the history and philosophy of science in learning optics is to support its arrangement in four fundamental theories in a unified picture of optics knowledge.

Picture Credits

- Fig. 7.5a Camera obscura. Illustration by Leonardo da Vinci from Gemma Frisius, *De Radio Astronomica et Geometrica*, 1545. Public Domain. https://commons.wikimedia.org/w/index. php?curid=51048791
- Fig. 7.5c God Aten and sun rays. Photo by Osama Shukir Muhammed Amin FRCP(Glasg). CC Attribution-Share Alike 4.0 International license. https://commons.wikimedia.org/wiki/ File:Akhenaten_with_one_of_his_girls,_detail_of_an_altarpiece_of_a_shrine._God_Aten_ and_his_cartouches_appear._C._1345_BCE._From_Amarna,_Egypt._Neues_Museum.jpg
- Fig. 7.5e Camera obscura. Drawing of 1910, Public Domain. https://commons.wikimedia.org/ wiki/File:Camera_obscura_1.jpg
- Fig. 7.6 *Annunciation* by Fra Angelico, 1432. Public Domain. https://commons.wikimedia.org/ wiki/File:La_Anunciaci%C3%B3n,_by_Fra_Angelico,_from_Prado_in_Google_Earth_-_ main_panel.jpg#mw-jump-to-license
- Figs. 7.7a and b and 7.8 The drawings from *Treatise of light* by Huygens, 1674, Figs. 20, 36, and 64. Public domain. https://www.gutenberg.org/files/14725/14725-h/14725-h.htm
- Fig. 7.10a and b Illustration from *Opticks* by Newton 1704. Book I, Part I. Fig. 18; Part II, Fig. 16. Public Domain. http://www.gutenberg.org/files/33504/33504-h/33504-h.htm
- Fig. 7.11a Illustration from *Opticks* by Newton 1704. Book II, Part I, Figs.1, 2, 3, 4. Public Domain. http://www.gutenberg.org/files/33504/33504-h/33504-h.htm
- Fig. 7.11c The Newton's rings. Illustration from The Britannica, 1st edition, 1771, Vol. 3, Plate CXVI, N. 39
- Fig. 7.11d Newton rings in a thin film of oil. Photo by John, The CC Attribution-Share Alike 2.5 Generic license. https://commons.wikimedia.org/w/index.php?curid=4081809
- Fig. 7.13 Frontispiece of *New Almagest* by Riccioli, 1651. Public Domain, https://commons. wikimedia.org/w/index.php?curid=16670323
- Portrait of Huygens by unknown author, 17th c. Public Domain. https://commons.wikimedia. org/w/index.php?curid=5505106
- Portrait of Newton by Sir Godfrey Kneller, 1689. Public Domain. https://en.wikipedia.org/ wiki/File:Sir_Isaac_Newton_(1643-1727).jpg

- Portrait of Ptolemy by unknown author, 1584, from *Popular Science Monthly*, Vol. 78, 1911, p. 316. Public Domain. https://commons.wikimedia.org/w/index.php?curid=19423714
- Fig. 7.14a Illustration from *Physics* by Taylor, 1941, p. 520.
- Fig. 7.14b Illustration from Lectures by Thomas Young, 1807, London's Royal Institution. Public Domain, https://commons.wikimedia.org/w/index.php?curid=39577646
- Fig. 7.18 The figure from A. Tonomura et al. *American Journal of Physics* 1989, Vol. 57, p. 117. The image by Bessazar, CCAttribution-Share Alike 3.0 Unported license. https://commons.wikimedia.org/wiki/File:Double-slit_experiment_results_Tanamura_2.jpg

References

Atwood, R. K., & Atwood, V. A. (1996). Preservice elementary teachers' conceptions of the causes of seasons. *Journal of Research in Science Education, 33*(5), 553–563.

Beaty, W. (1987). The origin of misconceptions in optics? *American Journal of Physics, 55*(10), 872–873.

Bendall, S., Goldberg, F., & Galili, I. (1993). Prospective elementary teachers' prior knowledge about light. *Journal of Research in Science Teaching, 30*(9), 1169–1187.

Berry, A. (1898/1961). *A short history of astronomy*. Dover.

Born, M. (1962a). *Einstein's theory of relativity*. Dover.

Born, M. (1962b). *Atomic physics*. Hafner.

Boyer, C. B. (1987). *The rainbow: From myth to mathematics*. Princeton University Press.

Cohen, R. M., & Drabkin, E. I. (1948). *A source book in Greek science*. McGraw-Hill.

Colin, P., & Viennot, L. (2001). Using two models in optics: Students' difficulties & suggestions for teaching. *American Journal of Physics, Physics Education Research Supplement, 69*(7), S36–S44.

Collingwood, R. G. (1956). *The idea of history*. Oxford University Press.

Crombie, A. C. (1959). *Medieval and early modern science*. Doubleday Anchor Books.

Cromer, A. (1993). *Uncommon sense*. Oxford University Press.

Cushing, J. (1994). *Quantum mechanics: Historical contingency and the Copenhagen hegemony*. University of Chicago Press.

Cushing, J. (1998). *Philosophical concepts in physics*. Cambridge University Press.

Darrigol, O. (2012). *A history of optics from Greek antiquity to the nineteenth century*. Oxford University Press.

Dedes, C. (2005). The mechanism of vision: Conceptual similarities between historical models and children's representations. *Science & Education, 14*, 699–712.

Descartes, R. (1637/1965). *Discourse on method, optics, geometry and meteorology. Second discourse – Of refraction*. Bobbs-Merrill.

Dirac, P. A. M. (1958). *The principles of quantum mechanics*. Calendon Press.

diSessa, A. (1993). Toward an epistemology of physics. *Cognition & Instruction, 10*, 105–225.

Driver, R., & Bell, B. (1986). Students' thinking and the learning of science: A constructivist view. *School Science Review, 67*, 443–456.

Duit, R., & Treagust, D. F. (1998). Learning in science – From behaviorism towards social constructivism and beyond. In B. Fraser & K. G. Tobin (Eds.), *International handbook of science education* (pp. 3–25). Kluwer.

Duit, R., Gropengießer, H., & Kattmann, U. (2005). Towards science education research that is relevant for improving practice: The model of educational reconstruction. In H. E. Fischer (Ed.), *Developing standards in research on science education* (pp. 1–9). Taylor & Francis.

Ecco, U. (2009). *The infinity of lists*. Rizzoli.

Einstein, A. (1952/1987). *Letters to Solovine: 1906–1955 (May 7, 1952)*. Open Road, Integrated Media.

Encyclopeadia Britannica. (1771/1979). *The First Edition*. Society of Gentlemen in Scotland.

Feher, E., & Rice, K. (1988). Shadows and anti-images: Children's conception of light and vision II. *Science Education, 72*(5), 637–649.

Feher, E., & Rice, K. (1992). Children's conceptions of color. *Journal of Research in Science Teaching, 29*(5), 505–520.

Feynman, R., Leighton, R., & Sands, M. (1964). *The Feynman lectures on physics*. Addison-Wesley.

Flavell, J. H. (1963). *The Developmental Psychology of Jean Piaget*. D. Van Nostrand.

Fresnel, A. (1866–1870). *Oeuvres complètes d'Augustin Fresnel* (3 vols). Imprimerie Impériale.

Galili, I., & Bar, V. (1992). Motion implies force. Where to expect vestiges of the misconception? *International Journal of Science Education, 14*(1), 63–81.

Galili, I., & Goldberg, F. (1996). Using a linear approximation for single-surface refraction to explain some virtual image phenomena. *American Journal of Physics, 64*(3), 256–264.

Galili, I., & Hazan, A. (2000a). Learners' knowledge in optics: Interpretation, structure, and analysis. *International Journal in Science Education, 22*(1), 57–88.

Galili, I., & Hazan, A. (2000b). The influence of historically oriented course on students' content knowledge in optics evaluated by means of facets-schemes analysis. *Physics Education Research, American Journal of Physics, 68*(7), S3–S15.

Galili, I., & Hazan, A. (2004). *Optics – The theory of light and vision in the broad cultural approach*. Science Teaching Center, The Hebrew University of Jerusalem.

Galili, I. (1996). Student's conceptual change in geometrical optics. *International Journal in Science Education, 18*(7), 847–868.

Galili, I. (2009). Thought experiment – Establishing conceptual meaning. *Science & Education, 18*(1), 1–23.

Galili, I. (2018). Physics and mathematics as interwoven disciplines in physics class. *Science & Education, 27*(1–2), 7–37.

Galili, I., Goldberg, F., & Bendall, S. (1991). Some reflections on plane mirrors and images. *The Physics Teacher, 29*(7), 471–477.

Gliozzi, M. (1965). *Storia della Fisica* (Vol. II). Storia della Scienze.

Gregory, R. L. (1979). *Eye and brain*. Princeton University Press.

Grimaldi, F. M. (1665). *Physico-Mathesis de lumine, coloribus, et iride*. Vittorio Bonati.

Guesne, E. (1985). Light. In R. Driver, E. Guesne, & A. Tiberghien (Eds.), *Children's ideas in science* (pp. 11–32). Open University Press.

Hakfoort, C. (1995). *Optics in the age of Euler. Conceptions of the nature of light, 1700–1795*. Cambridge University Press.

Hecht, E. (1998). *Optics*. Addison-Wesley.

Herzberger, M. (1966). Optics from Euclid to Huygens. *Applied Optics, 5*(9), 1383–1393.

Holton, G. (1985). *Introduction to concepts and theories in physical science* (Second edition revised by S. G. Brush). Princeton University Press.

Huygens, Ch. (1690/1912). *Treatise on light: In which are explained the causes of that which occurs in reflection & in refraction, and particularly in the strange refraction of Iceland crystal*. McMillan. In French: Huygens, Ch. (1992). Traité de la lumière. Dunod.

Kepler, J. (1611/2000). *Optics: Paralipomena to Witelo and the optical part of astronomy*. Green Lion Press.

Kipnis, N. (1991). *History of the principle of interference of light*. Birkhauser Verlag.

Kofka, K. (1925). *The growth of mind*. Harcourt, Brace &.

Koyré, A. (1943). In G. Holton (1952). *Introduction to concepts and theories in physical science* (pp. 21–22). Addison-Wesley.

Kuhn, T. S. (1962/1970). *The structure of the scientific revolution*. The University of Chicago Press.

Kwan, A., Dudley, J., & Lantz, E. (2002). Who really discovered Snell's law? *Physics World, 15*, 64.

Langley, D., Ronen, M., & Eylon, B. (1997). Light propagation and visual patterns: Pre-instruction learners' conceptions. *Journal of Research in Science Teaching, 34*, 399–424.

Levrini, O., Bertozzi, E., Gagliardi, M., Grimellini-Tomasini, N., Pecori, B., Tasquier, G., & Galili, I. (2014). Meeting the discipline-culture framework of physics knowledge: An experiment in Italian secondary school. *Science & Education, 23*, 1701–1731.

Lindberg, D. C. (1968). The theory of pinhole images from antiquity to the thirteenth century. *Archive for History of Exact Sciences, 5*(2), 154–176.
Lindberg, D. C. (1976). *Theories of vision from Al-Kindi to Kepler*. The University of Chicago Press.
Linn, M. C., Clark, D., & Slotta, J. D. (2003). WISE design for knowledge integration. *Science Education, 87*, 517–538.
Lipson, H. (1968). *The great experiments in physics*. Oliver & Boyd.
Losee, J. (1993). *A historical introduction to the philosophy of science*. Oxford University Press.
Mach, E. (1913/1926). *The principles of physical optics. An historical and philosophical treatment*. Dover.
McCloskey, M. (1983). Naïve theories of motion. In D. Genter & A. L. Stevens (Eds.) *Mental models* (pp. 299–324). Lawrence Erlbaum. Intuitive Physics, *Scientific American, 248*(4), 122–130.
Middleton, W. E. K. (1963). Note on the invention of photometry. *American Journal of Physics, 31*(2), 177–181.
Mihas, P. (2008). Developing ideas of refraction, lenses and rainbow through the use of historical resources. *Science & Education, 17*(7), 751–777.
Miller, A. I. (Ed.). (1986). *Frontiers of physics: 1900–1911. Selected essays*. Birkhauser.
Minnaert, M. (1940). *Light and colour in the open air*. Bell.
Minstrell, J. (1992). Facets of students' knowledge and relevant instruction. In R. Duit, F. Goldberg, & H. Niedderer (Eds.), *Research in physics learning: Theoretical issues and empirical studies* (pp. 110–128). IPN.
Morison, I. (2008). *Introduction to astronomy and cosmology*. Wiley.
Newton, I. (1687/1999). *The principia. Mathematical principles of natural philosophy* (B. Cohen & A. Whitman, Trans.). University of California Press.
Newton, I. (1704/1952). *Opticks*. Dover.
Osborne, J. F., Black, P., Meadows, J., & Smith, M. (1993). Young children's (7–11) ideas about light and their development. *International Journal of Science Education, 15*, 89–93.
Panofsky, W. K. H., & Phillips, M. (1962). *Classical electricity and magnetism*. Addison-Wesley.
Penrose, R. (1997). *The large the small and the human mind*. Cambridge University Press.
Piaget, J., & Garcia, R. (1989). *Psychogenesis and the history of science*. Columbia University Press.
Piaget, J. (1970). *Genetic epistemology*. Columbia University Press.
Popper, K. R. (1975). *Objective knowledge*. Clarendon Press.
Posner, G. J., Strike, K. A., Hewson, P. W., & Gertzog, W. A. (1982). Accommodation of a scientific conception: Toward a theory of conceptual change. *Science Education, 66*, 211–227.
Ptolemy, C. (1940/1948). Refraction. In M. R. Cohen & I. E. Drabkin (Eds.), *A source book in Greek science* (pp. 271–281). McGraw-Hill Book.
Ramadas, J., & Driver, R. (1989). *Aspects of secondary students' ideas about light*. University of Leeds, Center for Studies in Science and Mathematics Education.
Rashed. (2002). p. 313.
Ross, J. (2008). *Fermat's complete correspondence on light. Dynamics*. http://science.larouchepac.com/fermat. Accessed 24 Sept 2015.
Russo, L. (2004). *The forgotten revolution: How Science was born in 300 B.C. and why it had to be reborn*. Springer.
Sabra, A. I. (1981). *Theories of light. From Descartes to Newton*. Cambridge University Press.
Sabra, A. I. (2003). Ibn Al-Haytham's revolutionary project in optics: The achievement and the obstacle. In J. P. Hodgedijk & A. I. Sabra (Eds.), *The enterprise of science in Islam, new perspectives* (pp. 85–118). The MIT Press.
Schnepps, M. H., & Sadler, P. M. (1989). *A private universe – Preconceptions that block learning* (Videotape). Harvard University/Smithsonian Institution.
Schwab, J. J. (1964). Problems, topics, and issues. In S. Elam (Ed.), *Education and the structure of knowledge* (pp. 4–47). Rand McNally.
Schwab, J. J. (1978). Education and the structure of the disciplines. In J. J. Schwab (Ed.), *Science, curriculum and liberal education*. The University of Chicago Press.

Selley, N. J. (1996). Children's ideas on light and vision. *International Journal of Science Education, 18*(6), 713–723; Towards a phenomenography of light and vision. *International Journal of Science Education, 18*(8), 836–845.

Shapiro, A. E. (1973). Kinematic optics: A study of the wave theory of light in the seventeenth century. *Archive for History of Exact Sciences, 11*, 134–266.

Shapiro, A. E. (1993). *Fits, passions, and paroxysms*. Cambridge University Press.

Skinner, B. F. (1968/2003). *The Technology of Teaching*. Skinner Foundation.

Smith, A. M. (1982). Ptolemy's search for a law of refraction: A case-study in the classical methodology of "saving the appearances" and its limitations. *Archive for History of Exact Sciences, 26*(3), 221–240.

Smith, A. M. (1996). Ptolemy's theory of visual perception: An English translation of the optics with introduction and commentary. *Transactions of the American Philosophical Society, 86*(2), 1–300.

Steadman, P. (2002). *Vermeer's camera*. Oxford University Press.

Steneck, N. H. (1976). *Science and creation in the middle ages*. University of Notre Dame Press.

Taylor, L. W. (1941). *Physics. The pioneer science*. Dover.

Tonomura, A., Endo, J., Matsuda, T., & Kawasaki, T. (1989). Demonstration of single-electron buildup of an interference pattern. *American Journal of Physics, 57*, 117.

Tyndal, J. (1877). *Six lectures on light*. Appleton & Co.

Weinberg, S. (2001). *Facing up – Science and its cultural adversaries*. Harvard University Press.

Westfall, R. S. (1962). The development of Newton's theory of color. *Isis, 53*(3), 339–358.

Westfall, R. S. (1989). *Mechanical science in the construction of modern science* (pp. 50–64). Cambridge University Press.

Wichmann, E. H. (1971). *Quantum physics. Berkeley physics course* (Vol. 4). Education Development Center.

Wick, D. (1995). *The infamous boundary: Seven decades of controversy in quantum physics*. Birkhauser.

Wolpert, L. (1994). *The unnatural nature of science*. Harvard University Press.

Wood, R. W. (1911). *Physical optics*. Macmillan.

Chapter 8
From Comparisons Between Scientists to Gaining Cultural Scientific Knowledge: Leonardo and Galileo

Compare in order to understand.

© Springer Nature Switzerland AG 2021, corrected publication 2022
I. Galili, *Scientific Knowledge as a Culture*, Science: Philosophy, History and
Education, https://doi.org/10.1007/978-3-030-80201-1_8

Abstract Physics textbooks often present disciplinary knowledge in the sequential order of topics under instruction. Such presentation is usually univocal, that is, isolated from alternative claims and contributions regarding the subject matter in the pertinent scientific discourse. Here, I argue that comparing and contrasting the contributions of scientists addressing the same or similar subjects would not only enrich the picture of scientific enterprise but also be especially appealing to the learner seeking genuine understanding of the concept considered. This approach draws on the historical tradition from Plutarch in the distant past to Koyré in the recent history and philosophy of science. It gains new support in seeking cultural content knowledge of the subject matter. Here, we address two prominent individuals of the Italian Renaissance, Leonardo and Galileo, in their dealing with the issues relevant to introductory science courses. Although both figures addressed similar subjects of scientific content, their products were essentially different. Considering this difference is educationally beneficial, illustrating the meaning of what students presently learn in mechanics, optics, and astronomy and what they perceive with regard to the nature of science and scientific knowledge.

Introduction

The benefits of using history and philosophy of science (HPS) in science teaching have been under continuous, comprehensive debate for a long time. The recently published compendium of such efforts[1] creates a panoramic view of the ways in which HPS can be involved in science education. For many educators, the question "should we?" has been replaced with "how?," "to what extent?," and "what in particular?". Among the main reservations in involving HPS-based materials in regular teaching is that "incorrect" and "obsolete" knowledge from the past would "definitely" confuse novice learners whose knowledge is fragile and immature, thereby impeding their learning. Indeed, even when they do make historical references, authors of science textbooks often provide a fragmentary picture in which it is mentioned that the contribution of the particular scholar was to a particular subject, but the disagreements and arguments with other scholars are ignored. In other words, the ecology of major views and perspectives as they appeared in scientific debate are usually overlooked. For example, it is usually mentioned that Newton was the first who understood the physics of tides, but his account is never compared with the attempt by anyone else to do same, Galileo's endeavor, for instance. Why is that? Very simple: Newton was right and Galileo was wrong... We praise Galileo's discovery of the mountains on the Moon but remain silent regarding his proof of the Moon's atmosphere—a careful selection of right and wrong, the same approach as used in addressing the history of science in an educational setting.

Contributions to the progress of disciplinary knowledge are simply organized: Galileo—falling, ballistic, and inertial motion; Newton, the laws of dynamics and

---

[1] Matthews (2014).

the law of gravitation; Maxwell, the equations of electromagnetism and so on. It is known, however, that nothing was invented in science in a vacuum, in isolation from the diachronic and synchronic disciplinary discourse, in an intensive debate of alternative views and claims, and in this way of creating a special space in which scientific knowledge is molded. The mechanism of knowledge construction might look somewhat similar to the loop of reciprocal adjustment between matter and space-time in the theory of relativity: all material objects determine the curvature of space-time, which determines the motion of each object, and that changes space-time metric and so on. Scientists with their views and ideas create the medium of knowledge that determines the meaning of the new ideas produced, which in turn changes the medium of knowledge and so on. In other words, to understand something in science, one needs to observe the debate of ideas.

Methodological variety is in a similar dynamical reciprocal relationship with scientific knowledge. The inductive–empiricist approach is frequent in introductory courses, while the deductive–rationalist is typical in theoretical physics. Neither of them is absolutely sufficient. Different approaches are complementary, and only together may they reliably describe knowledge creation possessing the rationality and objectivity of science.

It might look preferable to be concise, accurate, and focused on distilled information regarding disciplinary knowledge. Historical studies may be threatened by obsolete concepts, values, and terminology. And yet, a strong cognitive constraint of meaningful learning was discovered. It asserts the need for conceptual variation, which means considering a space in which our cognition constructs understanding through *comparison* with alternatives—the space of variation.[2] In fact, this was and is the way in which collective knowledge emerges in discourse. The similarity of knowledge growth, the individual (among students) and collective (history of science),[3] hints to the universal mechanism of making sense by human cognition as stated in genetic epistemology.[4]

Both required aspects, the conceptual plurality and relevance, apparently suggest using science history in regular teaching.[5] The claim is not new. In general terms, it was already suggested by Mach who wrote that "historical investigation not only promotes the understanding of that which now is, but also brings new possibilities before us".[6] The challenge, however, is that even if many agree that history content would be beneficial, one needs more than that. In fact, history presents an ocean of facts, events, and stories, attracting interest but possibly leading astray. Deliberate

---

[2] Marton and Tsui (2004). In fact, it corresponds to the very old educational tradition in Talmudic schools where each claim is considered in a discourse of very different perspectives. It was adopted by medieval Christian scholars, such as Thomas Aquinas, who presented his claims in debate with alternatives.

[3] Recapitulation Kofka (1925, p. 44).

[4] Piaget (1970) and Piaget and Garcia (1989).

[5] For example, Welch (1973), Wandersee (1986, 1990), Matthews (1990), Stinner and Williams (1993), Galili and Hazan (2000b), Coelho (2010), and Stein et al. (2015).

[6] Mach (1883/1989, p. 316).

selection following a disciplinary rationale is crucial. Such selection and curricular adaptation require a certain conceptual framework that would guide navigation. For this, DC structuring of the fundamental theories in terms of nucleus–body–periphery[7] can provide the required guidance.

Here, we will deal with the relevant plurality of historical conceptions and compare the research approaches in several activities of Galileo Galilei, the pioneer of modern science,[8] and Leonardo da Vinci, a celebrated engineer, artist, inventor, and a curious scientist.[9] The teaching, including comparison of the elicited fragments in their disciplinary, philosophical, and methodological aspects, can facilitate meaningful learning of science through engaging and enjoyable activity.[10]

Comparison as a Method

Alexandre Koyré (1892–1964), a renowned historian and philosopher of science, used comparison as a powerful genre in analyzing the disciplinary and philosophical knowledge of Galileo and Kepler, Aristotle and Plato, and Newton and Descartes.[11] Such comparison made his inferences especially clear and elucidating, contrasting the often dry citing in textbooks, ignoring the nature of inquiry and learning. The science heroes usually appear sequentially, as if running in a narrow corridor or climbing up the ladder of progress of cumulative contributions, each placing his/her own brick. In contrast, in the eyes of Koyré (and quite in the tradition of Galileo's dialogues), it was in the hot debate of alternative approaches and research paradigms that knowledge revolutions took place and modern science emerged and that the new method and theories were established and have been shared and maintained by the scientific community ever since. Similarly, Arthur Miller addressed the next scientific revolution of the twentieth century. He applied comparative analysis to the epistemology of Einstein and Poincaré,[12] shedding light on their accomplishments, which are often confused.[13]

Comparison as a method is commonplace in history, literature, philosophy, and classical education.[14] Yet, in the disciplinary teaching of science, it is often lacking. In contrast, the framework of *discipline–culture* addresses the polyphony of ideas and the dialogical nature of the scientific discourse. The periphery content provides

---

[7] Chapter 6.

[8] Koyré (1943a).

[9] White (2000) and Capra (2007).

[10] Galili (2015).

[11] Koyré (1943a, 1943b, 1968).

[12] Miller (1981, 1984, 1986).

[13] For example, Darrigol (2000) and Granek (2006).

[14] In history, Plutarch (1989). In philosophy, comparison is standard in teaching: Parmenides versus Democritus, Plato versus Aristotle, Descartes versus Bacon, and so on.

rivals to the fundamentals of the nucleus, challenging them. In accordance, the knowledge of the learners becomes cultural content knowledge (CCK) rather than regular content knowledge (CK). One may consider CCK as corresponding and enriching Shulman's pedagogical content knowledge (PCK).[15] It corresponds to the specific mode of thinking of scientists and their hidden epistemological curriculum.[16]

Knowledge becomes cultural due to comparison with alternative perspectives. Blaise Pascal expressed a similar perception:[17]

> We have three principle objects in the study of truth: one to discover it when it is sought; another to demonstrate it when it is possessed; and the third, to discriminate it from the false when it is examined.

In this statement, one may relate *discover* to the content of nucleus, *demonstrate*, to the implications comprising the body knowledge, while *discriminating* the "truth" from the alternatives—the periphery of the DC structure facilitating such discrimination through comparison.

Leonardo and Galileo: Comparisons of Selected Aspects

Two prominent figures in the intellectual history of science, Leonardo and Galileo, were active about 100 years apart, during the transition into modern science: Leonardo, at its beginning, and Galileo, at its apex. However, while Galileo is widely recognized as a key figure in this process, the evaluation of Leonardo's contribution is different. Historians often consider him as a polymath: artist, engineer, inventor, but not exactly a scientist.[18] Others, however, argue that Leonardo was a scientist,[19] although they emphasize the very particular nature of his research style and contribution. It is educative to consider this controversy and its causes.

Galileo is often considered to be the *first* scientist.[20] This nomination, however, contradicts the term "transition" that presumes science also existed before Galileo. Considering him to be the first in "modern science" instead, it is important to see in what sense modern science was different. Such understanding emerges from the comparison with Leonardo with respect to the knowledge and to the method. We will address several subjects: mechanics, optics, and astronomy.

---

[15] CCK (Galili 2012); CK and PCK (Shulman 1986).

[16] Redish (2010).

[17] Pascal (1910, p. 454).

[18] For example, Pederson and Pihl (1974) and Dijksterhius (1986).

[19] White (2000), Atalay (2004) and Capra (2007).

[20] For example, MacLachlan (1997).

The Account of Motion

In the procession of scholars who accounted for motion, Leonardo and Galileo took different positions, both preceding the Newtonian theory. Let us touch on several representative points. We read in Leonardo's Notebooks (LNB):[21]

> Simple impetus is that which moves the arrow or the dart through the air. ... Impetus is the impression of movement transmitted by the mover to the movable thing. Impetus is a power impressed by the mover on the movable thing.
>
> All violent movements grow feebler the more they are separated from their cause.

This and other similar statements clearly indicate that Leonardo shared the medieval account of motion—dispensable impetus—that replaced the Aristotelian perspective.[22] Yet, Leonardo was not consistent. He also liked the Aristotelian mechanism of violent motion—antiperistasis, an intricate account by which the medium both resists and supports the motion of a projectile at the same time. Neglecting the criticism by other scholars before him,[23] Leonardo added antiperistasis to impetus, fusing the two conceptions into a hybrid:[24]

> Impetus is a power transmitted from the mover to the movable thing and maintained by the wave of the air within the air which this mover produces; and this arises from the vacuum which would be produced contrary to the natural law if the air which is in front of it did not fill up the vacuum, so causing the air which is driven from its place by the aforesaid mover to flee away.

Galileo, in turn, took a different stance. He refused to talk about the nature of motion and its causes (dynamics) while arguing:[25]

> The present does not seem to be the proper time to investigate the cause of the acceleration of natural motion concerning which various opinions have been expressed by various philosophers, some explaining it by attraction to the center, others to repulsion between the very small parts of the body, while still others attribute it to a certain stress in the surrounding medium which closes in behind the falling body and drives it from one of its positions to another. Now, all these fantasies, and others too, ought to be examined; but it is not really worthwhile. At present it is the purpose of our Author merely to investigate and to demonstrate some of the properties of accelerated motion (whatever the cause of this acceleration may be).

Avoiding theorizing ("fantasies") about motion, he curiously included Aristotelian antiperistasis ("a certain stress in the surrounding medium which closes in behind the falling body") from the past and the "attraction to the center," which

---

[21] MacCurdy (1938/1955, pp. 550, 557).

[22] The post-Aristotelian account of motion was introduced by Hipparchus and Philoponus of Hellenistic science and followed by Buridan and Oresme—medieval science. These scholars refuted the necessity of an outside mover (the air) for motion. It was the account by impetus—the "force of motion" (e.g., Pederson and Pihl 1974, pp. 237–241; Dijksterhius 1986, pp. 179–185).

[23] Hipparchus, Philoponus, Buridan in Cohen and Drabkin (1948, pp. 221–223), Clagett (1959).

[24] *LNB (pp. 509–510).*

[25] Galilei (1638/1914, p. 166).

would lead, no less, than to Newton's gravitation theory later. Galileo avoided dealing with the imposed ("violent") motion under the influence of an observed external agent and focused on the "spontaneous" and "natural" motion.

Lacking the theoretical framework of dynamics, Galileo employed broad colloquial terminology in his description of motion. Addressing the ball rolling along the wedge he wrote:[26]

> The speed reaches a maximum along a vertical direction, and for other directions diminishes as the plane diverges from the vertical. Therefore, the impetus, ability, energy, [l'impeto, il talento, l'energia] or, one might say, the momentum [il momento] of descent of the moving body is diminished by the plane upon which it is supported and along which it rolls.

For the contemporary reader, these terms, all in general, nonscientific use, should be well defined in physics and distinguished before being considered. Only after Newton, energy and momentum, synonyms for Galileo, were separated and became two basic characteristics of motion in mechanics. For Galileo, impetus coincided with momentum and was close in use to speed.

Yet, addressing motion, Galileo made very significant progress. While Leonardo made a short comment (important, but it remained undeveloped):[27]

> Every impression [ability of motion] tends to permanence or desires permanence.

Galileo discovered a special kind of motion—an *inertial* one. He wrote:[28]

> ...we may remark that any velocity once imparted to a moving body will be rigidly maintained as long as the external causes of acceleration or retardation are removed, a condition which is found only on horizontal planes; for in the case of planes which slope downwards there is already present a cause of acceleration, while on planes sloping upward there is retardation; from this it follows that motion along a horizontal plane is perpetual; for, if the velocity be uniform, it cannot be diminished or slackened, much less destroyed.

It was the third kind of natural motion, after the first two he mentioned (upward and downward), which, unlike them, was "perpetual". Before this claim of Galileo, it was a prerogative of superlunary space to maintain such motion. Now, this motion descended to the ground.

With the help of an inclined plane, popular in the medieval studies of statics, Galileo measured the natural inclination of an object to fall, while interpreting its spontaneous descent along the inclined plane as a constrained but still natural motion, a kind of redirected falling (Fig. 8.1). Though not explicitly named, he used here the medieval "positional" gravity—the fraction of natural weight, which causes the rolling down of an object along the inclined plane.[29] Leonardo was just there, in

---

[26] *Discorsi* stands for Galilei (1638/1914, p. 181).

[27] *LNB (p. 550).*

[28] Galilei (1638/1914, p. 214).

[29] For example, Jordanus de Nemore of the thirteenth century (e.g. Pedersen and Pihl 1974, p. 211), Moody and Clagett (1952, pp. 123–124). In our words, $W \sin\alpha$, with W—weight and α—the slope of the plane.

the medieval statics. He saw the inclined plane as a mechanical tool of pragmatic value—to reach force advantage: a smaller weight holds a bigger one (Fig. 8.1a).[30]

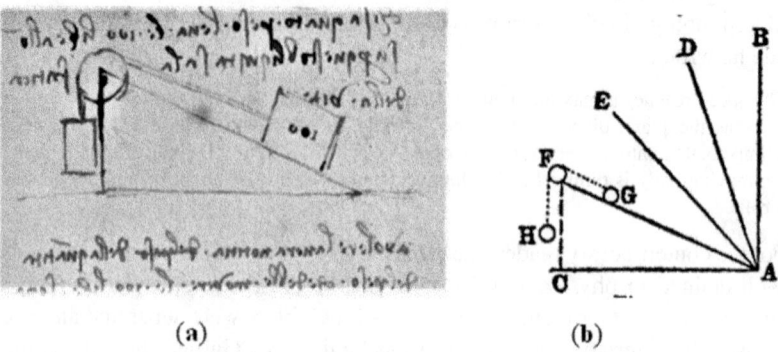

(a)   (b)

Fig. 8.1 (a) Leonardo's sketch of two masses connected over the edge. (b) Figure from the *Discorsi* (Galilei, 1638/1914, p. 181) illustrating the thought experiment of Galileo

Galileo's thought is much farther, in the same setting, he performs a thought experiment. By measuring the counter-weight H which prevented the descent of G along the slope (Fig. 8.1b), Galileo concluded that the magnitude of H decreased when the slope approached the horizontal plane:[31]

> Finally, on the horizontal plane the momentum vanishes all together: the body finds itself in a condition of indifference as to motion or rest; has no inherent tendency to move in any direction, and offers no resistance to being set in motion.

The *indifference to motion*—the term introduced by Galileo—replaced the indispensable impetus in inertial motion, while the impetus in falling increased and in ascending decreased. However, Galileo identified the horizontal motion as circular inertial motion equidistant from the center of the Earth because:[32]

> For just as a heavy body or system of bodies cannot of itself move upwards, or recede from the common center toward which all heavy things tend, so it is impossible for any body of its own accord to assume any motion other than one which carries it nearer to the aforesaid common center… Hence, along the horizontal, by which we understand a surface, every point of which is equidistant from this same common center, the body will have no momentum whatever. (Emphasis added)

This path "equidistant from this same common center," Galileo previously ascribed to the sunspots.[33] For Galileo, the universe remained finite and spherical,

---

[30] Leonardo da Vinci, Codex Forster fol. 087r.

[31] Galilei (1638/1914, p. 181).

[32] Galilei (1638/1914, p. 181).

[33] The first time that Galileo mentioned *circular inertia* was about 25 years earlier regarding the motion of sunspots observed by him (Galilei, 1613/1957, p. 113).

and the planets moved in circular orbits.[34] In that closed spherical universe, the claim of perpetual inertial motion along a straight line would be an absurdity.

Does a self-supporting, infinitely preserved motion that "maintains itself" regardless of any agent external or internal, present a refutation of the impetus dogma? In a way, yes… Yet, he still had no understanding of causes for motion change. As we see, he rejected considering causes regarding motion, as Aristotle did in his physics. Besides preserving the special role of the Earth's center, Galileo added that a body on the horizontal plane "*offers no resistance to being set in motion.*" Why is that? This point required a more advanced understanding. After Newton, we know that every material body does resist to being *set in motion* by its inertial mass and retains such resistance to any change in its state of motion.

In fact, parallel to the claim of *indifference to motion* by Salviati, Sagredo (the unbiased intellectual) raised the idea of a *resistance to motion*, the original idea of inertia:[35]

> (*What*) *I see in a movable body is the natural inclination and tendency it has to an opposite motion…. I said 'internal resistance', because I believe that this is what you meant and not external resistances, which are many and accidental.* (*Emphasis added*)

Salviati (Galileo) cautiously joined this speculation:

> *I wonder whether there is not in the movable body, besides a natural tendency in the opposite direction, another intrinsic and natural property which makes it resist motion.* (*Emphasis added*)

Galileo did not expand on this, apparently struggling with the contradiction of the two opposite tendencies (in a way similar to the Aristotelian antiperistasis, the air causing *and* resisting the mo*tion of a projectile*). *The idea of resistance to motion was already launched by Kepler* in his concept of inertia.[36] His inertia resistance was vehemently rejected by Newton in the *Principia:* instead of resistance to motion, he stated the resistance to any change in motion state—increasing or decreasing in speed or changing direction.[37] Neither Leonardo nor Galileo was there with inertia, despite the essential progress each of them made. It was Descartes who in his *Principles of Philosophy* made great progress, the recognition of motion—rectilinear and uniform—as a natural state, and so requiring no support at all.[38] Galileo's account was different from Leonardo's, but for both, motion was still a process and not a state. In this, both differed from the account of classical mechan-

---

[34] Galileo knew about Kepler's progress regarding elliptical orbits of planets and rejected this result as unacceptable, just as he did regarding the idea of attraction at a distance (e.g., Drake 1999a, p. 342).

[35] Galilei (1632/1953, p. 213).

[36] Chapter 4.

[37] The inherent *resistance to motion* introduced by Kepler for celestial objects appears as a misconception in class (Galili and Tseitlin 2003). Newton explicitly denied it and wrote (Cohen 1971): "I do not mean Kepler's force of inertia by which bodies tend toward rest, but the *force* of remaining in the same state whether of resting or of moving."

[38] Descartes (1644/1982, p. 59).

ics. Both scrutinized the account of motion, and on that track, Leonardo's views clearly lagged behind those of Galileo.

Falling Bodies

Falling presented a special case of motion—the "natural" motion for Aristotle. The historical discourse on falling—how and why objects fall—played a central role in the transition from medieval to early modern science.[39] Galileo's law of falling bodies is commonly learned in introductory physics courses.

Leonardo also considered falling. He was apparently committed to the simple observation of falling of objects in a medium. Quite in accordance with Aristotle, Leonardo wrote:[40]

> Among bodies of varying substance and of similar shape that which has most weight descends most rapidly.

In contrast, Galileo, in his *Discorsi*, refuted the Aristotelian claim that the swiftness of falling was proportional to the bodies' weights.[41] Teachers often quote the famous "thought experiment" in this regard.[42] In fact, this mental account displayed the logical inconsistency of the Aristotelian theory regarding falling. However, since no other theory was available, no other theoretical prediction was available. Unlike common claims, Galileo did not insist that his thought experiment *proved* that all bodies fall equally fast.[43] Despite the readiness of Simplicio—a peripatetic philosopher—to accept such logical refutation of "the heavier – the faster", Galileo proceeded to the empirical realm to support the new claim:[44]

> It is clear that Aristotle could not have made the experiment, yet he wishes to give us the impression of his having performed it when he speaks of such an effort as one which we see.

---

[39] For example, Gliozzi (1965), Koyré (1978), Dijksterhius (1986) and Lindberg (1992).

[40] LNB (p. 514).

[41] Galilei (1638/1914, p. 107).

[42] This thought experiment was adopted by Galileo from Benedetti (Gliozzi 1965; Koyré 1978; Dijksterhius 1986).

[43] Thought experiment may show only what the theory within which it was constructed prescribes. It may confirm the theory or reveal a contradiction in it (Galili 2009). Galileo's thought experiment of falling objects showed the internal contradiction in the Aristotelian theory. It could suggest looking for a new theory. A real experiment is required to confirm a new theory. It was introduced later by Newton.

[44] Galilei (1638/1914, p. 110), but also in Galileo's first work on mechanics (Galilei 1590/1969, p. 371). In his criticism of Aristotle regarding falling, Galileo followed Hipparchus, Philoponus, Benedetti, and Stevin who observed the discrepancy of the Aristotelian account and experimented with falling (e.g., Dijksterhius 1970).

Lacking a new theoretical framework, Galileo had to appeal to experiment and going to the limit of decreasing resistance.[45] A detailed elaboration of falling in a medium of decreasing density led Galileo to his conclusion:[46]

> Having observed this I came to the conclusion that in a medium totally devoid of resistance, all bodies would fall with the same speed.

The *empirical* law stating the independence of falling speed from a body's weight is known as Galileo's law. Within the Newtonian theory of gravitation, the law appeared to be only *approximately* correct for regular bodies.[47]

For Leonardo, falling was a natural motion with acceleration:[48]

> The heavy thing descending freely gains a degree of speed with every stage of movement.
> The heavy body which descends, at each degree of time acquires a degree of movement more than the degree of the time preceding, and similarly a degree of swiftness greater than the degree of the preceding movement. Therefore, at each doubled quantity of time the length of the descent is doubled and also the swiftness of the movement. It is here shown that whatever the proportion that one quantity of time has with another, the one quantity of movement will have the same with the other and similarly one quantity of swiftness.

Leonardo stated arithmetical ("pyramidal", in his words) proportion, which is the linear dependence *between time and the distance covered ("movement") and the speed ("swiftness"). His result implies distances covered in sequential time intervals to be in ratio 1:2:3:4:5....*[49] *With regard to swiftness, it was a correct statement, not so with regard to the distances covered. The fact that Leonardo states the same kind of growth for the speed and distance which cannot coexist indicates lack of a formal account of accelerated motion and any experimental bases for such. It looks as if his account here drew solely on a visual impression and the idea of universal validity of simple linear increase regardless of measurement.*

In contrast, Galileo applied a sophisticated empirical method of measurement to the ratio of distances covered during sequential time periods.[50] *He considered an object rolling down a slope that presents the constrained falling, slower than a free*

---

[45] This was a remarkable recourse. In other cases (such as regarding pendulum isochronicity), Galileo showed quite a different attitude to the experiment, preferring convincing reasoning and considering an ideal case, free from impeding factors in a real experiment which may mask the "true" phenomenon.

[46] Galilei (1638/1914, p. 116).

[47] Galileo's law of falling and the equivalence principle are often confused (Lehavi and Galili 2009). While the latter is unconditional, the acceleration of the two masses approaching each other—"fall"—depends on their masses. In the falling of regular ("small") objects to the ground, one may neglect the mass dependence, Galileo's law. However, in the astrophysical context of comparable masses, that law does not hold. The approximate nature of Galileo's law is a good lesson on the importance of having a theory for the full account, which may surpass certain observation.

[48] LNB (pp. 551, 570).

[49] *For example, Kemp (2006, pp. 116, 127).*

[50] *For example, Drake (1999a, p. 317), Chapter 2.*

falling and allowing direct measurement of the distances passed. His result differed from linear growth:[51]

> ...*it is clear that if we take any equal intervals of time whatever, counting from the beginning of the motion, such as AD, DE, EF, FG, in which the spaces HL, LM, MN, NI are traversed, these spaces will bear to one another the same ratio as the series of odd numbers, 1, 3, 5, 7; for this is the ratio of the differences of the squares of the lines [which represent time], ...or we may say [that this is the ratio] of the differences of the squares of the natural numbers beginning with unity.*

Leonardo, while drawing on observation, could catch the increasing gap between two sequentially dropped objects. He stated correctly:[52]

> *If two bodies of equal weight and the same shape fall one after the other from the same height in each degree of time, the one will be a degree more distant than the other.*

Though a rather subtle effect for those who consider the difference between two quadratic functions, it became observable to the acute eye of Leonardo who ascribed to the increasing gap linearity, this time a correct claim.[53]

An interesting case of falling was addressed by both Leonardo and Galileo. It was an imaginary scenario of a body thrown into the tunnel perforating the Earth's globe. Both scholars could have read Albert of Saxony's account of it.[54] *Leonardo wrote:*[55]

> *And if one should make a hole which was with its diameter or indeed its center the diameter of the world, and there were thrown there a weight, the more it were to move, the greater would its weight become. So, having arrived at the center of the earth which has only the name and it being itself equal to nothing, the weight thrown would not find any resistance at this center but would rather pass and then return.*

Addressing the same thought experiment, in his *Dialogo,* Galileo *used it to criticize the traditional separation in the falling motion. He equated the situation to that of an object rolling down the plane and ascending immediately after that on the rising slope of another plane, in other words, performing a pendulum-like oscillatory motion:*[56]

> *Having arrived at the centre ... Let me see you find an external thrower who shall overtake it once more to throw it upward. And what is said thus about motion through the centre is also to be seen up here by us. For the internal impetus of a heavy body falling along an inclined plane which is bent at the bottom and deflected upward will carry the body upward also, without interrupting its motion at all. A ball of lead hanging from a thread and moved from the perpendicular descends spontaneously, drawn by its internal tendency; without*

---

[51] Galilei (1632/1953, p. 176).

[52] LNB, p. 576.

[53] *Using modern notifications, the distances covered by the sequentially dropped bodies* $S_1 = 1/2gt^2$ *and* $S_2 = 1/2g(t - t_0)^2$ *(t_0 denotes time delay and g denotes free fall acceleration) yield for the separation:* $S_1 - S_2 \propto t$.

[54] Clagett (1959, p. 566).

[55] LNB, p. 592.

[56] Galilei (1632/1953, p. 236).

pausing to rest it goes past the lowest point and without any supervening mover it moves upward.

Both scholars refuted the scenario that would correspond to the account by Aristotle—a full stop at the center of the world. Galileo argued that by comparing with the rolling down the plane followed by climbing uphill, the motion which is continuous and does not stop in the middle. Leonardo did not explain the same predication. In this setting, Galileo made the first step to dismiss the traditional dichotomy, which was never questioned after Aristotle—natural versus violent motion. Yet, the full integration of the two was only achieved by Newton.

Gravity–weight

The context of falling and movement in general brought to the scene the concept of weight. As common in medieval science, Leonardo's view preserved much influence of the Aristotelian framework, which identified weight with tendency to fall. Leonardo's view related weight to spontaneous falling (or ascending, due to its counterpart—levity). *Like other medieval scholars, Leonardo stated that weight is not a force:*[57]

> Weight is corporeal and force is incorporeal. Weight is material and force is spiritual.... Force never has weight, although it often performs the function of weight. The force is always equal to the weight which produces it.

Weight increases in the object during its falling:[58]

> The heavy body which has a free descent with every degree of movement acquires a degree of weight.

This increase was the accidental gravity,[59] *which stemmed from the Aristotelian physics—the more the weight, the faster the fall—expanded to the perception of gravity/weight, a kind of operational definition. Indeed, the greater speed causes an increased difficulty of stopping the motion, similar to the* heaviness in holding an object. Natural weight was considered as the effective cause for falling. T*he falling body increased its swiftness* due to two weights, natural and accidental.

Yet, Leonardo went even farther. He related weight to motion in any direction, not only falling but also including horizontal and rotational:[60]

> Every heavy substance when it moves horizontally has only weight in the line of its movement. This is shown by the first part of the movement made by the ball from a mortar, this movement being in a horizontal direction.

---

[57] *LNB (p. 511).*

[58] *LNB (p. 516, 571).*

[59] *For example, Clagett (1959, pp. 561–564).*

[60] LNB (pp. 551).

In this, Leonardo anticipated Huygens' claims regarding the weight perceived in rotational system:[61]

The revolving movement made rapidly by the weight round the fixed point of its axis will have so much more heaviness in this weight as this revolving movement is more rapid.

Leonardo kept with the idea relating weight/gravity to seeking its natural state of rest in the arrangement of the setting of basic elements (water, air, and fire), which provides stability to physical reality. Yet, he stipulated weight by the *motion* toward such a state:[62]

No element possesses gravity or levity in its natural state. Gravity and levity are caused by one element being drawn into another.
Gravity is a power created by movement which transports one element into another by means of force, and this gravity has as much life as is the effort made by this element to regain its native place.
Gravity is a certain accidental power which is created by movement and infused into one element which is either drawn or pushed by another...

Leonardo's weight and levity ceased their existence with reaching by an object the state of rest. This view was rare even at that time:[63]

... this weight is ended with the movement of penetration which it made.
No simple element has gravity or levity in its own sphere.

Galileo's views on weight were very different. Firstly, he dismissed the concept of levity by the following logic:[64]

The experiment of inflated leather bottle of Aristotle proves conclusively that air possesses positive gravity and not, as some have believed, levity... for if air did possess this quality of absolute and positive levity, it should on compression exhibit greater levity and, hence, a greater tendency to rise; but experiment shows precisely the opposite.

Galileo's operational definition of weight was the accepted definition for years to come:[65]

The definition of heavy and light bodies given on the basis of their motion is not a good one. For while the heavy or the light body is moving, it is neither heavy nor light. That is, the heavy is that which presses down on something... Hence the heavy body, while exerting weight, does not move. This is clear if you have a stone in your hand. The stone will then press down so long as the hand resists its weight; but if the hand moves down with the stone, the stone will no longer press down in the hand. Therefore, a better definition will be "the heavier is that which remains at rest under the lighter".

---

[61] LNB (pp. 573), Chapter 5.
[62] LNB (pp. 588, 630).
[63] LNB (p. 285, 589).
[64] Galilei (1638/1914, p. 78).
[65] Galileo, Opere I (412–413), in Drake and Drabkin (1969, p. 383).

Since his early work on mechanics,[66] *for Galileo, weight was a feature of objects manifested in their heaviness to be measured by weighing.*[67] *It had* a clear operational meaning: weight corresponds to the heaviness perceived by holding an object or weighing it. The operational commitment of Galileo to see weight as a perceived faculty is seen in considering the weight of an object immersed in a medium:[68]

> *...as has been often remarked, the medium diminishes the weight of any substance immersed in it by an amount equal to the weight of the displaced medium.*

In his last major work, Galileo returned to weight in his famous debate on the speed of falling different objects. His view surpassed the medieval view:[69]

> SALV. *... Note that it is necessary to distinguish between heavy bodies in motion and the same bodies at rest. A large stone placed in a balance not only acquires additional weight by having another stone placed upon it, but even by the addition of a handful of hemp its weight is augmented six to ten ounces according to the quantity of hemp. But if you tie the hemp to the stone and allow them to fall freely from some height, do you believe that the hemp will press down upon the stone and thus accelerate its motion or do you think the motion will be retarded by a partial upward pressure? One always feels the pressure upon his shoulders when he prevents the motion of a load resting upon him; but if one descends just as rapidly as the load would fall how can it gravitate or press upon him? Do you not see that this would be the same as trying to strike a man with a lance when he is running away from you with a speed which is equal to, or even greater, than that with which you are following him? You must therefore conclude that, during free and natural fall, the small stone does not press upon the larger and consequently does not increase its weight as it does when at rest. ... We infer therefore that large and small bodies move with the same speed provided they are of the same specific gravity.*

Clearly, the commitment to measurement, the pivotal commitment of the scientific method, came to the fore, even if it was not a standard definition of weight, but definitely its essential indication. Yet, Galileo refrained from speculation regarding any theoretical account of weight and its cause.[70] While Leonardo did ask himself why the Moon possessing earthy ground did not fall for its weight,[71] Galileo ignored this question at all. For him, the Moon, though of similar to the Earth substance, performed circular motion, natural for the world of heavens in which weight does not cause falling.

Only Newton, in his Principia of 1687, equated the weight of an object to the gravitational force exerted on it, the force varying with distance—the next but not

---

[66] Galilei (1590/1969, p. 383).

[67] Galilei (1638/1914, p. 83).

[68] Galilei (1638/1914, p. 81).

[69] Galilei (1638/1914, pp. 63–64).

[70] Galilei (1638/1914, p. 166).

[71] His speculation was that local set ground-water-air-fire at the Moon provides stability to the Moon similar to the Earth (Richter 1952/2008, p. 52). In a way, it anticipated the later vision of Copernicus who split the world to areas of local attraction around celestial objects.

the final stage of the weight concept development.[72] *Earth, Moon, and all other planets do fall in the meaning of a free gravitational motion. Galileo and Leonardo, each in his own way, preceded the Newtonian understanding, but both remained very far from it.*

Projectile Motion

Both scientists addressed projectile motion—another central topic in mechanics. While both were dissatisfied with the Aristotelian approach to such motion as violent, composed of rectilinear and circular components,[73] Leonardo and Galileo still differed greatly regarding their accounts of that subject.

Leonardo, an observer of superior skills, understood that a water jet leaving an open pipe reproduces projectile motion. Drawing on his amazingly acute vision, he made sketches of such motion. To experiment with the trajectory, he imagined a special device launching water jets at different angles (Fig. 8.2a). The ability to vary the angle provided a controlled variable—the key element of the experiment as introduced in the scientific revolution much later, in the seventeenth century. Observation of the trajectory variation brought Leonardo very close to recognizing the trajectory as a parabola, but he did not do that. It was not a parabola; it was a realistic path of a projectile. Fully committed to reality, Leonardo's eye caught that the water trajectory stream was slightly asymmetrical. That was due to air resistance. Leonardo depicted trajectories of solid projectiles as lines for mortar shells (Fig. 8.2b) and a kind of stroboscopic image for cannon balls (Fig. 8.2c).[74] Leonardo did not try to produce any theoretical account; the motion was depicted holistically (without resolution to components). Rooted in acute observations, his results could bring some practical benefits. But no observations, even very precise, could provide an understanding and account of the physics behind the phenomenon. The experimentation of Leonardo provided a realistic trajectory of a projectile and replaced the common at that time schematic depiction of motion drawing on the impetus theory (Chap. 2, Fig. 2.9b). Galileo made the breakthrough.

---

[72] Chapter 5, *Galili (2001)*.

[73] For example, Dibner (1974, p. 181), Chapter 2.

[74] These sketches by Leonardo might suggest similarity with stroboscopic photos, but in fact, they were different. They rather depicted the path in perception by an observer. The true stroboscopic image would reveal segments of equal duration distinguishing between the segments with different speed of the projectile.

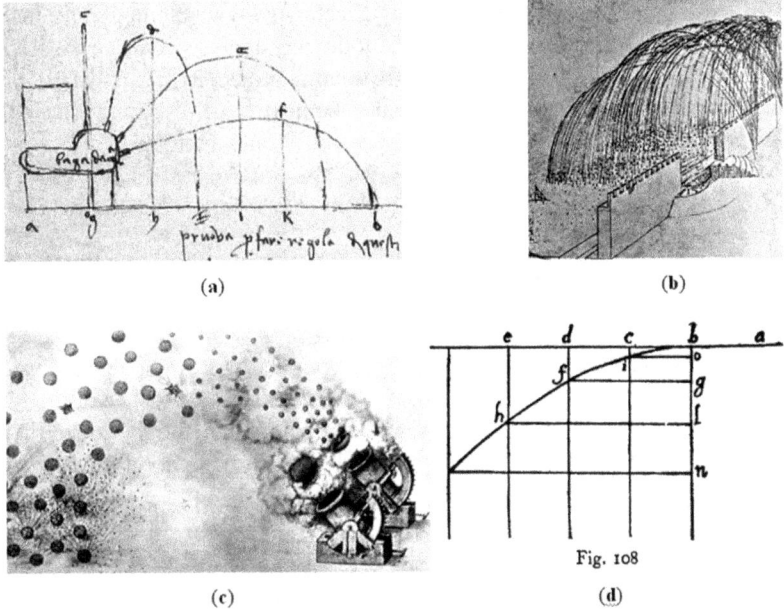

(a)   (b)

(c)   (d)

Fig. 8.2 (a) Leonardo depicted the apparatus investigating dependence of the water trajectory on the launch angle (Leonardo da Vinci, *Paris Manuscript C*, fol. 7r, Marcolongo (1956, p. 490)). (b) Leonardo depicted the ballistic motion of mortar shells (Leonardo da Vinci, *Codex Atlanticus*, fol. 72r, Calvi (1956, p. 284)), and (c) flying cannon balls (Leonardo da Vinci, *Codex Atlanticus*, fol. 9v-a, Calvi (1956, p. 281)). (d) Galileo's drawing in which he constructed the parabolic trajectory of a projectile was composed of equidistant horizontal segments *bc*, *cd*, *de*, and the vertical segments *bo*, *og*, *gl*, and *ln*, in the ratio of 1:3:5:7... The emerged line *bifh* was a parabola (Galilei (1638/1914, p. 249))

In contrast to Leonardo, Galileo's approach was theoretical. His major discovery was "seeing" the projectile motion as a composition of two: the horizontal uniform and vertical uniformly accelerated. The Oxford scholars of the Merton School (the "Calculators") previously treated both types of motion as separate abstract cases.[75] Galileo was the first to construct a parabola trajectory as a composition of the two simultaneous motions, and this progress allowed him to exactly predict the condition for the maximal distance and demonstrate the feature of angle duality (two angles of launch causing the same distance).[76] Unlike Leonardo, Galileo's theory totally neglected air resistance. His theory focused on "the phenomenon" as an idealized effect, similar to a thought experiment.[77] The advantage of such a rationalistic approach was a clear conceptual understanding of the considered phenomenon and made its further investigation possible. Since the two motions—the horizontally

---

[75] The account of motion by the Oxford Calculators was further developed by Oresme in Paris and Casali in Bologna in the fourteenth century (e.g., Pederson and Pihl 1974, pp. 222–225 and Clagett 1959, pp. 218–219; 235–246; 332–3, 382–391). They were apparently known to Galileo as they were taught and discussed in Padua (Clagett 1959, p. 644).

[76] Galilei (1638/1914, p. 276).

[77] McAllister (1996) and Galili (2009).

uniform inertial and the vertical uniformly accelerated—were empirically investigated by Galileo, their composition and the following analysis could be theoretical, the new science at its best (Fig. 8.2d). Yet, in treating projectiles, as in that of falling, Galileo did not proceed beyond kinematics. The next step—dynamics (the causal explanation by active forces)—was possible only through drawing on Newtonian theory.

Astronomy, Surface Illumination

Leonardo knew that the Moon reflected sunlight, and therefore its sphere is observed from the Earth in phases.[78] Yet, he also saw that actually a whole disc of the Moon is visible to us, but a part of it is seen as much dimmer. Leonardo explained that the faint part of the Moon disc can be observed only due to the light reflected from the Earth in the first place (Fig. 8.3):[79]

> *The moon has no light of itself, but so much of it as the sun sees it illuminates. And of that illuminated part we see as much as faces us. And its night receives as much light as our waters lend it as they reflect upon it the image of the sun, which is mirrored in all those waters that face the sun and the moon.*

His words indicated him holding the conception of moving image of the observed object to the eye of the observer (Chap. 3). He wrote:[80]

> *But the images of the objects conveyed to the pupil of the eye are distributed on this pupil exactly as they are distributed in the air.*

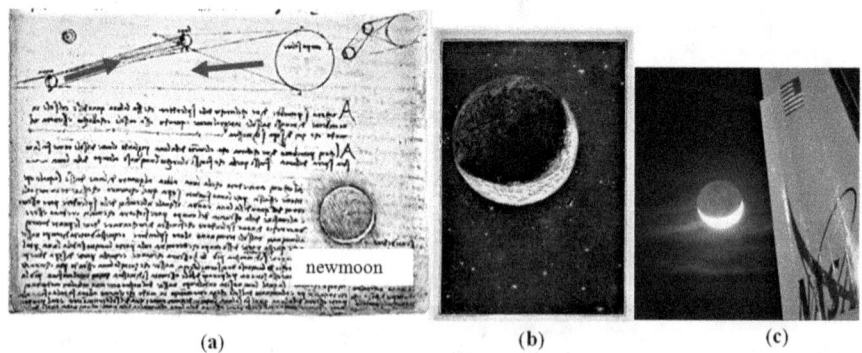

(a)   (b)   (c)

Fig. 8.3 (**a**) In his astronomical study (Leonardo da Vinci, *Codex Leicester* 2A, fol. 2r), Leonardo depicted the situation in which the sequential reflections of the Sun from the Earth to the Moon is the greatest—the new moon (arrows and notification are added). (**b**) The new moon observed by the astronauts on the way to the Moon in the novel by Jules Verne (1874). Was the picture faithful to reality? Compare with (**c**) the new Moon. (Picture credit: NASA)

---

[78] This knowledge is very old. Aristotle used it as an example of the second type of inductive reasoning based on the specific regularity in Moon phases: the crescent Moon always faces the Sun with its convex side (Fig. 10.12b).

[79] LNB (p. 291); Helden (1989, p. 22).

[80] Quoted in Richter, 1952/2008, p. 51).

Galileo did not miss this phenomenon too and explained the subtler aspect of the best timing for its being observed—during the new moon:[81]

Salviati:... the earth repays it by reflecting the solar rays when the moon most needs them, giving a very strong illumination—as much greater than what the moon gives us, it would seem to me, as the surface of the earth is greater than that of the moon. ... a certain baffling light which is seen on the moon, especially when it is horned, comes from the reflection of the sun's light from the surface of the earth and the sea; and this light is seen most clearly when the horns are the thinnest. For at that time the luminous part of the earth that is seen from the moon is greatest, in accordance with your conclusion a little while ago that the luminous part of the earth shown to the moon is always as great as the dark part of the moon which is turned toward the earth. Hence when the moon is thinly horned and conse- quently in large part shadowy, the illuminated part of the earth seen from the moon is large, and so much the more powerful is its reflection of light.

This was correct qualitative account. Kepler, in his *Astronomia Pars Optica* of 1604, paved the way for photometry[82] and empowered scientists to make numerical estimations of light intensity proving its decrease as the square of distance $-1/r^2$.

Galileo, a pioneer of telescopic observations could surpass Leonardo and discov- ered that the lunar terminator—the line separating the illuminated and non- illuminated areas—was a broken line:[83]

...the boundary which divides the part in shadow from the enlightened part does not extend continuously in an ellipse, as would happen in the case of a perfectly spherical body, but it is marked out by an irregular, uneven, and very wavy line.

He interpreted this as evidence of the irregular surface, mountains, and valleys. In accordance, in his sketches, Galileo interpreted bright spots in the dark area of the disc as where the peaks of the tall mountains, next to the terminator, met the sunlight. Moreover, drawing on the rectilinear light expansion and simple geometry, as well as the well-known, after Aristarchus, ratio of the Moon and the Earth diam- eters, Galileo could estimate the height of the mountains from their visual penetra- tion into the region not illuminated directly by the Sun.[84] His result was not very precise, he exaggerated the height of the mountains, but it was the first ever scien- tific estimation of the object.

Furthermore, Galileo identified the dark areas in the illuminated part of the Moon (Fig. 8.4a) as valleys, the areas that were smoother and lower than the surrounding surface. He identified them as natural water basins—seas. Leonardo shared the false idea of water presence on the Moon but made a different identification. In his view, the water areas looked brighter.[85] Apparently, he imagined the rippled surface reflecting light specularly from multiple spots, as happens when we observe sea surfaces illuminated by the sunlight in windy and sunny weather (Fig. 8.4c, d).

---

[81] Galilei (1632/1914, p. 67).

[82] This law was later elaborated by Bouguer, the founder of photometry in the eighteenth century (Wolf, 1961, p. 167).

[83] Galilei (1610/1989, p. 44).

[84] Galilei (1610/1989, p. 52).

[85] LNB (p. 291); Drake (1957, p. 34).

He wrote:[86]

The skin, or surface, of the water that makes up the sea of the Moon, is always ruffled, little or much, more or less; and this roughness is the cause of the proliferation of the innumerable images of the Sun, which are reflected in the ridges and concavities, and sides and fronts, of the innumerable wrinkles.

This note indicated Leonardo's belief that the Moon possesses its own atmosphere (wind a cause of ripples of water surface). Galileo was convinced of that. He brought "evidence" for the Moon's atmosphere—the apparently enlarged illuminated part of the Moon disc as observed in comparison with the nonilluminated part of the disc.[87]

In this context, Leonardo talked about "multiple images" of the Sun, which follow diffusive reflection from the rippled surface of water on the way to the observer's eye (Fig. 8.4c).[88] In his sketch, Leonardo also showed the single image of the Sun, which would be observed in the absence of ripples. Galileo, in his turn, considered both solid and water surfaces. He understood the need of diffusive reflection from the rough ground of the planets in order to be observed by us. He wrote:[89]

... why the Moon is not smooth, I reply that and all the other planets are inherently dark and shine by light from the Sun. Hence they must have rough surfaces, for if they were smooth as mirrors no reflection would reach us from them and they would be quite invisible to us.

Comparing the roughness of solid and water surfaces Galileo concluded, in contrast with Leonardo, that brighter areas should be solid and the dimmer indicated water. After Galileo, we call the big dark areas "Moon seas". Although incorrect with respect to water presence, this statement remained valid for indicating comparatively flat areas of the lunar landscape, which are closer to a flat mirror. These areas were used later in determining where the Apollo mission modules should land on the Moon.

Teaching may use the contrasting interpretations of Galileo and Leonardo, both making sense in physics, for better understanding of diffusive and specular reflections of light from any surface, and often confused by students.[90] Both Leonardo and Galileo refuted the Aristotelian view of celestial bodies as perfect spheres, the discovery often ascribed to Galileo and rarely mentions Leonardo.

In this context, there is also the aspect of accuracy of sketches in science. Compare Galileo's sketches with much more detailed drawings of the Moon by Scheiner (Fig. 8.4b).[91] Scheiner's drawing informed about numerous details of the Moon surface. Yet, Galileo's sketches were of schematic and focused character,

---

[86] Quoted in Capra (2007, p. 224).

[87] This confusion is related to the vision of an illuminated area against the dark background.

[88] Leonardo da Vinci, *Codex Arundel*, fol. 25r

[89] Galilei (1623/1957, p. 263)

[90] Demonstration of an illuminated mirror remaining dark often surprise students (Galili and Hazan 2000a).

[91] Lacking a telescope, Leonardo could not draw the lunar surface. There is, however, speculation regarding Leonardo's assembling a telescope (Argentieri 1956, pp. 416–426). His note might be addressing magnification of vision in an apparatus: "...here [with the device] only one star is seen, but it will be large. Therefore, the Moon will be seen larger and its spots in a more defined form."

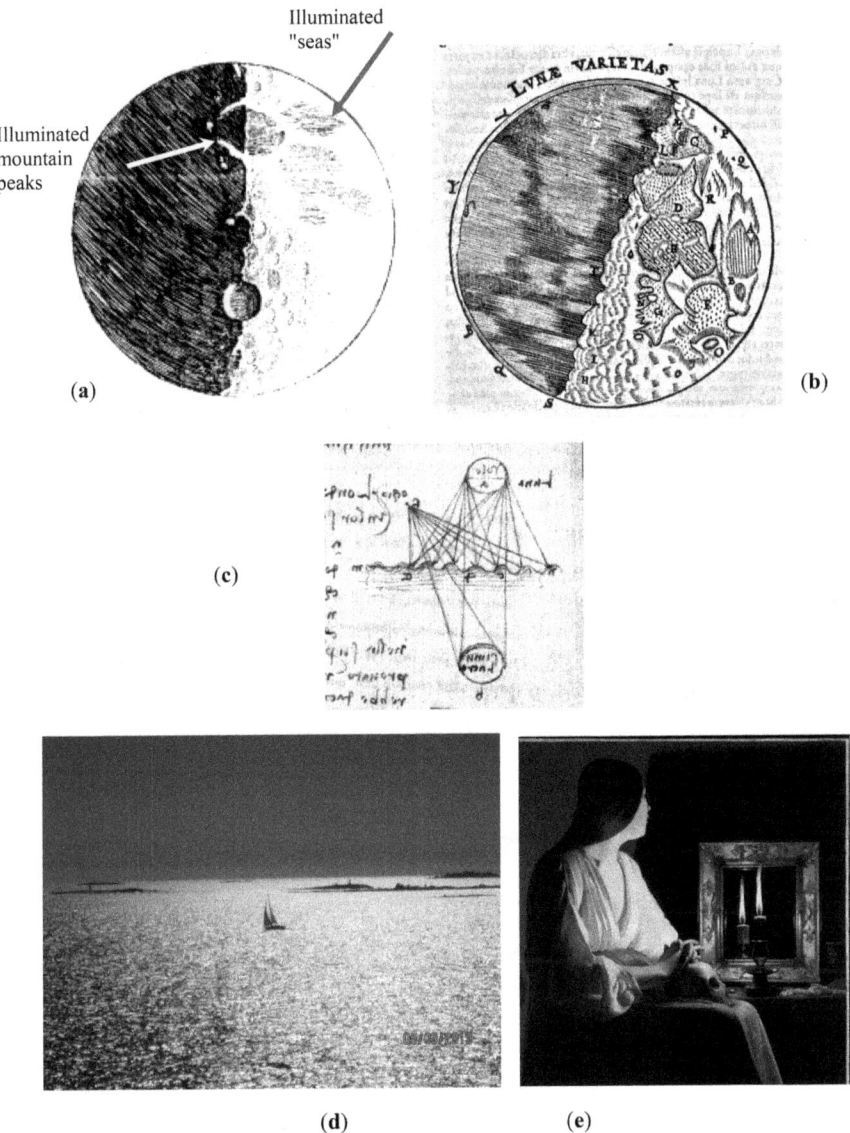

Fig. 8.4 (**a**) One may compare Galileo's sketch of the Moon (Galilei (1610/1892, p. 48)) with (**b**) Scheiner's sketch (Scheiner (1614, p. 58)) to appreciate the contrast with Galileo's drawing. (**c**) In his sketch, Leonardo showed a rippled surface that explained his identification of the bright areas observed on the Moon as "seas" (arrows added) (Leonardo da Vinci, *Codex Arundel*, fol. 25r, Emanuelli (1956, p. 206)). (**d**) A supportive illustration of Leonardo's understanding—multiple specular reflection of the sunlight from the sea surface in windy weather (Photo by the author). (**e**) The painting by La Tour (c. 1640) depicting the mirror surface being illuminated solely by a small light source (https://en.wikipedia.org/wiki/Magdalene_with_Two_Flames). This picture illustrates the reasoning of Galileo

lacking other details besides those useful in his qualitative argumentation and in supporting a numerical estimation of the height of the mountains. He understands what he sees. Light spots are shown near the terminator line by Galileo but not by Scheiner.

In his scientific drawings, Leonardo presents the preference of general descriptions including reliable documentation of details, which often left without attention of a superficial observer as normal in everyday life experience. Leonardo's vision is different; it is a vision of a researcher seeking the complexity in details, perhaps before it is even realized, still unknown. See how his extraordinarily acute vision and superior hand depicted intricate details of the water stream as if furnishing subsequent penetrating descriptions and analyses of the complex phenomenon. Leonardo depicted the turbulent motion created in water meeting with obstacles at various angles (Fig. 8.5).[92] There was practically no scientific knowledge of liquids in motion at that time. Yet, he clearly identified this reality as an extremely rich phenomenon. His extraordinarily accurate drawings were beyond sketches for amusing and impressing nonprofessional viewers. In fact, he documented details of the stream formation, which were investigated scientifically much later in the advanced exploration of fluid dynamics, in construction of bridges and dams, ships design, aerodynamics and development of flying apparatus, and ecological studies. If these were considered about the time of their production, scrutinized scientifically, these sketches could change the speed of scientific and technological progress. It did not happen, however.

Fig. 8.5 Leonardo's sketches of a water stream passing obstacles at different angles and the water pouring into a pool depicted the structure of turbulent motion and the creation of vortices (c. 1509)

---

[92] Leonardo da Vinci, *Codex Windsor*, fol. 12660r, v. https://commons.wikimedia.org/wiki/Leonardo_da_Vinci#/media/File:Studies_of_Water_passing_Obstacles_and_falling.jpg, Zammattio (1974, p. 193).

Leonardo foresaw mechanics to be "a paradise of mathematics" at the time which lacked the essential mathematical tools for a scientific analysis, especially of liquid motion. His exquisite artistic perception detected similarity of the water vortices to the hair curls and recognized the *interference* of two factors or motions as causing curly motion:[93]

> *Observe that the motion of the water curls resembles that of hair, which has two motions, one depending on the weight of the hair, the other on the direction of the curls; likewise the water has its swirling eddies, depending partly on the force of the principal flow, partly on the motion of impact and rebound.*

One might reveal the goal-dependent nature of scientific drawings. They may draw on understanding and focus on the essential features missed by others, represent an idea, and highlight and isolate the details to be investigated (Galileo's sketches of the Moon). In contrast, they may serve as documenting the details of certain phenomena, revealing them and suggesting their analysis (Leonardo sketches of fluid motion). Both types are complementary and indispensable.

Shadow, Optics

Yet, drawings may also reveal methodology. Such were the numerous studies of light and shadow produced by Leonardo. Those drawings, sometimes superficially identified as unfinished works, in fact, uncover an accomplished solution of certain problem. Leonardo was a true pioneer in representing illumination. In fact, his approach anticipated the future photometry and calculus introduced much later. Leonardo developed a semiqualitative procedure to construct partial shadow (penumbra), the gradually increasing illumination on the way from the axis behind the object facing an extended light source. Thus, Leonardo depicted the construction of the shadow pattern behind a disc facing a spherical light source (Fig. 8.6).[94] He showed shadow areas created by light cones from separate points of the light source.

In the course of construction, Leonardo first showed the simple case of just two light cones meeting the screen. Then, he proceeds to three and four cones, suggesting the resultant pattern in each case. In a great pedagogical style, Leonardo revealed his scientific understanding of penumbra, important for understanding illumination behind planets. Leonardo's approach to illumination in optical systems included tracing numerous individual rays arriving at a certain area:[95]

> *...shadows have in themselves various degrees of darkness, because they are caused by the absence of a variable amount of the luminous rays...*

---

[93] Gibbs-Smith (1978, p. 92). In that, Leonardo anticipated the vision of Huygens and Hooke in the seventeenth c. and Young and Fresnel, in the nineteenth c.

[94] Leonardo da Vinci, *Codex Atlanticus*, fol. 187r; Argentieri (1956, p. 413).

[95] Capra (2007, p. 222).

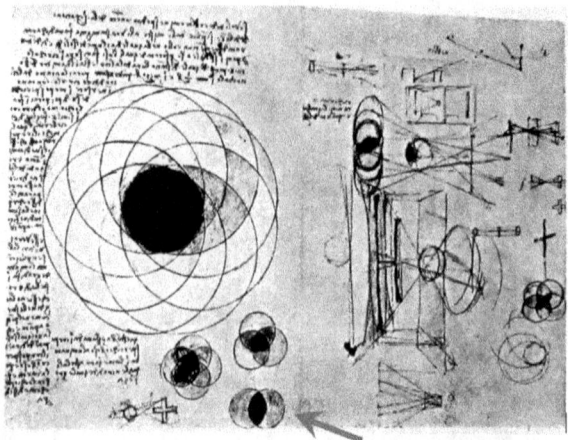

Fig. 8.6 Leonardo's sketch of the construction of complex shadow. At the bottom of the drawing, Leonardo illustrates the auxiliary cases of overlapping of two, three, and four light cones (pointed by an arrow)—a useful pedagogy

This way Leonardo tried to comprehend the light activity of the curved mirrors through meticulous following of the light rays. For example, lacking any analytical account of light to calculate focus distance (and related problems), he traced innumerable rays in their individual specular reflection in order to learn about the light reflected by the concave mirror, perhaps seeking for the focus of rays (Fig. 8.7a). Through such manipulation, he found such curve (Fig. 8.7b). We know that it is parabola, did he? In his investigation, he could find that the curvature of the concave mirror surface is in some inverse proportion to its radius (Fig. 8.7c).

Galileo, in turn, also uses light rays in his qualitative reasoning of natural phenomena. While discussing illumination of the Earth surface in his *Dialogo*, he illustrates the functional dependence of illumination on the angle of incidence:[96]

> *You must know, then, that a given surface receives more or less illumination from the same light according as the rays of light fall upon it less or more obliquely; the greatest illumination occurs where the rays are perpendicular.*

This way Salviati (Galileo) explained seasons by using a semiquantitative demonstration (Fig. 8.7b) in which light flux encountered a surface. The number of parallel rays stopped by the considered area varies with the angle of the area inclination toward the flux. This statement foresees Lambert's cosine law for illumination intensity established about a hundred years later, in 1729.[97] Yet, Galileo corrected the erroneous explanation of seasons by Copernicus[98] and excluded his ad hoc postulated "third movement" of the Earth—the annual precession of its axis.

---

[96] Galilei (1632/1953, p. 80).

[97] Wolf (1961, p. 167).

[98] Copernicus (1543/1952, p. 530).

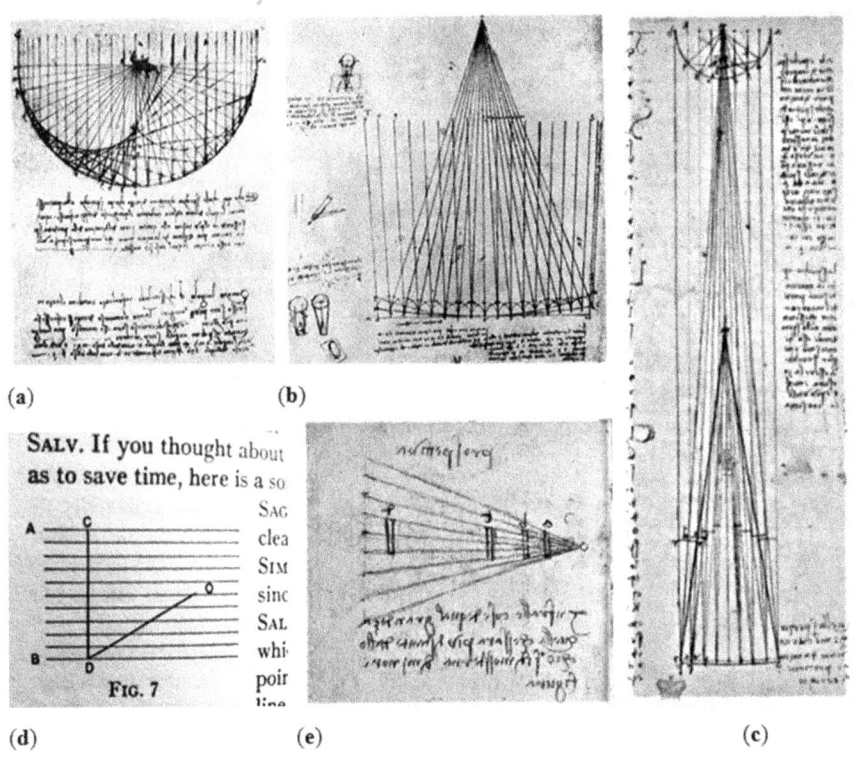

(a)   (b)

(d)   (e)   (c)

Fig. 8.7 (**a–c**) In studying light reflection from concave mirrors, Leonardo meticulously traced multiple rays to find out the impact of the light flux (Leonardo da Vinci, *Codex Arundel,* fol. 86v, 87r; Argentieri (1956, p. 417)). (**d**) Galileo's sketch in *Dialogo* supporting his reasoning (Galilei (1632/1953, p. 80)) of the angular dependence of illumination of the screen CD facing the ray flux AB. (**e**) Leonardo's comparative sketch demonstrating dependence of the focus mirror on its radius (Leonardo da Vinci, *Codex Forster* II, fol. 16v)

Where was Leonardo in this regard? Not too far away. Using the same technique of light rays, he investigated the highly important for art and science dependence of illumination (related to the quantity of rays stopped by a surface) on its distance to the light source (Fig. 8.7e). It was only a qualitative demonstration; no mathematical form was obtained. Much later, in 1609, Kepler was able to reach the famous inverse squares dependence in its mathematical form – $1/r^2$, known in photometry under the name of Bouguer who stated it in 1729.[99]

---

[99] Wolf (1961, p. 168). Currently, the laws of photometry (Bouguer's and Lambert's) disappeared in the school optics, which seriously impedes the quality of students' knowledge, for instance, regarding seasons (Galili and Lavrik 1998).

Problems of Statics

Both scholars joined the tradition of medieval scholars of investigating mechanical problems in the realm of statics.[100] Among the problems of interest, one finds the question of deformation of material objects, that is to say, their "resistance to breaking" under a load. In this field of basic engineering, Leonardo was a pioneer, and his contribution could have been significant if shared with others.[101] Leonardo's ideas are known from numerous but laconic claims dispersed among haphazard collections of his notes and even more numerous drawings often lacking explanations beyond empirical statements. They were very difficult to appreciate for the lack of details. Such was the idea of a self-supporting bridge (Fig. 8.8a),[102] which requires no nails or ropes to hold it together. It drew on a brilliant invention of a construction distributing and transferring weight stress among its identical components penetrating each other in a chain structure creating an arch.

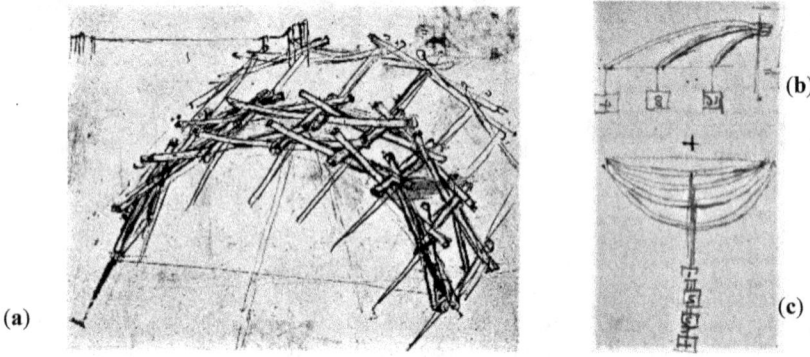

Fig. 8.8 (a) Leonardo's drawing of a self-supported wooden bridge. (b) Leonardo's drawing of the experiment of the loaded beam, fixed at one end. (c) Similar experiment with the beam fixed at both ends

Though not explained, justified, or discussed, the sketches were accompanied with praising the features of the proposed constructions:[103]

I can construct bridges very light and strong and suitable to be carried very easily, with which to pursue and at times flee from the enemy, and others, solid and indestructible by fire or assault, easy and convenient to transport and place in position.

---

[100] Clagett (1959) and Pederson and Pihl (1974).

[101] Uccelli (1956), Zammattio (1974) and Kurrer (2008).

[102] Leonardo da Vinci, *Codex Atlanticus*, fol. 23v-a; Calvi (1956, p. 297).

[103] Quoted in Gibbs-Smith (1978, p. 33).

Leonardo did perform experimentation to investigate the deformation and strength of structures designed to sustain weight. Treating a beam fixed at its end and stressed along its length, Leonardo drew inferences regarding the dependence of the deformation on the beam length and the load magnitude (Fig. 8.8b). Similar investigation was performed with the beam fixed at both ends (Fig. 8.8c).[104] Drawing on experience-based intuition and collected data, Leonardo tried to generalize in the form of scaling predictions such as in the following:[105]

> If a rod projected by one hundred thickness [100 times its diameter] from a wall carries ten pounds, how much will be borne by a hundred similar rods similarly projecting and bound together into a unit? – I say that if [a rod of] hundred thickness will carry ten pounds, the [bundle of] five thickness will carry ten times as much as the [rod] of 100 [thickness], and if ab is the [bundle] of five thickness, the hundred rods will carry 20,000 pounds.

However, this type of prediction, lacking any justification (within a theory of elasticity) and dealing with an unclearly chosen experimental setting, was often destined to be incorrect, even if pointing to the relevant parameters of the problem.[106]

Galileo addressed the issue of strength in material constructs in his *Dialogue Concerning Two New Sciences,* that is, being already a mature researcher.[107] Of the two new sciences presented, the first one was statics. There, Galileo heralded a new approach: mathematical modeling. A system of real objects was described by a theoretical construct—a model—which acted in accordance with certain principles, assumptions, and the empirically established parameters characterizing particular materials. The principle used by Galileo was the well-established rule of a lever, plus a number of empirically known characteristics of material objects. Galileo applied the model to structures and inferred regarding their characteristics—a procedure in the coming modern science:[108]

> Hitherto we have demonstrated numerous conclusions pertaining to the resistance which solids offer to fracture. As a starting point for this science, we assumed that the resistance offered by the solid to a straight-away pull was known; from this base one might proceed to the discovery of many other results and their demonstrations; of these results the number to be found in nature is infinite.

As Leonardo before him (Fig. 8.8b), Galileo looked for the deformation of a loaded rod fixed at its one end—a cantilever beam (Fig. 8.9a)—but instead of appealing solely to experiment, Galileo built a model, scrutinized it, solve a cluster of related problems employing the lever principle (Fig. 8.9b), and seek the form of transferring the stress along the axis of the curved beam (Fig. 8.9c). He was able to deduce the dependence of the critical length causing the fracture of the beam. Although many of the obtained results were only approximately correct, as checked

---

[104] Leonardo da Vinci, *Codex Forster II*, fol. 089r; Uccelli (1956, p. 271).

[105] Uccelli (1956, p. 271).

[106] Uccelli (1956) and Kurrer (2008).

[107] Galilei (1638/1914, Day Two).

[108] Galilei (1638/1914, pp. 149–150).

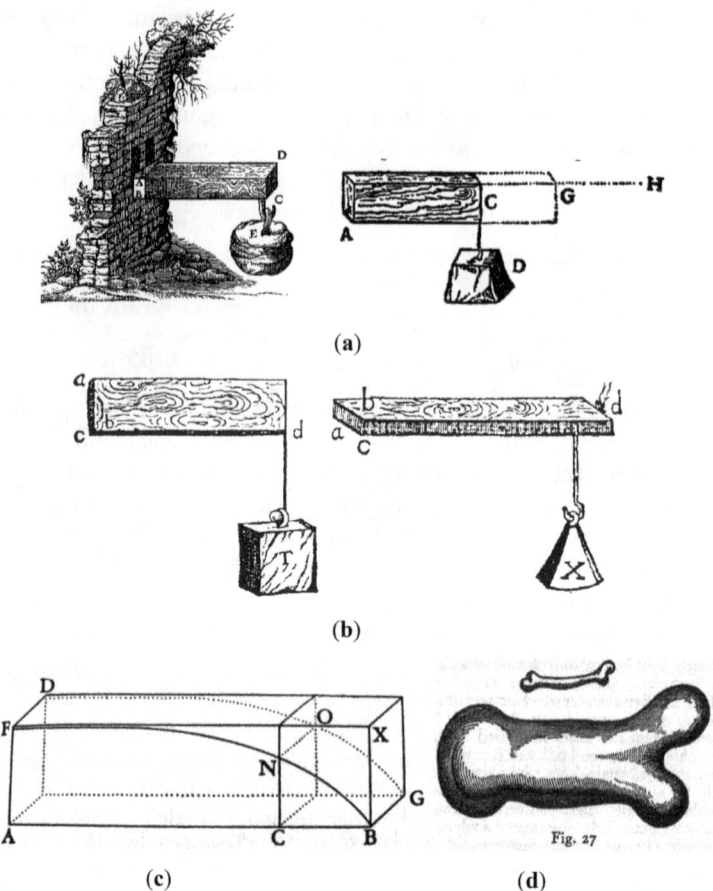

Fig. 8.9 Galileo's drawings in his *Discorsi*, the Second Day. (**a**) The problem of a solid prism ABCD fastened onto a wall at the end AB is presented and considered in looking for the maximal load and length of the beam prior to its fracture; (**b**) Galileo's sketches illustrating the claim of the higher resistance to fracture of the beam of greater width: weight T is bigger than weight X; (**c**) Galileo's mathematical model of the parabolic shape of the loaded beam. (**d**) Galileo's drawing depicted a regular human leg bone and the bone that would correspond to that of a giant

by modern methods, Galileo succeeded in providing some correct results that became famous.[109]

Galileo refuted the claim of Leonardo:[110]

Any support of twice the proportion of a smaller one will carry twice as great a load as a smaller, each of the two being in one piece and solid.

---

[109] Kurrer (2008).

[110] Leonardo da Vinci, *Codex Atlanticus*, fol. 46r; Uccelli (1956, p. 269).

While proceeding in this investigation, Galileo demonstrated that any increase of the linear dimensions of a creature (its height, depth, and width) would cause the corresponding increase of its volume (and thus the weight) in cubic proportion. Clearly, the cross-sectional area of bones (and so their resistant power to the sustained weight) increased only in quadratic proportion would not withstand the new weight. Hence the bones and the whole skeleton would crash. The required increase of dimensions that could prevent the whole body from collapse would dramatically change the proportions of the human body, given the same material of bones (Fig. 8.9d). This excludes the reality of the speculative depictions of mythological giants being big but preserving human proportions. Galileo generalized:[111]

> ...you can plainly see the impossibility of increasing the size of structures to vast dimensions either in art or in nature; likewise, the impossibility of building ships, palaces, or temples of enormous size in such a way that their oars, yards, beams, iron-bolts, and, in short, all their other parts will hold together; nor can nature produce trees of extraordinary size because the branches would break down under their own weight...

Galileo drew a bone of a small animal being enlarged in order to show that the bone of the larger animal violates proportional enlargement of the smaller one. Giants *of our proportions* (as well as microorganisms of our dimensions) cannot exist when keeping with the same type of materials. This and other scaling arguments continue to serve as a guiding method in paleontology in the interpretation of remnants of dinosaurs.

Although both scholars, Leonardo and Galileo lacked an inclusive theory incorporating elasticity, solid dynamics, materials, and calculus; their methods and results laid a foundation for such a theory. Leonardo's studies illustrate for us the need for experimentation. His works and ideas greatly surpassed in veracity of the parallel studies by other scholars. Yet, they were of limited worth in providing knowledge of the relevant phenomena. This was in contrast to Galileo, who clearly pointed to the essential feature of the new science—mathematical modeling and account drawing on basic principles. It was much more than the actual calculations and his solved problems of statics. Galileo never wrote formulas, but his long, verbal reasoning could be reduced to such. He creatively bridged the experience-based descriptions with the theoretical method. It was the future of science.

Pendulum as a Device and as a Phenomenon

Both Leonardo and Galileo studied the pendulum, but the difference between the approaches they employed was remarkable and therefore deserves attention. For Leonardo, a pendulum was an item in the ocean of mechanical devices he imagined

---

[111] Galilei (1638/1914, p. 130).

and depicted impressively creatively but only briefly mentioned or theoretically explored.[112] In contrast, for Galileo, the pendulum was one of the central theoretical tools—a simple device that manifests the phenomenon that he used for revealing and exploring the principles governing motion[113] and thus could be considered as experimental inquiry. Leonardo conceived the pendulum motion as composed of descending ("natural") and ascending ("accidental") motions of different nature. This fully corresponded to the Aristotelian grand separation between natural and violent motions. In accordance, for Leonardo, a pendulum supported the rule: accidental motion is always weaker (velocity is decreasing) than the natural one (velocity increasing):[114]

> Violent movement the more it is exerted the more it grows weaker: natural movement does the opposite. Accidental motion will always be shorter than the natural.

This feature causes spontaneous decay of successive oscillations (Fig. 8.10a).[115] Leonardo's sketch of a swinging pendulum depicts a series of bob positions in a sequence of moments (Fig. 8.10b).[116] This impressive image may create a feeling of reality in a sequence of moments. Importantly, the moments are in half-a-period intervals. The picture is not instead a stroboscopic photo of a single swing, which would show more crowded locations at the extremes where the speed is lower (Fig. 8.11c). In Leonardo's sketch (Fig. 8.10b), the denser images are next to the lowest point of equilibrium, where the velocity is the highest. Moreover, he connected the points of extreme deviation with a special line showing convergence of oscillations to the equilibrium point. There is great visual appeal of the drawing, as is usual with Leonardo.

Leonardo "sees" the decay of the pendulum and only mentioned that for smaller arks the difference between the two motions decreases (it is indeed difficult to observe) leading, in effect, to the isochronicity:[117]

> The smaller the natural motion of a suspended weight, the more the following accidental motion will be equal in length.

Leonardo proceeded with pragmatic goal. If decaying is natural, for any reason, he suggested its "correction" using a contrivance that could make the pendulum oscillations permanent and stable (Fig. 8.10d).[118] He replaced a free pendulum with an oscillator driven by an external agent moving along the guiding sinusoidal groove. This step would be inconceivable for Galileo who saw in pendulum motion nothing but constrained natural falling—a subject of his investigation.

---

[112] Bedini and Reti (1974).

[113] Matthews (1994/2015, 2000).

[114] LNB (p, 517).

[115] Leonardo da Vinci, *Codex Madrid II* fol. 147 r; Bedini and Reti (1974, p. 262).

[116] Leonardo da Vinci, *Codex Madrid II*, fol. 147r; Bedini and Reti (1974, p. 263).

[117] Leonardo da Vinci, *Codex Madrid II*, fol. 147r; Bedini and Reti (1974, p. 262).

[118] Leonardo da Vinci, *Codex Madrid II*, fol. 8r; Bedini and Reti (1974, p. 258). Leonardo included an escape device to control weight support in order to prevent oscillations from decay.

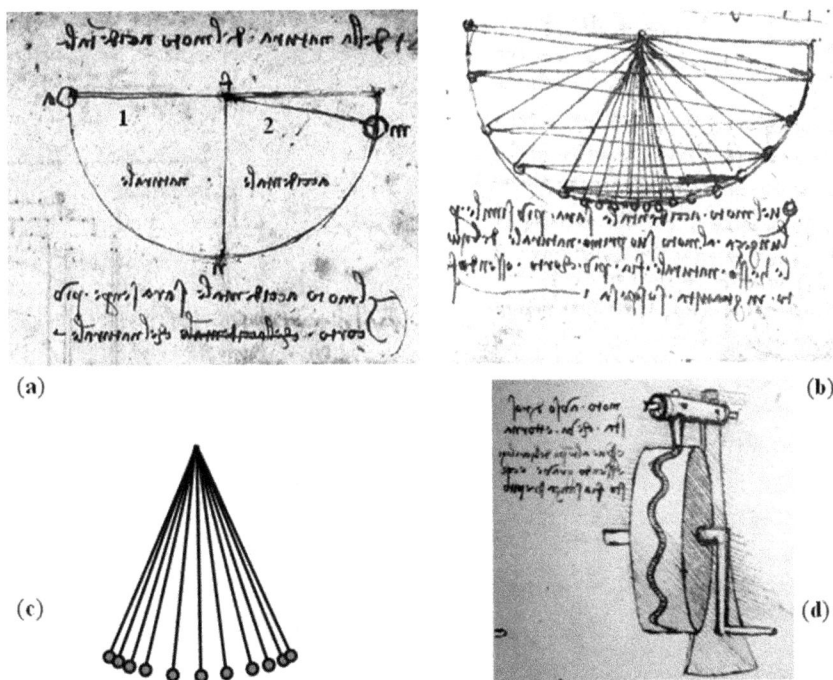

Fig. 8.10 (**a**) Leonardo's drawing of a swinging pendulum during a single swing: the amplitudes are different, and the motion is separated to two parts, 1 and 2 (labeled). (**b**) Leonardo's drawing of a swinging pendulum during an amount of time long enough for a full decay of oscillations. (**c**) A stroboscopic image of a pendulum depicted to contrast sketch **b**. (**d**) Another contrast emerges in juxtaposing Leonardo's sketch of a non-decaying pendulum forced to follow a waving groove on a spinning disc with an external mover

While Leonardo's thinking was focused on making oscillations non-decaying, Galileo analyzed the *theory* of pendulum motion and, through it, the regularity of falling as natural phenomenon. The pendulum was a tool for decoding natural laws. Figure 8.11a[119] illustrates such use. Galileo discovered regularity: no matter which pass the pendulum bob makes, it always ascends to the *same* height—ignore the decay. This regularity holds with regard to not only the pendulum but also the inclined plane with various slopes (again, the resistance impeding motion is ignored). Galileo's ultimate interest was the regularity of a free (vertical) fall.[120] The goal was to infer regarding the motion in its ideal, unperturbed form. Here, the interests of the two scholars reflected two worlds of preference—engineering and science.

---

[119] Galilei (1638/1914, p. 171).

[120] Galilei (1590/1960, p. 65).

In his *Dialogo* (Fourth Day), Galileo raised the claim for *isochronicity* of the pendulum motion, its period independent of its initial position. He stated isochronicity of pendulum swings regardless the angular span:[121]

> It must be remarked that one pendulum passes through its arcs of 180°, 160° etc. in the same time as the other swings through its 10°, 8°, etc. … if two persons start to count the vibrations, the one the large, the other the small, they will discover that after counting tens and even hundreds they will not differ by a single vibration, not even by a fraction of one.

The proof came later, in his *Discorsi* (Third Day). There, Galileo considered the bodies descending along the chords AC, DC, EC, FC, and GC of a circle and stated the same time for the descent along each of the chords—"the law of chords" (Fig. 8.11b).[122] Alas, this isochronicity valid for the chords does not hold for the descent along the corresponding arcs. Galileo's thought was: if the descent is independent on the chosen chord and on a very small chord it "practically" coincides with the arc, then the times of pendulum oscillations along different arcs are also equal—isochronicity. That was wrong. No matter how small the chord is, it never becomes the arc… Indeed, it was not obvious…

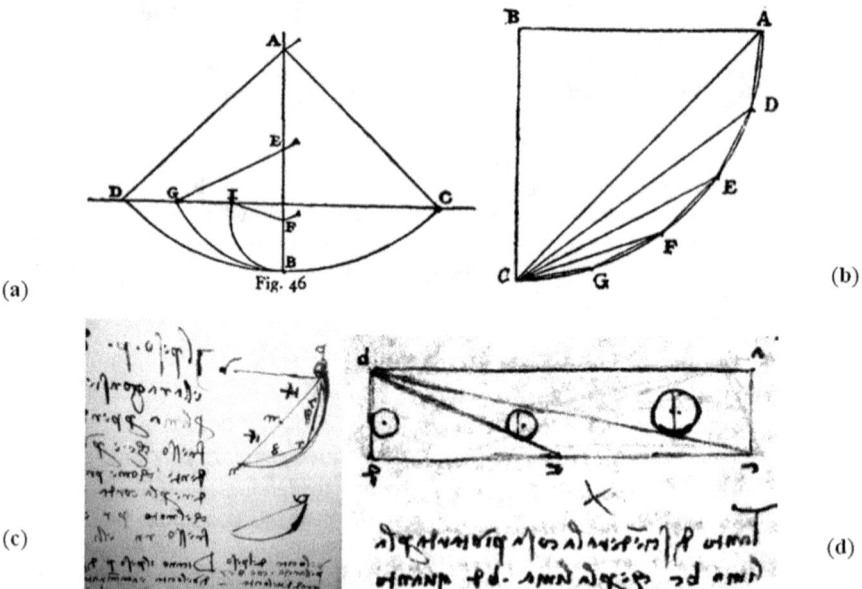

(a)   Fig. 46   (b)

(c)   (d)

Fig. 8.11 (**a**) Galileo's experiment with the constrained pendulum motion revealing the law of motion of general validity. (**b**) This drawing supported Galileo demonstration of the law of chords. The claim was that rolling down along the chords CA, CD, CE, CF, and CG takes the same time was correct. Getting shorter and shorter, the chords indeed approach the arc but not in the kinematic sense. Geometrical closeness of the chords and the arc could mislead Galileo in his claim of isochronicity regardless of the amplitude of oscillations. (**c** and **d**) Leonardo's sketches illustrating his considering a ball rolling down along the chords and along the inclined planes of different lengths

Leonardo considered a similar context of the travel time of a body descending along chords in comparison with arcs (Fig. 8.11c)[123] but claimed the opposite advantage of the descend time. He also considered descent along inclined planes of the same height but different slopes (Fig. 8.11d).[124] In the latter case, he observed that the time of descent is proportional to the length of the planes, so the natural falling was the fastest way to reach the ground.

Perhaps inspired by this fact, Leonardo considered descent along the different paths in a circle (Fig. 8.11c) and wrote:[125]

> *Every heavy substance not held back out of its natural place desires to descend more by a direct line than by an arc. This is shown because every body, whatever it may be, that is away from its natural place which preserves it, desires to regain its first perfection in as brief a space of time as possible; and since the chord is described in a less time than the arc of the same chord it follows from this that every body which is away from its natural place desires to descend more speedily by a chord than by an arc.*

Taken verbatim, this is an erroneous claim. The descend along the chord is slower than along its arc. Both are not "natural" motions, and so the object path is not an extreme trajectory. In a way, Leonardo felt that his reasoning regarding might not hold when he proceeded with a rather obscure inference regarding the motion of a balance:

> *...the movement of gravity in the balance is not entirely natural. This is evident from the fact that the arms of this balance as they descend describe an arc, and as a consequence curved lines. The second is that the heavy movement in the arm of the balance which descends is not entirely violent, since in this manner it acquires in its descent natural movement. The third is that the heavy movement in the balance is half-way between the natural and the violent.*

Leonardo did not go further, while Galileo did. Being limited to geometry and proportional reasoning (neither dynamics nor calculus were available) made Galileo's demonstration of isochronicity cumbersome.[126] Indeed, one needs Newtonian dynamics to reveal that isochronism holds for the chords but it does *not* hold for a pendulum.[127] Motion along the curve remains different, even if chords visually approach the arc. Yet, Galileo succeeded in revealing the functional dependence of the pendulum period as proportional to the square root of its length, but he

---

[123] Leonardo da Vinci, *Codex on the Flight of Birds*, Turino: the Biblioteca Reale.

[124] Leonardo da Vinci, *Codex Atlanticus*, fol. 338r.h. Milano: the Biblioteca Amhroriana.

[125] LNB (p. 513). The translation of Leonardo's statement into English seems being not the best possible.

[126] In classical mechanics this claim receives a simple proof shown in the school physics class.

[127] For example, Ariotti (1968), MacLachlan (1997). The problem of pendulum isochronicity was solved in 1673 by Huygens with cycloid, the true *brachistochrone*, which provides the quickest descent between two points in space. The same solution was proved by Leibniz, Newton, and Bernoulli later, in 1697 (Matthews, 2000, pp. 124–133).

could not receive the full form (including the coefficient of 2π) and failed to see that isochronicity was limited in validity to small angles only. Different approximations provide deviation from the pendulum isochronicity of about 16% for the angles around 90°.[128] For the periods of approximately a second, this was a small disparity that, even if it were detected, could be ignored by Galileo or attributed to air resistance.

The idea that the laws of nature could be masked by contingent impediments was popular in medieval science in explaining the obvious deviations from the Aristotelian tenets. In their interpretation of the collected data, physicists always face a choice: either the theoretical framework used is incorrect, or the appearance is modified by other factors. To resolve this problem, physicists go both ways. They never rush with decisive falsification but proceed with analysis, experimentation, and the development of a theory.

Consider discovery of neutrino. The violation of energy conservation in the spontaneous nucleus decay was reported in the 1930s. Bohr speculated that energy conservation did not hold in the microworld, whereas Pauli suspected a missed product (a particle) that could save energy conservation.[129] In the 1950s, neutrino— the missed product of the process—was discovered. The need to break with one of the basic tenets of physics was rejected.

Furthermore, facing the Michelson—Morley experiment with its zero result for the Earth's motion through ether, most physicists tried to keep with the old theory and modify Maxwell equations.[130] Einstein, in contrast, assumed that the Maxwell equations were correct but indicated a need for a new theory of mechanics. He was right, not the others. The theory of relativity entered the stage.

School teaching usually rounds corners. It is upon the teacher to depict the dual strategy in science activities. The collected data are always limited in accuracy, which leaves the mentioned two options. Galileo failed to find the full account of pendulum motion. Nevertheless, he tried and partially succeeded, while Leonardo did not try beyond observation and manipulation with technical devices. For science, that was insufficient.

Summary of the Comparison

Our comparison of Leonardo and Galileo has addressed several of their common interests. The subjects considered here are conceptually important and relevant to the school curriculum. The comparison between the two scientists stimulates cultural knowledge enforcing the disciplinary knowledge of the subjects.

---

[128] For example, Lima and Arun (2006).

[129] Fermi (1934).

[130] For example, Miller (1981, 1986, 1996).

The summary of Table 8.1 comprises six clusters of the chosen topics mentioned in the first column from the left. The way the physics curriculum accounts for these items is in the first column from the right, whereas the two central columns outline the accounts by Leonardo and Galileo. Cultural knowledge presumes the ability to affiliate each of the elements of knowledge to the disciplinary areas of nucleus, body, or periphery of the pertinent theory—the classical mechanics and optics.

What Could Be Learned?

Leonardo was an exceptionally talented and skillful scholar. His products in art, science, and engineering were innovative, attractive, and interesting to the wider public now, far beyond experts and professionals. The power of his imagination, the scope of his curiosity, and the quality of his products were so high that the interest in him caught a wide public of various interests. Through him, many people distant

Table 8.1 Comparative summary of Leonardo–Galileo accounts of several subjects in physics

1. Account of motion	Leonardo	Galileo	Modern science
	Impetus causes violent motion combined with antiperistasis (hybrid conception).	Natural motion was investigated.	All-inclusive account (kinematics–dynamics).
	Natural and violent motion.	Kinematic account of motion (no dynamics).	Motion with uniform velocity establishes a natural state of a body.
	Velocity increase in natural motion and decrease in accidental.	The third type of natural motion is horizontal and uniform.	Inertial motion means preservation of the state of motion.
	The impetus/power of the violent motion is dispensable.	Natural falling presents a uniformly accelerated motion.	Physical equivalence of inertial observers.
	Friction impedes motion.	Compound motion of projectiles produces parabolic trajectory.	Inertial mass of a body is the measure of its resistance to the change of states of motion.
	Friction with the surface of support is proportional to the pressing force on the support[a].	Physical equivalence of rest–uniform motion.	Motion characteristics: energy and momentum. Force determines acceleration.
		Horizontal means equidistant from the center (circular inertia).	Friction with the plane of support is proportional to the pressing force on the support.
		Friction masks the "true" natural phenomena.	

(continued)

Table 8.1 (continued)

2. Falling bodies and weight	Falling is natural motion.	Dismissing the separation of natural and violent motions.	Universal account of motion.
	Falling is weight proportional.	Falling is weight independent (friction neglected).	Free falling is the motion solely under gravitational attraction.
	Natural falling accelerates with the speed and distance increase in linear proportion with time.	Natural falling: Constant acceleration, speed grows in linear proportion whereas distance—in square proportion with time.	Free falling of an object to the ground is independent of its mass—an excellent approximation for regular objects.
	Body thrown upward is in accidental motion with linearly decreasing speed.	Weight is an inherent property of objects.	Close to the Earth's surface the free fall is at a constant acceleration—an excellent approximation for regular objects.
	Gravity–levity conceptual duality.	Weight is the heaviness perceived and measured by weighing.	(Newton) Weight splits to true (the gravitational force on the body and apparent (weighing result). Alternatively: Weight is the result of standard weighing.
	Weight is one of the causes of motion.	Within medium weight depends on the density of the medium.	An object in a free fall is weightless effectively or truly (weight definition dependent).
	Weight is related to spontaneous motion.	In falling, weight is not "realized" for the inability to press on the support.	
	Gravity/weight is related to the direction of motion, power/push.		
	Accidental/positional weight.		
3. Projectile motion	Visual account, close to parabola path (slightly deformed).	Decomposition of projectile motion to two: the horizontal inertial, uniform and the vertical, uniformly accelerated.	Dynamic account of motion under gravitational force.
	Empirical investigation by water jets at various angles.	Parabolic trajectory was derived.	Parabolic trajectory as approximation of the elliptic trajectory (Kepler's laws).
	Projectile motion is natural and violent. Interplay of gravity and impetus.	Theoretical kinematic account: Full control of parameters.	Taking into account air/medium resistance produces various trajectories.
	Projectile motion is always considered in medium.	Air resistance was neglected.	

(continued)

Table 8.1 (continued)

4. Astronomy of the Moon /optical studies	Qualitative account based on naked eye observations.	Quantitative account of mountains height based on the telescope observations of the Moon.	Quantitative photometry, the law of Bouguer–Lambert (angular and distance dependence) explains the observed Moon's illumination and all other illumination patterns.
	Moon and Earth are similar celestial bodies (possessing water and atmosphere).	Moon and Earth are similar celestial bodies (mountains, water, and atmosphere).	Moon and Earth possess different environments (no water, no atmosphere, and no soil on the Moon).
	Moon is illuminated by the light reflected from the Earth.	The dark areas on the Moon disc are seas (specular reflection of light from water surface).	Specular and diffusive reflection of light from the Moon surface.
	The bright areas on the Moon disc are seas (diffusive reflection of light from water surface).	Angular dependence of surface illumination using light rays' flux (semiqualitative account)—explanation of seasons on the Earth.	Moon seas are plain areas, which are observed as dark areas on the Moon disc.
	Vision/illumination through the images of images transferred by light rays.		The account of optical systems and phenomena by the theory of light flux and rays.
	Light behavior and shadow area construction account in mirrors by tracing flux of light rays.		
5. Structures and statics	Experimental investigation of elastic deformation and strength of the loaded beams, ropes, devices.	Theoretical account of strength of the loaded beams using lever-based mathematical model, empirical investigations.	The account by the theory of elasticity, providing a model for each specific case.
	Structures of distributed support.	Scaling claims supported by mathematical modeling.	Calculus-based modeling.
	Speculative scaling claims with no theory. (often incorrect)		Scaling method within the area of validity.
			Tension distribution calculated and tested.

(continued)

Table 8.1 (continued)

6. Pendulum	Pendulum as a mechanical devise producing oscillatory motion.	Pendulum oscillations present constrained falling and illustrate the laws of natural falling.	Pendulum motion is treated by the laws of dynamics.
	Pendulum oscillations naturally decay.	Each pendulum swing presents a single process.	Mathematical pendulum provides a model of harmonic oscillator.
	Pendulum motion is neither natural nor enforced: the downward part of a swing is natural while the upward part of it is accidental due to the accrued speed.	Pendulum demonstrates speed independence on weight in falling.	The swing period of the mathematical pendulum is weight independent.
		Swing time is proportional to the square root of the pendulum length.	Isochronicity of the mathematical pendulum holds as an approximation, good for small amplitudes.
		The claim of pendulum isochronicity experimentally and theoretically established investigated (limited accuracy).	Period of the mathematical pendulum oscillations at small angles is $= 2\pi \sqrt{L/g}$.

ᵃFor example, Truesdell (1968, p. 9) quote from *Codex Atlanticus* (72v. b)

from science and engineering learned about the structure of old and new mechanical devices, motion transmission, and regularities of light and fluid phenomena. Through Leonardo, fans of engineering and science discovered that their products could be artistically beautiful and affectively impressing. There is, therefore, a special educational benefit in bringing Leonardo to the science class, to learn about his vision of nature, and to perceive his passion of exploration. In many aspects, Leonardo's life and activities were unusual. In a way, history provides us with a unique experiment. We may see what might happen if one changes the traditionally recognized features and rules of inquiry, which we use to proclaim as the scientific method. How far can we twist it? What is the price that we may pay in changing them?

Firstly, we may learn a lot about observation. It seems very difficult to compete with Leonardo in the accuracy of documentation of the observed. At the same time, we may learn from his experience that no precise observation of natural phenomena, however accurate it is, can replace the theoretical account of reality. Observation furnishes and stimulates understanding, but reliable scientific knowledge requires theoretical guidance leading to experimentation based on the known physical principles and a mathematically shaped account. Leonardo was a Renaissance individual: art, specifically painting, was for him the "queen of all sciences" providing the "means of obtaining knowledge and communicating it."[131] Leonardo defined his method of knowledge production as:

1) Close observation;
2) Repeated testing of the observation from various viewpoints;
3) Drawing the object or phenomenon so skillfully that it became a "fact" which all the world could see, or could grasp with the aid of brief explanatory notes.

He did exactly that in his numerous studies documented in numerous notes and sketches rather than studies focused on a single topic. The combination of his superior acute vision and space perception served Leonardo in producing drawings, which became a highly valuable contribution to anatomy, physiology,[132] and solid geometry.[133] Yet, this program is essentially limited, often not surpassing description and intuition. Any inquiry cannot manage without relying on what is already known about the subject. In considering many issues, Leonardo could take advantage from knowing the studies of others who had investigated similar and related subjects in the recent and distant past.[134] That could accelerate his work tremendously. In science and engineering, Leonardo worked completely alone: he had no discussants, nor assistants versed in science. Not only did he not publish his writings or lecture them to students, he encoded his writings to prevent sharing their content and so could not receive any reflection of his *scientific* claims, either critical or confirming. A lack of professional dialogue was his real misfortune and a gross impediment in his activities.

In art, it was very different. There, he was and has remained forever a celebrity of the first magnitude, whom we oblige for essential innovations that have become emblematic of the Renaissance culture. Yet, Leonardo's secrecy and isolation in scientific studies placed him close to the status of a sage or mage rather than a scholar immersed in the academic discourse. In fact, however, he was far from esoteric or religious views and debates; Galileo was just a cleric in comparison to him. Isolation may befit an artist but not a scientist. One cannot make a more convincing case of the importance of the diachronic and synchronic scientific discourse for scholars. And yet, his bright mind was so creative, his senses were so acute that he did reach achievements in mechanics, optics, solid geometry, and anatomy, though seemingly below his potential. In their scientific progress, people managed without

---

[131] Wallace (1966, p. 104).

[132] Favaro (1956).

[133] Crowell (1999, pp. 123–126; Pacioli, 1509).

[134] Even though he lacks formal education, Leonardo made a great effort to compensate for that vacuum by self-education. He purchased, and somehow it is not known how he familiarized himself with a rather rich collection of science books (Capra 2007, pp. 155–156). He learned geometry from Luca Pacioli (Capra 2007, p. 95) to whom he contributed unbelievably exact and aesthetically superior drawings of polyhedra in Luca's *Divina Proportione*. Yet, Leonardo's knowledge was fragmentary. He was not exposed to fundamentals as happens to students at universities.

Leonardo. His numerous inventions were reintroduced by others, and his achievements were reinvented (for instance, his laws of sliding friction).[135]

Leonardo's scientific claims known from dispersed collections (codices) are difficult to consume. They were laconic, nondiscursive, often lacking comprehensive explanation, made as notes to himself, associative comments rather than valid scientific statements. In the numerous sketches he left behind, we may recognize questions he asked which were answered many years later. For instance, look at two drawings in which Leonardo suggested the way to learn about the pressure in liquid (Fig. 8.12a, b).[136] These sketches indicate that Leonardo realized that the pressure in the fluid increases with the depth of immersion or the height of the water column above certain point. He understood that this pressure determined the speed of the water erupted from a U-tube or ejected from the vessel through the opening in its wall and that speed determines the shape of the water jet. The elaboration is so clear and so suggestive for demonstration of the law of fluid dynamics known as Torricelli's law, discovered in 1643 ($v = \sqrt{2gh}$). Today it is in the curriculum of middle school. Then, in the fifteenth century, it was an unsolvable problem for lacking any mathematical account of the fluid theory. Similarly, he depicted that identical water jets leave the vessels with different cross-sectional areas which indicates the independence of pressure and ejected speed of the water on this the cross-section area of the vessel containing the liquid (Fig. 8.12c).[137] In all these settings, Leonardo was so close to stating law-like regularities but could not proceed to the clear laws for lacking mathematical models depicting the behavior of mater, liquid, or solid in motion. These so strongly suggestive settings and depictions could guide other researchers, if Leonardo's ideas were known and publically discussed. But they were not...

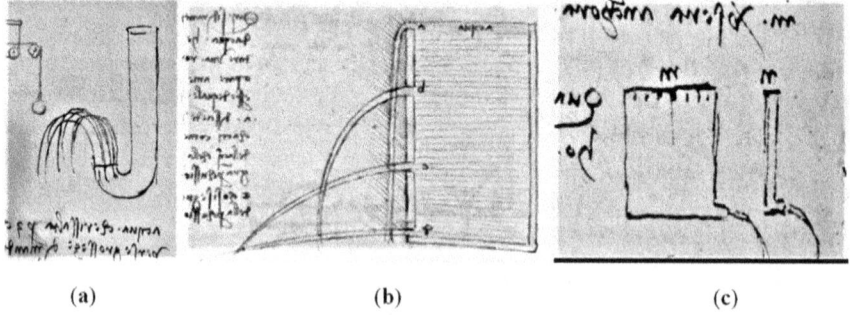

(a)   (b)   (c)

Fig. 8.12 (a) Leonardo's drawings revealing his considering the physics of fluids. (a) The height of the water jet leaving the U-tube increases with the height of water column in the other parts of the tube. (b) The shape of the water jet leaving the opening in the vessel depends on the distance of the opening from the water surface. (c) The identical shape of the water jet leaving the vessels with different cross-sectional areas indicates independence of the water pressure and speed in the outgoing jet on the area of the vessel

---

[135] Leonardo's laws of slide friction (*Codex Atlanticus*, 72v, b) were rediscovered by Amontos in 1699 (Truesdell 1968, pp. 9–12).

[136] Leonardo da Vinci, *Codex Forster II*, fol. 102v, *Codex Madrid I*, fol. 134v.

[137] Leonardo da Vinci, *Codex Forster II*, fol. 117v.

The variety of the imagined by Leonardo technical devices remained as ideas rather than projects, and many of them were never realized. Some ignored the aspects crucial for their technical realization (such as the required energy resources, constraints of the laws of hydrostatics crucial in underwater activity, and the laws of optics in optical devises).[138] The absence of mathematical modeling and lack of basic tools for precise measurement (especially of time) present striking deficiencies in Leonardo's toolkit in comparison with Galileo.[139]

Galileo provided a striking contrast. He was knowledgeable both in the old Hellenistic and the recent medieval science. He learned a lot from Archimedes (buoyancy, lever, pumps, and mathematics). From the Merton School in Oxford, he adopted physical and mathematical concepts: the instant velocity and acceleration, pure uniform and uniformly accelerated motions, mean-speed theorem, etc. From Casini and Oresme, Galileo adopted the graphical representation of Merton's kinematics developed in Paris and brought to Padua by Casini.[140] The new space-time accounts of motion became the central tool for his discoveries—inertial horizontal and vertical accelerated natural motions, their synthesis in the full account for projectiles, and the removal of the 2000 years old dichotomy of natural and violent motion.

The major revolt against the old physics was his method of quantitative account of natural phenomena and experiments.[141] He introduced experiments with controlled variables, which upgraded mere holistic descriptions and observations. Theoretical modeling was his major tool. He believed in the world designed in lawlike regularities—*the Book of Nature*—a clear Platonic idea. Yet, he considered these basic regularities as being masked by accidental factors, features of materials, and environmental constrains, friction, for example. In accordance, he tried to address idealized situations, which would reveal the laws of Nature in their "pure" form, as they were written in the Book of Nature. It was this activity that he considered as *experiments*.[142]

Galileo relied on authorities from the past and present. However, at the same time, he did not hesitate to criticize them—Aristotle, Jesuit scholars, and his university colleagues. From his first days as a lecturer in Pisa, he was continuously involved in intensive debates with his fellow scientists on the nature of motion, falling, Moon environment, comets, Sun spots, buoyancy, solid structures, pendulum, projectiles, and, of course, on the celestial world. Addressing controversy was the axis of his activities illustrating the central feature of science of all times.

His major books Galileo presented dialogues with other views, a vivid debate he skillfully constructed between imaginary discussants. It was a very effective and

---

[138] Canestrini (1956).

[139] It is not only due to the gap in time. Much could be learned from Ptolemy, Archimedes, Alhazen, and among others from the past if learned by Leonardo.

[140] Pederson and Pihl (1974, pp. 219–225), Clagett (1959), Hannam (2009), Uritam (1974), and Dijksterhius (1986).

[141] Koyré (1943a, 1968a).

[142] McAllister (1996).

appealing form of explanatory power, far from a dry, distilled, disciplinary presentation. Galileo was widely educated and taught at universities. He was surrounded with academics and intellectuals exploring a variety of scientific topics. Yet, unlike others, not only did he publish his books and essays, he introduced in them the colloquial Italian, Vulgate, instead of Latin, presented academic canon. This step removed the barrier for the wider public beyond academy to know about his claims. His style was often casual, and his writings could resemble a free-flowing conversation (sometimes "shapeless and undisciplined"[143]) rather than a rigor demonstration. He could mix a variety of topics, associations from various areas of knowledge. Yet, Galileo possessed a great pedagogical talent and even an entertaining skill for disseminating his ideas among curious intellectuals and enlightened amateurs outside the academy. It caused his ideas quickly spread and become influential throughout Europe. He became famous for heralding the new science, and his name became a symbol of the scientific revolution that changed science forever.

While Leonardo often refrained from revealing his reasoning beyond depiction and claims, never discussed his views with opponents or students, Galileo, when he adopted the heliocentric view of Copernicus, started an entire program for its demonstration. Being challenged with critic and rivals highly instigated Galileo's research agenda. He interpreted his discovery of the moons of Jupiter as evidence against the unique status of the Earth possessing satellites. Another Earth's uniqueness was removed in his depiction of the Moon's surface. He identified the phases of Venus testifying to its planetary motion around the Sun. However, the climax was his dispute with cardinal Roberto Bellarmine on the truth versus modeling that sent Galileo to look for the phenomenon, which would unequivocally testify for the Earth's motion, diurnal, and annual as the incontrovertible truth.[144] He thought he had found it—the ocean tides—and in 1632, decided to violate the arrest of talking about Copernicus theory, he complied with in 1616, and published his *Dialogo*.[145] Albeit his tides argument soon appeared entirely erroneous; it manifested the scientific method—displaying a theoretical account of certain phenomenon, the tides, as the evidence supporting the central theory—the heliocentric system of the world. The dispute with alternatives was recognized as the essence of the scientific method, which requires both theoretical and empirical components. Galileo's account for tides was dynamical. In that, he broke his approach of a mere kinematic account of motion[146] and criticized the different account by Kepler.[147] It appeared to be a pre-

---

[143] Sharratt (1994, p. 153).

[144] Cardinal Roberto Belarmine, the central figure of the Catholic institution of Inquisition, was the most formidable and intellectual discussant and opponent of Galileo in the fundamental problem of the Earth motion (Sharratt 1994, pp. 114–116, Finocchiaro 1989, p. 30).

[145] Galilei (1632/1953, The Fourth Day). Originally, the *Dialogo* included tides in its title was the *Dialogue on the Ebb and Flow of the Sea* (Drake 1999b, p. 23; Finocchiaro 1989, pp. 119–133).

[146] Galilei (1638/1914, p. 166) wrote: "Salv. The present does not seem to be the proper time to investigate the cause of the acceleration of natural motion concerning which various opinions have been expressed by various philosophers."

[147] Drake (1978, p. 42).

mature attempt and a wrong explanation.[148] So, the problem waited for the theory of Newton that truly explained the tides.

In contrast with Leonardo, who relied on visual perception, Galileo did not unconditionally yield to the observed but proceeded to its deeper analysis, a mathematical demonstration combined with illustrative experiments. It is important to see the complex and nontrivial role that Galileo ascribed to experiment.[149] While in one place, he answered by:[150]

Did you [Salviati/Galileo] do an experiment? – No, I do not need it, as without any experience I can affirm that it is so, because it cannot be otherwise. … I without experiment am certain that the effect will follow as I tell you, because it is necessary that it should.

In another place, however, Galileo argued:[151]

…the fact that all human reasoning must be placed second to direct experience. Hence, they will philosophize better who give assent to propositions that depend upon manifest observations than they who persist in opinions repugnant to the senses and supported only by probable reasons…

On the surface, Galileo may look inconsistent… Yet, in fact, this is the nature of the scientific method, which prevents a simple identification of a scientist either as a rationalist or empiricist philosopher. Apparently, Galileo adopted the dialectical approach of complementarity, which had proved itself in scientific inquiry. In his activities, one observes consolidation of the new scientific method: an amalgam of both approaches kept in balance by the practitioners. Physics research may look like pure empiricism in some cases (as it appears in some introductory textbooks) and pure rationalism in others (as it appears in the advanced theoretical physics).[152] Galileo illustrated this complementarity—the scientific method is pluralistic. Missing this perspective leads to the confusion of "a poor teacher of elementary physics who seeks a simple picture for his students."[153] Koyré summarized Galileo's approach ironically but essentially:[154]

The new science is for him an experimental proof of Platonism.

---

[148] Indeed, Galileo explained the tidal motion of seas by the difference in summarizing of the diurnal and annual movements (Drake 1999b, pp. 23–36). This effect, in eyes of Galileo, would be a physical evidence of the annual motion of the Earth. During the 24 hours, however, the annual motion is practically rectilinear and uniform in comparison with the diurnal rotation. This is regardless other inconsistences widely elaborated (the wrong periodicity and spring tides). And yet, Galileo's effect was not an absurd (Drake 1999b, p. 34, Ch. 4). The diurnal change of velocity due to summarizing the two *rotational* movements implies the change in the centrifugal force on the ocean water (velocity changes at the same radius), causing a single pair of water ebb and flow diurnally. Lacking dynamics of Huygens' Vis centrifuga provided later, Galileo could not produce this explanation, even if it would remain not the true explanation of the ocean tides:

[149] Segre (1991, p. 47).

[150] Galilei (1632/1953, p. 145).

[151] Galilei (1613/1957, p. 118).

[152] Einstein (1949/1979, pp. 683–684).

[153] Cohen (1993), Chapter 9.

[154] Koyré (1968, p. 43).

Behavioristic Comparison

We have summarized the comparison of Leonardo and Galileo in their contribution to the content knowledge of mechanics and optics (Table 8.1). Such comparison facilitated our perspective on how our knowledge grew and what is its meaning. In that, we "stand on the shoulders of giants," as articulated by Bernard of Chartres[155] in the medieval philosophy. Discussing the differences enhanced the understanding of the essence of the subject matter. Therefore, in parallel, we may expand our inquiry asking about behavioristic and methodological differences between the two outstanding scholars, and this way enhance another domain of our knowledge about science—its epistemological aspect (Table 8.2).

Table 8.2 Features of the scientific activities of the two scholars in juxtaposition

Leonardo	Galileo
Self-educated in science, received occasional tuition in mathematics	University formal education
	Knowledge of scientific heritage
Solitary studies	Active participation in the scientific community discourse
No assistance, no pupils (in science)	Several assistants–pupils–followers
Never taught science	University professor in mathematics and philosophy
Freelancer	Court scientist
Secrecy, codifying own writings	Using Italian (Vulgate) in publications, seeking wide public exposure
Mixture of notes on different topics, often not organized, collection of comments often not explained	Genre of dialogue with alternative views (scientific culture)
Insularity	Participation in public discourse on scientific topics
No published scientific studies	Several published scientific treatises including nonformal narrative, widely discussed
Laconic comments regarding philosophical topics	Comprehensive addressing of issues in religion and philosophy of science (epistemology)
Experimental inquiry of specific devices	Experiments with controlled parameters to elicit general principles of natural organization
Mechanical, hydrodynamic, optical modeling	Mechanical and optical modeling
Numerous ideas of various technological devices	Attempts to reduce factors (friction, external influence) which mask natural "true" natural regularities. Thought experiments
Using mathematics: Simple proportion, linear functional dependence, geometrical considerations	Using mathematics: Simple and quadratic proportion/functional dependence
	Geometrical approach to demonstration in physics
	Mathematical modeling of physical systems
	Graphical method of functional dependence
Perfect artistic depiction (documentation)	Use of illustrative pictures of physical settings
Accurate drawing of objects and phenomena ("stroboscopic" depiction)	Schematic scientific drawings
Schematic scientific drawings	"Ear for music" (using pulse rate measuring time in observation)

---

[155] Crombie (1959, p. 27).

This table suggests that juxtaposition of different approaches could be elucidative. The statement from each of the columns emphasizes the strong difference between the two in the distance to the scientific method, as we understand it today. Such a comparison, considering what was indeed missed by each of the two scholars, can facilitate a productive discussion on the nature of science, its methodology, and interaction with society. The recognition of the range of cognitive preferences and variation of individual skills of researchers can further emphasize the still unavoidable requirement in doing science, the science commitment to objectivity, need of empirical evidence, mathematical proficiency, theory, modeling, and so on. Some of these are rather obvious for interpretation, but none of them can be unique and univocal; there are plenty of aspects open for clarification, while the other will require a deeper consideration. We will expand on these items in the general discussion on the nature of science in the next chapter.

Our comparison emphasized what was lacking in the scientific activities of Leonardo and Galileo. This approach—revealing what they missed—may remind of the apophatic approach to understanding of the legacy of the two great scholars. It creates a cultural perspective, which refines and enriches the oversimplified image of a genius who just added this or that discovery and whom we all are, therefore, obliged for the progress of science. In our view, even if it is common to recognize Galileo among the leaders of the scientific revolution of the seventeenth century, which produced modern science, ignoring the image of Leonardo as a scientist impoverishes the overall picture of science and its method in education. Leonardo and Galileo both present the culture of science and help each generation to understand the nature of science in its variety.

Picture Credits

- Self-portrait of Leonardo da Vinci, 1513, Turin Royal Library. Public Domain, https://commons.wikimedia.org/w/index.php?curid=59570
- Portrait of Galileo by Justus Sustermans, 1640. Public Domain. https://commons.wikimedia.org/w/index.php?curid=230543
- Fig. 8.1a Leonardo's sketch of two masses in balance over an edge. *Codex Forster II*, fol. 087r. Public Domain. https://www.vam.ac.uk/articles/explore-leonardo-da-vincis-notebook-codex-forster-ii#?c=&m=&s=&cv=174&xywh=-504%2C-638%2C10940%2C7548
- Fig. 8.1b Illustration from Galilei, 1638/1914, *Discorsi*, p. 181. Public Domain
- Fig. 8.2a Water jets experiment. Sketch by Leonardo da Vinci. *Paris Manuscript C*, fol. 7r. Public Domain. https://commons.wikimedia.org/wiki/Category:Manuscrit_C_(RMN)#/media/File:Leonardo_da_Vinci_-_19-547105.jpg
- Fig. 8.2b Mortars firing into a fortress. Sketch by Leonardo da Vinci, 1504, *Codex Atlanticus*, fol. 72r. Public Domain. https://www.ambrosiana.it/en/opere/atlantic-codex-codex-atlanticus-f-72-recto/
- Fig. 8.2c Cannon shells, sketch by Leonardo da Vinci, 1518, *Codex Atlanticus*, fol. 33r. Public domain. https://commons.wikimedia.org/wiki/File:Codice_Atlantico_-_009r_original.jpg
- Fig. 8.2d Drawing of parabola. Illustration from Galilei, *Discorsi*, 1638/1914, p. 249. Public Domain
- Fig. 8.3a Astronomical sketch by Leonardo da Vinci, *Codex Leicester* 2A, Fol. 2r. Public Domain. https://commons.wikimedia.org/w/index.php?curid=1685117

- Fig. 8.3b Picture of the Moon. Illustration in Jules Verne, *From the Earth to the Moon,* 1874. New York: Schribner, Armstrong & Co. Public Domain
- Fig. 8.3c The New Moon. Picture credit: NASA
- Fig. 8.4a Moon sketch by Galilei, *Siderius Nuncius,* 1610/1989, p. 44. Public Domain. www.rarebookroom.org/Control/galsid/index.html
- Fig. 8.4b Illustration in Scheiner, 1614, p. 58. Public Domain. https://astronomy.wikia.org/wiki/Telescope_construction?file=Galileo_scheinermoon.bmp.jpg
- Fig. 8.4c Leonardo da Vinci's sketch of light reflection. *Codex Arundel,* fol. 25r. Public Domain. http://www.bl.uk/manuscripts/Viewer.aspx?ref=arundel_ms_263_f001r
- Fig. 8.4e Magdalene by La Tour, 1640, Metropolitan museum of art. Public Domain. https://commons.wikimedia.org/w/index.php?curid=153615
- Fig. 8.5 Studies of water by Leonardo da Vinci, 1509, *Codex Windsor,* fol. 12660r, v. Public Domain. https://commons.wikimedia.org/w/index.php?curid=59577
- Fig. 8.6 Shadow pattern construction. Leonardo da Vinci sketch, *Codex Atlanticus,* fol. 187r. Public Domain. Reproduced in Argentieri (1956, p. 413). https://www.codex-atlanticus.it/#/
- Fig. 8.7a, b, c Light reflection in curved mirrors. Leonardo da Vinci's sketches, *Codex Arundel,* fol. 87r. 86v. Public Domain. http://www.bl.uk/manuscripts/Viewer.aspx?ref=arundel_ms_263_f001r#
- Fig. 8.7d Illustration from *Dialogo* by Galilei, 1632/1953, p. 80. Public Domain
- Fig. 8.7e The distance dependence of light illumination. Leonardo da Vinci's sketch, *Codex Forster II,* fol. 16v. Public Domain. https://www.vam.ac.uk/articles/explore-leonardo-da-vincis-notebook-codex-forster-ii#?c=&m=&s=&cv=31&xywh=-1%2C-81%2C5518%2C3807
- Fig. 8.8a Self-supporting bridge. Leonardo da Vinci's sketch, *Codex Atlanticus,* fol. 23v. (69r) Public Domain. https://www.flickr.com/photos/prof_richard/48807160781/in/photostream/.
- Fig. 8.8b, c Loaded beam. Leonardo da Vinci's sketchs, *Codex Forster,* fol. 089r. Public Domain. https://www.vam.ac.uk/articles/explore-leonardo-da-vincis-notebook-codex-forster-ii#?c=&m=&s=&cv=178&xywh=0%2C-374%2C9845%2C7280
- Fig. 8.9a-d Illustrative drawings from Galilei, *Discorsi,* 1638/1914, p. 114, 129, 131, 116, 140. Public Domain
- Fig. 8.10a Pendulum. Leonardo da Vinci sketch, Leonardo da Vinci's sketch, *Codex Madrid* I, fol. 147r. Public Domain. http://leonardo.bne.es/index.html
- Fig. 8.10b Decaying Oscillations. Leonardo da Vinci's sketch, *Codex Madrid* I, fol. 147r. Public Domain. http://leonardo.bne.es/index.html
- Fig. 8-.10d Enforced pendulum. Leonardo da Vinci's sketch, *Codex Madrid* I, fol. 8r. Public Domain. http://leonardo.bne.es/index.html
- Fig. 8.11a Illustration from Galilei, *Discorsi,* 1638/1914. p. 167. Public Domain
- Fig. 8.11b Illustration from Galilei, *Discorsi,* 1638/1914. p. 232. Public Domain
- Fig. 8.11c Descending on a chord. Leonardo da Vinci's sketch, *Codex on the Flight of Birds.* Public Domain. Turino: the Biblioteca Reale. https://airandspace.si.edu/exhibitions/codex/codex.cfm#page-6-7
- Fig. 8.11d Descending on slopes. Leonardo da Vinci's sketch, *Codex Atlanticus,* fol. 338r., h. Public Domain. https://www.codex-atlanticus.it/#/
- Fig. 8.12a Water jet leaving a tube. Leonardo da Vinci's sketch, *Codex Forster II,* fol. 102v. Public Domain. https://www.vam.ac.uk/articles/explore-leonardo-da-vincis-notebook-codex-forster-ii#?c=&m=&s=&cv=205&xywh=471%2C488%2C8204%2C6067
- Fig. 8.12b Water jets leaving a vessel at different heights. Leonardo da Vinci's sketch, *Codex Madrid I,* fol. 134v. Public Domain. http://leonardo.bne.es/index.html
- Fig. 8.12c Water jet leaving to vessels of different cross areas. Leonardo da Vinci's sketch, *Codex Forster II,* fol. 117v. Public Domain. https://www.vam.ac.uk/articles/explore-leonardo-da-vincis-notebook-codex-forster-ii#?c=&m=&s=&cv=235&xywh=-1%2C-354%2C9845%2C7280

References

Argentieri, D. (1956). Leonardo's optics. In E. Vollmer (Ed.), *Leonardo da Vinci* (pp. 405–436). Reynal.

Ariotti, P. (1968). Galileo on the isochrony of the pendulum. *Isis, 59*(4), 414–426.

Bedini, A. S., & Reti, L. (1974). Horology. In L. Reti (Ed.), *The unknown Leonardo* (pp. 240–263). McGraw-Hill.

Calvi, I. (1956). Military engineering and arms. In E. Vollmer (Ed.), *Leonardo da Vinci* (pp. 275–306). Reynal.

Canestrini, G. (1956). Leonardo's machines. In *Leonardo da Vinci, collected studies* (pp. 493–507). Reynal.

Capra, F. (2007). *The science of Leonardo*. Anchor Books.

Clagett, M. (1959). *The science of mechanics in the middle ages*. The University of Wisconsin Press.

Coelho, R. L. (2010). On the concept of force: How understanding its history can improve physics teaching. *Science & Education, 19*, 91–113.

Cohen, I. B. (1993). A sense of history in science. *Science & Education, 2*(3), 251–277. [1950, *American Journal of Physics*, 18, 343–359].

Cohen, R. M., & Drabkin, E. I. (1948). *A source book in Greek science*. McGraw-Hill.

Copernicus, N. (1543/1952). *On the revolutions of the Heavenly spheres*. Encyclopedia Britannica.

Crombie, A. C. (1959). *Medieval and early modern science*. Doubleday Anchor Books.

Crowell, P. R. (1999). *Polyhedra*. Oxford University Press.

Darrigol, O. (2000). *Electrodynamics from Ampere to Einstein*. Oxford University Press.

Descartes, R. (1644/1982). *Principles of philosophy*. D. Reidel.

Dibner, B. (1974). Machines and weaponry. In L. Reti (Ed.), *The unknown Leonardo* (pp. 166–189). McGraw-Hill.

Dijksterhius, E. J. (1970). *Simon Stevin: Science in the Netherlands around 1600*. Martinus Nijhoff.

Dijksterhius, E. J. (1986). *The mechanization of the world picture. Pythagoras to Newton*. Princeton University Press.

Drake, S., & Drabkin, I. E. (1969). *Mechanics in sixteenth-century Italy*. The University of Wisconsin Press.

Drake, S. (1957). *Discoveries and opinions of Galileo*. Doubleday & Company.

Drake, S. (1978). *Galileo at work. His scientific biography*. The University of Chicago Press.

Drake, S. (1999a). Galileo: Scientific method and philosophy of science. In *Essays on Galileo and the history and philosophy of science* (Vol. I). University of Toronto Press.

Drake, S. (1999b). Galileo: Scientific method and philosophy of science. In *Essays on Galileo and the history and philosophy of science* (Vol. II). University of Toronto Press.

Einstein, A. (1949/1979). Autobiographical notes. In P. A. Schilpp (Ed.), *Albert Einstein: Philosopher-scientist*. Harper.

Emanuelli, P. (1956). da Vinci's astronomy. In E. Vollmer (Ed.), *Leonardo da Vinci* (pp. 205–208). Reynal.

Favaro, G. (1956). Anatomy and biological sciences. In E. Vollmer (Ed.), *Leonardo da Vinci* (pp. 363–372). Reynal.

Fermi, E. (1934). Versuch einer Theorie der β-Strhlen. I. *Zeitschrift für Physik, 88*, 161–177.

Finocchiaro, M. A. (1989). *The Galileo affair*. University of California Press.

Galilei, G. (1590/1960). *De Motu*. In I. E. Drabkin & S. Drake (Eds.), *On motion & on mechanics; comprising*. The University of Wisconsin Press.

Galilei, G. (1610/1892/1989). The Sidereal Messenger. In Le Opere di Galileo Galilei 3, p. 48. University of Chicago Press, . Also Galilei, G. (1653). Nuntius Sidereus. In P. Gassndi (Ed.), *Institutio Astronomica, Juxta Hypotheses Tam Veterum quam Recentorium* (pp. 18–20). London: Facobi Flesher.

Galilei, G. (1613/1957). Second letter on sunspots. In S. Drake (Ed.), *Discoveries and opinions of Galileo*.

Galilei, G. (1623/1957). *The Assayer*. Translated by Stillman Drake. In *Discoveries and opinions of Galileo* (pp. 237–238). Anchor Books.

Galilei, G. (1632/1953). *Dialogue concerning the two chief world systems – Ptolemaic and Copernican*. University of California Press.

Galilei, G. (1638/1914). *Dialogue concerning two new sciences*. Dover.

Galili, I., & Hazan, A. (2000a). Learners' knowledge in optics: Interpretation, structure, and analysis. *International Journal in Science Education, 22*(1), 57–88.

Galili, I., & Hazan, A. (2000b). The influence of historically oriented course on students' content knowledge in optics evaluated by means of facets-schemes analysis. *Physics Education Research, American Journal of Physics, 68*(7), S3–S15.

Galili, I., & Lavrik, V. (1998). Flux concept in learning about light: A critique of the present situation. *Science Education, 82*, 591–613.

Galili, I. (2001). Weight versus gravitational force: Historical and educational perspectives. *International Journal of Science Education, 23*(10), 1073–1093.

Galili, I. (2009). Thought experiment – Establishing conceptual meaning. *Science & Education, 18*(1), 1–23.

Galili, I. (2012). Promotion of content cultural knowledge through the use of the history and philosophy of science. *Science & Education, 21*(9), 1283–1316.

Galili, I. (2015). From comparison between scientists to gaining cultural scientific knowledge: Leonardo and Galileo. *Science & Education, 25*(1), 115–145.

Gibbs-Smith, C. (1978). *The inventions of Leonardo da Vinci*. Charles Scribner's Sons.

Gliozzi, M. (1965). *Storia della Fisica* (Vol. II). Storia della Scienze.

Granek, G. (2006). Poincaré's light signaling and clock synchronization thought experiment and its possible inspiration to Einstein. In J. M. Alimni & A. Fuzfa (Eds.), *Albert Einstein century international conference* (pp. 1095–1102). American Institute of Physics.

Hannam, J. (2009). *Gods philosophers. How the medieval world laid the foundations of modern science*. Icon Books.

Van Helden, A. (1989). Introduction. In G. Galilei (Ed.), *The sidereal messenger*. The University of Chicago Press.

Kemp, M. (2006). *Leonardo da Vinci, the marvelous works of nature and man*. Oxford University Press.

Kofka, K. (1925). *The growth of mind*. Harcourt, Brace & Co.

Koyré, A. (1943a). Galileo and the scientific revolution of the seventeenth century. The philosophical review. *Journal of the History of Ideas, 52*(4), 333–348.

Koyré, A. (1943b). Galileo and Plato. *Journal of the History of Ideas, 52*(4), 400–428.

Koyré, A. (1968). Newton and Descartes. In *Newtonian studies* (pp. 85–155). University of Chicago Press.

Koyré, A. (1968a). *Metaphysics and measurement: Essays in scientific revolution*. Harvard University Press.

Koyré, A. (1978). *Galileo studies*. Harvester Press.

Kurrer, K.-E. (2008). The history of the theory of structures. From arch analysis to computational mechanics. .

Lima, F. M. S., & Arun, P. (2006). An accurate formula for the period of a simple pendulum oscillating beyond the small angle regime. *American Journal of Physics, 74*(10), 892.

Lindberg, D. C. (1992). *The beginnings of the Western Science*. The University of Chicago Press.

MacCurdy, E. (Ed.). (1938/1955). *The notebooks of Leonardo da Vinci*. G. Brazilller.

Mach, E. (1883/1919/1989). The science of mechanics, a critical and historical account of its development. : Open Court.

MacLachlan, J. (1997). *Galileo Galilei. First physicist*. Oxford University Press.

Marcolongo, R. (1956). da Vinci's mechanics. In E. Vollmer (Ed.), *Leonardo da Vinci* (pp. 483–492). Reynal.

Marton, F., & Tsui, A. B. M. (2004). *Classroom discourse and the space of learning*. Lawrence Erlbaum.

Matthews, M. R. (1990). Ernst Mach and contemporary science education reforms. *International Journal of Science Education, 12*(3), 317–325.

Matthews, M. R. (1994/2015). *Science teaching. The contribution of history and philosophy of science.* Routledge.

Matthews, M. R. (2000). *Time for science education: How teaching the history and philosophy of pendulum motion can contribute to science literacy.* Plenum Press.

Matthews, M. R. (Ed.). (2014). *International handbook of research in history, philosophy and science teaching.* 3 Volumes. Dordrecht, The Netherlands: Springer.

McAllister, J. (1996). The evidential significance of thought experiments in science. *Studies in History and Philosophy of Science, 27*(2), 233–250.

Miller, A. I. (1981). *Albert Einstein's special theory of relativity: Emergence (1905) and early interpretation (1905–1911).* Addison-Wesley.

Miller, A. I. (1984). *Imagery in scientific thought: Creating 20th-century physics.* Birkhauser.

Miller, A. I. (Ed.). (1986). *Frontiers of physics: 1900–1911. Selected essays.* Birkhauser.

Moody, E. A., & Clagett, M. (Eds.). (1952). *The medieval science of weights.* The University of Wisconsin Press.

Pacioli, L. (1509). *Divina proportione.* . https://archive.org/details/divinaproportion00paci

Pascal, B. (1910). Of the geometrical spirit. In *Blaise Pascal: Thoughts, letters, and minor works* (The Harvard Classics) (Vol. 48). Collier & Son.

Pedersen, O., & Pihl, M. (1974). *Early physics and astronomy.* McDonald & Janes.

Piaget, J., & Garcia, R. (1989). *Psychogenesis and the history of science.* Columbia University Press.

Piaget, J. (1970). *Genetic epistemology.* Columbia University Press.

Plutarch. (1989). *Parallel lives, Volume VII. Demosthenes and Cicero. Alexander and Caesar.* Loeb Classical Library, Harvard University Press.

Redish, E. F. (2010). Introducing students to the culture of physics: Explicating elements of the hidden curriculum. *AIP Conference Proceedings, 1289,* 49.

Richter, I. (1952/2008). *Leonardo da Vinci notebooks.* Oxford University Press.

Scheiner, C. (1614). *Disquisitiones Mathematicae.* Eder for Elisabeth Angermaria.

Sharratt, M. (1994). *Galileo decisive innovator.* Blackwell.

Shulman, L. S. (1986). Those who understand: Knowledge growth in teaching. *Educational Researcher, 15,* 4–14.

Stein, H., Galili, I., & Schur, Y. (2015). Teaching new conceptual framework of weight and gravitation in the middle school. *Journal of Research in Science Teaching, 52*(9), 1234–1268.

Stinner, A., & Williams, H. (1993). Conceptual change, history, and science stories. *Interchange, 24*(1 & 2), 87–103.

Truesdell, C. (1968). *Essays in the history of mechanics.* Springer.

Uccelli, A. (1956). The science of structures. In E. Vollmer (Ed.), *Leonardo da Vinci* (pp. 261–274). Reynal.

Uritam, R. A. (1974). Medieval science, the Copernican revolution, and physics teaching. *American Journal of Physics, 42*(10), 809–819.

Verne, J. (1874). *From the earth to the moon.* Schribner, Armstrong & Co..

Wallace, R. (1966). *The world of Leonardo.* Time-Life.

Wandersee, J. H. (1986). Can the history of science help science educators anticipate students' misconceptions? *Journal of Research in Science Teaching, 23*(7), 581–597.

Wandersee, J. H. (1990). On the value and use of the history of science in teaching today's science: Constructing historical vignettes. In D. E. Herget (Ed.), *More history and philosophy of science in science teaching* (pp. 278–283). Florida State University.

Welch, W. W. (1973). Review of the research and evaluation program of Harvard project physics. *Journal of Research in Science Teaching, 10*(4), 365–378.

Wolf, A. (1961). *A history of science, technology & philosophy in the 18th century* (Vol. 1). Harper.

Zammattio, C. (1974). The mechanics of water and stone. In L. Reti (Ed.), *The unknown Leonardo* (pp. 190–215). McGraw-Hill.

Chapter 9
A Refined Account of Nature of Science

Everything should be made as simple as possible, but not simpler.

Albert Einstein

The truth is never univocal.

Andrei Sakharov

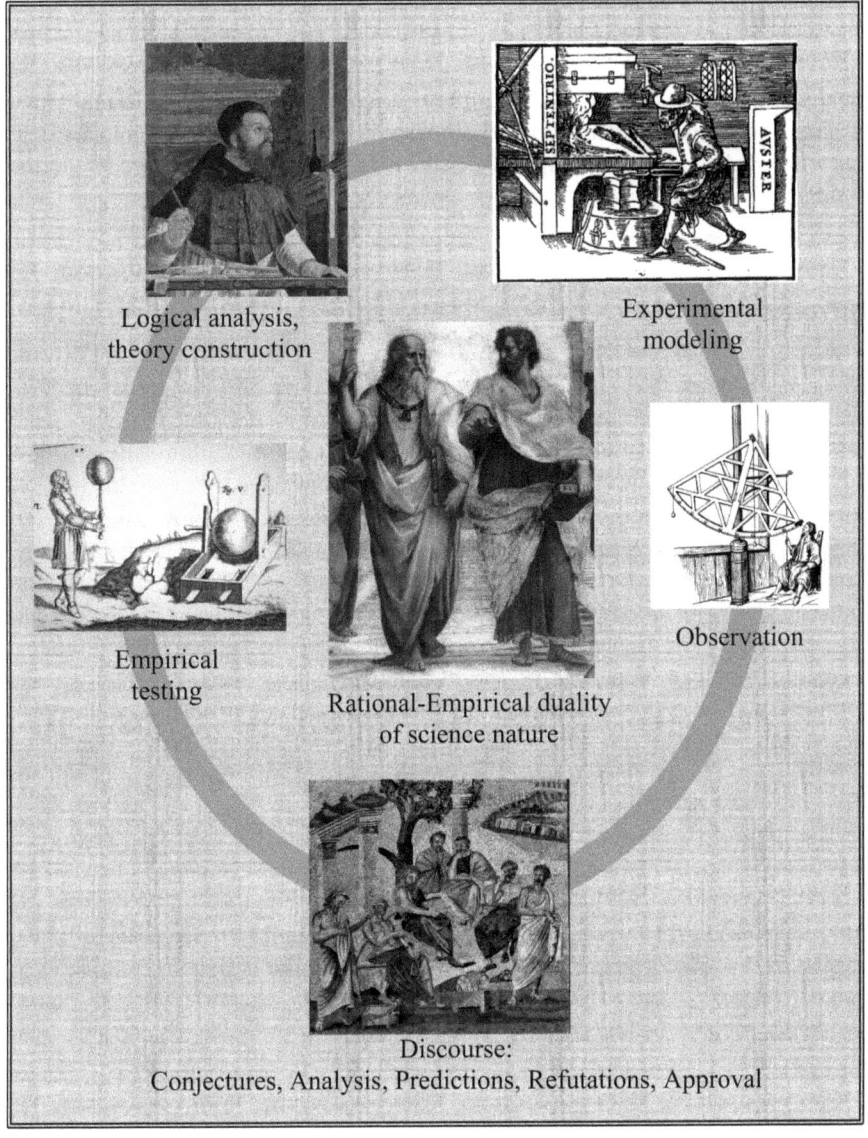

Logical analysis, theory construction

Experimental modeling

Empirical testing

Rational-Empirical duality of science nature

Observation

Discourse: Conjectures, Analysis, Predictions, Refutations, Approval

© Springer Nature Switzerland AG 2021
I. Galili, *Scientific Knowledge as a Culture*, Science: Philosophy, History and Education, https://doi.org/10.1007/978-3-030-80201-1_9

Abstract   The nature of science (NOS) has become a popular topic of science education research. This is also due to its short list of features, which essentially revised traditional views on the subject and provided simple answers to the complex questions regarding the status of scientific knowledge. In this chapter, the list of NOS features is reconsidered. It is stated that this list, as is, may harm, impede, and mislead the understanding of science. Yet, the refinement of its claims may enrich or sometimes reverse them. The suggested analysis calls for addressing the range of variation in each particular aspect of NOS. The approach of discipline–culture provides a useful platform for such reconsideration clarifying the questions common in the teaching of science. Cultural content knowledge naturally incorporates conceptual variation referring to the structure of knowledge and plurality of scientific methodology while drawing on the history and philosophy of science. The implication of such revision highlights the major role of science educators who face veracity of polar claims while constructing science curricula. The cultural approach protects their disciplinary content from oversimplification and provides the necessary philosophical defense of the traditional NOS claims as required for genuine understanding of science.

Introduction

We normally do things under numerous assumptions and preconditions of which we might not be aware. For many years, this was the situation with regard to the nature of scientific knowledge and its method in public education. Traditionally, teachers use to present science disciplines as the objective knowledge about nature, its laws unknown before their construction ("discovery") in research through using the scientific method. Nature of science (NOS) was often tacitly presumed, in a way, "known to all," as put by Newton addressing time and space. And exactly as with time and space, in our days, those fundamentals, originally taken as obvious in science classes, were reconsidered. In education research, an intensive discourse regarding the latter started by the end of the last century. The previously ignored complexity of NOS became a subject for discussion outside of science philosophy. Not surprisingly therefore, the issues of philosophy of science usually conceived naively, if at all, by class teachers were reconsidered. The curricular pendulum started to move in the opposite swing.

The progress in science—the scientific revolution in the twentieth century—dragged after its revolutionary changes in the philosophy of science. It was just a matter of time before this wave of the new worldview and perspectives reached the curricular discourse. Joseph Schwab was among the pioneers. He argued for the need of the knowledge about knowledge to enter curricular design in science education. He introduced the notions of *substantive* and *syntactic* structures of disciplinary knowledge.[1] The *substantive* structure contained the basic concepts and

---

[1] Schwab (1964, 1978); Shulman (1986, 1987).

principles, which constitute a discipline (conceptual knowledge), while the *syntactic* structure informed about the ways, methods, and tools through which truth and the validity of knowledge were established. Substantive and syntactic knowledge of the curriculum context corresponded to *ontology* and *epistemology* in the philosophy of science. By using specific terminology, Schwab indicated the specific context of education implying specific forms, vocabulary, and level of presentation. A new discourse was established. Years later, it was given the title of "Nature of Science"—a special type of inquiry regarding the features and status of elements of the scientific knowledge, their validity, and reliability.

It appeared that the inclusion of NOS raised the requirement to involve several domains of knowledge and that caused a special complexity to be realized and dealt with (Fig. 9.1). A simple suggestion to consult philosophers can be illuminating for the individuals with no specific backgrounds (as science teachers often are) and appeared to be often not practical at the same time. It can even introduce obscurity for the nontrivial preferences required to take. This is because of polar differences of the pertinent claims in different areas.[2] Science teacher discovered the need to familiarize oneself with a range of ideas before any decision. In other words, considering NOS requires a new type of knowledge, knowledge about knowledge, which can be identified as discipline–cultural knowledge of the subject matter. Teacher training programs have to develop a specific proficiency in perspective teachers of science.

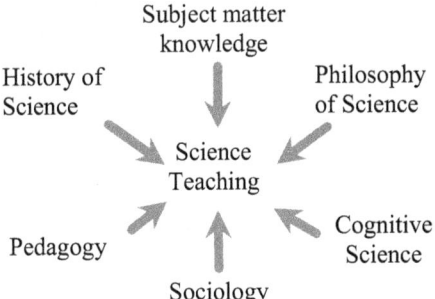

Fig. 9.1 Areas of competence required from science educators to comprehend NOS

Background of the Debate

The issues of NOS arise each time when learners and teachers ask themselves "What kind of knowledge does science represent?", "How it is different from other knowledge?", "What makes knowledge scientific?", and "How should one present it in teaching?" These questions define the unique role and challenge faced by science educators—the need to synthesize different aspects and resolve their controversy.

---

[2] Consider, for example, the philosophy of science of Glasersfeld (1992, 1995) versus Bunge (1973) or versus van Fraassen (1980, 2008).

Vygotsky in the 1930s provided a helpful clue.[3] He pointed to the fact that scientific knowledge represents the culture of a particular society, and there are two ways to learn about it—*from inside* and *from outside*.

Scholars in each academic area engage in a specific, long-term, comprehensive discourse. One may say that they continuously learn their subject matter *from inside*. This experience creates their vision of NOS throughout all their professional lives. In contrast, students of science become acquainted with scientific knowledge *from outside*. In school classes, they take short-term courses composed of selected topics. The resulting knowledge is unavoidably superficial, fragmental, and often pragmatic and lacking the idea of the overall structure.

Vygotsky illustrated the difference between the two learning types by contrasting the learning of one's mother tongue with that of a foreign language. The knowledge of science concepts was equated with that of a foreign language.[4] Students have trouble with the concepts and features of science and its method. If they draw on naive views and experience, there is no chance of success. One needs specific support from *the outside*, a clear and distinct account of theories, laws, models, principles, and experiments—the code of science. Yet, even basic knowledge of this kind, provided by HPS, is often lacking in teacher training programs,[5] pushing teachers to reinvent the fundamentals. This is an inevitably shaky ground.

Furthermore, unlike the disciplinary content possessing clear standards of correctness, the features of NOS emerge in a range including opposite claims, different ideas, and plural methodologies, which often puzzles consumers in science education.[6] The challenge is twofold: with respect to the normative features and regarding the way to teach them. One should start from the first aspect—the normative features of NOS and their clarification. Considering this aspect, a short list of NOS features introduced by Norman Lederman and his collaborators can attract attention also for its title "Consensus view."[7] It declared that:

1. Scientific knowledge, owing to scientists' theoretical commitments, beliefs, previous knowledge, training experiences, and expectations, is unavoidably *subjective*.
2. The distinction between *theories and laws*. Laws are statements or descriptions of the relationships among observable phenomena. Theories, by contrast, are inferred explanations for observable phenomena.

---

[3] Vygotsky (1934/1986).

[4] Vygotsky (1934/1986, pp. 190–208).

[5] This is often the situation in many countries. Kampourakis (2017) pointed to the problem in a wider scope including faculty members in science departments.

[6] Clough and Olson (2004).

[7] We coded these items as L1–L5: L1, for Lederman et al., 2002; L2, for Lederman, 2004; L3, for Lederman, 2007; L4, Lederman et al., 2014; and L5, Lederman et al., 2015a, b. The list by Lederman and colleagues has been promoted since 1998 (Lederman et al. , 1998; Abd-El-Khalick et al., 1998). Erduran and Dagher (2014) mentioned other lists containing similar features.

3. The development of scientific knowledge involves human imagination and creativity.
4. The empirical nature of scientific knowledge. The distinction between observation and inference.
5. The social and cultural embeddedness of science. Scientific knowledge affects and is affected by the various elements and intellectual spheres of the culture.
6. The tentative nature of scientific knowledge: scientific knowledge is never absolute or certain. This knowledge, including "facts," theories, and laws, is inherently tentative or subject to change.
7. Myth of the scientific method (the absence of unique recipe-like stepwise procedure).

The context of this list may serve us as a stage in presenting critique of the subject. Even widely criticized,[8] the list (L-list) remains frequently discussed in studies, and its teaching is considered.[9] The L-list presents the only account of NOS in the *Encyclopedia of Science Education*,[10] which practically justifies the label of *consensus* view. The L-list reconstructed the traditional image of science[11] and of its method[12] presenting a strong shift. It challenges teachers and researchers. Let us list the reservations made to the L-list provided in science education studies.

First: It was mentioned that the all-sweeping claims about tentativeness of scientific knowledge and its subjectivity were oversimplified, unproductive, and inadequate.[13] They missed any difference between established and controversial ideas, naïve, and sophisticated tentativeness.

Second: The idea of a list of univocal statements regarding philosophical issues is not appropriate in light of the lack of consensus in the philosophy of science.[14] The list was criticized for its philosophical pitfalls, the superficial addressing of much-debated epistemological issues.[15]

Third: The list authors' widely used "assumption that NOS learning can be judged and assessed by the students' capacity to identify some declarative statements" is false. With regard to NOS standards, "it is unrealistic to expect students, or trainee teachers, to become competent historians, sociologists or philosophers of science." "Teachers should aim for a more complex understanding of science."[16] Matthews urged having "modest goals" when teaching about NOS.

---

[8] Millar (2000), Elby and Hammer (2001), Osborne et al. (2003), Clough (2007); Wong and Hodson (2009, 2010), Allchin (2011, 2013, 2017), Matthews (2012), Hodson (2014); Hodson and Wong (2017), Bazzul (2017), Berkovitz (2017), Dagher and Erduran (2017), and Wallace (2017).
[9] McComas (1998), Niaz (2009), Abd-El-Khalick (2012), Duschl & Grandy (2013), Erduran (2014).
[10] Gunstone (2015).
[11] Merton (1973).
[12] Hempel (1966).
[13] Elby and Hammer (2001), Matthews (2012).
[14] Alters (1997) examined the views of 187 members of the US Philosophy of Science Association.
[15] Matthews (2012).
[16] Matthews (2012), Allchin (2011).

Fourth: Several scholars mentioned that the L-list missed a panoramic image of science, which would include the features of mathematization, technology, modeling, etc.[17] This trend of thought produced the idea of the *Whole Science*.[18] Its author, Allchin, suggested science be addressed in three dimensions, 10 subdimensions, and 41 categories. However, this multitude makes it unfeasible to address the subject in an informative and clear way, at least in a regular teaching context.

Furthermore, Allchin argued that the knowledge of epistemological subtleties such as the differences between theory and law, model and principle, etc. is secondary in education. Instead, students need the ability to evaluate reliability of information and familiarize the practical meaning of science and critical thinking. He suggested performing "practical epistemic analysis" while learning historical cases of knowledge production, revealing tentativeness, and other numerous features.

Fifth: With the same intention to depict NOS beyond the L-list, an inclusive multidimensional approach of Family Resemblance (FRA) was suggested.[19] FRA rejects defining science. It tries "to overcome essentialism, the necessary and sufficient conditions" of being scientific. FRA considers *"science as a cognitive-epistemic system of thought and practice* on the one hand and *science as a social-institutional system* on the other." *Each of these two big aspects of science can be characterized in three dimensions of meaning.*

In effect, FRA becomes elastic and blurs the demarcation between science and nonscience. In contrast, cognitive studies advise emphasizing *difference* rather than *resemblance*, in grasping any subject matter by the learners.[20] To understand NOS, it is essential to emphasize the *difference* between astronomy and astrology, chemistry and alchemy, and science and creationism rather than their resemblance, which definitely exists (especially if science is vaguely defined as has been just mentioned). The abundance of misleading nonscientific publications makes it crucial to realize what science is in contrast to what it is not. For that, one needs to combine resemblance (agreement) with difference (disagreement) as practiced since medieval science philosophy,[21] but making the difference especially clear. They, in the past, did want to elicit the essential aspects of knowledge, clearly and distinctly.

Sixth: There was an effort to elicit NOS features from practicing scientists.[22] Hodson and Wong revealed features of NOS, which only partially coincided with the L-list. The researchers found that practitioners may hold fuzzy knowledge with regard to the status of basic concepts of law, theory, model, and principles that, as the authors inferred, did not affect successful functioning in research. This is reminiscent of another claim: scientists' views about science are rather productive than

---

[17] Matthews (2012).

[18] Allchin (2011, 2013, 2017).

[19] Irzik and Nola (2011, 2014).

[20] Marton and Pang (2006, 2013).

[21] Losee (1993).

[22] Wong and Hodson (2009) investigated 13 individuals from different domains of science. Their study can be representative of this trend of investigation.

correct.[23] Osborn[24] reminded Lakatos' extreme metaphor comparing scientists' knowledge about science to that of fish about hydrodynamics.

Hodson and Wong admitted that their sample was small and included practitioners who faced a number of questions about NOS for the first time. Not to forget, philosophy of science is not a requirement of scientists' training.[25] It is not clear, however, how one could expect inclusive and deep truth about NOS, beyond "it seems to me," on behalf of anybody without specific study and relevant background. Apparently, the context of science education is different from that of "doing" science. Education is different; it requires a more representative account of NOS. Observing the variety in their results and in those of others who also asked scientists,[26] Hodson and Wong dismissed the idea of a unique list of tenets. In their latest study,[27] they criticized the L-list as superficial and misleading. They called for the study of NOS through learning "from scientists, about scientists and with scientists."

Seventh: In still another study,[28] an alternative to the L-list, a list of six distinctive "styles of reasoning," was suggested. It mentioned essential characteristics of science having emerged from the cognitive history of science and having been shared across its different domains. His list, however, rather addressed the realm of methodology and ignored the essential features of science, the status of its constructs and its specific tools, needed for a representative image of science.

Eighth: Facing the complexity of the NOS subject and the controversy of perspectives, some researchers suggested inverting the items of Lederman's list from claims into questions, thus guiding a corresponding discussion and inquiry.[29] The following realizes this suggestion.

Alternative Perspective

Clarification may start from recognizing the advantage of the L-list. Indeed, a compact presentation of NOS features is not only valuable, but it is essential in making feasible its inclusion in the teaching–learning agenda in a practical sense. Yet, the critique of the L-list is too strong to be ignored. Therefore, the analysis will follow the items of the L-list, often drawing on the platform of discipline–culture[30] in considering the structure and meaning of knowledge components (theories, principles,

---

[23] Elby and Hammer (2001).

[24] Osborne (2017).

[25] Not in the past, it is true today.

[26] Osborne et al. (2003).

[27] Hodson and Wong (2017).

[28] Osborne (2017).

[29] Clough and Olson (2004), Clough (2007).

[30] Chapter 6.

laws, and models). Despite the frequent disparagement of "philosophical subtleties" by many practitioners,[31] education philosophers insist on the central role of addressing the meaning and status of scientific constructs, the structure of knowledge, and its epistemology.[32] Clear complementarity of the two approaches to scientific knowledge—the worldview versus the practical importance—has accompanied science from its dawn.[33] For a striking example, one may compare the epistemology of Newton as expressed in his *Principia*[34] with the Marxist analysis of Newtonian physics by Hessen.[35] This opposition is permanent: holistic conceptual understanding versus practical problem solving, theory-based ("worldview") versus modeling-based ("practical science") curricula orientation, realist versus constructivist understanding of theoretical terms, nominal versus operational concept definitions, and so on. It is a challenge for science teachers to be able to illustrate and passionately convince their students in the deficient quality of the knowledge, lacking these "subtleties": warning against the ignorance regarding the big picture of science and remaining with a puzzle of knowledge pieces, fragmentary instrumental understanding of occasional phenomena, and some technological gadgets focused merely on their usefulness.

To enrich the quality of education, NOS should be explicit in science curricula rather than left for discovery by the consumers. We believe that a *compact* account of NOS, highlighting its features, is the most effective way of appealing to a greater audience. Such was the L-list in the recent past. This list may serve as a starting point for its refinement, elaboration, and modification, while identifying the rationale of its claims. Where can guidance in this process be found?

Schwab[36] asked his famous "Who knows?" in relation to the structure of the disciplinary curriculum. He answered "Nobody," pointing to the enormous variety of parameters in considering substantive and syntactic knowledge. Indeed, looking for a single authority may confuse. Numerous bright minds from the distant and recent past held different and even contradictory views on science as if stating the "deep truth" (Niels Bohr) in this regard.[37] Indeed, similar to quantum objects, NOS understanding requires complementarity. Thus, in addressing science, philosophers

---

[31] Nobody checked it reliably among science practitioners, but it seems plausible and corresponds to the claim by Allchin (2011). It, however, contradicts the posture of Newton, Einstein, Heisenberg, Bunge, and Weinberg, for instance, as expressed in a variety of circumstances.

[32] Matthews (2009).

[33] One may start by pointing to the opposition of agenda between the Hellenic and Hellenistic sciences, in general terms, and of Aristotle versus Archimedes, in particular.

[34] Newton (1687/1999, pp. 27–30; 43, 585–589). For that confession and the incomparable product of *the Principia*, Edmond Halley wrote about Newton (p. 26): "No closer to the gods can any mortal rise."

[35] Hessen (1933).

[36] Schwab (1978).

[37] Bohr (1949, pp. 199–241) chose the claim "opposites are complementary" for his coat of arms. The deep truth was defined as the statement for which the opposite is also true. For instance, both statements "electron is a particle" and "electron is a wave" are true statements and present a deep truth.

are often cited in pairs: Aristotle–Plato, Descartes–Bacon, Galileo–Newton...[38] Each of these luminaries stressed a certain facet of science often (actually, always) problematic if taken univocally and literally.

Literal understanding may mislead. "Anything goes" by Feyerabend in his *Against Method* critique does not mean a lack of any methodology. *How the Laws of Physics Lie* by Cartwright does not mean that the laws of physics are untrue. "Science without laws" by Giere does not imply that one can manage without laws in physics. A close view shows a mere figurative nature of this rhetoric. Specific understanding is required. Thus, van Fraassen defined scientific knowledge as *anti-realistic* (being empirically verified but not literally considered).[39] Many practitioners would disagree with his label "anti-realist" taken verbatim in different sense but agree with his *constructive materialism* being introduced to their meaning. The difference between conceptual and material realisms is often not known to science consumers for whom the claim of science being anti-realistic may sound an oxymoron.

Similarly, with regard to educational psychology, Piaget, Skinner, Vygotsky, and others provided cognitive perspectives with apparently contradictory curricular implications. Educational practice suggests teachers draw on all of these in making pedagogical decisions in each and every educational context.

As already mentioned, asking practitioners of science about philosophical generalization regarding science is problematic. They may express views dependent on their specific expertise and background, which vary tremendously in the kind of experience, interest, success, and failure. HPS is normally not a requirement of science students, and so, unfortunately, its knowledge is usually a product of autodidactic effort among contemporary scientists. Therefore, episodic interviews without serious individual analysis prior to answering may provide biased, superficial, and misleading views that the individual would change after a more serious consideration.[40] Einstein's comment "Don't listen to their words, fix your attention on their deeds" may be helpful but not sufficient. In fact, practitioners are often not familiar with the epistemological discourse but may quickly be introduced into it being challenged by the claims regarding NOS, especially of extreme kind.[41]

What to do? One may listen mainly to those scientists who devoted special attention to such reflection and their time and specific effort. As it is well known, highly valuable information of this kind was provided by the prominent scientists who analyzed and articulated the features of scientific knowledge and its methodology— Galileo, Descartes, Newton, Duhem, Einstein, Bohr, Heisenberg, and Weinberg, to name but a few. In other words, asking "the horses" is a good idea, but for some questions, one should choose which horses to ask. Seemingly, for the questions regarding NOS, one should ask the charioteer.

---

[38] Losee (1993).

[39] Feyerabend (1993, p. 241); Cartwright (1983); Giere (1988); and van Fraassen (1980, pp. 8, 12).

[40] Einstein (1934/2011b).

[41] Wong and Hodson (2009); Kampourakis (2017).

Facing these considerations, science educators are better recognizing the range of argumentation in the cluster of univocal claims with respect to epistemological issues. The refinement is required in this sense. It should involve the variation of perspectives and their conceptual width where one may point to a certain feature as central and argue for such preference, while still preserving the range.

Our analysis will contextualize these claims in physics and mention, where relevant, the domains of chemistry and biology. This preference is justified by the increasing complexity of the systems, the number of their irreducible components (dimensions of freedom), and the factors of influence.[42] It is from this perspective that the following considerations arise.

Refinement of the NOS Features

The L-list of NOS features was not hierarchical. Therefore, in order to simplify our treatment, we have made a change in the original order—the forth claim regarding the *subjective* nature of scientific knowledge is addressed here first, after making it reversed—*objective* knowledge—both for its central importance and decisive implications for the rest of the features of science.

The Feature of Objectivity

The first and most problematic feature of the L-list is the claim of the *subjectivity* of the scientific knowledge. This claim initially appeared in L1 in the form "The theory-laden nature of scientific knowledge." Yet, in L2 and later, it appeared as "Scientific knowledge is subjective *and* theory laden," bringing subjectivity to the fore as in the following clear-cut statements:

> All these background factors form a mind-set that affects the problems scientists investigate and how they conduct their investigations, what they observe (and do not observe), and how they make sense of, or interpret their observations. (L3)

> Scientific knowledge, owing to (based on) scientists' theoretical commitments, beliefs, previous knowledge, training, experiences, and expectations, is unavoidably subjective. (L4)

Actually, the authors made two claims—subjectivity of the knowledge *and* its being theory-laden.[43] Firstly, they consider idiosyncratic features of knowledge

---

[42] Schwab (1978).

[43] Several authors, while citing the list, corrected this point without even mentioning the inaccuracy of the original claim. They stated subjectivity *as* being theory-laden. Here, we address both aspects separately as the L-lists deserve. The claim of *subjectivity of physics knowledge* contradicts the situation in introductory physics textbooks around the globe. We never saw such a claim in hundreds of physics textbooks examined and in equally numerous supporting materials.

production that make, in their view, scientific knowledge subjective. Indeed, people of different backgrounds and mindsets have constructed scientific knowledge. However, being a cultural product does not necessarily implies being subjective. Being objective normally presumes being independent of a subject (producer), independent of a voluntary change, being constrained by reality, fitting it, functioning independently of the creator, all together, as if it is "factual." Such status is traditionally ascribed to scientific knowledge.[44]

There is a special mechanism, a procedure by which fragments of scientific knowledge become objective. It includes (a) the hypothetico-deductive method; (b) the multisteps, iterative, reciprocal theoretico-empirical, many-faceted refinement, and verification; and (c) the intersubjective agreement reached through the cumulative dialogue of scholars causing multiple refinement and self-correction.[45] Scientific knowledge becomes an object possessing its own life.

In a sense, this procedure refines the statement by Russell: "The question whether objective truth belongs to human thinking is not a question of theory, but a practical question."[46] Kuhn cited the common view: "Objectivity enters science… through the processes by which theories are tested, justified, or judged."[47] He defined "subjective" as being a "matter of taste" and illustrated it by the view of Einstein on quantum mechanics, who ascribed to nature certain features as he believed they are. Kuhn mentioned that: "Einstein was one of the few, and his increasing isolation from the scientific community in later life, shows how very limited a role taste alone can play in theory choice."[48] Quantum mechanics was repeatedly proved as objectively representing the reality of the microworld.

Secondly, the claim of being theory-laden is separate and extremely important in itself since scientific knowledge is indeed *theory-based*. Physicists consider a big structure of science as composed of a few fundamental theories—widely inclusive clusters of internally coherent knowledge elements, related and hierarchically organized.[49] The use of the notion *theory* in this context is different from the meaning of theory as a counterpart to *practice*, *experiment*, and *experience*. Here, it is neither a synonym for *abstract* nor *hypothetical*, a consideration, making sense explanatory

---

[44] Losee (1993), Longino (1990), Couvalis (1997), Godfrey-Smith (2003, pp. 6, 229), and Scheffler (2009, p. 7).

[45] Some scholars do not grant scientific knowledge more than being intersubjective, which literally means being a product of agreement among scholars, community ("conventionalism"). We consider this feature insufficient, since *being agreed* does not presume *multiple empirical many-staged verifications* on which such an agreement draws in science and which present the core requirement of objectivity (Popper, 1975). The procedure of reaching objectivity must include all three aspects—a, b, and c—together.

[46] Russell (2009, p. 169). In this, Russell mirrored Marx's second thesis on Feuerbach in 1845.

[47] Kuhn (1977, p. 326).

[48] Though Kuhn (1977, pp. 336–337) stated that the adopted theory may preserve some idiosyncratic features. He rejected, however, the claim that he deprived science of objectivity in its "standard application", that is, as opposed to the "matter of taste," which implies being subjective and undiscussable.

[49] Heisenberg (1959/1971); Bunge (1967a, 1973); Weizsacker (1985/2006).

model. In contrast, theory in science can signify a different construct, which is fundamental for science. This use as a structural whole is close to that described by Giere[50] and is shared by science and the philosophy of science.[51] In fact, Heidegger *defined science* as a theory of actual reality[52] implying natural science to be a *theory of nature*. Though the latter definition may include any organized thinking about reality (such as the traditionally developed rules of constructing ships), the addressing theory as a very inclusive cluster of knowledge elements, related in a unifying picture of the world, is fundamental in science. In any case, we need to exemplify this meaning.

Theories establish science as a "systematic knowledge of subject matter" also in science curriculum.[53] However, a mere dependence on a theory does not imply subjectivity of science. This is because scientific theories can be objectively verified.[54] Such are the fundamental theories of physics: classical mechanics, thermodynamics, classical electrodynamics, quantum physics, and the general theory of relativity—all objective theories of nature (Fig. 9.2).

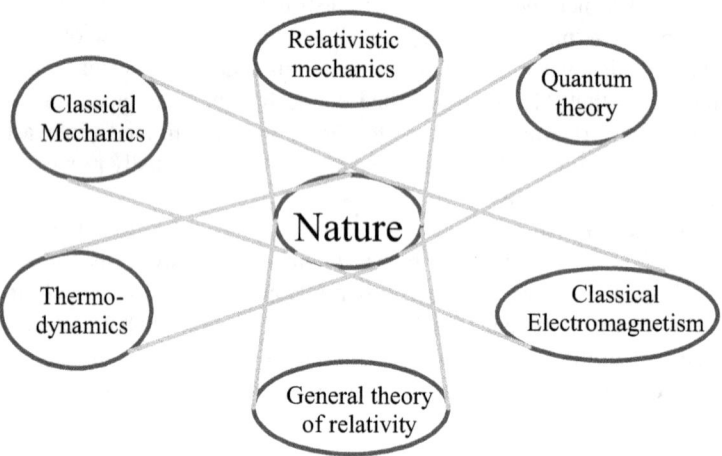

Fig. 9.2 Fundamental theories of physics objectively represent the actual reality of nature in its different but unifying aspects, the "Pictures of the World"

In teaching the subject of NOS, there is a need to depict how human knowledge can be objective and how it may lose its idiosyncratic status and becomes impersonal. After being developed by, say, Aristotle or Newton, a theory is detached from

---

[50] Giere (1985, p. 16; 1999, pp. 97–99).

[51] Chalmers (1976, Ch. 7, 8).

[52] Kockelmans (1985, p. 162).

[53] Schwab (1978).

[54] Couvalis (1997, p. 12), Hempel (1983).

them and proceeds with its independent life. Popper epitomized this fact by introducing the concept of "the third world"—a virtual space of cultural products and collective knowledge. In that world, scientific theories present the objective knowledge, free of supernatural and voluntary references. He argued:[55]

> ...it is possible to accept the reality or (as it may be called) the autonomy of the third world, and at the same time to admit that the third world originates as a product of human activity.

Scientific theories are often credited with the virtue of being "universal." The latter corresponds to the idea of being invariant and deserves clarification. Indeed, a scientific theory holds everywhere, at any time and regardless of who applies it. Yet, each scientific theory depicts nature only in certain aspects, perspective, creating a partial, specific picture, valid in a well-defined domain of parameters. Therefore, we have several fundamental theories, but none of them is all inclusive.

The superficial claim that scientific knowledge is subjective breaks with the tradition and norm of science and science education. It commonly surprises science teachers and contradicts with the values, which they normally hold and represent in science classes, in possible contrast with some content of other disciplines such as humanities (history, philosophy, literature, arts, etc.). In the vision of different philosophers, however, science ranges from the objective to a socially determined construct. To prevent confusion, this implies the need for teachers to explicitly determine the meaning of objectivity and be ready to argue for that in class instruction addressing the nature of science.

Popper clarified that only objective knowledge can be a subject of critical discourse and progress.[56] Historians assert that practitioners normally consider scientific knowledge objective and distinguish it from other types of knowledge systems practiced in society.[57] Physicists, however, seldom talk about objectivity, apparently taking it to be a presumed norm.[58] Still, when asked, they normally firmly argue for the objectivity of collective knowledge being proved quantitatively in a variety of experiments by numerous independent scholars and by the striking success of technology, no matter who uses it, correct predictions, and no matter who makes them. All this has nothing to do with recognized limits of power and of areas of applicability of each scientific product. Scientists illustrate the acquired objective knowledge of the natural world by the ability of the quantum field theory to reach the amazing accuracy of up to 10^{-11}—an evidence of the highest validity.[59] Weinberg argued:[60]

---

[55] Popper (1975, Part 4, p. 159).

[56] Popper (1975). In contrast, one may see such areas as religion where *to be subjective* is presented as a goal of mature knowledge (Kierkegaard, 2009). The same can be said with regard to the art of teaching science but not to the knowledge which pretends to be scientific.

[57] Holton (1985), Shapin (1996); Nozick (2000), and Agazzi (2014).

[58] di Francia (1976); Wong and Hodson (2009).

[59] For instance, in calculating the magnetic moment of electron Feynman (1985/2014, p. 7), Sokal & Bricmont (1998, p. 57).

[60] Weinberg (2001, pp. 91–92).

We [scientists] believe in an objective truth that can be known, and at the same time we are always willing to reconsider, as we may be forced to, what we have previously accepted.

That is, being subject to correction, improving, and replacement does not change the status of objectivity. He proceeded:[61]

I have come to think that the laws of physics are real because my experience with the laws of physics does not seem to me to be very different in any fundamental way from my experience with rocks. For those who have not lived with the laws of physics, I can offer the obvious argument that the laws of physics as we know them work, and there is no other known way of looking at nature that works in anything like the same sense.

His summary was:

. . . there is an essential element in science that is cold, objective, and nonhuman . . . the laws of nature are as impersonal and free of human values as the rules of arithmetic . . .

Yet, *Stanford Encyclopedia of Philosophy* starts from the definition:[62]

It [scientific objectivity] expresses the idea that the claims, methods and results of science are not, or should not be influenced by <u>particular perspectives</u>, value commitments, community bias or personal interests, to name a few relevant factors. (Emphasis added)

The reference to a "particular perspective" as a feature of subjectivity (similar to that stated in L3) is not the best expression here. The fundamental physics theories do provide *particular perspectives*, which still constitute objective knowledge. They guide our successful interaction with reality, the planning and production of industrial products, making correct predictions. Within the scientific perspective, empirical data are interpreted and attain specific meaning.[63]

Popper followed Kant in his understanding of objectivity:[64]

... scientific knowledge should be justifiable, independently of anybody's whim: a justification is 'objective' if in principle it can be tested and understood by anybody. 'If something is valid', he [Kant] writes, 'for anybody in possession of his reason, then its grounds are objective and sufficient.'

Carnap was more precise:[65]

...knowledge is objective insofar as it is limited or constrained by certain conditions (the facts) over which arbitrary will has no control.

The idea of science as the objective knowledge of nature was introduced in the epistemological evolution, which replaced myths with the idea of cosmos—the universe organized in a stable order, independent of any personal will. For that, Carnap praised Democritus:[66]

---

[61] Weinberg (2001, p. 151).

[62] Reiss (2014).

[63] We address here neither the truth of the theory nor the certainty of scientific knowledge but its objective nature as impersonal and involuntary.

[64] Popper (1959/2002, p. 22).

[65] Weinberg, J. (1936, p. 215).

[66] Carnap (1971, p. 206) .

> *Democritus, for example, regarded the regularities of nature as completely impersonal, not connected in any way with divine commands. ... this step from the personal necessity of divine commands to an impersonal, objective necessity was a great step forward.*

Russell summarized his rejection of subjectivity of scientific knowledge in the form difficult to compete with:[67]

> *There is a widespread philosophical tendency towards the view which tells us that Man is the measure of all things, that truth is man-made, that space and time and the world of universals are properties of the mind, and that, if there be anything not created by the mind, it is unknowable and of no account for us. This view, if our previous discussions were correct, is untrue; but in addition to being untrue, it has the effect of robbing philosophic contemplation of all that gives it value, since it fetters contemplation to Self.*

However, within meaningful education, there is a special need to address the difficulties of recognizing objectivity as the essential virtue of scientific knowledge and illustrate them in the context of science curriculum.

Deception by Perception

The first challenge to objectivity is deception by senses. It is good to recollect that Greek philosophers talked about contemplation rather than observation in natural philosophy. In other words, their looking was interwoven with analysis of the observed. In 1670, Newton worried about the validity of his claims regarding colors. What troubled him was whether colors obtained behind a prism were innate to sunlight.[68] Newton realized that white color can be reproduced by combining some (at least three), not necessarily all, colors of the spectrum.[69] The reason for this was shown later by Young, who in 1801, introduced a theory of color *perception* based on the three basic colors that corresponded to three types of color receptors in the human eye.[70] According to that understanding, Young separated the *objective* (observer independent) color from the *subjective* (psychological, observer dependent) one. Similarly, nobody relies on the sense perception of heat to measure temperature or on vision for detection of light. Generally, science does not rely on senses and observation, but transcends it through using instruments, which probe, measure, and monitor objective reality in a variety of ways. Perceptual evidence of physical phenomena is neutralized or refined, normally being replaced by instrumental measurement prior to inferences within the construction of objective knowledge. Strengthening of the *objectivity* by the transition from observation to measurement was behind the scientific revolution of the seventeenth century.[71] We have illustrated it considering the difference between Leonardo, as the apex of the

---

[67] Russell (1912/1990).

[68] Shapiro (1984).

[69] Mach (1913/1926, p. 97).

[70] RGB in modern terms.

[71] Koyré (1968).

premodern science, and Galileo, as the pioneer of the modern science.[72] This same *revolution of the objectivity of scientific knowledge* is traditionally and continuously imitated by teachers in our science classes starting from the very beginning—the elementary school.

Representation versus Interpretation

One can distinguish between the representation and content depicting specific features of the *objective* reality and its interpretation. Consider the law of light reflection (the equality of the coplanar angles of incidence and reflection —a mathematical statement). Over the course of history, this law was given different interpretations. While its empirical content remained, its interpretations varied from Heron's idea of nature seeking the shortest path, to Fermat's interpretation of the path *extremal* in time. Then, Newton, keeping in mind light particles, interpreted the same form of the law as conservation of the momentum component parallel to the surface, while Huygens[73] proved the same regularity of the light path by interference among the secondary waves. Feynman[74] transferred the wave account to the interference between the waves along the multiple paths involved in the virtual motion of a single photon. Some of the interpretations were correct, some were not, but they all were different.

Over time, the objective empirical law of reflection obtained accounts in three theories of light (geometrical optics, physical optics, and quantum optics)—all three are correct in the realms of their validity.[75] The mathematical form of the relationship between physics quantities was not only more compact and numerically accurate but also neutralized various unsupported speculations. Newton understood this advantage and tried his best in providing physical theory with mathematical form.[76] He wrote:[77]

> But truly with the help of philosophical geometers and geometrical philosophers, instead of the conjectures and probabilities that are being blazoned about everywhere, we shall finally achieve a natural science supported by the greatest evidence.

The mathematical account of motion,[78] internal energy, heat and work, entropy, electric and magnetic fields, and quantum wave function, all revealed aspects of certain objective content of empirical reality whose meaning were clarified within the established theory. Even more, two different mathematical forms could appear

---

[72] Chapter 8, Table 8.2.

[73] Huygens (1690/1912, pp. 22–28).

[74] Feynman (1985/2014, pp. 37–47).

[75] Chapter 7 .

[76] Shapiro (1984).

[77] Newton (1670). We quote from the article by Shapiro (1984) depicting Newton's Optical Lectures of 1670–1672.

[78] Chapters 2 and 8.

later as equivalent in representing the same objective reality, as happened with the accounts of quantum phenomena by Heisenberg and Schrodinger within the quantum theory. They were different in form but equivalent in the objective essence of reality.[79]

Scientists joke that equations are smarter than people in their coverage of objective reality. Dirac's equation in relativistic quantum mechanics described the electron and unexpectedly provided additional states with negative energy. They led imagination of scientists to interpret them as an evidence for antiparticles of electrons—positrons, which were never observed until then. Since then, the *objective* existence of antielectrons was experimentally proved: antiparticles were discovered.

Objectivity versus Correctness

It is common to equate objectivity and correctness. While both definitely present the goal of scientific knowledge, these characteristics do not coincide. Showing that a certain scientific theory was imprecise does not imply making it subjective. This is because being adopted by science presumes certain impersonal, independent claim about nature, its phenomena. The same claim could be later recognized as limited, less accurate, less effective, even erroneous, and replaced by other knowledge.

The geocentric theory of the solar system by Ptolemy and geostatic theory by Tycho Brahe were dismissed by Galileo, Kepler, and Newton. Yet, these theories were objective, providing accounts of a certain appearance, independent of anyone's will. They provide account of the reality by anybody who used them. Thus, geocentric models remained valid, and they are still used in everyday navigation depicting motion in the geocentric frame of reference.[80,81] With the coming of Newtonian dynamics, Kepler's laws were corrected: the point at rest in the Sun–planet system is not the Sun but the center of masses of the two objects. This correction does not change the objective nature of Kepler's laws.

Scientists suggest different working models of reality, which vary in correctness and accuracy, providing accounts of certain aspects of nature. In the theory of electromagnetism, we use models of fluids and gas for the electrical current.[82] Each model is successful in some aspects but fails in others. Both are objective, however. Ampere introduced microscopic electrical currents, never observed, but effective.[83] They supported mathematical accounts and explanations for several physical phe-

---

[79] Dirac (1958, pp. 108–116).

[80] A very simple example: the claim that the "Sun is rising in the East" is an objective claim. Its correctness, however, depends on the frame of consideration, geocentric or heliocentric, both legitimate.

[81] We use the notions of theory and model within the discipline–culture framework (Chap. 6). As such, theory includes models of different kinds and affiliation (Fig. 6.7). Thus, the geocentric theory includes various models.

[82] Knight (2013, p. 877); Steinberg and Wainwright (1993).

[83] Darrigol (2000).

nomena. Over time, the limits of validity of this theory were revealed, a new theory of quantum electromagnetism was introduced, but the theory of classical electromagnetism has remained effective and objective. The model of a current within an atom faithfully explains an orbital magnetic moment, though there is no current in the simple sense in an atom. In considering atom as quantum object, the classical model of current is incorrect. Thus, scientific correctness is distinct from objectivity. The opposite of falsity is correctness, not objectivity.

Context Dependence

Scholars have indicated the essential difference between the regime of thought of the researchers in which they develop knowledge and that where they argue for its adoption. While knowledge construction may involve subjective ideas and beliefs and reflect personal ideas, environment, and skills, the subsequent adoption of that knowledge requires argumentation and empirical verification of an objective nature.[84] Reichenbach insisted that scientific epistemology and method cannot deal with the act of discovery but only with the procedure of justification. The latter does not, or at least need not, involve subjective factors. Laudan introduced a parallel pair, *pursuit* and *acceptance*, which is not different in meaning.[85]

Losee traced the recognition of the *two contexts* to John Hershel, who was impressed by Ampere's new theory of electromagnetism.[86] Ampere did not draw on any specific data or inductive scheme but on his heuristic invention of circular currents in the magnetized matter, which were much later proved to be real. Similarly, Maxwell derived his fundamental equations of electromagnetism using the heuristic model of elastic medium, which appeared to be merely imaginary.[87]

After theories are introduced, they detach from their creators in the process of justification, analysis, development, and testing by other researchers on the way to becoming collective objective knowledge—an item in the third world of Popper.[88] Such justification is not a single ("crucial") experiment but includes a variety of

---

[84] Reichenbach (1938, pp. 6–7, 381–382). Goodman (1968/1976, p. 251) put it as follows: "Indeed, in any science, while the requisite objectivity forbids wishful thinking, prejudicial reading of evidence, rejection of unwanted results, avoidance of ominous lines of inquiry, it does not forbid use of feeling in exploration and discovery, the impetus of inspiration and curiosity, or the cues given by excitement over intriguing problems and promising hypotheses." Nersessian (1992) termed this stage as *the context of development*.

[85] Laudan (1977), Godfrey-Smith (2003, pp. 108–109).

[86] Losee (1993, pp. 121–126).

[87] Nersessian (1992); Darrigol (2000, pp. 149–151). In contrast, Duschl and Grandy (2013) stated that the two contexts might be interwoven: "What occurs in science is neither predominantly the context of discovery nor the context of justification but the intermediary contexts of theory development and conceptual modification." However, even their being interwoven does not dismiss the high validity of recognizing the two aspects of knowledge creation as different with respect to the status of objectivity.

[88] Popper (1975, 1978).

experiments by different researchers. It is a long process of rational elaboration, which, in the view of Toulmin, invalidates the idea of *revolution* in science in favor of its *evolution*.[89]

For instance, Einstein's theory of relativity was not justified by Michelson–Morley or any other single experiment, but by a whole list of experiments and following implications continuously verified Einstein's theory through experience.[90] The process of justification included the contest of six parallel theories: the ether theories of Lorentz and the ballistic theories of Ritz.[91] Einstein's theory surpassed Fresnel's explanation of ether drag in accuracy.[92] The subjective inquiry of Einstein, his mental images of rods, and clocks in a spacious lattice were left behind.[93] From 1902 to 1909, Einstein's inquiry was out of the academic mainstream,[94] while the process of justification moved him to the center of scientific discourse, theoretical and experimental, eventually providing his theories with the status of objective knowledge about space-time. Perhaps the most convincing justification of objectivity of the theory of relativity was the atomic power technology and weaponry developed independently by many researchers in several very different countries.

Neglecting context can mislead. In our refinement of the L-list, it is appropriate to mention Hodson[95] who quoted Mitroff's depiction of science in terms of *particularism, solitariness, interestedness*, and *nonrationality* as better characterizing the reality than the *universalism, disinterestedness*, and *rationality* proclaimed by Merton.[96] In close look, however, one may reveal that the argumentation provided by Mitroff rather addressed the aspects appeared in the context of inquiry.

Objectivity of Modern Physics

Modern physics, the theory of relativity and especially the quantum theory changed the status of the *observer* in physics theories, bringing it to the fore. Galileo, Kepler, Descartes, and Newton addressed reality presuming no intrusion by an observer, who was considered as a perturbation factor commonly causing distortion of the appearance of its law-like regularity. Such disturbance could be taken into account

---

[89] Toulmin (1972, p. 105).

[90] Panofsky and Phillips (1962, p. 240) illustrated the process of justification of the special theory of relativity. In event, five of the rival theories successfully accounted for the zero result of the Michelson–Morley experiment, but only Einstein's theory could explain all 13 different experiments performed by different researchers in different environments. Actually, criticism of the Einstein theory of relativity never stopped.

[91] Miller (1981, pp. 25, 280; 1986, p. 2).

[92] French (1971, pp. 131–132).

[93] Einstein and Infeld (1938), Miller (1984), and Granek (2006).

[94] Miller (1996).

[95] Hodson (2011, pp. 111–112).

[96] Merton (1973).

and ignored in an ideal environment and "correct perspective."[97] This vision became a subject of cardinal revision in the twentieth century. Multiple observers (frames of reference) and quantum measurement raised the doubt of subjectivity—"all is relative," "all depends on an observer," and so on. It appeared, however, that the new physical theories remained objective in the strict sense.

The theory of relativity draws on the strict rules of invariance, relating between all the accounts by different observers. A new type of absolute quantities was introduced and became central in the theory of relativity, those preserved under the Lorentz transformations. In the quantum theory, an observer may choose the kind of macroscopic apparatus and physical environment, which determine possible states of a micro-object (as in classical mechanics). The new and subtle feature of micro-world reality—the *probability* of measurement results—is observer-*independent*. As Dirac explained, that probability is made by Nature and not by the observer.[98] Modern physics thus remains objective in providing knowledge about reality, though different from the classical account.[99]

Summarizing our refinement of this item in the L-list, we argue for the objective nature of scientific knowledge, though subjective elements may be present in knowledge construction and ongoing research. Einstein expressed this aspect as follows:[100]

> Science as something <u>existing and complete</u> is the most objective thing known to man. But science in the <u>making</u>, science as an end to be pursued, is as subjective and psychologically conditioned as any other branch of human endeavor...

Though the objectivity of science is a philosophical and scientific commonplace and is the norm when defining scientific knowledge, there is a need for refinement of this claim in order to prevent its misinterpretation in education. In the contemporary environment of postmodern deconstruction, the question of the validity of the fundamentals of science frequently emerges. Despite possible confusion, this is a positive development for stimulating clarification of sometimes complex settings of social and individual knowledge creation. It leads to better understanding. People tell about Niels Bohr who showed satisfaction when realizing existence of a hidden mistake in a problem solution, exclaiming "now we have a chance to genuinely understand the problem."

---

[97] For example, the struggle with friction by Galileo (Chap. 2) and the rejection of the account by a non-inertial observer by Huygens (Chap. 4).

[98] Bohr (1949, p. 223) and Cushing (1994, p. 178).

[99] There is extensive discussion of the objective nature of quantum mechanics (Heisenberg, 1967; Popper, 1967; Bunge, 1967a; Cushing, 1994; Agazzi, 2014).

[100] Einstein (1934/2011a, p. 57).

Scientific Laws and Theories

The following item of the L-list deals with the concepts of theory and law. L2 determined:

> Laws are statements or descriptions of the relationships among observable phenomena. Boyle's law, which relates the pressure of a gas to its volume at a constant temperature, is a case in point. Theories, by contrast, are inferred explanations for observable phenomena (e.g., kinetic molecular theory provides an explanation for what is observed and described by Boyle's law). Scientific models are common examples of theory and inference in science.

The provided definition of laws is deficient, being not fully representative. Firstly, as Mach stated:[101]

> All general physical concepts and laws, the concept of a ray, the laws of dioptrics, Boyle's law and so on, are obtained by idealization.

It is rather evident that in practice, we have a more or less acceptable approximation of the claimed functional dependence and there is often the need of asymptotic consideration to state the law in its ideal form. Thus, Galileo never observed *simultaneous* falling of bodies with different masses (experimentation in a vacuum was not available) but derived his law through asymptotic consideration.[102] A similar procedure is required to reach the law of inertia and others.

Secondly, the concepts related through physics laws are not necessarily observable/measured directly as in the Boyle law. Such are, for instance, *energy* in the law of its conservation, *entropy* in the Boltzmann law, *electromotive force* in the Kirchhoff laws, and *internal energy* in the first law of thermodynamics. In quantum mechanics, wave function is a nonobservable quantity, and it determines observable ones in the law-like rules for transition probabilities, particle interference, etc. "Theoretical laws" are theory products, and they are not solely about observables.[103] Moreover, besides concepts, the *units* used to measure physical quantities do not draw any more on the directly measured *kg, m,* and *sec.* They all have been elicited through sophisticated *theoretical* accounts from the world constants h, c, and e considered now as fundamental. Therefore, the L-list claim about the nature of laws in science has to be corrected.

The meaning of the term "theory" in science is twofold. Only the first has been addressed in the L-list that of the opposition theory-practice. This meaning is common in everyday use, but it is not unique. The list ignores the meaning of theory in science as an inclusive cluster of coherent and hierarchically arranged knowledge elements.[104] Starting from Plato, creating theories became an agenda of scientific exploration. From then on, they present the fundamentals of science. The ancient Greek term *theory* is elucidating the meaning of the term in science. It emphasizes

---

[101] Mach (1976, p. 140).

[102] Galilei (1638/1914, p. 116).

[103] Carnap (1971, p. 305), *Wilczek* (2004).

[104] Kuhn (1969) used the terms *constellation* or *disciplinary matrix* when he addressed a theory.

the replacement of reality by its representation as a *theory*, as if it happens in a *theater*. As mentioned already, Heidegger defined science (the whole of science!) as a theory of what is actually real[105] implying natural science to be a theory of nature.

Indeed, knowing the theory helps the understanding of its laws. Even so, to define a theory solely as an explanation is not sufficiently representative. We derive the law of mechanical energy conservation and the law of pendulum from Newton's second law, but this derivation does not explain those laws conceptually. Empirical laws such as Hooke's law of elasticity, da Vinci's law of sliding friction, and Ohm's law of electrical circuits are not explained by the theories of mechanics and electricity, but belong to them, respectively.

Each fundamental theory of physics presents a "picture of the world" from a specific perspective (Fig. 9.2). Each such theory is a very inclusive cluster, incorporating numerous elements—principles, laws, concepts, models, phenomena explanations, solved problems, experiments, and specific apparatus—in a self-consistent system.

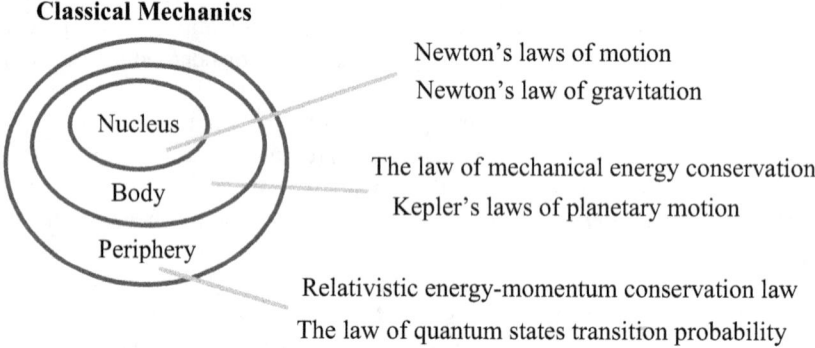

Classical Mechanics

Nucleus

Body

Periphery

Newton's laws of motion
Newton's law of gravitation

The law of mechanical energy conservation
Kepler's laws of planetary motion

Relativistic energy-momentum conservation law
The law of quantum states transition probability

Fig. 9.3 Examples of law affiliation in the theory of classical mechanics structured as a discipline–culture

Beyond consistency, a scientific theory is also hierarchically ordered. One may identify fundamental laws, principles, and basic concepts—the *nucleus* of a theory.[106] Other elements subdued to the nucleus would present its *body*. For instance, if the nucleus includes Newton's Law of Gravitation, the body includes Kepler's laws of planetary motion, which are derived from the law of gravitation, and if the nucleus includes Newton's Laws of Motion, its body includes the work–energy theorem and the law of conservation of mechanical energy (Fig. 9.3).

---

[105] Kockelmans (1985, p. 162).

[106] Placing principles in the nucleus of a theory is a striking feature of Einstein's Theory of Special Relativity (Einstein, 1949/1979, pp. 52–53, Darrigol, 2000, p. 383).

As already mentioned, the important feature of a physical theory is that it is not valid universally but only within a well-defined range of parameters (length, time, mass, particles, and type of interaction)—the area of validity. Thus, classical mechanics is not valid for black holes where the theory of general relativity holds. Yet, it does *not* imply that physical theories (and laws) "lie"[107] nor that physics knowledge presents a "patchwork," as Cartwright expressed.[108] Giere wrote:[109] "Close inspection, I think, reveals that they are neither universal nor necessary – they are not even true." Physics laws do not "lie" but are valid, each in the particular area of validity.[110] The patchwork metaphor is inappropriate because the theory of general relativity is valid in the area where classical mechanics holds but not vice versa. Newton's law of gravitation works on the leaning tower in Pisa but not in quasars, while Einstein's theory of gravitation holds in both cases. Quantum mechanics works in both micro- and macroworlds, while Newtonian mechanics does not hold in the microworld. In short, the scenario of a simple division of areas of validity does not properly represent the nature of science. In Einstein's view, the splitting of one picture of the world to several theories is "not a matter of fundamental principle."[111]

Consider the domain of optics. To account for image creation in a simple microscope, one may use geometrical optics, or the theory of light rays, and correctly predict the location and size of the image. However, the gradually increasing magnification blurs the image when it approaches the limit of resolution, indicating the invalidity of the ray theory and the need for the wave theory. The wave theory not only provides the new accounts but also reproduces all the results of the ray theory regarding image creation.[112] Furthermore, by gradually reducing the light flux to the very low intensities, one distorts the image again. This time, it splits into light spots,[113] indicating leaving the area of validity of the wave theory and the need for yet another light theory, the quantum theory of light, and the theory of photons. Thus, the meaning of *universality* with respect to scientific laws and theories is of a specific kind.

The relationship between fundamental theories can be represented by placing the nuclei, say, of Einstein's theory of relativity and quantum theory in the periphery of classical mechanics and the nuclei of the ray and photon theories in the periphery of the wave theory of light. This model alternatively represents the traditional claim[114] that a more advanced theory (such as special relativity) subsumes, under ceteris paribus reservations, its predecessor (such as the theory of classical mechanics), and

---

[107] Cartwright (1983).

[108] Cartwright (1994).

[109] Giere (1988, p. 128).

[110] Heisenberg (1948).

[111] Einstein's (1918/2002).

[112] This approach is termed Abbe optical theory (Hecht, 1998, pp. 602–604).

[113] Tipler (1987, pp. 184–186) and Serway et al. (2005, pp. 180–182). The reference to the original experiment by Taylor (1909) is rare (Rabinowitz, 2017).

[114] Nagel (1961).

visualizes *incommensurability* of the theory fundamentals in parallel with *commensurability* of certain numerical accounts of the two theories.[115]

Organization of knowledge elements in the tripartite structure—nucleus, body, and periphery—can represent the normal scientific practice and "visualize" the dynamics of theories' exchange in scientific revolutions.[116]

Theory structure incorporates laws.[117] When laws belong to a nucleus (such as Newton's laws in classical mechanics), they are used as principles.[118] Theoretically established laws of the body of knowledge may be explained in the sense of being reduced to (or derived from) the elements of nucleus (e.g., energy conservation, laws of projectiles, planetary motion laws, and the law of pendulum). Empirical laws, such as da Vinci's law of sliding friction and Hooke's law of elasticity, are affiliated to the body knowledge of mechanics but are not explained by it. Such elements as friction, elasticity, nonconservative forces, and Ohm's laws are formally irreducible to the nuclei axioms of mechanics and electromagnetism. They appear as *emergent* properties, empirical laws. They indicate the conceptual incompleteness of the particular theory but do not prevent its validity in the certain area of parameters.

We have elaborated on the law–theory relationship in the context of physics, but one can extend this relationship for natural sciences in general. While physics deals with the "simplest" material systems, it projects its knowledge to chemistry and biology treating more complex systems (greater number of parameters). Chemical laws (such as the law of periodicity in the interatomic interaction activities of elements) can be demonstrated from the electronic structure of atoms (physics), while the regularities in biological cells are explained using molecular genetics (biology).

We teach theories in class not in the form that these theories were historically introduced. Indeed, Newtonian mechanics did not include energy, Mendeleev's periodic law did not draw on electronic structure, and Darwin did not justify selection of species by genetic rules. Moreover, chemistry knowledge includes historically achieved *empirical* laws, which appeared to have exceptions and so are not universal.[119] However, the pattern of discipline–culture for a *fundamental* theory structure remains valid, together with the idea of a scientific law being included into a theory. Chemistry curriculum may affiliate the periodic law and the matter conservation principle with its nucleus. Then, the specific regularities of chemical reactions can be ascribed to the body of knowledge, and the alternative classifications of elements (by Meyer in groups of valences, by Lavoisier according to unbroken in reactions substances, including light and caloric, and the alchemic "principles"[120]) will be identified with the periphery.

---

[115] Chapter 6.

[116] Chapter 6.

[117] Duhem (1905/1982).

[118] Giere (1995).

[119] Dagher and Erduran (2014); Erduran (2014).

[120] For example, Pattison Muir (2004, Ch. 5).

Moreover, placing to the periphery, the *Table of Affinities*, which was the central element of chemistry—the theory of substances—of the eighteenth century,[121] may by comparison elucidate students' understanding of the nature of chemistry as a scientific discipline and the meaning of the scientific revolution, and the strong paradigmatic shift, which brought the chemistry that we possess and so widely practice nowadays. This curricular step would culturally enrich chemistry students bridging them to the ideas of Descartes and Newton regarding interaction, symbolism, and the concepts and principles of chemical reactions contrasting those they learn today in isolation and so promoting their meaningful learning.[122]

In biology, one may also arrange the theory of evolution in the tripartite structure, placing the Darwinian natural selection in the nucleus, while the alternative claims of Lamarck regarding the nature of selection (and the ideas of catastrophes and creationism) can be affiliated to the periphery.[123] Similarly, the presentation of molecular biology as a fundamental theory would have the central dogma of DNA-based mechanism of transferring information of heredity in the nucleus of the theory, specific examples of such mechanism, in its body, and the alternative ways of information transfer, epigenetics—in the periphery. One may consider the representation including periphery as a recommended "dissolving dominance" of the central dogma valid for teaching.[124]

The character of "laws" may change with the increase in system complexity (additional dimensions of freedom), allowing exceptions in chemistry and even more so in biology. Complex biological systems result in stochastic emergence over a huge time span. Fluctuations may be preserved in specific environmental conditions of isolated pockets. This, however, does not imply the dismissing of laws in biology for their wide validity providing at least a probable account.[125] Presenting the theory–laws relationship within a DC structure can support displaying NOS in biology classes, bridging their students to other sciences. One may then use such discussions in considering the similarities and differences within the family of natural sciences. The latter is a rare event in science classes.

To summarize, regarding this item of the L-list, the difference between law and theory should be revised in science teaching. One may do it at various levels of simplicity while avoiding oversimplification. Such revision is feasible in high school instruction as was demonstrated in the first attempts with regard to optics and mechanics.[126]

---

[121] Encyclopeadia Britannica (1771/1979, Vol. 2, pp. 66, 166–168).

[122] Partington (1962, pp. 52–55) and Hudson (1992, pp. 49, 102).

[123] Greighton (1999, p. 3810), Garvey (2007), and Frank-Kamenetskii (2013).

[124] Allchin (2013).

[125] Garvey (2007) and Frank-Kamenetskii (2013).

[126] Levrini et al. (2014) and Goren and Galili (2018).

The Involvement of Creativity and Imagination

In L2, the authors stated:

> *Even though scientific knowledge is, at least, partially based on and/or derived from obser-*
> *vations of the natural world (i.e. empirical), it nevertheless involves human imagination*
> *and creativity*

Although this statement regarding NOS looks simple and obvious, a refinement may make it more informative, less trivial, and specific for science. Einstein stated that concepts and theories are "free inventions of the human spirit."[127] Yet, he specified the place for imagination among other activities, which included *testing* the products of imagination—the invented constructs—in order to verify whether they fit the objective world on the way to their adoption or refutation (Fig. 9.4).

Fig. 9.4 The cycle of concept introduction and verification in scientific practice as explained by Einstein

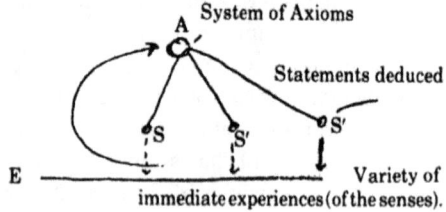

For Einstein, imagination is guided by reality and participates in a self-correction cycle of knowledge production: the intuitive move from experience (E) to creating fundamental axioms (A) and the rational deduction to statements (S), checking these statements against experience and returning to further consideration. Is this process reminiscent of the Aristotelian research cycle[128] and the procedure of deduction from phenomena which Newton contrasted against "hypothesizing" on possibilities?[129] This circle continuously repeats itself as an iterative procedure, making it spiral, that is, developing. In it, human imagination and creativity are empirically tested against the objective reality in a self-regulative process of scientific theory creation. In this, the imagination in science (destined to be meticulously tested) is different from that in poetry where it remains a mere fantasy appealing to spiritual cognition. One may ask about the engine that put this wheel into motion. Aristotle thought it was curiosity about Nature specific for humans, but it can also be a simple advantage of possessing an account for a situation, a pragmatic need for a problem to be solved. The range is determined by the knowledge already possessed by an individual and community.

---

[127] Einstein (1952/1987).

[128] Losee (1993, p. 6).

[129] Shapiro (2004, p. 188).

At the same time, however, one may and should realize the need to complement human imagination and creativity with the equally correct idea of the "economy of thought" suggested by Ernst Mach,[130] implying scientific knowledge as a product to be consumed without repeating an in-depth analysis in each case, regardless imagination. This reality allows a whole industry of problem solving and actually enables action in science (as well as in technology). Thus, using Kepler's laws of planetary motion for solving problems in astrophysics and space technology does not require the ability of their consumers to derive them from the first principles of Newtonian theory. One can design a flying apparatus without knowing the proof of the Bernoulli and Stocks equations from theoretical aerodynamics; one can dismiss a supposed perpetuum mobile without the ability to demonstrate the laws of thermodynamics. Of course, we are talking here about the routine application and not about tackling with the novel situation. Imagination and creativity do not remove the economy of thought and automatic repetitive use of scientific algorithms. Scientists can proceed with their research only because they can combine both aspects in their inquiry. They may draw heavily on the authority and the known tools until they fail; then, when they face a novel problem, their imagination and creativity move to the fore.

Within this expanded perspective, there is a continuous spectrum to be exposed in science education (Fig. 9.5), precluding an extreme view of any kind and showing the importance of both *equally* important virtues in describing NOS.

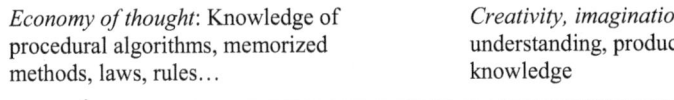

Economy of thought: Knowledge of procedural algorithms, memorized methods, laws, rules…

Creativity, imagination: In-depth understanding, production of new knowledge

Fig. 9.5 The broad variation of cognitive activity with regard to scientific knowledge in different contexts. Each scientific activity (or product) can be "located" on this axis as combining the two complementary virtues in a dialectic symbiosis

The Empirical Nature of Scientific Knowledge (Observation–Inference)

The statement of L1 that scientific knowledge is empirically based is also correct and expands the previous point of the list to its counterpart—practice. Einstein continued to guide:[131]

> *Pure logical thinking can give us no knowledge whatsoever of the world of experience; all knowledge about reality begins with experience and terminates in it.*

---

[130] Mach (1976, p. 354).
[131] Einstein (1934/2011b).

Yet, in the context of education, this claim is not sufficiently representative. There are other nonscientific domains of knowledge, which are also empirically supported and application oriented, but that was not science. For thousands of years, people have followed the paths of planets, registered the level of water in rivers, practiced traditional healing, and developed a wide variety of crafts. However, these systems of rich empirical knowledge did not constitute science. What was missed in this knowledge that precluded it to be considered science? What is special for science with regard to empirical support?

The essential feature of scientific knowledge is its arrangement in accordance with especially inclusive, abstract, and conceptual systems—*theories* of nature. Scientific knowledge is empirical but nonetheless theoretical. They create a symbiosis. The same scholar proceeded:[132]

> *...on principle, it is quite wrong to try founding a theory on observable magnitudes alone. In reality the very opposite happens. It is the theory which decides what we can observe.*

Thus, theory projects onto observations and vise-versa. Yet, we still need to distinguish between *observation* and *inference*, which are interwoven, but differ in logical status. Their difference is often subtle and is not always adequately presented in classes. For instance, one may see in a school curriculum the following instruction:[133]

> *A feature of the work in both the middle and senior classes is that pupils will investigate falling objects. They will discover that objects fall because of the force of gravity. They will measure force by constructing their own spring balances.*

It could be even more inspiring for the kids, however, to be informed that there is no way to "discover that objects fall because of the force of gravity" regardless of the time invested by them in practical exploration. By the way, ironically, the first critique of the "discovery" of the gravitational force instead of its *invention* as an abstract tool was due to the Irish philosopher Berkeley.[134] In fact, for thousands of years, observation of falling and the studies of gravity did not compel using *force* as in other settings. The opposite was true: being natural *excluded* force as an agent of violence and imposition. Galileo rejected the idea of gravitational attraction.[135] Look how people naturally managed without believing that force caused falling. Thomas Aquinas, being inspired with the Aristotelian perception (natural motion does not require force), wrote in the twelfth century:[136]

> *A thing moved by another is forced if moved against its natural inclination; but if it is moved by another giving to it the proper natural inclination, it is not forced; as when a heavy body is made to move downwards by that which produced it, then it is not forced. In like manner God, while moving the will, does not force it, because He gives the will its own natural inclination.*

---

[132] Heisenberg (1971, p. 63).

[133] Primary School Curriculum of Ireland, Science. Teacher Guidelines (1999, p. 13).

[134] Popper (1962, p. 109).

[135] Galilei (1638/1914, p. 166).

[136] Aquinas (1267/1952, pp. 541–542).

And in fact, many of our students would subscribe to these words. Their misconceptions, as judged by classical mechanics, repeat the understanding, which reigned until Newton.

The force of gravitation was introduced, *invented,* by Newton as a part of his theory in the seventeenth century. Students may better appreciate this fact if they are at least informed that Aristotle and Einstein also explained gravity (peripheral elements of the DC-framed classical mechanics). Each did it in his own way, and both explanations were different to the Newtonian force. Just as theoretical and experimental activities are deeply interwoven, so too are observation and inference. Observation in science often depends on a particular fundamental theory,[137] and involvement of tacit knowledge in the scientific observation is unavoidable. Kuhn discusses the same point when considering the pair of perception–interpretation of something, cemented together by the tacit knowledge of the explorer.[138]

Summarizing this refinement, the nature of scientific knowledge combines empirical and theoretical aspects manifested in the distinction between observation and inference. Their connection is not dichotomous; it is symbiotic. Einstein called it "the eternal antithesis of the two inseparable constituents of human knowledge, Experience and Rationale, within the sphere of physics."[139] In a classroom, their continuous reciprocal influence should be contextualized in teaching with concrete examples of the iterative process of knowledge construction. This symbiosis epitomized in artistic images could be shown in science classes as the logos of science, its nature.[140]

The Social and Cultural Embeddedness of Scientists and Science

The list proceeded to another important point. The text in L2 reads:

> Science as a human enterprise is practiced in the context of a larger culture and its practitioners (scientists) are the product of that culture. Scientific knowledge affects, and is affected by, the various elements and intellectual spheres of the culture in which it is embedded.

Indeed, scientific knowledge is a cultural–social product.[141] However, there is a need to refine this claim in regular teaching, its limits, in order to distinguish scientific knowledge from other numerous knowledge systems produced by society. The central norm of scientific knowledge is its being objective, being reduced to a few inclusive fundamental theories based on empirical verification, reproduction, and

---

[137] Kuhn (2000, pp. 246–247).

[138] Kuhn (1969, pp. 197–198).

[139] Einstein (1934/2011a).

[140] Appendix 1.

[141] Longino (1990).

successful predictions. Different social environments definitely influence the manner in which scientists live, think, and function. Environment affects the agenda of the studies in form and content. Even so, despite the social impact, the essence of the products has to be objective in order to be scientific. This norm originally caused the split of science from mythology in ancient Greece, and it allowed continuity of the scientific discourse, dealing with the *same* fundamental questions and scientific problems across different societies and at different times.[142]

Consider the scientific knowledge of light. It was continuously constructed through the societies of Classical Greece, the Hellenistic civilization, Muslim and Christian medieval worlds, seventeenth century Europe, and so on. Yet, despite the striking social differences, the development of pertinent knowledge continued as a single process.[143] Addressing the period of so called Islamic Science in that regard, Weinberg mentioned that they (the Muslim, Christian, and Jewish scholars of that time) saw themselves as refiners of Hellenistic thought, and they promoted science much farther.[144] They "were not doing Islamic science. They were doing science."[145] A very similar claim was made by another Nobelist, Leon Lederman, who stated categorically:[146]

> We believe that there is only one science, not Western, not indigenous, not even Maori. Its origins may be traced to the Ionian Greek civilization, and it flowered in Europe in the seventeenth century.

Weinberg related the cultural independence of physics laws to their objective nature:[147]

> One of the things about laws of nature like Maxwell's equations that convinces me of their objective reality is the absence of a multiplicity of valid laws governing the same phenomena, with different laws of nature for different cultures.

Similar examples can be provided from mechanics and astronomy developed in very different cultural milieu and by the researchers holding a variety of worldviews. A vast body of literature documented the tragic reality of Soviet science during Stalin's regime and the brutal repression of the social environment, including physical elimination, torturing, and imprisonment of numerous scientists.[148] In spite of the often inhuman conditions, the scientists there managed to produce results universally valid, regardless of the nightmare they faced, while being committed to the universal scientific norms for argumentation and creating objective, socially independent products.[149] By contrast, when the subjective social demand penetrated

---

[142] The further complexity of this claim is briefly addressed in Appendix 2.
[143] Lindberg (1976).
[144] Al-Khalili (2010).
[145] Weinberg (2015, p. 70).
[146] Lederman, L. (1998, p. 132).
[147] Weinberg (2001, p. 158).
[148] Gorelik and Frenkel (1994), Gorelik and Bouis (2005), and Ginzburg (2005).
[149] Josephson and Sorokin (2017).

scientific content, as happened in the Lysenko case (Lamarckian paradigm) where Stalin destroyed the opponents from the new science of genetics, the product was pseudoscience, not science.[150] In an interesting parallel with education, the entrance of social, subjective factors may cause *pretending* social behavior and pseudo-conceptual understanding on behalf of students who try to succeed avoiding understanding.[151]

Scientific knowledge was produced standing on the shoulders of scientists who were active in different societies regardless of social factors such as values and ideology. In Soviet Russia, genetics and cybernetics were considered to be bourgeois pseudoscience or "capitalistic" products. The development of these areas of objective knowledge was thus suppressed in Russia for many years but eventually came there too.[152] Once developed in one social environment, "science swept across the world as a forest fire."[153] Scientists may hold different views on the meaning of the established laws, but none replace $E = mc^2$ with $E = mc^3$, regardless of any social factor.

Besides the social–political aspect, there are other aspects of social influences on science, its content, and practice, including the practical and religious demands of each society throughout all periods of history.[154] Another special issue is the social impact on scholars *within* scientific communities.[155] Besides the subjective voluntary pressure as exemplified by scientists in the Soviet Russia, this impact could be also interpreted in the objective terms of political economy within the Marxist perspective. Thus, it was claimed that Newton's *Principia* was actually the answer to the needs of England in constructing canals and locks, problems of chronometry in navigation, etc.[156]

Our argument states the social independence of the scientific content in its essence of nonideological nature. It is this aspect of NOS to be emphasized in introductory science education: the *objective meaning of the scientific* content is *independent* of *social environment* being determined by nature itself and regardless possible interpretations of the individuals producing it. Without such refinement, the claim of scientific knowledge as dependent on social environment might be grossly misleading.

---

[150] Birstein (2001).

[151] Vinner (1991, 1997).

[152] We limit our discussion to science, arguing using scientific theories, but technology is not different in this perspective. Think about the striking difference, in all aspects of social environment and ideology, between the USA, USSR, China, Pakistan, and North Korea. Despite the differences, practically the same scientific and technological products were created—the atomic weapon and rocketry, for instance.

[153] Needham (2004, p. 231).

[154] Lindberg (2002); Al-Khalili (2010).

[155] Kuhn (1962/1970); Latour (1987).

[156] Hessen (1933, pp. 30, 62); Hessen (2009, pp. 41–102). The Marxist interpretation of Newtonian science contrasted with Hessen's own tragic life, being executed in 1936 (Gorelik, 1995, pp. 54–75).

Adoption of the social constructivist perspective on scientific knowledge as merely being "socially constructed" but neglecting its independent objective essence may threaten science education by reducing science to ideology and pedagogy to mere propaganda.[157]

The Tentative Nature of Scientific Knowledge

The next point of the L-list was not less strong, however. In L1, the text relating to it was especially short and categorical as if informing about factual data: "scientific knowledge is tentative." L2 and L3 expanded and finalized that scientific knowledge is "never absolute or certain." It is not difficult to trace this claim to the Popperian perspective[158] described as follows:[159]

> Scientific theories, for him [Popper], are not inductively inferred from experience, nor is scientific experimentation carried out with a view to verifying or finally establishing the truth of theories; rather, all knowledge is provisional, conjectural, hypothetical – we can never finally prove our scientific theories, we can merely (provisionally) confirm or (conclusively) refute them.

This perspective belongs to the dialogue on the continuous and unlimited progress of scientific understanding of nature taken holistically, across domains and theories. It looks as if the knowledge is addressed there in the *metaphysical* sense with respect to the *ultimate all-inclusive* truth about nature—the *gnosis*.[160] To expand on this concept, it is interesting to bring the view of the famous theologian, cardinal Bellarmine, who was an opponent of Galileo in his famous affair regarding heliocentric system of the world. In 1615, Bellarmine expressed his position:[161]

> ... I say that it seems to me that Your Paternity and Mr. Galileo are proceeding prudently by limiting yourselves to speaking _suppositionally and not absolutely_, as I have always believed that Copernicus spoke. For there is no danger in saying that, by assuming the earth moves and the sun stands still, one saves _all the appearances_ better than by postulating eccentrics and epicycles; and that is sufficient for the _mathematician_. However, it is different to want to affirm that _in reality_ the sun is at the center of the world and only turns on itself without moving from east to west, and the earth is in the third heaven and revolves with great speed around the sun...(emphasis added, IG)

---

[157] Slezak (1994).

[158] Popper (1959/2002, p. 22). We refer to Popper in this regard and not to other philosophers of the past, such as Hume, who expressed a similar and even more famous criticism in his replacing the cause–effect necessity relationship in science with experience based on mere probability. Popper addressed science in a more inclusive and mature way.

[159] Thornton (2016).

[160] In religious Judeo-Christian literature, the knowledge of ultimate truth is labeled *gnosis*. In a similar sense, in Soviet Russia, university students learned *gnoseology*, which replaced *epistemology* in other countries.

[161] Finocchiaro (1989, p. 67).

In a sense, the position of Bellarmine in theology share features with the position of Popper in philosophy of science. Both do not grant people to know the *absolute truth* about nature as a product of people ("mathematician") exploration—science, Bellarmine—in favor of Holy Scriptures, no matter any evidence, even "saving all the appearances," while Popper was in favor of other more advanced theories to come.[162]

This philosophical perspective is distant from science. Teaching and practicing science addresses knowledge in the essentially different sense of *episteme*—the *rational understanding* within a well-defined framework of assumptions, created and structured in a scientific discourse using the specific method, a subject for demonstration (theoretical and experimental).

The Popperian negation of "being proved" as a characteristic of science products clearly contradicts the scientific practice, the normal activity of science[163] that presents the agenda of science education. Leon Lederman stated in presenting a novel educational program:[164]

> We believe that science is tentative at its frontiers, but that there is a large body of knowledge that is objective truth for all practical purposes.

It is, however, not a divorce from philosophy but rather the recognition of a complex relationship.[165] Bunge wrote: "What is obvious to the practitioner of a science may be problematic to its philosopher."[166] Why, then, not ignore philosophy? Bunge answered: "Ignore all philosophy and you will be the slave of one bad philosophy"[167] and elaborated:[168]

> Physics cannot dispense with philosophy, just as the latter does not advance if it ignores physics and the other sciences. ... Science and sound (i.e., scientific) philosophy overlap partially and consequently they can interact fruitfully. Without philosophy, science loses in depth; and without science philosophy stagnates.

The clarification of the difference between episteme and gnosis actually displays the importance of philosophy for science education and its difference from "doing" science.

In fact, there is no need to wait until the scientific theories we teach are falsified and replaced by more advanced, but also "provisional," theories, as Popper saw the situation. Scientific knowledge is already organized in a family of fundamental

---

[162] In philosophy, the position of Popper is not too distant from the old tradition to reject the human ability to comprehend the very truth of the external reality, which always remains to be "things in themselves," as Kant put it. This perspective of empirical or transcendental idealism (Stang, 2018) does not deprive people from successful account of *appearances*.

[163] Weinberg (2015, p. 24).

[164] Lederman, L. (1998, p. 132).

[165] Russell (1912/1990).

[166] Bunge (1973, p. 28).

[167] Bunge (1967b, p. 261).

[168] Bunge (2000, p. 461).

theories, none of which is all-inclusive in validity.[169] Each such theory represents a specific aspect of reality, valid within a certain area of parameters. In the tripartite structure of a theory—a discipline–culture—the limits of the validity of its nucleus are indicated by its periphery. The status of *proven* regarding a scientific statement implies its coherence with the nucleus of a certain adopted theory, being verified empirically and theoretically. To call the myriad of products of classical mechanics, thermodynamics, and electrodynamics "tentatively correct" presents a misleading assertion, erasing the meaning of being erroneous and incorrect. All the discoveries and inventions rewarded by a Nobel Prize were *proven* to be true. Thus, the Nobel Prize in Medicine of 1945, awarded for the discovery of penicillin and its curative effect that saved the lives of millions and proved the theory of immunology. Likewise, the Nobel Prize in Physics of 2017, given for the observation of gravitational waves, added another proof of correctness of the Theory of General Relativity by Einstein established a century before.

In exact similarity, to *explain* a natural phenomenon scientifically means to display its coherence with the tenets of a certain fundamental theory. This is a working norm in a scientific laboratory and in school science classes. At the same time, researchers often, and even normally, deal with tentative knowledge in the context of exploration of a new subject, while they draw on the great amount of knowledge previously established as certain.

Science textbooks and teachers tell their students that Aristotle proved the spherical shape of the Earth, and Archimedes proved the law of the lever and the law of buoyancy. Galileo proved that Venus rotates around the Sun and not around the Earth. Pascal and Boyle proved the existence of atmospheric pressure and a vacuum. Newton proved the oblateness of the Earth globe and that tides are due to lunar attraction. All these facts are taught not as *tentative* but as *certain* proved results derived through drawing on scientific theories and verified by corresponding experiments and predictions. The certainty of science reached tremendous accuracy on molecular, atomic, subatomic, and cosmic scales. How can a teacher declare that "all scientific knowledge is tentative" while presenting numerous factual discoveries and products of our knowledge about nature? Is the word "tentative" sufficiently representative *in the context of education* for the numerical accuracy of 10^{-11} reached in the experimental approval of the advanced physics theories?[170]

No doubt it is not easy to establish certainty. It may require much time and effort, and often complex technology and a variety of experiments carried out by independent research groups. During the ongoing inquiry, which can take years, the scientific claims under investigation could be addressed as *tentative*. Therefore, in the context of education, to avoid confusion and to distinguish science from other areas, one should present the *variation* in tentativeness between its extremes: from an

---

[169]This perspective may resolve the confusion of those who do recognize the progress of science but do not see it approaching the truth about nature (Kuhn, 1962/1970, p. 170). The approach of science is not linear but multifaceted in different aspects of truth revealed in greater and deeper extent by several fundamental theories.

[170]Feynman (1985/2014, p. 7).

argumentative hypothesis to certitude (Fig. 9.6). Each scientific result or claim should be considered within this span from hypothetical and tentative to certain and accurate. The atomic structure of matter may illustrate the historical change of knowledge status along the time axis, from a speculative hypothesis at the dawn of science to the objective fact to be taught as such nowadays.

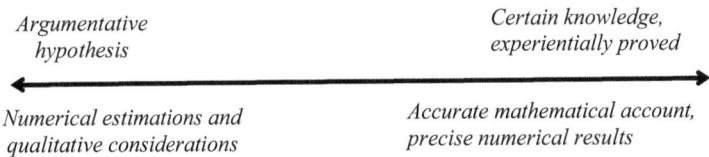

Fig. 9.6 The span of variation of tentativeness of scientific claims in the context of science education

Let us further exemplify through the knowledge of the electron as an elementary particle. An electron, within the theory of classical electromagnetism, presents a particle with certain, clearly determined characteristics (mass, charge, and spin). Within the quantum theory, the electron obtains a new account. In the latter, the electron is a *quanton* (quantum particle[171]), which in addition to the previously known, possesses new features corresponding to the nature of the quantum world (i.e., includes nonclassical "wavity" manifested in diffraction and interference). Physics, as a subject of education, does not address concepts, such as electron, in their absolute universal amorphous sense, but within specific fundamental theories. Indeed, physics theories, classical and quantum, were produced in historical sequence. Hence, one may see them holistically as a development of knowledge, its progression. Yet, in each of the two fundamental theories, classical and quantum, the electron presents a well-defined object explaining a vast amount of natural phenomena and revealing new facts about nature, not merely tentatively but with certainty.[172] Heisenberg ascribed the status of certainty to the fundamental theories of physics, such as classical and quantum mechanics. To emphasize this status, he labeled them "closed theories," that is, they are not a subject for any change of their nuclei.

The perspective, which does not recognize the theory-based structure of science, may bring the learner to the claim of the all-tentative nature of knowledge, which makes sense with respect to *gnosis* but not as *episteme*. It is the latter that is adopted and continuously practiced in science. Discipline—culture structure of theories displays the scientific knowledge as an episteme.

---

[171] Lévy-Leblond (2001).

[172] Bokulich (2008, pp. 29–48).

The Scientific Method

Finally, the L-list considers the topic of epistemological importance—the scientific method. L1 states:

> *One of the most widely held misconceptions about science is the existence of the scientific method.*

In the following text, the authors moderated this extreme claim by explaining:

> *The myth of the scientific method is regularly manifested in the belief that there is a recipe-like stepwise procedure that all scientists follow when they do science. This notion was explicitly debunked: There is no single scientific method that would guarantee the development of infallible knowledge.*

The claim looks consistent with the previous items of the list: if all scientific knowledge is merely tentative, subjective, socially monitored, and never proven without mentioning that the opposite at least partially present too, then who can talk about a method? This negation might be highly misleading regarding the specific knowledge (*techne*)—method—of production of scientific knowledge (*episteme*). In fact, however, techne has been continuously developed and meticulously refined throughout the history of science and its philosophy. The method specified the meaning of logical proof, measurement, mathematization, modeling, scaling, systematic observation, data processing, and specific types of experiment through which scholars act and produce reliable knowledge. The scientific method is what makes knowledge scientific and distinguishes it from other kinds of knowledge.[173]

Addressing the scientific method often alludes (especially in the introductory context) to a certain sequence: (a) defining the problem; (b) gathering background information; (c) forming a hypothesis, (d) making observations, (e) testing the hypothesis, and (f) drawing conclusions.[174] This way the numerous research projects for MSc and PhD degrees are organized in the universities around the world. It determines the structure of proposals for research grants in science in all countries. Within this framework, these proposals are considered and criticized by other scientists—experts—who decide to adopt them or possibly reject for violation or shortcomings in the applied method. This is a requirement in science practice. May we ignore this reality in considering NOS? Even if one considers the L-list as a merely schematic, one-sided comment, its refinement in the broad educational context is an obvious requirement. Science teachers must introduce the approach of science, which is specific and different from other approaches used in nonscientific problem solving and in other areas of knowledge—interpersonal relations, "common sense," intuition, traditional healing, religion, politics, magic, and other areas representing subjective knowledge. They all may use science, as they often do, but science has no ambition, knowledge, and appropriate tools to reign there by applying indisputable

---

[173] Losee (1993, pp. 120–136), Gower (1997), Lakatos (1999, pp. 19–108), Betz (2011). See Appendix 3 for an emblematic image of the scientific method in the Renaissance.
[174] Hempel (1966, p. 11).

prescriptions. The violation of this kind is named scientism, which is widely resisted. Feyerabend explained:[175]

> *However, the profits [of science] should not be imposed; they should be examined and freely accepted by the parties of the exchange.*

This warning, however, is not about the deficiency of the *scientific method*. We cannot manage without it in any serious problem. If scientism fails, its reverse, anti-scientism, fails even faster. Bunger exlained:[176]

> *Why has anti-scientism failed? Arguably, it failed because it condemned and spurned the scientific method, which has characterized all of the scientific achievements since the Scientific Revolution.*

A close look reveals that denying a "recipe-like stepwise procedure" does not imply a lack of method, but calls for proficiency of its application. In the much quoted critique of the scientific method in *Against Method*, Feyerabend wrote:[177]

> *I argue that all rules have their limits... I do not argue that we should proceed without rules and standards.*

Scientists normally practice "methodological pluralism"[178] showing an amalgam of philosophical approaches. Feyerabend's "anything goes,"[179] as a logo, may represent naïve confusion in observing the plural nature of the method applied in the context of inquiry. Einstein elaborated:[180]

> *He [the scientist] therefore must appear to the systematic epistemologist as a type of unscrupulous opportunist: he appears as realist insofar as he seeks to describe a world independent of the acts of perception; as idealist insofar as he looks upon the concepts and theories as free inventions of the human spirit (not logically derivable from what is empirically given); as positivist insofar as he considers his concepts and theories justified only to the extent to which they furnish a logical representation of relations among sensory experiences. He may even appear as Platonist or Pythagorean insofar as he considers the viewpoint of logical simplicity as an indispensable and effective tool of his research.*

Importantly, scientific knowledge itself serves as a tool for achieving new products and so presents a methodology.[181] Lakatos considered fundamental physics theories, such as Newtonian mechanics, as a *methodology* (the rules to solve problems rather than a theory of the world), therefore calling it a *scientific research program.*[182] Rephrasing Engels' maxim of 1876 in *Dialectics of Nature* regarding the synergy of human hand and labor, we may say: *scientific knowledge is not only the product of a certain method, but equally true is that the scientific method is the product of scientific knowledge.* The role of a specific method is especially striking in modern

---

[175] Feyerabend (1993, p. viii).

[176] Bunge (2015).

[177] Feyerabend (1993, p. 231).

[178] Feyerabend (1999b, p. 216).

[179] Feyerabend (1999a, p. 296).

[180] Einstein (1949/1979, pp. 683–684).

[181] Popper (1962).

[182] Lakatos (1980).

physics, quantum, and relativistic. Heisenberg defined scientific knowledge as nothing but the knowledge that obtains its substance and meaning through specified tools, meaning the use of macroscopic apparatus,[183] while Bohr expanded to the principle constrain of using *macroscopic* terms for comprehension of the *microscopic* world.[184] Method of knowledge construction became, thus, a central issue/problem in the adequate understanding of nature.

In specifying the scientific method, we draw on the inclusive, empirically tested fundamental theories structured hierarchically and on the specific mechanism of knowledge production. Even if other areas share some features of the scientific approach to problem solving,[185] which is true, science teachers should strengthen the *specific* method of science. As already mentioned not once, it is the difference from other human endeavors rather than commonality that makes science unique and indispensable. It is worth to discover that difference, to contrast it with other approaches in science class.

The features of the scientific method include *mathematization* of the account of reality (seeking numerical accuracy), theory-based *modeling*, many faceted *experimentation* with controlled parameters, testing through *prediction and reproduction*, and *statistical analysis*—all highly developed in science. One may consider as proof of this method, the fact that its violations lead to the fiasco of pseudoscience—the products of the activity in which the method was violated this or other way (e.g., astrology, alchemy, Lysenkoism, cold fusion, creationism, and Feng Shui).[186] Each of these cases was analyzed and contrasted with the scientific methodology. Historically, science and pseudoscience were highly interwoven. Astrology and mysticism went a long way with astronomy. The claim of the mystic correspondence and relationship between macrocosmos (the world) and microcosmos (the human organism) was considered scientific for centuries and shared by leading scholars.[187] The historical offshoots of science contributed in observation, experimentation, terminology, data classification and conceptualization,[188] and their separation from science illustrated the inadequacy of "anything goes."[189]

---

[183] Heisenberg (1959/1971, p. 81).

[184] Bohr (1949, p. 209).

[185] McComas (1998).

[186] See Kuhn (1957) and Lakatos (1998) with regard to astrology; Read (1995), for alchemy; Birstein (2001), for the Lysenko case; Huizenga (1993), for cold fusion; Roob (2001a, 2001b), for mysticism; and Matthews (2019), for Feng Shui.

[187] Indeed, who constructed astrological horoscopes for emperors, military leaders and nobility?—Not the same scholars who contributed so much to science progress?—Tycho, Galileo, and Kepler were experts in astrology. They all "personalized their gifts for their patrons by means of their patrons' horoscopes" (Rutkin, 2001, pp. 154–155). Seemingly, it was due to the introduction of telescope by Galileo and Kepler that astronomy and astrology began to go separate ways (Boxer, 2020, Ch. 9).

[188] Hoskin (1997), Glashow (1994), and Lindberg (1992).

[189] As a striking contra argument against "anything goes," we bring the special issue of pseudo and even anti-science, the violation of ethics in investigations by Nazi scientists in their medical experiments. It led to the establishment of the Nuremberg code to regulate scientific experimentation (https://encyclopedia.ushmm.org/en).

The scientific method arranges an inquiry not as a protocol procedure, but as logically related structure of components integrated in a continuous inquiry. Imagination and creative leaps introduce uniqueness to each scientific inquiry while preserving its circularity. Therefore, newcomers can join a research project at any stage of the loop even if the process was started by others, at another time and at another place. This is an implication of science being a culture and students of science as joining this culture, performing *enculturation*.

Mathematization appeared to be among the most distinctive features of the scientific method first proclaimed by Plato as the way to reveal truth, providing it with a highly precise expression. The claim of Galileo that mathematics is the language in which science is written in the grand book of Nature became emblematic.[190] Its concise symbolism though insufficient to express comprehensive meaning still allowed essentially compact codification of scientific claims and ideas. It created a new space in which scientific knowledge became a subject of extremely effective manipulation and problem solving including precise numerical results and hinting on the new ways of conceptual understanding. From the very beginning, mathematics shadowed the scientific enterprise,[191] until it became a required component of the method of modern science.[192] History informs us about mathematization as a tool for breakthrough achievements of science. Mathematization changed scientific disciplines.[193] It presents an indicator of maturation, providing clarity and distinctness of scientific statements, which through becoming quantified, are useful in practice and easier to critique their content and falsify certain elements.

Modeling is another central feature of the scientific method.[194] The replacement of a real system with its conceptual and mathematical models enables its effective representation and treatment through simplifying both the subject matter (to be simpler than reality) and its investigation (treatment by means of less than a full-scale theory). Within the conceptual structure of discipline–culture, one may identify three basic kinds of models—paradigmatic, working, and heuristic—ascribing them to nucleus, body, and periphery of a theory, respectively (Fig. 6.7). As a clarifying example, one may compare the models of Leonardo (suggestive, testing, and

---

[190] Galilei (1623/1957, in Drake, 1957, pp. 235–236).

[191] Thales of Miletus, commonly considered the first natural philosopher (Russell, 1959, p. 16). His contribution mixed natural science and mathematics: reductionist theory ("all things are made of water"), the demand for a proof (beyond plausible claims) and mathematical account in problem solving (using scaling method).

[192] Neugebauer (1993), for Babylonian and Egyptian science; Berry (1898/1961) and Dreyer (1953), for Hellenic astronomy; Russo (2004), for Hellenistic science; Pedersen and Pihl (1974), for medieval science; and Kepler (1621/1972) and Gorham et al. (2016), for modern science, Galili (2018).

[193] Striking changes of this kind were observed in quantum chemistry, physical chemistry, and molecular biology.

[194] Hestenes (1992), Giere (1995).

scaling) and models of Galileo (theoretical, mathematical, and idealizing).[195] Though different in nature, both modeling served as a pivotal tool of methodology.

Within any problem solving, the scientific method prescribes "gathering information" and selecting a proper theory to work with, searching for similar problems previously discussed and solved. Solving problems starts with copying others and learning and applying the known methods in a new context. Kuhn stated normal science as "puzzle resolving" within certain paradigms, following standard prescriptive and prohibitive rules and norms.[196] However, this picture is not sufficiently representative, as he mentioned himself. It lacks the problems, which could not be solved within the known paradigm. In this case, the scientific norm is not to forget them, but accumulate them in the periphery of the DC structure and preserve them as a subject for consideration in the fundamental research. This methodology invites new tools to be developed and approaches to be suggested leading to the new paradigms.

Teaching the scientific method may start from simple cases of a logical Hempel stated loop procedure including observation and empirical verification of a spontaneously emerged hypothesis.[197] Yet, seeking adequacy in representation of scientific methodology, a teacher can proceed to more complex cases, such as asking where the hypothesis for solution may come from. The circularity of investigation process refines the originally looking simple method and reveals to students its additional components and aspects. The recognized features of investigation should be raised to the status of a universal approach—the scientific method. The reference should be to the vast success of science due to the method applied to the numerous problems and phenomena, solved and explained. With regard to other subjects, it is an effective pedagogy to consider violations of the scientific method leading to failures. Such strategy will encourage the understanding of *necessity* to keep with specific method rather than relying on luck, natural skill of improvisation, and advanced equipment, all important but very often insufficient if the method is neglected.

Summary of the Performed Refinement

Table 9.1 summarizes the results of the performed refinement of the L-list. In most cases, the univocal tenets were replaced by a range of variation, including opposites, thus providing a more adequate account of the basic NOS features.[198]

---

[195] Chapter 8.

[196] Kuhn (1962/1970).

[197] Khan Academy (2017).

[198] Beyond the comparison with the original L-list, this summary can be considered as a polemic with Table 9.1 (Erduran, 2014), which is compared between the consensus view and the tenets of logical positivism.

Table 9.1 Summary of refinement for the L-list of NOS features

N	L1–L5 lists	Suggested refinement
1	Scientific knowledge, owing to scientists' theoretical commitments, beliefs, previous knowledge, training experiences, and expectations, is unavoidably *subjective*.	Scientific knowledge is *objective* as collective knowledge essentially independent of personal will and values. Scientific inquiry may include subjective elements in form and interpretation. Exclusion of subjective aspects takes place in the context of knowledge justification, through the continuous many-faceted experimental verification and reproduction, iterative steps of investigations, and discourse of scholars.
2	The distinction between *theories and laws*. Laws are statements or descriptions of the relationships among observable phenomena. Theories, by contrast, are inferred explanations for observable phenomena.	Scientific knowledge is composed of fundamental *theories*— inclusive clusters of various knowledge elements, coherent, and hierarchically organized. Each theory is valid in a certain area of validity. A fundamental *theory* is structured in a *nucleus* (basic concepts and principles) and a *body* containing derived and associative elements of knowledge coherent with the nucleus. *Periphery* is added to include the elements of pertinent knowledge at odds with the nucleus. *Laws* present reproducible functional relationship among the characteristics of natural phenomena, stable in a variety of settings and valid in certain areas of validity. Laws are normally affiliated to a particular theory. Laws of theoretical or empirical kind differ in status in accordance with the area of their affiliation in the theory structure—nucleus, body, or periphery.
3	The development of scientific knowledge involves human imagination and creativity.	The progress of scientific knowledge draws on a continuous disciplinary discourse within the scientific community. It involves imagination and creativity drawing on the previously developed knowledge taken as given. The rate of invented versus given knowledge varies while both are necessary.
4	The empirical nature of scientific knowledge. The distinction between observation and inference.	Scientific knowledge is essentially theoretical and draws on empirical verification. Observation and inference differ, but they are intrinsically interwoven, influencing each other, as are the empirical and theoretical aspects of scientific knowledge.
5	The social and cultural embeddedness of science. Scientific knowledge affects and is affected by the various elements and intellectual spheres of the culture.	Science is a cultural product. Social and cultural environments affect scientists in their life, activities, agenda, forms, and ways of knowledge production. Yet, the structure of the scientific knowledge in its essence, conceptual and operational meanings, is independent of social and cultural factors and environment. Being committed to objectivity makes science products and knowledge valid across different societies and be independent of them.
6	The tentative nature of scientific knowledge; scientific knowledge is never absolute or certain. This knowledge, including "facts," theories, and laws, is inherently tentative or subject to change.	Science continuously advances, producing knowledge of different status of validity and reliability ranging from hypothetical and tentative to certain and accurate products Making claims in science presumes demonstrating their coherence with a certain theory, theoretical and/or experimental verification using scientific methodology All fundamental scientific theories are empirically proven and valid in certain areas of basic parameters (such as length, time, and mass in physics).

(continued)

Table 9.1 (continued)

N	L1–L5 lists	Suggested refinement
7	Myth of the scientific method (absence of unique recipe-like stepwise procedure).	The scientific method possesses plurality of specific norms, procedures, prescriptive, and prohibitive rules of formal and ethical nature. Scientific inquiry unfolds in a spiral sequence of repeating components. It includes the hypothetico-deductive cycle, experimentation, modeling (idealization), a mathematical account, logical rules, rules of qualitative, and quantitative analysis of the collected data. Fundamental theories serve as methodological programs of knowledge production. Scientific method distinguishes science from any other activity, and its violations result in nonscientific products.

Some Inferences

The review of the L-list showed that behind each of its items there is a nontrivial facet of science and a range of meaning. Any univocal claim, even if trying to find the so called "golden middle," is inevitably nonrepresentative, missing the important aspects of NOS. Refinement, stating range of variation, stipulating conditions, and area of validity appear to be essential.

Indeed, the L-list succeeded in capturing central epistemic aspects of scientific knowledge—the status of knowledge, its features, dependence on the producers and their environment, and the method of construction. As was illustrated by concrete examples, the adequate account of these issues requires conceptual variation often including opposites—objective–subjective, tentative–certain, socially dependent–independent, creative–algorithmic, theoretical–empirical, unique–plural, and so on. The range reveals the complex reality and furnishes adequate understanding. Univocal claims regarding NOS mislead and reduce the quality of education. Such is the issue of objectivity of scientific knowledge. If the context of inquiry or justification is ignored, any inference becomes ambiguous and misrepresenting. Specific knowledge is required to consider NOS; intuition and "commonsense" are by no means sufficient. Therefore, investigating the views of students and individual views on NOS without their pertinent background does not provide valid answers beyond informing about the situation in a particular student/teacher population.[199]

Rather than claiming overall tentativeness of scientific products, one should specify the changing status of such products in the course of history. The cultural content knowledge provides necessary background for such analysis in different domains of scientific knowledge. A desirable educational goal is to reveal the progress in the status of certain knowledge element following its affiliation in terms of nucleus–body–periphery on the way to certainty within a fundamental theory. This process characterizes the scientific knowledge as an episteme. Instead of presenting the final product, the range of change in tentativeness is to be displayed in school curriculum.

---

[199] Bronowski (1967), Cromer (1993), and Wolpert (1994).

Peripheral knowledge (emphasizing "what it is not" next to "what it is") is important for considering NOS. It provides variation of the concept in the negative sense, expanding beyond resemblance, which stipulates meaningful learning.[200] Addressing contrasting claims in teaching imitates scientific discourse, which involves *controversy*, the understanding evolving through comparison and refutation.[201] Students will adopt the habit to consider a particular framework—a specific theory to choose—in any scientific account. They will begin to ask—In what sense? Under what conditions? Within what theory?—instead of running to manipulate with some formalism they remember from class instruction or seen in books.

The important aspect is the reciprocal relationship within the complementary pairs of NOS features.[202] The symbiosis of observation *and* inferences, theoretical (conceptual) *and* experimental (practical) considerations, and rationalist *and* empiricist approaches characterizes the scientific method and reveals the iterative back and forth evolutionary progress, which establishes the scientific knowledge.[203] Being the mechanism of reaching objectivity, this process elucidates science progress in terms of evolutionary epistemology.[204] It clarifies the content of the naturalized philosophy of science[205] and the relationship between science and mathematics.[206] It is in this sense that science is *cumulative* in maintaining diachronic discourse.

Implications in Teaching

Duschl and Grandy reviewed relevant studies and stated the advantage of long-term courses ("weeks and months long") adopting *naturalized* epistemology in order to teach NOS.[207] In our view, dealing with isolated practices of "doing" science can be informative about the context of inquiry but is only fragmentally suggestive regarding the norms, easily missing the big structure of scientific knowledge. Teaching of

---

[200] Marton and Pang (2006, 2013).

[201] Sequeira and Leite (1991), Scheker and Niedderer (1996), Kalman and Adulls (2003), Kipnis (2010), Schwartz et al. (2011), Braga et al. (2012), Levrini et al. (2014), and Goren and Galili (2018).

[202] In humanities, Buber proclaimed: "In the beginning is relation" (Buber, 1958, p. 18).

[203] It is different from zoological (Darwinian) evolution (Toulmin, 1972, pp. 140–141) and rather goes with historical materialism, which clarifies Popper's dialectical account of science progression by conjectures and refutations. The *reciprocal evolution* changes both partners in the process, the researchers, and their knowledge. To depict this duality, Paget introduced two notions, different for both partners, assimilation and accommodation.

[204] This way Mach (1883/1989), Einstein and Infeld (1938), Taylor (1941), and Glashow (1994) depicted the history of science. That was the method of learning science (Crombie, 1959, p. 27, 1996, p. 7, Lindberg, 1992, p. 193).

[205] Godfrey-Smith (2003, pp. 149–150).

[206] Galili (2018).

[207] Duschl and Grandy (2013).

NOS illustrated through the subject matter of a regular course looks as preferable and more feasible goal in school education.

In particular, teachers may infuse short discussions on this or that aspect of NOS relevant to the particular subject matter. The list of Table 9.1 may guide this activity, and conceptual *excursus* provides the required HPS materials for training prospective teachers of science.[208] A special opportunity of students to reveal NOS content is a summary lecture. It was developed as a *delay organizer* of the knowledge organized in a discipline–culture manner.[209] A summary lecture addresses NOS aspects of Table 9.1 integrated with the content of the course reviewed and provides a special perspective on the learned before. Such a summary upgrades and maturates the knowledge constructed in the course.[210]

Given this or other way of teaching, one can stimulate students' thinking of NOS though posing specific questions raised by the teacher and bringing students toward the issues of Table 9.1:

1. How are humans, each a *subjective* individual, able to create *objective* knowledge about the world? How can such knowledge be preserved through different societies?
2. How could tentative initial accounts of experiments or phenomena transform into certain and accurate scientific knowledge?
3. How does scientific knowledge arrange its numerous components (theories, laws, concepts, models, principles, experiments, and explanations) in one coherent body of knowledge?
4. In what sense is scientific knowledge universal, and in what sense is it not?
5. In science, can we imagine an experiment without any theory behind it? Have we any theory without a reference to experiment?
6. Could any observation constitute a science? If yes, exemplify such. If no, what should be added to produce an item of scientific knowledge?
7. How does the cultural social environment influence scientific activities and products? What are the limits of that influence?
8. What are the constraints on the individual creativity and imagination in establishing new *scientific* knowledge of nature?
9. Count the specific rules, tools, and activities in the creation of scientific knowledge. Are there restrictions, formal and ethical, on the way of knowledge construction? What does not count as science?

---

[208] Chapters 1, 2, 3, 4, and 5.

[209] The suggestion of *delay* organizer was inspired by the *advance* organizer introduced by Ausubel (1968, 2000, pp. 11–12).

[210] Chapter 7; Levrini et al. (2014) and Goren and Galili (2018).

Conclusion

The L-list deconstructed the traditional account of science and its method as depicted by Merton (1973) and Hempel (1966). At the same time, the L-list articulated the locus of central features representing NOS. There is need to resolve this controversy with the traditional understanding through reconsideration and refinement of the items of the L-list, drawing on the historical development of science knowledge and the philosophy of science taken in a sufficiently broad perspective, rather than fragmentary claims, which could be polar different. Discipline–cultural perspective on scientific knowledge could be of effective support in this approach. It introduces the structure and hierarchy of scientific knowledge as episteme, makes explicit its epistemological foundation (its method), and emphasizes its *objective* nature as *the major difference* from nonscience. The advantage of this perspective is in presenting science features in the range of their variation, emphasizing the dialogical nature and contrasting between different conceptual accounts. All these present a subject of explicit teaching and imply a conforming change in teacher training programs. NOS concepts cannot be left for individual discovery and intuitive inferences.

A special challenge of science education is to prevent *oversimplification* and losing the general picture of science (Einstein's epigraph to the chapter). In considering NOS tenets, one *cannot be univocal* (Sakharov's epigraph to the chapter).

Appendix 1

Art provides appealing images of the symbiotic relationship of reason and experience as the essential feature of knowledge and method in science.[211] Such is Rafael's renowned collective portrait of *The School of Athens* (1501). The fresco in the Vatican has in its focus two figures of the founders of natural philosophy, Plato and Aristotle, who present through their gestures reason (rationalism) and experience (empiricism) as their principles of knowledge. Their symbiosis represents the genus of science (Fig. 9.7a).

A similar image from the Far East is on display in the National Museum of Seoul. It employs the idea of the up and down complementarity of the world and the knowledge about it in a no less appealing way (Fig. 9.7b). The two interwoven figures relate the Earthly and Heavenly origins, possibly interpreted as ratio and experience. The right-angle tool symbolizes the Earth, considered to be of rectangular shape by Eastern scholars in the past. It therefore corresponds to the pointing downward by Aristotle. The compasses represent rationality related to the heavens, considered to be of round shape. It therefore corresponds to the pointing upward by Plato. The Korean image may become even stronger, if one interprets the interwoven figures as showing their symbiosis in a reciprocally supported emergence in the iterative evolution (subsection *Some Inferences*).

---

[211] Chapter 10; Galili (2013).

(a)   **(b)**

Fig. 9.7 (**a**) A fragment of Rafael's *The School of Athens* (1501) in the Vatican (*arrows are added to both figures to symbolize the dual origin of knowledge*).; (**b**) Fuxi and Nüwa—the gods symbolizing the dual origin of everything in the picture of the seventh century in the National Museum of Korea in Seoul (*Arrows are added to both figures next to the tools symbolizing heavens and Earth—compass to produce circle and right angle to produce square*)

Appendix 2

The essential independence of social environment does not dismiss the question why, despite the international nature of the scientific enterprise at the present time; it was invented in Greece and nowhere else; why modern science was developed in Europe 2000 years later, but not in other places. Why did the long intellectual tradition of interest in nature in China and India not produce a similar outcome? This question is known as the Needham Question.[212] As a possible answer, one may mention that human history, being one whole, is composed of interacting components and that prevents isolated paths of development, which could need more time for independent growth. It is enough for one trend in some place of the world to develop faster than in others, for any reason, that its products would influence scholars in other societies preventing alternative or similar inventions, thus imposing the adoption of the products on others in the very different cultural environment. Adoption of discoveries, repeating and copying methods, and learning from others

---

[212] Needham (2004), Sivin (2005), and Gorelik (2012, 2018).

are all much easier than constructing original concepts, solutions, and theories. This is something that we all know in education, technology, and science. Strong interaction integrates people's ideas and undermines cultural originality, establishing a collective universal mode of scientific knowledge. Science is international in its nature.

Appendix 3

The theories' contest in cosmology—Copernicus versus Tycho (with the victory of the latter!)—is depicted on the front page of the 1651 book by Riccioli (Fig. 9.8). In science education, it can serve as an emblematic image for its explicit reference to the scientific method as was defined already in Biblical times. Specifically, the method required *numeration*, *measurement*, and *weighing* as its central NOS activities—"Numerus, Mensura, Pondus." This triad was adopted from the first century *Book of Wisdom of Solomon* (11:21), signifying the supernatural revelation (the name of God in Hebrew presents the ultimate origin).

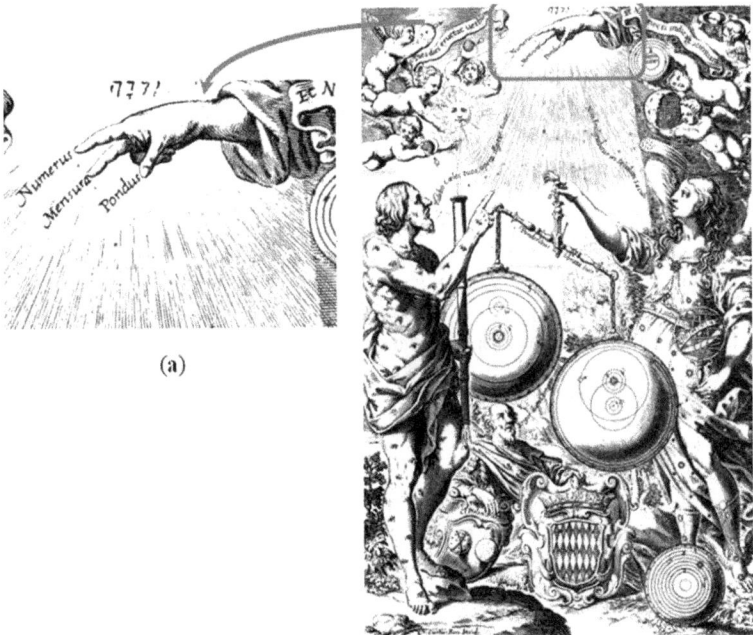

(a)

Fig. 9.8 The front page of the book by Riccioli (1651). (*a*) The enlarged fragment specifying the scientific method

Picture Credits

- Fragment from *St. Augustine in His Study* by V. Carpaccio, 1502. Web Gallery of Art: Image. Public Domain. https://commons.wikimedia.org/w/index.php?curid=9772902
- Fragment from illustration in the digitalized book, O. Guericke, Experimenta Nova, 1672, Amstelodami: Janssonius. The Herzog August Bibliothek Wolfenbüttel. http://diglib.hab.de/wdb.php?dir=drucke/34-5-phys-2f. Public Domain. https://commons.wikimedia.org/w/index.php?curid=18292849
- Plato conversing with pupils. Roman mosaic, first c. BCE, at the Museo Nazionale Archeologico, Naples. Photo by Jebulon, CC0 1.0. Public Domain. https://commons.wikimedia.org/w/index.php?curid=45896144
- Seventeenth c. Observatory. Illustration from Hogben, L. 1938. *Science for the Citizen*. London: George Allen & SL Unwin, Fig. 150, p. 236. Public Domain
- Illustration from W. Gilbert, *De Magnete*, 1600, London: Peter Short. Republished, 1900, London: Chiswick Press, Book 3, Ch, XII, p.139. Public Domain. Posner Memorial Collection in Electronic Format (cmu.edu)
- Fig. 9.7a Fragment of Rafael's *The School of Athens*, 1501, the Vatican Museum. Web Gallery of Art. Public Domain. https://commons.wikimedia.org/w/index.php?curid=75881
- Fig. 9.7b Fuxi and Nuwa painting, seventh c. by Anonymous Korean artist. The collection of the National Museum of Korea. Picture credit: the Korea Open Government, License Type I: Attribution; © the National Museum of Korea https://www.museum.go.kr/site/eng/relic/represent/view?relicId=435
- Fig. 9.8 The front-page in Riccioli, *New Almagest,* 1651. Public Domain. https://commons.wikimedia.org/w/index.php?curid=16670323

References

Abd-El-Khalick, F. (2012). Examining the sources for our understandings about science: Enduring conflations and critical issues in research on nature of science in science education. *International Journal of Science Education, 34*(3), 353–374.

Abd-El-Khalick, F., Bell, R. L., & Lederman, N. G. (1998). The nature of science and instructional practice: Making the unnatural natural. *Science Education, 82*(4), 417–436.

Agazzi, E. (2014). Objectivity as a replacement for truth in modern science. In *Objectivity and its contents* (pp. 1–10). Springer; and *Scientific objectivity and its contexts* (pp. 54–55). Springer.

Al-Khalili, J. (2010). *Pathfinders. The golden age of Arabic science*. Penguin Books.

Allchin, D. (2011). Evaluating knowledge of the nature of (whole) science. *Science Education, 95*(3), 518–542.

Allchin, D. (2013). *Teaching the nature of science. Perspectives and resources*. SHiPs.

Allchin, D. (2017). Beyond the consensus view: Whole science. *Canadian Journal of Science, Mathematics and Technology Education, 17*(1), 18–26.

Alters, B. J. (1997). Whose nature of science? *Journal of Research in Science Teaching, 34*, 39–55.

Aquinas, T. (1267/1952). *Summa Theologica*. Encyclopaedia Britannica.

Ausubel, D. P. (1968). *Educational psychology: A cognitive view*. Holt, Rinehart & Winston.

Ausubel, D. P. (2000). *The acquisition and retention of knowledge: A cognitive view*. Springer-Science.

Bazzul, J. (2017). From orthodoxy to plurality in the nature of science (NOS) and science education: A metacommentary. *Canadian Journal of Science, Mathematics and Technology Education, 17*(1), 66–71.

Berkovitz, J. (2017). Some reflections on "going beyond the consensus view" of the nature of science in K–12 science education. *Canadian Journal of Science, Mathematics and Technology Education, 17*(1), 37–45.

Berry, A. (1898/1961). *A short history of astronomy*. Dover.

Betz, F. (2011). Origin of scientific method. In *Managing science, Innovation, technology, and knowledge management* 9, 21. Springer.

Birstein, V. (2001). *The pervasion of knowledge. The true story of soviet science*. Westview Press.

Bohr, N. (1949). Discussion with Einstein on epistemological problems in atomic physics. In P. A. Schilpp (Ed.), *Albert Einstein: Philosopher-scientist* (pp. 199–241). Harper Torchbooks.

Bokulich, A. (2008). *Reexamining the quantum-classical relation. Beyond reductionism and pluralism*. Cambridge University Press.

Boxer, A. (2020). *A scheme of heaven*. Profile Books.

Braga, M., Guerra, A., & Reis, J. C. (2012). The role of historical-philosophical controversies in teaching sciences: The debate between biot and ampere. *Science & Education, 21*, 921–934.

Bronowski, J. (1967). *The common sense of science*. Harvard University Press.

Buber, M. (1958). *I and thou*. Charles Scribner's Sons.

Bunge, M. (1967a). *Quantum theory and reality*. Springer.

Bunge, M. (1967b). *Foundation of physics*. Springer.

Bunge, M. (1973). *Philosophy of physics*. Reidel Publishing Company.

Bunge, M. (2000). Energy: Between physics and metaphysics. *Science & Education, 9*(5), 457–461.

Bunge, M. (2015). In defense of scientism. *Free Inquiry, 35*(1), 24–31.

Carnap, R. (1971). *Philosophical foundations of physics. An introduction to the philosophy of science*. Basic Books.

Cartwright, N. (1983). *How the laws of physics lie*. Clarendon Press.

Cartwright, N. (1994). Fundamentalism vs the patchwork of laws. *Proceedings of the Aristotelian Society, 93*(2), 279–292.

Chalmers, A. F. (1976). *What is this thing called science?* The Open University Press.

Clough, M. P. (2007, January). Teaching the nature of science to secondary and post-secondary students: Questions rather than tenets, *The Pantaneto Forum*, Issue 25, http://pantaneto.co.uk/issue-25/

Clough, M. P., & Olson, J. K. (2004). The nature of science: Always part of the science story. *The Science Teacher, 71*(9), 28–31. Reprinted in Koulaidis, V., Apostolou, A., & Kampourakis, K. (Eds.) (2008). *The nature of sciences: Didactical approaches* (pp. 287–296).

Couvalis, G. (1997). *The philosophy of science. Science and objectivity*. Sage Publications.

Crombie, A. C. (1959). *Medieval and early modern science*. Doubleday Anchor Books.

Crombie, A. C. (1996). *Science, art and nature in medieval and modern thought*. The Hambledon Press.

Cromer, A. (1993). *Uncommon sense*. Oxford University Press.

Cushing, J. (1994). *Quantum mechanics: Historical contingency and the Copenhagen hegemony*. University of Chicago Press.

Dagher, Z., & Erduran, S. (2014). Laws in biology and chemistry: Philosophical perspectives and educational implications. In M. Matthews (Ed.), *International handbook of history, philosophy and science teaching* (pp. 1203–1233). Springer.

Dagher, Z. R., & Erduran, S. (2017). Abandoning patchwork approaches to nature of science in science education. *Canadian Journal of Science, Mathematics and Technology Education, 17*(1), 46–52.

Darrigol, O. (2000). *Electrodynamics from ampere to Einstein*. Oxford University Press.

Di Francia, G. T. (1976). *The investigation of the physical world*. Cambridge University Press.

Dirac, P. A. M. (1958). *The principles of quantum mechanics*. Clarendon Press.

Drake, S. (1957). *Discoveries and opinions of Galileo*. Doubleday & Company.

Dreyer, J. L. E. (1953). *A history of astronomy from Thales to Kepler*. Dover.

Duhem, P. (1905/1982). *The aim and structure of physical theory*. Princeton University Press.

Duschl, R. A., & Grandy, R. (2013). Two views about explicitly teaching nature of science. *Science & Education, 22*(9), 2109–2139.

Einstein, A. (1918/2002). *Principles of research. The collected papers of Albert Einstein: The Berlin years, 1918–1921* (pp. 42–45). Princeton University Press.

Einstein, A. (1934/2011a). Address at Columbia University, New York, January 15. In A. Einstein (Eds.), *Essays in science*. Open Road Integrated Media.

Einstein, A. (1934/2011b). On the method of theoretical physics. In *Essays in science*. Open Road.

Einstein, A. (1949/1979). Autobiographical notes. In P. A. Schilpp (Ed.), *Albert Einstein: Philosopher-scientist*. Harper.

Einstein, A. (1952/1987). *Letters to Solovine: 1906–1955 (May 7, 1952)*. Open Road, Integrated Media.

Einstein, A., & Infeld, L. (1938). *Evolution of physics*. Cambridge University Press.

Elby, A., & Hammer, D. (2001). On the substance of a sophisticated epistemology. *Science Education, 85*(5), 554–567.

Encyclopeadia Britannica. (1771/1979). The First Edition. Society of Gentlemen in Scotland.

Erduran, S. (2014). Beyond nature of science: The case for reconceptualizing 'science' for science education. *Science Education International, 25*(1), 933–111.

Erduran, S., & Dagher, Z. R. (2014). *Reconceptualizing the nature of science for science education*. Springer.

Feyerabend, P. (1993). *Against method*. Verso.

Feyerabend, P. (1999a). *Knowledge, science and relativism philosophical papers* (Vol. 3). Cambridge University Press.

Feyerabend, P. (1999b). Rationalism, relativism and scientific method. In P. Feyerabend (Ed.), *Knowledge, science and relativism*. Cambridge University Press.

Feynman, R. (1985/2014). *QED. The strange theory of light and matter*. Princeton University Press.

Finocchiaro, M. A. (1989). *The Galileo affair. A documentary history*. University of California Press.

Frank-Kamenetskii, M. (2013). Are there any laws in biology? Comment on "how life changes itself: The read–write (RW) genome" by James Shapiro. *Physics of Life Reviews, 19*, 328–330.

French, A. (1971). *Newtonian mechanics*. Norton.

Galilei, G. (1623/1957). *The Assayer*. Translated by Stillman Drake. In *Discoveries and opinions of Galileo* (pp. 237–238). Anchor Books.

Galilei, G. (1638/1914). *Dialogue concerning two new sciences*. Dover.

Galili, I. (2013). On the power of fine arts pictorial imagery in science education in science education. *Science & Education, 22*, 1911–1938.

Galili, I. (2018). Physics and mathematics as interwoven disciplines in physics class. *Science & Education, 27*(1–2), 7–37.

Garvey, B. (2007). *Philosophy of biology*. Acumen.

Giere, R. N. (1985). Philosophy of science naturalized. *Philosophy of Science, 52*, 331–356.

Giere, R. N. (1988). *Explaining science. A cognitive approach*. University of Chicago Press.

Giere, R. N. (1995). The sceptical perspective: Science without laws of nature. In F. Weinert (Ed.), *Laws of nature: Essays on the philosophical, scientific and historical dimensions* (pp. 120–138). Walter de Gruyter.

Ginzburg, V. L. (2005). *About science, myself and others*. Institute of Physics Publishing.

Glasersfeld, E. (1992). A constructivist view of learning and teaching. In R. Duit, F. Goldberg, & H. Niedderer (Eds.), *Research in physics learning: Theoretical issues and empirical studies* (pp. 29–40). IPN.

Glasersfeld, E. (1995). *Radical constructivism: A way of knowing and learning*. The Falmer Press.

Glashow, S. L. (1994). *From alchemy to quarks. Physics as liberal art*. Brooks.

Godfrey-Smith, P. (2003). *An introduction to the philosophy of science. Theory and reality*. The University of Chicago Press.

Goodman, N. (1968/1976). *Languages of art. An approach to a theory of symbols*. The Bobbs-Merrill Company.

Gorelik, G. (1995). Moscow, physics, 1937. In V. A. Kimanev (Ed.), *Tragic destiny*. Nauka.

Gorelik, G. (2012, April 6). How the modern physics was invented in the 17th century, part 1: The Needham question. *Scientific American*.

Gorelik, G. (2018). Hessen's explanation and the Needham question, or how Marxism helped to put an important question but hindered answering it. *Epistemology and Philosophy of Science, 55*(3), 153–171.

Gorelik, G., & Bouis, A. W. (2005). *The world of Andrei Sakharov. A Russian physicist's path to freedom*. Oxford University Press.

Gorelik, G., & Frenkel, V. Y. (1994). *Matvei Petrovich Bronstein and Soviet theoretical physics in the thirties*. Birkhauser Verlag.

Goren, E., & Galili, I. (2018). A summary lecture as a delay organizer of students' knowledge of mechanics – A Discipline-Culture Approach. Proceedings of the 11th Conference of the European Science Education Research Association (ESERA), Dublin.

Gorham, G., Hill, B., Slowik, E., & Waters, C. K. (Eds.). (2016). *The language of nature. Reassessing the Mathematization of natural philosophy in the seventeenth century*. University of Minnesota Press.

Gower, B. (1997). *Scientific method. An historical and philosophical introduction*. Routledge.

Granek, G. (2006). Poincare's light signaling and clock synchronization thought experiment and its possible inspiration to Einstein. In J. M. Alimni & A. Fuzfa (Eds.), *Albert Einstein century international conference* (pp. 1095–1102). American Institute of Physics.

Greighton, T. E. (1999). *Encyclopedia of molecular biology*. Wiley.

Gunstone, R. (Ed.). (2015). *Encyclopedia of science education*. Springer.

Hecht, E. (1998). *Optics*. Addison-Wesley.

Heisenberg, W. (1948). Der Begriff Abgeschlossene Theorie in Der Modernen Naturwissenschaft. *Dialectica, 2*(3–4), 331–336. Quoted in Popper (1962).

Heisenberg, W. (1959/1971). *Physics and philosophy. The revolution in modern science*. Harper.

Heisenberg, W. (1967). *Quantum mechanics and objectivity*. Martinus Nijhoff.

Heisenberg, W. (1971). *Physics and beyond*. Harper & Row.

Hempel, C. G. (1966). *Philosophy of natural science*. Prentice Hall.

Hempel, C. G. (1983). Validation and objectivity in science. In R. S. Cohen & L. Laudan (Eds.), *Physics, philosophy and psychoanalysis essays in honor of Adolf Grilnbaum* (pp. 73–100). Reidel Publishing Company.

Hessen, B. M. (1933). *Socio-economical roots of Newton's mechanics*. GTTI.

Hessen, B. M. (2009). The social and economic roots of Newton's *Principia*. In G. Freudenthal & P. McLaughlin (Eds.), *The social and economic roots of the scientific revolution. Texts by Boris Hessen and Henryk Grossmann* (Boston studies in the philosophy of science) (Vol. 278). Springer.

Hestenes, D. (1992). Modeling games in the Newtonian world. *American Journal of Physics, 60*, 732–748.

Hodson, D. (2011). Looking to the future. In *Building a curriculum for social activism*. Sense Publishers.

Hodson, D. (2014). Nature of science in the science curriculum: Origin, development, implications and shifting emphases. In M. R. Matthews (Ed.), *International handbook of research in history, philosophy and science teaching* (pp. 911–970). Springer.

Hodson, D., & Wong, S. L. (2017). Going beyond the consensus view: Broadening and enriching the scope of NOS-oriented curricula. *Canadian Journal of Science, Mathematics and Technology Education, 17*(1), 3–17.

Holton, G. (1985). *Introduction to concepts and theories in physical science* (2nd ed. revised by S. G. Brush). Princeton University Press.

Hoskin, M. (1997). *The Cambridge illustrated history of astronomy*. Cambridge University Press.

Hudson, J. (1992). *The history of chemistry*. The Macmillan Press.

Huizenga, J. R. (1993). *Cold fusion: The scientific fiasco of the century*. University of Rochester Press.

Huygens, C. (1690/1912). *Treatise on light: In which are explained the causes of that which occurs in reflection & in refraction, and particularly in the strange refraction of Iceland crystal*. Macmillan.

Kierkegaard, S. (2009). *Concluding unscientific postscript to the philosophical crumbs*. Cambridge: Cambridge University Press.

Irzik, G., & Nola, R. (2011). A family resemblance approach to the nature of science for science education. *Science & Education, 20*, 591–607.

Irzik, G., & Nola, R. (2014). New directions for nature of science research. In M. R. Matthews (Ed.), *Handbook of historical and philosophical research in science education* (pp. 999–1021). Springer.

Josephson, P., & Sorokin, A. (2017). Physics moves to the provinces: The Siberian physics community and Soviet power, 1917–1940. *British Journal for the History of Science, 50*(2), 297–327.

Kalman, K. S., & Adulls, M. W. (2003). Can an analysis of the contrast between pre-Galilean and Newtonian theoretical frameworks help students develop a scientific mindset? *Science & Education, 12*(8), 761–772.

Kampourakis, K. (2017). History and philosophy of science courses for science students. *Science & Education, 26*, 611–612.

Kepler, J. (1621/1972). *Epitome of Copernican astronomy* (p. 845). Britannica.

Khan Academy. (2017). *Scientific method.* Retrieved December 3, 2017, from https://www.you-tube.com/watch?v=N6IAzlugWw0

Kipnis, N. (2010). Scientific controversies in teaching science: The case of Volta. *Science & Education, 10*, 33–49.

Knight, R. D. (2013). *Physics for scientists and engineers* (3rd ed.). Pearson.

Kockelmans, J. J. (1985). *Heidegger and science.* University Press of America.

Koyré, A. (1968). *Metaphysics and measurement: Essays in scientific revolution.* Harvard University Press.

Kuhn, T. S. (1957). *The Copernican revolution. Planetary astronomy in the development of Western thought.* Harvard University Press.

Kuhn, T. S. (1962/1970). *The structure of the scientific revolution.* The University of Chicago Press.

Kuhn, T. S. (1969). Postscript 1069. In Kuhn, T. (1970). *The structure of the scientific revolution.* The University of Chicago Press.

Kuhn, T. S. (1977). Objectivity, value judgement, and theory choice. In T. S. Kuhn (Ed.), *Essential tension* (Selected studies in scientific tradition and change) (pp. 320–339). The University of Chicago Press.

Kuhn, T. S. (2000). *The road to science structure.* The University of Chicago.

Lakatos, I. (1980). Falsification and the methodology of scientific research programmes. In J. Worrall & G. Currie (Eds.), *Imre Lakatos philosophical papers* (*The methodology of scientific research programs*) (Vol. 1, pp. 8–101). Cambridge University Press.

Lakatos, I. (1998). Science and pseudoscience. In M. Curd & J. A. Cover (Eds.), *Philosophy of science. Central issues* (pp. 20–26). Norton.

Lakatos, I. (1999). Lectures on scientific method. In I. Lakatos & P. Feyerabend (Eds.), *For and against method.* The University of Chicago Press.

Latour, B. (1987). *Science in action: How to follow scientists and engineers through society.* Harvard University Press.

Laudan, L. (1977). *Progress and its problems.* University of California Press.

Lederman, L. (1998). A response. *Studies in Science Education, 31*, 130–135.

Lederman, N. G. (2004). Syntax of nature of science within inquiry and science instruction. In L. B. Flick & N. G. Lederman (Eds.), *Scientific inquiry and nature of science* (pp. ix–xviii). Kluwer Academic Publishers.

Lederman, N. G. (2007). Nature of science: Past, present, and future. In S. K. Abell & N. G. Lederman (Eds.), *Handbook of research on science education* (pp. 831–879). Erlbaum.

Lederman, N. G., Abd-El-Khalick, F., Bell, R. L., & Schwartz, R. S. (2002). Views of nature of science questionnaire: Toward valid and meaningful assessment of learners' conceptions of nature of science. *Journal of Research in Science Teaching, 39*(6), 497–521.

Lederman, N. G., Abd-El-Khalick, F., & Schwartz, R. (2015a). Measurement of NOS. In R. Gunstone (Ed.), *Encyclopedia of science education* (pp. 704–708). Springer.

Lederman, N. G., Bartos, S. A., & Lederman, J. S. (2014). The development, use, and interpretation of nature of science assessments. In M. R. Matthews (Ed.), *International handbook of research in history, philosophy and science teaching* (pp. 974–978) Springer Dordrecht.

Lederman, N. G., Schwartz, R., & Abd-El-Khalick, F. (2015b). Conceptualizing the construct of NOS. In R. Gunstone (Ed.), *Encyclopedia of science education* (pp. 694–698). Springer.

Lederman, N. G., Wade, P. D., & Bell, R. L. (1998). Assessing understanding of the nature of science: A historical perspective. In W. McComas (Ed.), *The nature of science in science education: Rationales and strategies* (pp. 331–350). Kluwer Academic Publishers.

Levrini, O., Bertozzi, E., Gagliardi, M., Grimellini-Tomasini, N., Pecori, B., Tasquier, G., & Galili, I. (2014). Meeting the discipline-culture framework of physics knowledge: An experiment in Italian secondary school. *Science & Education, 23*, 1701–1731.

Lévy-Leblond, J.-M. (2001). On the nature of quantons. *Science & Education, 12*(5), 495–502.

Lindberg, D. C. (1976). *Theories of vision from Al-Kindi to Kepler.* The University of Chicago Press.

Lindberg, D. C. (1992). *The beginnings of the Western science.* The University of Chicago Press.

Lindberg, D. C. (2002). The Western reception of Arabic optics. In R. Rashed (Ed.), *Encyclopedia of the history of Arabic science* (Vol. 2, pp. 363–371). Routledge.

Longino, H. (1990). *Science as a social knowledge. Values and objectivity in science inquiry.* Princeton University Press.

Losee, J. (1993). *A historical introduction to the philosophy of science.* Oxford University Press.

Mach, E. (1883/1919/1989). *The science of mechanics, a critical and historical account of its development.* Open Court.

Mach, E. (1913/1926). *The principles of physical optics. An historical and philosophical treatment.* Dover.

Mach, E. (1976). *Knowledge and error. Sketches on the psychology of enquiry.* D. Reidel.

Marton, F., & Pang, M. F. (2006). On some necessary conditions of learning. *The Journal of the Learning Sciences, 15*(2), 193–220.

Marton, F., & Pang, M. F. (2013). Meanings are acquired from experiencing differences against a background of sameness, rather than from experiencing sameness against a background of difference: Putting a conjecture to test by embedding it into a pedagogical tool. *Frontline Learning Research, 1*(1), 24–41.

Matthews, M. R. (2009). Teaching the philosophical and worldview components of science in science. *Science & Education, 18*, 697–728. M. R. Matthews (Ed.). *Science, worldviews and education.* Springer.

Matthews, M. R. (2012). Changing the focus: From nature of science (NOS) to features of science (FOS). In M. S. Khine (Ed.), *Advances in nature of science research* (pp. 3–26). Springer.

Matthews, M. R. (2019). *Feng Shui: Teaching about science and pseudoscience.* Springer Nature.

McComas, W. F. (1998). The principal elements of the nature of science: Dispelling the myths. In W. F. McComas (Ed.), *The nature of science in science education: Rationales and strategies* (pp. 53–70). Kluwer.

Merton, R. K. (1973). *The sociology of science: Theoretical and empirical investigations.* University of Chicago Press.

Millar, R. (2000). Science for public understanding: Developing a new course for 16–18 year old students. *Melbourne Studies in Education, 41*(2), 201–214.

Miller, A. I. (1981). *Albert Einstein's special theory of relativity: Emergence (1905) and early interpretation (1905–1911).* Addison-Wesley.

Miller, A. I. (1984). *Imagery in scientific thought: Creating 20th-century physics.* Birkhauser.

Miller, A. I. (1996). *Insight of genius.* Copernicus. Springer.

Nagel, E. (1961). *The structure of science.* Harcocoart, Brace and World.

Needham, J. (2004). *Science and civilization in China* (Vol. 7, Part 2). Cambridge University Press.

Nersessian, N. J. (1992). How do scientists think? Capturing the dynamics of conceptual change in science. In R. Giere (Ed.), *Cognitive models of science* (Minnesota studies in the philosophy of science) (pp. 3–45). University of Minnesota Press.

Neugebauer, O. (1993). *The exact sciences in antiquity.* Barrens & Noble.

Newton, I. (1670). Optical lectures. In A. Shapiro (1984). *Newton's optical lectures.* Cambridge University Press.

Newton, I. (1687/1999). *The principia. Mathematical principles of natural philosophy* (B. Cohen & A. Whitman, Trans.). University of California Press.

Niaz, M. (2009). *Critical appraisal of physical science as a human enterprise: Dynamics of scientific progress.* Springer.

Nozick, R. (2000). The objectivity and the rationality of science. In J. H. Fetzer (Ed.), *Science, explanation, and rationality: Aspects of the philosophy of Carl G. Hempel* (pp. 287–308). Oxford University Press.

Osborne, J. (2017). Going beyond the consensus view: A response. *Canadian Journal of Science, Mathematics and Technology Education, 17*(1), 53–57.

Osborne, J., Collins, S., Radcliffe, M., Millar, R., & Duschl, R. (2003). What "ideas-about-science" should be taught in school science? A Delphi study of the expert community. *Journal of Research in Science Teaching, 40*(7), 692–720.

Panofsky, W. K. H., & Phillips, M. (1962). *Classical electricity and magnetism*. Addison-Wesley.

Partington, J. R. (1962). *A history of chemistry*. Macmillan.

Pattison Muir, M. M. (2004). *The story of alchemy and the beginnings of chemistry*. The Project Gutenberg eBook.

Pedersen, O., & Pihl, M. (1974). *Early physics and astronomy*. McDonald & Janes.

Popper, K. R. (1959/2002). *The logic of scientific discovery*. Routledge.

Popper, K. R. (1962). Theories as instruments. In *Conjectures and refutations. The growth of scientific knowledge*. Basic Books.

Popper, K. R. (1967). Quantum mechanics without "the observer". In M. Bunge (Ed.), *Quantum theory and reality* (pp. 7–44). Springer.

Popper, K. R. (1975). *Objective knowledge*. Clarendon Press.

Popper, K. R. (1978). Three worlds. *The Tanner lecture on human values*. The University of Michigan. http://www.tannerlectures.utah.edu/lectures/documents/popper80.pdf. Accessed 24 Sept 2015.

Primary School Curriculum of Ireland, Science. (1999). *Teacher guidelines*. Government Publications. https://docplayer.net/1871672-Primary-school-curriculum-science-social-environmental-and-scientific-education-teacher-guidelines.html

Rabinowitz, M. (2017). *Examination of wave-particle duality via two-slit interference*. https://arxiv.org/pdf/physics/0302062. Retrieved 14 Dec 2017.

Read, J. (1995). *From alchemy to chemistry*. Dover.

Reichenbach, H. (1938). *Experience and prediction: An analysis of the foundations and the structure of knowledge*. University of Chicago Press.

Reiss, J. (2014). *Scientific objectivity*. https://plato.stanford.edu/entries/scientific-objectivity/. Retrieved on August 16, 2017.

Roob, A. (2001a). *Alchemy & mysticism*. Taschen.

Roob, A. (2001b). *The hermetic museum: Alchemy & mysticism*. Tachen.

Russell, B. (1912/1990). *The problems of philosophy*. Hackett Pub.

Russell, B. (1959). *Wisdom of the west*. Crescent Books.

Russell, B. (2009). Dewey's new logic. In R. E. Egner & L. E. Denonn (Eds.), *The basic writings of Bertrand Russell*. Routledge.

Russo, L. (2004). *The forgotten revolution: How science was born in 300 B.C. and why it had to be reborn*. Springer.

Rutkin, H. D. (2001). Celestial offerings: Astrological motifs in the dedicatory letters of Kepler's *Astronomia Nova* and Galileo's *Sidereus Nuncius*. In W. R. Newman & A. Grafton (Eds.), *Secrets of nature. Astrology and alchemy in early modern Europe* (pp. 133–172). The MIT Press.

Scheffler, I. (2009). *Words of truth. A philosophy of knowledge*. Willey-Blackwell.

Scheker, N., & Niedderer, H. (1996). Contrastive teaching: A strategy to promote qualitative conceptual understanding of science. In D. F. Treagust, R. Duit, & B. I. Frazer (Eds.), *Improving teaching and learning in science and mathematics* (pp. 141–151). Teachers College Press.

Schwab, J. J. (1964). Problems, topics, and issues. In S. Elam (Ed.), *Education and the structure of knowledge* (pp. 4–47). Rand McNally.

Schwab, J. J. (1978). Education and the structure of the disciplines. In J. J. Schwab (Ed.), *Science, curriculum and liberal education*. The University of Chicago Press.

Schwartz, D. L., Chase, C. C., Oppezzo, M. A., & Chin, D. B. (2011). Practicing versus inventing with contrasting cases: The effects of telling first on learning and transfer. *Journal of Educational Psychology, 103*(4), 759–775.

Sequeira, M., & Leite, L. (1991). Alternative conceptions and history of science in physics teacher education. *Science Education, 75*(1), 45–56.

Serway, R. A., Moses, C. J., & Moyer, C. A. (2005). *Modern physics*. Thomson, Brooks/Cole.

Shapin, S. (1996). *The scientific revolution*. The University of Chicago Press.

Shapiro, A. E. (1984). Experiment and mathematics in Newton's theory of color. *Physics Today, 37*(9), 34–42.

Shapiro, A. E. (2004). Newton's "experimental philosophy". Newtonianism: Mathematical and 'experimental'. *Early Science and Medicine, 9*(3), 185–217.

Shulman, L. S. (1986). Those who understand: Knowledge growth in teaching. *Educational Researcher, 15*, 4–14.

Shulman, L. S. (1987). Knowledge and teaching: Foundations of the new reform. *Harvard Educational Review, 57*, 1–22.

Sivin, N. (2005). Why the scientific revolution did not take place in China—Or didn't it? htstp://ccat.sas.upenn.edu/~nsivin/scirev.pdf

Slezak, P. (1994). Sociology of scientific knowledge and scientific education. *Science & Education, 3*, 265–294.

Sokal, A., & Bricmont, J. (1998). *Fashionable nonsense. Postmodern Intellectuals' abuse of science*. Picador.

Stang, N. (2018). *Kant's transcendental idealism*. https://plato.stanford.edu/archives/win2018/entries/kant-transcendental-idealism

Steinberg, M. S., & Wainwright, C. L. (1993). Using models to teach electricity – The CASTLE project. *The Physics Teacher, 31*, 353–357.

Taylor, G. I. (1909). Interference fringes with feeble light. *Proceedings of the Cambridge Philosophical Society, 15*, 114–115.

Taylor, L. W. (1941). *Physics. The pioneer science*. Dover.

Thornton, S. (2016). Karl Popper. In *Stanford encyclopedia of philosophy*. Retrieved on August 18, 2017, https://plato.stanford.edu/entries/popper

Tipler, P. A. (1987). *Modern physics*. Wort Publishers.

Toulmin, S. (1972). *Human understanding*. Clarendon Press.

van Fraassen, B. C. (1980). *The scientific image*. Clarendon Press.

van Fraassen, B. C. (2008). Scientific representation: Paradoxes of perspective. Clarendon Press

Vinner, S. (1991). The role of definitions in teaching and learning mathematics. In D. Tall (Ed.), *Advanced mathematical thinking* (pp. 65–81). Academic Publishers.

Vinner, S. (1997). The pseudo-conceptual and the pseudo-analytical thought processes in mathematics learning. *Educational Studies in Mathematics, 34*(2), 97–129.

Vygotsky, L. (1934/1986). *Thought and language*. The MIT Press.

Wallace, J. (2017). Teaching NOS in an age of plurality. *Canadian Journal of Science, Mathematics and Technology Education, 17*(1), 1–2.

Weinberg, J. R. (1936). *An examination of logical positivism*. Kegan Paul, Trench, Trubner & Co.

Weinberg, S. (2001). *Facing up – Science and its cultural adversaries*. Harvard University Press.

Weinberg, S. (2015). *To explain the world: The discovery of modern science*. Harper Collins Publishes.

Weizsacker, C. F. (1985/2006). *The structure of physics*. Springer.

Wilczek, F. (2004). Whence the *force* of F = ma? *Physics Today, 57*(12), 10.

Wolpert, L. (1994). *The unnatural nature of science*. Harvard University Press.

Wong, S. L., & Hodson, D. (2009). From the horse's mouth: What scientists say about scientific investigation and scientific knowledge. *Science Education, 93*(1), 109–130.

Wong, S. L., & Hodson, D. (2010). More from the horse's mouth: What scientists say about science as a social practice. *International Journal of Science Education, 32*(11), 1431–1463.

Chapter 10
On the Power of Fine Arts Pictorial Imagery in Science Education

Gravure on the front page of the book *Ars Magna Lucis et Umbrae* (The Great Knowledge of Light and Shadow) by Athanasius Kircher published in 1650.

> *Drawing the object or phenomenon so skillfully that it would become a "fact" which all the world could see, or could grasp with the aid of brief explanatory notes.*

Leonardo da Vinci

Abstract This chapter illustrates the use of pictorial artistic images in teaching scientific concepts and the nature of science. One example is Giotto's fresco of the stigmatization of St. Francis. Its interpretation associates an artistic image of a philosophical idea with plane mirror features facilitating the expression of the idea. Specific implications of this case for teaching optics are suggested. Other pictorial images are discussed in the intention to suggest them for using in the teaching of science at school. These images can facilitate hermeneutic reconsideration addressing the meaning and nature of scientific knowledge, its specific features in forms especially appealing to people for their aesthetic value, and imagination and the surprising discovery of aspects easily missed in a disciplinary instruction of scientific technicalities. These aspects are of holistic importance in public education. Finally, science presents an image of reality in human mind. This image may also be expressed in artistic form.

Introduction

Science education mediates between socially constructed scientific knowledge and society itself. Various tools are available to a teacher serving as a mediator and trying to represent the knowledge of science. They range from the formal analytical to the visual graphical. According to developmental psychology, the stage of concrete perception precedes abstract conceptual thinking in individual growth. This fact explains quantitative and qualitative changes in the learning materials textbooks used at different levels of teaching. Typically, the more advanced the level, the less abundant the use of pictorial illustrations, which are often considered as less specific and fuzzier in meaning than the propositional representation of knowledge.

In the course of learning, graphs, diagrams, and sketches become increasingly more and more important, gradually replacing pictorial images as visualizations of knowledge claims. Eventually, abstract analytical form, equations, and formulas supersede all other forms,[1] pushing aside other visualized accounts.[2] In general, the advanced texts of physics seek the generality of abstract claims, free of concrete images. In fact, this feature distinguishes formal texts from popular ones.

The original version of this chapter was revised: We have to update the erratum correction on page 463, 464 and 468. The correction to this chapter is available at https://doi.org/10.1007/978-3-030-80201-1_12

---

[1] For example, Landau and Lifshitz (1960) on physics, and Whittaker (1910/1951) on physics history, are free of any visual images.

[2] For instance, lines of fields and Feynman diagrams are replaced with analytical terms.

If so, one may ask whether geometry, which manipulates visually concrete symbolism, is inferior to algebraic symbolism. Descartes[3] seemingly thought that it was, and invented analytic geometry as a more inclusive representation of geometrical knowledge. Even so, Descartes's own writings in physics[4] included plenty of illustrative sketches, which became emblematic of his work.[5]

This topic of visual concrete versus abstract analytical was once raised in the discussion regarding mathematical curricula in public schools. The renowned Russian mathematician, Arnold,[6] reflected on this debate with his colleagues in France. Arnold rejected the stigma of primitiveness sometimes ascribed to visual images and geometry in mathematics. Earlier, Poincare[7] identified two kinds of mathematicians and students: "some prefer to treat their problems 'by analysis', as they say, others 'by geometry'." Maxwell had addressed the same when he stated that three types of mind existed among practitioners in physics. Those of the first type of master mathematical symbols have pure quantities with which they manipulate physics knowledge. Others, he continued, prefer to follow geometrical forms for the same purpose. Still, others are the consumers who enjoy solving concrete quantitative problems. Maxwell concluded:[8]

> *For the sake of persons of these different types, scientific truth should be presented in different forms, as should be regarded as equally scientific, whether it appears in the robust form and the vivid coloring of a physical illustration, or in the tenuity and paleness of a symbolical expression.*

Visual images played a crucial role in Maxwell's major contribution—the theory of electromagnetism.[9] Moreover, those scientists who resisted his theory because it drew upon the visual model of the hypothetical media under tension and motion[10] paid a heavy price by neglecting the theory, which became the classical theory of electromagnetism. Thus, de Broglie wrote regarding Duhem:[11]

> *It seems his researches on electromagnetism were less happy, for he always had a great hostility towards Maxwell's theory and preferred Helmholtz' ideas, which are quite forgotten today. His deep antipathy with regard to all pictorial models prevented him, moreover, from understanding the importance of the Lorentz theory of electrons, then in full development, and rendered him as unjust as he was shortsighted about the rise of atomic physics, then in its beginning.*

Psychologists describe the two hemispheres of the human brain as being responsible for different activities: the left one processes analytic-verbal information, logic, numbers, and abstraction, whereas the right hemisphere operates with holistic imagery, metaphor, and music.[12] The brain thus requires complimentary activities. The

---

[3] Descartes (1637/1965).

[4] Descartes (1644/1982).

[5] Baigrie (1996a, b).

[6] Arnold (1998, 2002).

[7] Poincare (1903).

[8] Maxwell (1890/1965, Vol. 2, 220).

[9] Miller (1984), Nersessian (1992).

[10] Maxwell (1890/1965, Vol. 1, pp. 451–513, 526–597).

[11] de Broglie (1953).

[12] Orenstein (1975), Shlain (1991).

curricular policy in education is to start with teaching pictorial presentations and slowly shifts to abstract concepts codified in analytical terms. However, schools do not allow visual means to be suppressed since both cognitive functions, holistic and analytical, are required to master scientific knowledge professionally or for literacy. The claim of progression from visual to abstract as reflecting maturation requires refinement since we know that at least some scholars preserve a clear cognitive preference for visual images.[13] Therefore, teaching addressing a wide public should preserve visual imagery at every educational level.

Within the broad topic of visualization, we will consider a special type of artwork, which illustrates, this way or that, certain scientific and philosophical ideas and sheds light on their meaning in a different context.

Such "pictures are vehicles for the storage, manipulation, and communication of information."[14] Thus, they bridge between science and art. Indeed, art deals with aesthetics, beauty, feelings, and values. On the surface, it is distant from the logically framed "laws of nature." Yet, art may be relevant to learning science, because understanding pictures goes beyond the reading of naïve mimetic representations of objects by the viewer.[15] Pictures present "symbol systems."[16] They signify and depict reality interwoven with ideas in a system-relative representation.[17] The symbols used by an artist in a certain piece of art can be borrowed from elsewhere, mystics, mathematics, science, etc. In such cases, they can provide an interpretation valid in an educational perspective.

Fig. 10.1 Fragment from the *Scenes of Nativity* by Vitale da Bologna (S. Salvatore church in Bologna). The arrow points to the geometrical structure looking as a symbol rather than an element of architecture. (Photo by the author)

---

[13] Allegedly, Einstein used to say: "If I can't picture it, I can't understand it."

[14] Lopes (2004, p. 7).

[15] Gombrich (1972, p. 27).

[16] Goodman (1968/1976, p. 5).

[17] Lopes (2004, p. 59).s

For example, let us take *Scenes of Nativity* by Vitale da Bologna from the fourteenth century (Fig. 10.1). This picture shows an event of Christian mythology, which is depicted in thousands of similar works. However, in this case, the eye of an observer may catch a strange detail in the center: a series of right isosceles triangles creating a form of a fractal (self-similar) structure. What is the meaning of this symbol given the time and style of painting much before the modern formalistic art of the twentieth century? It is in striking difference to the rest of elements.

The art literature that provided descriptions of the picture did not help me. Yet, when I asked my friend, a teacher of mathematics, he told me right away: "Oh, this is a cluster of right-angled isosceles triangles which infinitely multiplies itself in similar smaller triangles just by drawing one line in each triangle – the altitude from the right-angle vertex causing the endless fractal branching."[18] My second question, about the meaning of this object in geometry, took him more time, and his answer impressed me: "The right isosceles triangle"—he said—"was the shape in which Pythagoreans discovered the incommensurability of certain distances which corresponds to so called irrational numbers." If so, one may infer that Vitale da Bologna drew a geometrical pattern that, for a person literate in geometry, indicated *irrationality* and *infinity,* or at least suggested to him this trend of thought for possible interpretation.[19] Was that image a powerful symbol for expressing the divine meaning that Christian philosophy ascribes to the event of Nativity? And if so, was this a case of art borrowing symbols for an expression from another realm of knowledge? How can we know if the artists do not talk? Let us take another case from scholars of that time who did.

Luca Pacioli in the fifteenth century renamed the Greek "golden ratio" (more precisely, *the division in extreme and mean ratio*) to be "*divine* proportion." In his *De Divina Proportione* he explained:[20]

> *3. Just like God cannot be properly defined, nor can be understood through words, likewise our proportion cannot be ever designated by intelligible numbers, nor can it be expressed by any rational quantity, but always remains concealed and secret, and is called irrational by the mathematicians, …*
>
> *4. The omnipresence and invariability of God is like the self-similarity associated with the divine proportion: its value is always the same and does not depend on the length of the line being divided or the size of the pentagon in which ratios of lengths are calculated.*

If Pacioli, a man of the Renaissance, made this transfer of meaning from mathematics to religious context, why could other educated people of that time, like Vitale de Bologna and Botticelli, all deeply equally immersed in the religious intellectual ecology, not do that in their media?

---

[18] This knowledge from Euclid's *Elements* had been available in Europe in Latin translations from Arabic since the twelfth and thirteenth centuries (Crombie, 1959, p. 44).

[19] Another later consumer of the same symbolism was Sandro Botticelli in his *The Adoration of the Magi* (the National Gallery, Washington, DC).

[20] Herz-Fischler (1987, p. 172), Livio (2002, p. 132). The golden ratio is represented by an irrational number: $\varphi = (1 + \sqrt{5})/2$.

The science educator may focus on the scientific content and still involve artistic images. A pictorial image may not immediately provide appreciation of a specific meaning, not the image Poincare described as "to perceive the whole of the argument at a glance."[21] Rather, it may require interpretation, possibly speculative, to be discussed in light of an educational goal, beneficial in itself. In the following, some examples of this are given, addressing the scientific content and certain aspects of the nature of scientific knowledge and its epistemology. These images can not only esthetically please and emotionally involve students but also provide an intellectually rich environment in science class. The inclusion of such images provides science students with the features usually ascribed solely to the students of humanities—cultural values. It may unfetter the self of students and benefit their taking science classes in a unique way.

Encoding Meaning into Artistic Images

The tradition of involving art started from the time science was first adopted for teaching at public schools. The pioneers drew on the values of liberal education in its broadest sense. Huxley wrote in his essay on science education:[22]

> There are other forms of culture besides physical science, and I should be profoundly sorry to see the fact forgotten, or even to observe a tendency to starve or cripple literary or aesthetic culture for the sake of science. Such a narrow view of the nature of education has nothing to do with my firm conclusion that a complete and thorough scientific culture ought to be introduced into all schools.

These claims were not specific and did not refine how or what to do. The idea of the art–science coexistence in science teaching simply fits the image of the harmonious personality. Huxley somewhat expanded on this idea when he mentioned that painting affected people differently since the perceived truth "depends entirely upon the intellectual world of the person to whom art is addressed." He wrote:[23]

> The intellectual knowledge we possess brings its criticism into our appreciation of works of art, and we are obliged to satisfy it, as well as the mere sense of beauty in colour and in outline. And so, the higher the culture and information of those whom art addresses, the more exact and precise must be what we call its "truth to nature."

People observing pieces of art not only enjoy their beauty but can be perceptually and conceptually engaged with the truth of nature. There is a rich intellectual benefit of the involvement of art in science education. Numerical proportions such as the golden section and Fibonacci numbers infused mathematical meaning into structures and pictures. The Egyptian pyramids, the Greek Parthenon, medieval cathedrals, and Renaissance paintings integrated certain mathematical meaning in their design.[24] Piero della Francesca and Leonardo de Vinci codified the numerical ratio

---

[21] Miller (1984, p. 233).

[22] Huxley (1882, p. 162).

[23] Huxley (1882, p. 178).

[24] For example, Lawlor (1982); Heilbronn (1998, pp. 235–241), Vincent (2001), Livio (2002).

in the depicted objects and their arrangements.[25] This way the artists tried to convey the idea of the universal metaphysical principle underling the cosmos. Contemporary observers are unaware of this content, and painters do not subscribe in any way to the call of Luca Pacioli in his *Divina Proportione* (1509) for the involvement of mathematics because:[26]

> *A work necessary for all clear-sighted and inquiring human minds, in which everyone who loves to study philosophy, perspective, painting, sculpture, architecture, music and other mathematical disciplines will find a very clear, subtle and admirable teaching and will delight in diverse questions touching upon a very secret science.*

The method of perspective avoided the fate of oblivion. Introduced in the *Optics* of Euclid,[27] it was rediscovered in painting and architecture through the efforts of Filippo Brunelleschi and Leon Battista Alberti, scholars of the Renaissance.[28] Perspective became a standard technique in art—the way to reproduce spacious reality on a flat canvas and on the curved vaults and cupolas. Such were Correggio's paintings on the curved surfaces in Parma.[29] At the same time, flat plane was perceived as spacious. Pozzo's creation of a spacious panorama on the flat ceiling, his virtual dome of 18 m in diameter in Saint Ignatius of Rome and the similar frescos in the University Church of Vienna amaze contemporary scholars.[30] Floors received depth and amazed crowds (Fig. 10.2). Masters of art challenged regular perception, surprised, confused, and transferred new meaning.

Fig. 10.2 Using perspective rules in the marble paving of floors in the Cathedral of Florence. (**a**) The floor in the central hall creates a view of a much higher level of the worship place (a virtual steep wall downward). (**b**) The floor fragment that creates a view of a cupola and may cause a sensation of losing one's balance. (Photo by the author)

---

[25] Lawlor (1982, pp. 62–63), Andersen (2007, pp. 65).

[26] Livio (2002, p. 131).

[27] Cohen and Drabkin (1948).

[28] Park (1997).

[29] Wind (2002).

[30] Andersen (2007, pp. 389–394).

Conversely, the impact of art on science was also remarkable. Art contributed through the aesthetic decoration of scientific documents in different realms. Artisans, driven by great fantasy, embellished astronomical maps, instruments, furniture, and scientific instruments (telescopes, mirrors, thermometers, etc.).[31] Artistic representations of the *contents* and the *structure* of scientific knowledge were preserved throughout history, being both informative and attractive for generations of consumers:[32]

> By admitting that it is possible to think in images, we come closer to the Renaissance view of how pictures actually function, and thus to how pictures achieved their authoritative structure. For a thought picture is not necessarily either 'true' or 'false', but instead occupies a realm of its own. It is a kind of mental construct that may (or may not) be subject to whatever passes for 'rational analysis' within the framework of the thinker's world. ... This business of thinking-in-pictures is clearly not science. ... And yet, it is not unrelated to science either.

Thinking in pictures is an extremely old way humans expressed themselves. Yet, one may see its clear surge in the time of Renaissance when arts are raised in status reaching the height of philosophy. Seeking the beauty of appearance of the observed form and its imagery became a commonplace. It was a great return (and so the name of the period, Renaissance—rebirth) to the classical Greece. Within this great ambition of the new vision of life, art entered to science and vice versa. In a way, the books of the period illustrate the new genre of art.

Margarita Philosophica, a textbook by Gregor Reisch from 1503, presents a good example. It included allegorical pictures describing the views and knowledge of that time.[33] Thus, the science of logic was represented by the drawing of Fig. 10.3a—*Typus Logice*. The picture invites a long observation and a gradual revealing of the basic elements of logic, its agenda, tools, and potential difficulties, which comprise a revealing scene of a young hunter marching toward his future endeavors—questions to be answered. Similarly, Fig. 10.3b—*The Tower of Learning*—represents the whole conception of liberal education at that time, its components, hierarchy, heroes, and values.

In this regard, it is appropriate to mention the artistic images of cover pages of the scientific books. In fact, the front pieces of books established a special artistic genre, which exists ever since. It was always a challenge for an artist to represent a book as a whole. The front page had to represent the book content, perhaps holistically, and possibly beyond, to the philosophical idea, perspective. Illustrations of such were brought in the previous chapters. Thus, Fig. 3.16 (Ch. 3) presented the front page of Witelo's textbook *Perspectiva* as was published in Basel in 1572. It presented Alhazen's optics to the Europeans. The artist, or the author, scanned the book content and chose the most impressive, as he thought, optical phenomena (the rainbow, concave mirror, and the light focusing by a mirror as was allegedly used

---

[31] For example, Fludd (1617), Cellarius (1661), Margotta (1968), Goodwin (1979), Miniati (1991), Whitfield (1995), Dekker (2004), and Evers (2006).

[32] Hall (1996, p. 36).

[33] Jung (1944/1993), Piltz (1981), and Roob (2001).

(a)   (b)

Fig. 10.3 (a) *Typus Logice*—The picture representing the structure of logic and its application in the form of pictorial allegory. (b) *The Tower of Learning* depicting the structure of liberal education in a pictorial allegory. (Reisch 1517)

against the Roman fleet in the siege of Syracuse). These images became emblematic for optics.

Another front page of science book was used as the front piece in Chap. 6. It was the cover page of Galileo's famous dialogue regarding the major systems of the world (Galilei, 1632/1953). The page represented the participants of the dialogue who reconsidered the previous views on the subject. The chosen set was, however, not dual (old versus new) but triple, and this was a special approach of Galileo's presentation, which made the presentation cultural. Seeking inclusiveness, the front page images also addressed philosophical issues of the nature of science. In the following, we will illustrate them too.

Leonardo da Vinci surpassed mere description by artistic means. He considered painting to be the queen of science and a tool of scientific research. His pictures "interrogated" the subject. To the researcher, he suggested:[34]

1. Close observation.
2. Repeated testing of observation from various viewpoints.
3. Drawing the object or phenomenon so skillfully that it would become a "fact," which all the world could see or could grasp with the aid of brief explanatory notes.

---

[34] Wallace (1966, p. 104).

Through this approach, he explored anatomy and the functioning of the human body.[35] His drawings supported the training of generations of medical students. However, his uniquely precise sketches revealed the inability of a true understanding of a theory. They skillfully depict stimulated thought but could not explain ballistic and pendulum motion, light behavior, turbulent stream, or various mechanical systems.[36] Science surpasses photographic observations even if it uses them extensively. The divine proportion between the limbs of a human body detected by the acute eye of Leonardo was a revealing fact, but it could not provide any causal doctrine beyond speculative claims.

In summary, science and art appear as two forms of an intellectual approach to nature that are complementary. Since a picture representing an object becomes a symbol for it,[37] art pieces may provide new symbols of scientific concepts furnishing their further account. In a way, this enriches art with an appeal beyond aesthetic and social associations, giving new meaning. Teachers may involve artistic images, which, though not created to be accessories of curricula, still provide an effective tool of intelligent expression of the subject matter taught in science class. The following case illustrates this.

The Case of Giotto – Displaying an Idea Through a Representative Image[38]

We first consider a specific work by Giotto di Bondone (1267–1337)— the stigmatization of St. Francis (1300). In the picture, Giotto connected the wounds of Christ with the stigmata of Francis through lines. He connected the left hand of one figure with the right hand of the other and likewise, the other hand and feet. The artist himself never explained his drawings,[39] and the question why this and not another manner of connection was chosen is thus open for interpretation, especially since most of the later depictions of the same event were drawn connecting the left hand with the left hand and so on—perhaps a more natural way. One may assume that his decision was taken from a certain design, which we intend to reveal.

Giotto was invited to commemorate St. Francis in the central basilica devoted to the saint in Assisi. Within this mission, it would be natural to address certain features emblematic of St. Francis. In fact, the traditional title of St. Francis was *The*

---

[35] Arnheim (1974, p. 159).

[36] For example, Reti (1974), Argentieri (1956), Calvi (1956), and Canestrini (1956).

[37] Goodman (1968/1976, p. 5).

[38] I am very obliged to the rector of the Basilica of Santa Croce, Padre Antonio Di Marcantonio, who allowed my approach to the discussed fresco and kindly supported my attempts to clarify the status of the Giotto fresco in the Basilica.

[39] Vasari tells a legendary, but illustrative, story about Giotto's great confidence in his drawings speaking for themselves (Vasari, 1991, p. 22). To receive the commission from the Pope, Giotto drew, with a single stroke, a simple circle whose perfect appearance was, however, sufficient to inform of his competence.

Mirror of Perfection. It appeared due to Brother Leo,[40] the close disciple of Francis, who published the account of his teacher's story under that title in 1227 and thus developed this metaphor, which has become commonplace ever since. Brother Leo pointed to the similarity of the images of St. Francis and of Jesus by introducing the mirror metaphor: starting from their similar appearance and extending to the level of philosophy. Indeed, Francis taught that every object presents a reflection of God. Giotto could not miss that reality.

One may imagine that in order to depict this feature of reflection, Giotto invented a special conceptual framework. Crucifixion was *mirrored* to stigmatization, the summative event for both persons. To express the mirror effect, Giotto connected the right hand with the left hand, the right leg with the left leg, as happens in a plane mirror.

The image in a plane mirror is very much like the original, but not the same. The symmetry of the mirror reflection changes in reality in only one of the three spacious dimensions—that perpendicular to the mirror. As a result, the parts of the object facing the mirror and located on the left remain on the left in the image, and those on the right remain on the right. However, in the mirror, the left *hand* becomes the right hand, and the right hand becomes the left (Fig. 10.4).[41]

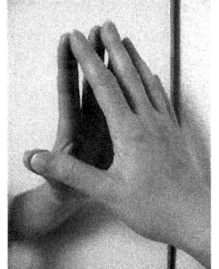

Fig. 10.4 Hand reflected in a flat mirror. (Photo by the author)

This way Giotto represented the metaphor of a mirror's reflection without actually showing a mirror (Fig. 10.5) by connecting the wounds of the two figures, one to one. This was unlike a number of earlier depictions of the stigmatization, such as by Belingueri Bonaventure,[42] who did not draw any lines. From then on, for centuries, many artists continued to draw lines to connect the two images.

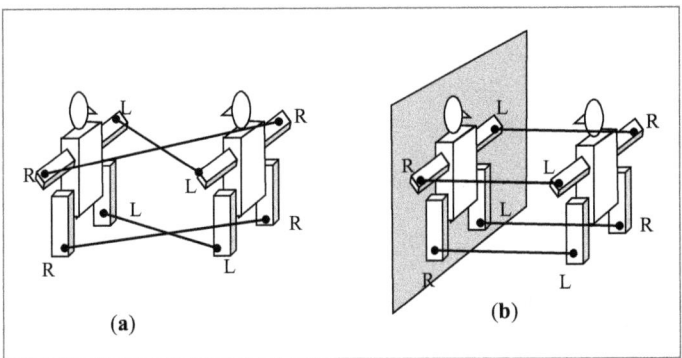

Fig. 10.5 (a) Connections made between right and left hands/feet for two persons facing each other. (b) Connections emphasizing mirror inversion of the image of a person facing the plane mirror

---

[40] Leo, Brother (1227/2010).

[41] In science, such a change is labeled as changing *chirality* as caused by the inversion of the axis perpendicular to the mirror surface.

[42] Bonaventure (1235).

Furthermore, Giotto connected the wound on the *right side* of the body of Jesus with the right side of Francis's chest—no inversion. One may speculate on the reason for that. Perhaps, Giotto wanted to further emphasize the mirror idea: only the hands and feet change their identification. Indeed, this interpretation might be rather complex for the viewers, especially those unfamiliar with the mirror reflection label of the saint. Given that Giotto never explained his pictures, his original idea was soon abandoned. Some artists preserved the tradition of Giotto, but most of others "corrected" the master, and thus destroyed the mirror idea.[43]

Giotto produced only three images of stigmatization: two frescos, in the Basilica of San Francesco in Assisi (1295) and in the Basilica of Santa Croce in Florence (1320), and a panel (1300), currently in the Louvre museum. In two of them, the connection between the wounds of Jesus and the stigmata of Francis were in keeping with the mirror principle correspondence (Fig. 10.5b). However, the fresco above the Bardi Chapel in Santa Croce (Fig. 10.6c) was different. The latter was produced some 20 years after the fresco in Assisi (Fig. 10.6a) and the panel in Louvre (Fig. 10.6b). Did Giotto change his approach?

The support for Giotto's original vision may come from the similar fresco located in the refectory of the same basilica. Giotto's closest student and collaborator at that time, for 24 years—Taddeo Gaddi—produced the fresco in 1330. Besides the fresco, Gaddi also produced a panel painting of the same event.[44] Both works by Gaddi exactly match Giotto's works in Assisi and Paris.[45] Taking seriously the pictures of Giotto and Gaddi, one could infer[46] that the fresco in Santa Croce was corrected during several sweeping perturbations in the basilica.[47] The providential opportunity to check the original was due to the restoration of the frescos. A close observation in 2010 confirmed the original conjecture of 2007. Visiting the fresco, located more than 10 meters above the floor, revealed the old lines of Giotto connecting the hands and feet wounds in the manner identical to the original works of Giotto and Gaddi (Fig. 10.7a). This finding placed a dilemma before the restoration crew.[48] What lines to enhance, those of the original or those of the correction?[49]

---

[43] Galili and Zinn (2007).

[44] It is currently located in the Fogg Art Museum of Harvard University in Cambridge, Massachusetts.

[45] Another small piece of the same type by Taddeo Gaddi is in Pinacoteca Nazionale di Bologna.

[46] Galili and Zinn (2007).

[47] Micheletti (2007) and Muray and Muray (1996, p. 199). Emma Micheletti (2007, p. 12) wrote: "Both [chapels in Santa Croce] were considerably restored during the nineteenth century by Gaetano Bianchi (who according to the custom of the time went so far as to restore or recreate faces (buildings and whole landscapes) robbing the great art of Giotto of much of its genuineness and original vigor."

[48] The earlier frescos of the fourteenth century that covered the inner walls of the basilica were almost entirely destroyed (Micheletti, 2007). The walls around Galileo's tomb still bear a few fragments of superior quality. This is an illustration of the different cultural norms that reigned in the past.

[49] In 2013, the newer lines were removed and the original ones by Giotto strengthened returning the fresco to its original appearance.

Fig. 10.6 (**a**) Stigmatization of St. Francis by Giotto before 2010. The fresco in Assisi. (**b**) The panel in Louvre, (**c**) The Fresco in Florence before restoration. (**d**) The stigmatization of Santa Catharine of Siena by Manetti. (One line of connection is emphasized in each of the pictures. IG)

(a)   (b)

Fig. 10.7 (a) A close view of the fresco *Stigmatization of St. Francis* in Santa Croce in Florence (c. 1325) in 2010. The later drawn line connects the hand wounds: right to the right. (Photo by the author) (b) The fresco after renovation (2014). One connection is emphasized to display the change. In the renovation, the lines of the later production were washed out, the original lines were strengthened

The artistic metaphor of Giotto allows one to interpret the emblem of Franciscans: two arms crossed, one representing the Deity and the other, St. Francis, both bearing signs of wounds. This is a pair of right and left hands as in an object-mirror image couple related by Giotto. "Similar but different" could be a new motto of the order.

Educational Implication

Consideration of the restoration of the fresco in Santa Croce may be used in the teaching of optical images in different settings. The teacher may inlay the interpretation of the fresco comparing Giotto's works with depictions by other artists who faced the same choice of correspondence. The idea of a mirror image poses a question: What does the mirror do to the image seen in it? Does it change left to right?[50] A comparison of Giotto's work with Manetti's (Fig. 10.6d), who unlike Giotto changed right to left also regarding the chest wound, may reveal an additional aspect of Giotto's strategy in hinting about mirror reflection: a mirror does not change left to right, but does change left hand to right hand.

The effectiveness of teaching might be enhanced by considering transformations of optical images observed in various optical settings in comparison and contrasting: mirror, convex lenses, and the camera obscura. The four cases (Fig. 10.8) together establish a space of variation for meaningful learning. A representative set that can promote students' understanding of the correspondence between an optical image and the object under comparison in different optical settings.

---

[50] Galili and Goldberg (1993).

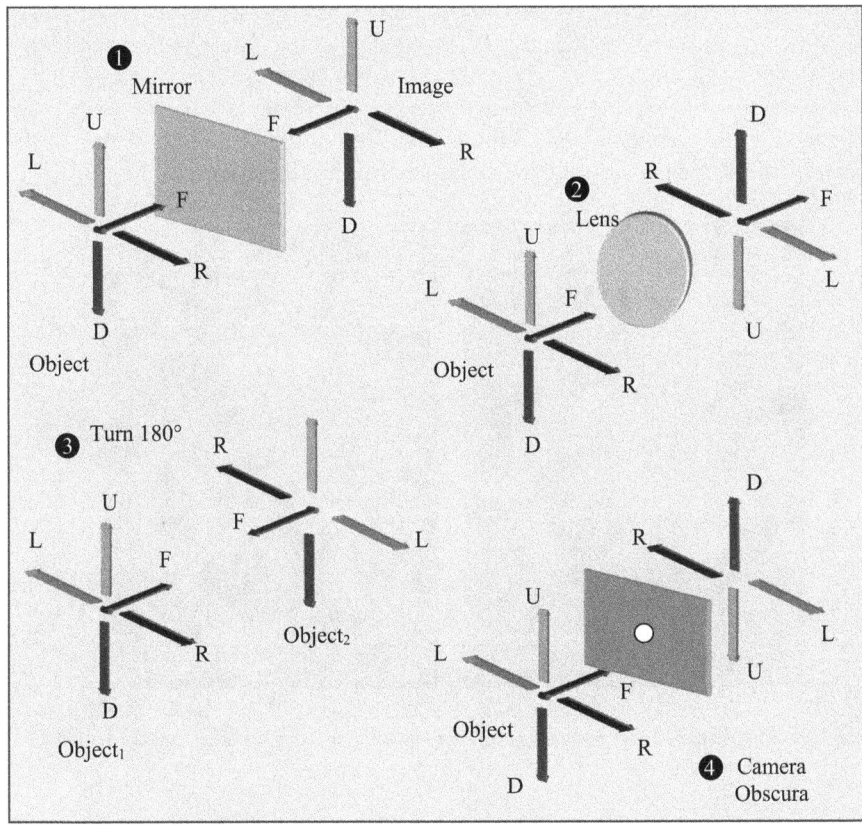

Fig. 10.8 Transformation of object-image dimensions in four basic situations: (1) in the plane mirror; (2) in the real image by the convex lens; (3) when two persons face each other; and (4) in camera obscura (no depth to the image). To share the assumed considerations of Giotto compare cases 1 and 3

An Extra Example in Learning About Optical Image

To further illustrate the features of mirror image in an unusual setting, one may address two sculpture images of the mythological Gorgon Medusa, which appear in the underground water cistern in Constantinople. The Byzantine Emperor, Justinian I, constructed the cistern in the sixth century, and the two sculptures support two columns. The strange feature is the orientation: one is upside down, and the other lies on its side (Fig. 10.9). In both cases, the face of Medusa is just above the water. The reason that might have guided the Byzantine builders could have been that Greek mythology ascribed to Medusa the power to petrify any observer who dared to look at her. To overcome this obstacle, Perseus, a legendary Greek hero, was equipped by the goddess Athena with a mirror shield so that he could look at Medusa's image while he approached and attacked her. This way he succeeded in beheading Medusa. But even the dead Medusa's head preserved its lethal power (used by Perseus in later adventures). The builders of the cistern in Constantinople played with the legendary ban on looking directly at Medusa's face by suggesting to the viewer to avoid looking directly at her but to observe her face indirectly by looking at the water just below the sculptures. The setting illustrates the feature of the horizontal mirror in inversing the image to the vertical direction.

Fig. 10.9 Two heads of Gorgon Medusa supporting columns in the water cistern in Istanbul. (Photos by the author)

Nature of Science (Scientific Knowledge)

Art provides images for understanding the core features of scholarship; the nature of science and scientific knowledge is among them. We may return here to the already mentioned images on the front pages of scientific books. In Chap. 9, we considered the front page of science book (*New Almagest* by Riccioli, 1651), which declared the principles of knowledge construction—*numeration*, *measurement* and *weighing*—making theory contest specific for science, that is, keeping with these principles from the old tradition especially flourished in Hellenistic period (Fig. 9.8). The image chosen to open this chapter is, however, different. It reproduces the front page of another book published in the same time—the famous optics book by Kircher (*Ars Magna Lucis et Umbrae*, 1650). Though both art pieces declared the same ultimate origin of the universe and its designer (the God's name is given in both pictures explicitly, written in Hebrew), the Jesuit scholar Kircher stated two other principles of knowledge production—the *divine authority* and *ratio*—clearly coming from the religious medieval rationalistic philosophy. The Renaissance philosophy of science was different. It fuses materialistic with rationalistic approaches.

We find such new understanding when we consider the fresco of Rafael in the Vatican—*The School of Athens* (Fig. 10.11a). In the best "perspectivist" style introduced by Renaissance artists, the fresco focuses on two *equally* emphasized figures by placing both around the point of convergence (vanishing point): Plato and Aristotle—the two founders of science as developed from the natural philosophy of Classical Greece. Each of the two heroes, by a simple hand gesture, represents the philosophical method stated to be the correct one for the objective account of nature: rationalism or empiricism (Fig. 10.10a). The two basic approaches to the scientific method are frequently presented as a complementary pair in normal scientific practice incorporating the *theoretical* and *experimental* components of the science curriculum.

(a)

(b)

(c)

Fig. 10.10 (**a**) Fragment from the fresco *School of Athens* by Rafael Sanzio (1511). Plato and Aristotle represent the epistemological core of scientific method combining the rationalist and empiricist philosophies (arrows added to emphasize the gestures). (**b**) The same picture when used in class teaching regarding the debates on the nature of light and motion between Descartes and Newton. (**c**) The univocal pointing to the authority in *The Disputation of the Holy Sacrament*

This picture by Rafael has a long tradition of interpretation, including the gestures of its main heroes.[51] In science class, however, the traditional interpretation
could be more specific. The physics teacher may, for instance, address the particular debate of the two competitive theories of light developed in the seventeenth
century.[52] This teaching, by addressing the cultural content knowledge about light,
may reveal to the students not only the ontological contest but also the epistemological one—the difference between the method of Descartes (and his adherent,
Huygens) and that of Newton (Fig. 10.10b). The gestures of Plato and Aristotle will
then be given concrete meaning. Plato's pointing upward signifies the transcendental and metaphysical origin of the ultimate knowledge—the gnosis. His renowned
pupil, Aristotle, challenges Plato pointing down, to the earth, a moderation of the
metaphysical ideas, suggesting the alternative foundation—human contemplation
and   experience   as   the   true   and   reliable   resource   of   theoretical
knowledge—empiricism.

The teacher may extend this discussion by pointing to another fresco by the same
artist, which faces the *Scuola* (Fig. 10.11a) from the opposite wall of the same
room—*Disputa* (*Disputation of the Holy Sacrament*) (Fig. 10.11b). Both being collective "portraits" of scholars, the two pictures are different in arrangement and
gesturing. The *Disputa* has a single central speaker, the scholar, showing a clear
gesture: he points up as if preaching to those who are around him (Fig. 10.10c). This
is in contrast to the pair of opposite references by Plato and Aristotle. The figures in
the *Disputa* are clearly arranged in vertical hierarchy. Actually, the picture does not
represent a dispute; there is no controversy there. The declared truth seems to be
fully revealed to the participants beyond any doubt. The personalities present clear
evidence of authority. In contrast, *The School of Athens* appears as a crowd, groups
of individuals sharing and exchanging thoughts, states of mind, learning from each
other. The principle unifying them is complex and dialogical. It is rather a method
of complementarity of subjects, perspectives, and approaches—a dispute about the
production of knowledge, as presented by the central couple. It is passionately
argued by Socrates and ignored by the cynical skepticism of Diogenes. In the lower
corners, correspondent to Plato and Aristotle, Pythagoras and Euclid are immersed
in problem solving. Plato and Aristotle's opposition of visions in natural philosophy
manifests itself in the opposition of abstract–numerical (Pythagoras) and material–
geometrical (Euclid) mathematical methods in correspondence (emphasized in
Fig. 10.11a). It will remain a dichotomy present among all consumers of science,
from young pupils in schools to scientists throughout history.

By juxtaposition of the two frescos facing each other, Rafael showed the essential epistemological difference between science and religion (regardless of content):
unlike religion, there can be no single authority in science maintained by a hetero-
geneous community, preserving its eternal dialectic discourse of opposites, in the
variety of realms, philosophy, mathematics, astronomy, ethics, and so on.

Furthermore, below the *School of Athens* fresco, Rafael makes much smaller,
more modest, and colorless frescoes depicting *The Debate on the Earth Sphericity*,

---

[51] For example, Gutman (1941), Haas (2012), Phelan (2002).

[52] For example, Gliozzi (1965), Chap. 7, Fig. 7.13.

Fig. 10.11 (**a**) *The School of Athens*. The dual nature of science is represented by the two figures (Plato– Aristotle) and their implications in numbers theory (Pythagoras) and geometry (Euclid) (**b**) *The Disputation of the Holy Sacrament*. The central religious figure is connected by a unique connection to it evidence—sacrament. Fragments to compare are emphasized

Siege of Syracuse, and the *Death of Archimedes*. Why just Archimedes out of all the many heroes and events in the history of science? Are they representative, emblematic cases of a common scientist–society relationship—challenging the "common-sense," vital need of scientific products to protect society, and finally, the absurdity of societal pay back when unique talents are destroyed by a brutal force unaware of the meaning of its deeds? Is it too far a stretch to a contemporary eye, in particular, to see in that scene of the murder of Archimedes, all those victims of persecution, suppression and destruction throughout history?[53] They were brought to trial by all forms of barbarous regimes simply for doing true science, for doing something that benefited all humankind. Again, we may only guess regarding the intentions of the artist, thinking in terms of associations, as here observing the panoramic image chosen by Rafael to represent science as a whole, and take advantage of this in revealing the nature of science relevant for a teaching setting. In any event, it is apparent that in this fresco composition in the Vatican, the ingenious artist showed himself also as a true intellectual.

Yet, before leaving the two frescos, one may look again at *the School of Athens* and reveal that, actually, it includes not only natural philosophers and mathematicians but also Socrates and a group of humanists who never analyzed unanimated nature but were active in the humanities (Socrates and his group are next to Plato). The name of the fresco is the *School*, which makes the gathering more inclusive. This aspect may suggest ascribing to all these scholars the concept of *episteme* in its "archaeological" meaning by Michel Foucault.[54] It incorporates a whole layer of thought characterizing a period of culture, its specific manner of thought. Here too, we may look to the second fresco across the room, which actually depicts another episteme quite in contrast to Athens. The episteme shared by Athens with the following history could be related to their making sense, analyzing and revealing through *contemplation* the underlying meaning and order of both nature *and* human society—the episteme of *Cosmos*. As to *the Disputa* on the opposite wall, it clearly symbolized a different episteme that reigned in a broad sense throughout medieval society and into the Renaissance. Thus, the fresco included not only saints and the doctors of the church (the higher level) but also practitioners, clerics, priests, and plain Christians (the lower level). This is a portrait of the episteme—the episteme of *Faith*. The two epistemes, Cosmos and Faith, are looking at each other celebrating the two gross products of human civilization of eternal validity.

More About Scientific Method

Raphael's *Scuola* is very powerful indeed and deserves thorough observation and discussion. It is not unique in its representation of the two poles of scientific epistemology, Platonic rationalism and Aristotelian empiricism. Consider a pair of pictures by Botticelli and Caravaggio (Fig. 10.12). Used together, they address the same epistemological opposition. Botticelli's *Calumny* shows the significance of the a priori basic claims of a theory over the false alternative. Two allegorical figures suggest that truth is apparently different from falsehood. The viewer cannot be deceived. Botticelli chose a nude figure to be the allegory of truth and contrasted it with falsehood completely covered from head to foot.

---

[53] There is possibly another image in the fresco, which echoes Archimedes. It is the image of Hypatia (just above Pythagoras), a great scholar of Alexandria, not only the only woman in the picture but also a martyr.

[54] Foucault (1972), Appendix 1 in Chap. 6.

Here, a discussion may arrive at the essential idea of the Greek founders of natural philosophy—the concept of Aletheia (the state of being evident). Nature is open to comprehension through careful observation combined with theorizing on the reality—contemplation. Nature is unconcealed.[55] Although criticized from the modern perspective of scientific "truth,"[56] such openness ("unclosedness") of nature is presumed as the fundamental precondition of science. The idea is similar to Einstein's vision as expressed in his "Subtle is the Lord, but malicious He is not."[57] It supports the optimism of all those scholars trying to reveal the objective truth about nature. But should the truth be also apparent? Botticelli, in art, had no other means to depict the contrast with falsehood. Yet, Plato's vision of truth, however, already surpassed obviousness. His idea of beauty presumed reasoning similar to mathematics, instead of using deceptive senses. Indeed, science does not rely on the "obvious." To return to the issue, the teacher using this picture of Botticelli may moderate by the list of scientific accounts used, obviously true and reliable obvious principles that appeared to be wrong (the old theories of motion, the old theories of vision, Descartes' theory of collisions, and so on…).[58]

(a)

(b)

Fig. 10.12 (**a**) A fragment of Botticelli's *The Calumny of Apelles* (1497) for possible illustration of Platonic rationalism. (**b**) Caravaggio's *The Incredulity of Saint Thomas* (1602) for possible illustration of the Aristotelian empiricism

Caravaggio's *The Incredulity of Saint Thomas* provides the image of contrast to Botticelli's *Calumny*. The artist shows extreme determination in seeking reliable evidence by Apostle Thomas. His finger does not merely point to the wound of

---

[55] Heidegger (1972, p. 70).

[56] Heidegger (1977, pp. 160–161, 166).

[57] Pais (1982/2005, pp. vi, 113).

[58] Chapters 1, 2, and 3.

Jesus but *penetrates into* it. The idea of seeking tactile evidence as if being the most reliable source for knowledge can be emblematic for the empiricist (positivistic) epistemology of science. The picture visualizes the in vitro approach of investigation[59] ("interrogation of nature," as put by Francis Bacon) as preferred to the in vivo approach (Greek contemplation of nature). The strong affective contrast with Bottichelli's *Calumny* calls for a second thought on a priori adopted principles, even if it appeared obvious. This is an epistemological claim. It may be exemplified by Mach's critique with respect to Newton's definition of mass as a "quantity of matter." *Operational* definitions of physics concepts were suggested as preferable to theoretical declaratives.[60] The operational definition draws on direct measurement of the observed, a kind of hands-on experience. The contraposition between the nominal and operational definitions is present in the school curriculum—the concepts of weight and force.[61] This issue returns us to the same split between Plato and Aristotle, resolved through complementarity.

Discussion of the nature of the knowledge of science as supported by artistic images may proceed to the issue of discovery versus invention, modern versus classical doctrines of objective reality and its comprehension. This may touch on the epistemological revolution that took place at the beginning of the twentieth century. What artistic image could one choose for that illustration?

(a)   (b)   (c)

Fig. 10.13 (**a**) The characteristic side of the Nobel Prize medal for physics (the medal of Albert Einstein). The scene represents *Scientia*, which discovers the features of *Nature* through uncovering her face. (*Names are emphasized*). (**b**) Science in action. Revealing the reason for phases of the Moon (Byzantine mosaic). (**c**) Schematic drawing of Pygmalion casting Galatea (Ovid's poem)

---

[59] In vitro is usually associated in biology with the dead subject of investigation observed under the microscope (imagine a mosquito between the glass plates).

[60] Mach (1883/1989, pp. 216–218). Similar cases of the operational definitions introduced are those of simultaneity in the relativity theory of Einstein (1905/1923) and the concept of weight in classical mechanics (Galili, 2001).

[61] Chapters 4 and 5.

To represent the epistemology of classical science, the teacher may use the image appearing on the Nobel Prize medal for physics (Fig. 10.13a). Two allegorical figures represent nature (natura) and science (scientia).[62] In an extremely gentle motion, scientia raises the veil, revealing the face of natura. The image represents the idea of science as revealing (discovering) the features of nature originally hidden. The uncovered features are independent of an observer/scientist—an observer. They objectively exist. This scenario corresponds to the realist philosophy. The claim is strong: scientia looks at nature from below, but natura, above, remains entirely indifferent. She looks forward ignoring her investigator. The veil, the finest possible material, hints of the utmost effort of the investigator to minimize the disturbance caused in discovering the objective truth about nature, which is that nature is both unique and beautiful as is. However, scientia does not passively gaze at nature. She contemplates: her look is interwoven with an analysis of the observed to be documented in the scroll that she holds in her other hand. Not every viewing of nature presents science and brings discovery. Until the twentieth century, this scenario could represent the ideal, somewhat iconic, even in the nineteenth century epistemological paradigm of science.

Addressing the scientific method, Aristotle, among others, described the induction of the second type and illustrated it by the inference that the Moon is an opaque body. It does not produce light but is lit only by the Sun.[63] This claim is illustrated in pictures one can meet in old mosaics decorating Byzantine monasteries (Fig. 10.13b).[64] In it, the saint points to a configuration of the Sun and the Moon. With only the Sun given as a light source, the crescent area of the Moon is lit. It is, then, shown that the lit area is always observed as convex toward the Sun. Thus, the basic feature of the Moon phases is explained.

To provide an idea of the modern scientific epistemology, one can use the illustration of the poem about Pygmalion and Galatea from Ovid's *Metamorphoses* (Fig. 10.13c). It may trigger the required image to be perceived. The story tells of Pygmalion, who produced a sculpture of the ideal woman and fell in love with it. Eventually, the goddess Aphrodite was gracious and animated the stone into a living woman. The new couple was happy. This story has inspired many artists. The original idea of Ovid was the human ability to compete and even challenge nature in the beauty of her creatures: humans possess a special "Pygmalion power."[65] Numerous artists have focused on the beauty of the human body.[66] The hammer and chisel are hardly seen, if at all, in those pictures. Only a few put the act of construction to the

---

[62] In other languages, the two female figures might become an inappropriate choice. Thus, in Hebrew, both terms are of masculine gender, reflecting a different conception of Nature in Judaism.

[63] For example, Losee (1993, p. 6).

[64] This type of picture is canonical and can be seen in different places as fresco or mosaic, e.g., the monastery of Troodos (Cyprus), fresco, or the Palatino Cathedral of Palermo (Sicily), mosaic. The latter is reproduced here.

[65] Gombrich (1972, pp. 93–115).

[66] For example, the pictures by Agnolo Bronzino (1530) and Jean-Léon Gérôme (1890).

fore.[67] It is that version of the image, together with the myth, that can be brought to the science class to support the debate of the human origin of the product (the sculpture was human made) but its objective nature (Galatea was a regular woman) despite being human-made. The parallel is a scientific theory that is also a human-made construct; it is not discovered as is, it is produced; nevertheless, it is objective—became a creature in its own right, representing the true aspect of reality. For example, Newton invented his theory of gravitation, he did not discover it, but it is still an objective theory, faithfully representing reality within a certain range of parameters (mass, length, and time). The idiosyncratic (religious) views of the Creator and the special circumstances of his activity and social environment (very much specific) did not change the essence of the theory used by all since then. The story of Pygmalion may symbolize the objective nature of science; the statue being a beautiful human construct did not divorce from the natural object it represents.

The Idea of Knowledge Organization

Chapter 6 described the special organization of scientific knowledge in a number of fundamental theories, each structured holistically using a triple cultural code, thus producing a discipline–structure. To illustrate it, one may consider several artistic samples clarifying the underpinning idea. In fact, the religious iconic tradition is rich, sharing the same idea of holistic representation of certain fundamental (this time, nonscientific) theories (dogma). Figure 10.14a illustrates an abundance of similar images on the walls of churches and art museums today. The content is organized in a triple code nucleus–body–periphery, in which the body content (paradise) presents the results of keeping within the principles of the nucleus (the worshiped figures) and the periphery shows what happens with the violation of these principles (hell).

Figure 10.14a is, however, not fully representative of the periphery of the DC structure, the nature of its elements. They could be more than violations. The periphery should include the alternative ideas, which are much beyond erroneous behavior of various kinds. Thus, the periphery of classical mechanics includes the basic ideas of relativity and quantum theories, not only obsolete and erroneous ideas. In terms of names, if Newton is in the nucleus, the periphery includes Einstein and Bohr. Such affiliations clearly change the image of the periphery.

Another artistic image better represents this idea. Consider a picture of the Last Supper from Christian tradition. The attention of the viewers is often focused on two figures, Jesus and Judas, the nucleus and periphery of Christian dogma. In the chosen picture of Fig. 10.14b, Judas is sitting across the table, isolated, facing Jesus,

---

[67] For example, Francisco José de Goya (1812–1820), http://search.getty.edu/gateway

Fig. 10.14a A typical presentation of the Christian dogma of the Last Judgment using the triple code (S. Lochner, 1435). The three areas of different status are marked with boundaries added to illustrate the idea of thematic separation

opposite him, in opposition to him. He was not depicted as ugly and deformed but as a figure of contrast, representing another ideology.[68]

An expressive presentation of the last supper is that by Leonardo da Vinci (Fig. 10.14c). The same codification, nucleus–body–periphery, is supported there by indicative gesturing of the figures. The first pair of gestures (Jesus versus the Apostles, 1–2 arrows) represents the relationship nucleus–body: the nucleus suggests and imposes on the body, and the body adopts and refers to the nucleus—a consonance. The second pair of Jesus and apostles versus Judas (3–4 arrows) represents nucleus/body versus the periphery relationship—a dissonance. These elements are similar to the Plato–Aristotle gesturing (Fig. 10.10) representing confrontation.

---

[68] Judas indicates the break of the new ideology from traditional dogma. In another version (Giovanni da Modena in San Petronio, Bologna, 15th c.), the periphery included Mohamed representing the later offspring of the monotheistic religion—Islam—not a less strong alternative to the nucleus.

Fig. 10.14b The Last Supper of the Christian tradition interpreted in the triple code. The three areas of the content are marked. The confronting dog and cat on the floor indicate the ideological confrontation between the two incommensurable worldviews of the nucleus and periphery. Rosselli (1480), the Sistine Chapel of the Vatican

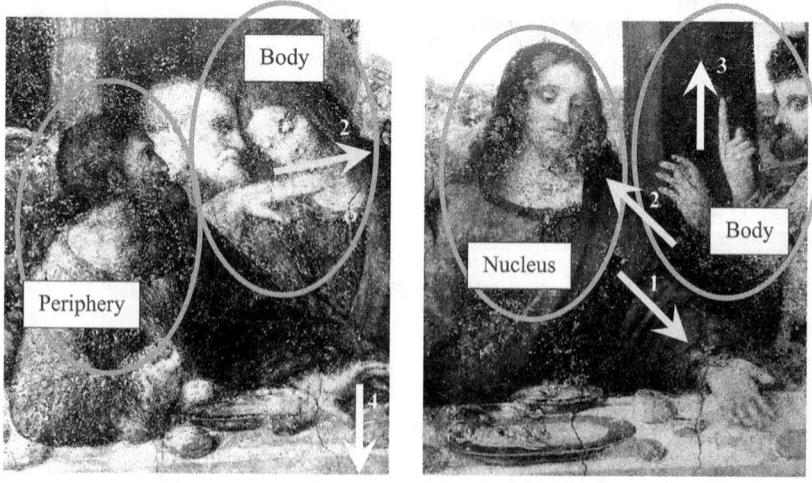

Fig. 10.14c Fragments from *The Last Supper* fresco by Leonardo da Vinci (Santa Maria delle Grazie, Milan, c. 1490). Nucleus, body, and periphery are identified with the images of the fresco. Two pairs of arrows represent the relationships between the areas

The Cumulative Nature of Science

One of the important aspects of scientific knowledge is its cumulative nature. It is much discussed and often discharged through the vision of scientific reality as a succession of revolutions each "incommensurable" with the previous understanding. Instead of refining the complex process of knowledge construction as one net, this conception may prevent appreciation of the cumulative nature of scientific knowledge, which is molded in a continuous diachronic scientific discourse. In it, each new piece of knowledge, whether confirming or refuting the previous one, essentially relies on history, conceptually and empirically, to share the major rules of the enterprise.[69]

To illustrate this claim using artistic images, one may use the Christian tradition, which was canonized by Bernard of Chartres, in the twelfth century as the basic approach to education and learning, especially in natural philosophy. For that, he used the famous metaphor to put the idea:[70]

We are like dwarfs on the shoulders of giants, so that we can see more things than them, and can see further, not because our vision is shaper or our stature higher, but because we can raise ourselves up thanks to their giant stature.

Fig. 10.15 Reproduction of the stained glass windows (1225–1230) in the Notre-Dame Cathedral in Chartres, France. It symbolically depicts the progress of knowledge. The four Evangelists, the authors of the New Testament, are sitting on the shoulders of the Major Prophets from the Old Testament. The enlarged fraction allows appreciating the artistic beauty as well as the important detail: all the figures wear spectacles, which were emblematic of scholarship

[69] Chapter 9.

[70] Crombie (1959, p. 27).

In the tradition of that time, his words were implemented in a visual image in the form of stained glass windows in the local cathedral. They became famous as an emblem of the idea of the cumulative nature of knowledge in learning and exploration (Fig. 10.15). This dictum became especially famous due to Newton, who in 1675 cited this old maxim:[71]

If I have seen further, it is by standing on the shoulders of giants.

Polyphony, Perspective, Learning Science

Is the produced scientific knowledge theoretically homogeneous? As we know, science is composed of several natural sciences (physics, chemistry, biology, etc.), and each of these is itself constructed from several fundamental theories. Physics includes classical mechanics, electromagnetism, thermodynamics, and classical and quantum theories, each addressing a different facet of reality, a range of phenomena. These theories produce several pictures of the world, several perspectives with which people represent objective reality using a specific account.

Apparently, a polyphony characterizes the subject. The artistic allegory of a choir of singers in Van Eyck's *Singing Angels* (1429) (Fig. 10.16a) could be a relevant image. The idea *one source. many voices* is apparent here. The angels sing from the same Book (Galileo's "Book of Nature"[72]), but their faces and emotions vary, although they share the same subject and melody.

Indeed, despite the differences, the angels share similarities. Differences and similarities make the group a fitting background to a discussion introducing the big picture of physics. Composed of different, sometimes contradictory, claims, fundamental theories, the disciplines still share much similarity and a family resemblance; they possess a common genus.[73] What is it?—Objectivity, a cumulative, distributed, and self-correcting nature, mathematical account, logical apparatus, and empirical verification could all be considered. This common–distinct nature matches the idea of a discipline–cultural curriculum, explicit with respect to the dialogical nature of science and manifested in the multiple, similar–different scientific accounts.[74]

---

[71] *From Newton's "Letter from Sir Isaac Newton to Robert Hooke."*

[72] Galilei (1623/1957).

[73] Wittgenstein (1953/2001), Tseitlin and Galili (2005), Irzik and Nola (2011).

[74] Chapters 1–5.

Fig. 10.16 (**a**) Van Eyck's *Singing Angels* (1429) from the Altarpiece in Ghent Cathedral. (**b**) Labyrinth on the floor (pointed by arrow) of the Cathedral in Amiens on the old gravure (See the gravure *Chartres Cathedral* (1750) by Jean Baptiste Rigaud). (**c**) The image of God—Creator—a fragment from the fresco on the ceiling of Capella Sistine by Michelangelo in the Vatican. After Creation, God observes from above his subjects underneath—a clearly advantageous position— and can guide their way in the labyrinth of life, "the way to Jerusalem"

Why does one need this complexity in science class? In answering that question, one may point to the labyrinth on the floors of several cathedrals (Fig. 10.16b). Why place a labyrinth in a temple? Is it a mere decoration? A possible explanation is that the labyrinth symbolizes the complexity of finding the way to survival out of confusing situations. Indeed, to find the way out of a labyrinth is difficult. In the Greek myth, the hero Theseus succeeded in using the trick of unraveling thread. Yet, there is another way out of confusion, looking onto the labyrinth from above. The church suggests such a perspective to guide people in their troubles. The old tradition labels the labyrinth in cathedrals as "The Road to Jerusalem," as if suggesting that the difficult pilgrimage to Jerusalem for the believer provides such a perspective, looking from above. That makes a difference. In science education, this might be compared with providing the learner with the big picture of scientific knowledge in a DC structure.

Finally, one may address the progression in scientific knowledge. Very often, the metaphor of climbing is used in addressing learning. This can be allegorically represented by using the canonical fresco depicting the *Divine Ascent* of those believers on the way to perfection overcoming barriers and driven with their faith (Fig. 10.17a). In a way, this imagery represents the similar progression of students of science in their way of knowledge construction and the science researchers, devoted individuals who have to overcome multiple barriers on their way. The image of ladder may symbolize the accumulative nature of these both activities, the idea of learning as gradual construction, piece by piece and step by step, and the path on which not all participators succeed.

(a)   (b)

Fig. 10.17 (a) *The Ladder of Divine Ascent.* Icon from the twelfth century. Monastery of St. Catherine in Sinai. (b) Synchronic debate *in Plato's Academy* as depicted in the Roman mosaic of the first c. from Pompeii (National Archaeological Museum, Naples)

The ladder metaphor is adequate for the lifelong labor and professional promotion of scientists but misses other important aspects of science nature. A complementary aspect is the community discourse. It is a social interaction of a specific type as depicted in the Roman mosaic in Pompeii (Fig. 10.17b). It is different from other social activities being an interaction of equal contributors. No figure is portrayed as superior to others, including the leader of the group, Plato himself. At the same time, the knowledge is constructed in a cumulative manner, building on the previously established ideas—diachronic discourse of scholars across different civilizations. This aspect may return to use the mentioned metaphors of ladder and sitting of the shoulders of giants (Fig. 15).

Modern Science

To appreciate the change, which has taken place in scientific epistemology in the twentieth century, one may turn to modern art. One of the central features of modernity is the role of the observer. The role of an observer in classical physics was related to the image on the Nobel Prize medal (Fig. 10.13a). The change of the role, moving from observer to a sculptor, as initially suggested (Fig. 10.13b), may be elaborated differently in the two modern theories—relativity and quantum.

The theory of relativity states that we account for reality through universal laws, which possess a form independent of the observer, or the frame of reference (covariant form).[75] While various aspects of reality, including time, length, and simultaneity, may appear differently to various observers, other characteristics, such as the speed of light, charge, mass, and present universal constants, that is to say, they remain the same for all observers. One may echo this dichotomy–unity in the world-view while interpreting some pictures by Picasso.

Picasso was impressed by the ideas of this new physics, which was widely discussed at that time. This may be the explanation behind the new genre in depicting objects he introduced. He tried to depict the perception of a certain subject from more than one perspectives simultaneously—*relativity*—but combining them in the same picture as if emphasizing the same personality—the *invariance* of the essence (Fig. 10.18a).[76] It was a smart image of Einstein's idea of the theory of relativity.

(a)   (b)

Fig. 10.18 (a) The sketch reproduces the idea of cubism—several pieces of an object from various perspectives combined in a single portrait (sketch by Guy Galili). (b) Portrait of Picasso by Juan Gris (1912)

Another feature of modern physics is the role of the observer in determining the state of the object of the atomic scale from the microworld.[77] The essential dependence of the micro-object interacting with macroscopic apparatus can be compared with the dependence of the appearance of the subject on the individual style of the

---

[75] Born (1962). School curricula are often silent on this important subject. The introduction of observer dependence is argued for school teaching of electromagnetism and mechanics (Galili and Kaplan, 1997, Stein and Galili, 2018).

[76] For example, Portrait of Dora Maar by Pablo Picasso. (https://en.wikipedia.org/wiki/Portrait_of_Dora_Maar).

[77] For example, Heisenberg (1959/1971, pp. 55–57).

artist, his or her perception. Taking that into account, the product is made intentionally different from a photographic reproduction (Fig. 10.18b). Yet, it remains the same object. We observe a portrait in the cubist style, and we are forced to think about the artist: What did he want to say by this and that deviation from reality? Was he positive, critical? And so on. This aspect becomes not less, but perhaps even more important than the object itself. Yet, we can recognize the person; the subject still possesses his/her identity, some essential traits. This manner of cubism may be related to the major feature of the account by quantum theory. In particular, the macroscopic environment determines the state of the microscopic object, more exactly, the range of results in the measurement. The obtained result is limited by the nature of the subject and environment: not everything can be changed and not in any way. The set of options is known to us in advance. At the same time, the subject preserves its fundamental features (mass, spin, charge, etc.).

The situation is not like an X-ray "portrait" of a human arm, for example. An X-ray image depicts what we cannot observe directly. Inspecting X-ray pictures, we never think about the technician who made them, as he did not change anything in the arm. In a sense, it is a classical image, seeking photographic similarity with the classical object. The cubist approach treats the object aggressively, moving the artist to the fore. This is somewhat similar to the role of the physics apparatus in quantum theory. Although the actual person posing for the portrait can be recognized by some essential traits, the portrait is a new object. The analysis of the roles of the artist, the portrait maker, and the person depicted can be compared to the role of environment determining the appearance of the quantum object.

(a)   (b)

Fig. 10.19 (a) Nike of Samothrace a Hellenistic sculpture of the Greek goddess of victory (the second century BC) displayed at the Louvre. (Photo by the author) (b) "Unique Forms of Continuity in Space" (1913) by Umberto Boccioni on display in several museums

A striking difference between modern and classical science is in their accounts of motion. In particular, classical motion, with a trajectory of an object in space, is dismissed by the quantum theory. Yet, the reality is complex. Indeed, there is a change of location of an object in time also in the microworld. However, an exact representation of microworld events by images in the macroworld is impossible. Physicists can depict the change of location by considering a tube (in so called phase space) with dimensions allowed by the principle of indeterminacy. This presentation replaced a linear trajectory in classical mechanics by a quasi-classical approximation of motion. An image of such "vague" motion could be a somewhat "diffused" figure of the modern sculpture by Umberto Boccioni in the twentieth century (Fig. 10.19b). It can be compared with an image of motion from classical Greece (Fig. 10.19a) with its clear edges outlining the shape of a moving body.

With quantum mechanics, the mentioned *epistemological* duality of rationalism–empiricism (Figs. 10.11 and 10.12) obtained a new manifestation—the renowned pair of physicists Einstein and Bohr. In quantum theory, duality became *ontological* too. It starts with the wave–particle nature of matter, *two* patterns of behavior of material objects in the microworld: as a particle and as a wave—complementarity. There are only *two* types of particles, fermions and bosons, possibly in *two* forms: matter and antimatter. There are *two* groups of observables with respect to the coex-

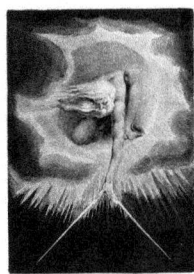

istence of certain features: commutating and anti-commutating. In short, *duality* appears to be a central feature of objective reality; the way it exists or, if one prefers, is designed. In a sense, matter appears in *two* facets, *two* ways of existence, two, and nothing else. This reality also was perceived from two basically different perspectives held by physicists during the twentieth century scientific revolution. The two groups could be represented by the renowned pair of protagonists, Einstein and Bohr (Fig. 10.20).

To supply an emblematic image, one may combine the picture by Blake of the God–Creator acting with the two-legged compasses with the photo of Einstein and Bohr walking together while maintaining their eminent dialogue about the way nature is.[78] It is with regard to quantum mechanics that one may realize that scientists do not just reflect and reveal but also construct and invent a special holistic account of our universe. Science *constructs* knowledge in a dialogue of ideas rather than *discovers* it. As with science, in science education, we *reconstruct* the scientific knowledge rather than *adopt* it, preferably by dialogue.

Fig. 10.20 Combining the picture of God–Creator by William Blake of 1794 with the photo of Einstein and Bohr in Brussels in 1927 in order to emblemize the dichotomy of possible approaches to reality held by the two prominent scientists

---

[78] Beller (1999).

The Beauty and Pleasure of Understanding

The appearance of artistic images with reference to disciplinary knowledge may surprise those who expect formal scientific content to be, as it normally is, represented by textual slides, equations, graphs, tables, and experiments. Artistic images change the atmosphere of discussions. The appeal of the artistically envisioned idea is challenging, being open to a variety of interpretations. When presented in class, students ask the lecturer to provide an interpretation in order to enable them to match the "authorized" vision with their own. This triggers discussion regarding the meaning of the subject matter. We observed such a reaction to the images we used in summary lectures following regular disciplinary courses in optics[79] and mechanics.[80] They caused excitement and enthusiastic reflection. Artistic images can do more; their power easily expands beyond facilitation of pragmatic issues of science. Let me proceed in the new direction.

People often use certain artistic images to represent abstract ideas: Botticelli's Primavera, love and beauty; Michelangelo's David, force, beauty, and courage; Carpaccio's St. Augustine, revelation; and Grunewald's crucifixion, suffering. All these are abstract concepts in humanities. With regard to science and its learning, images may do the similar; they may facilitate introduction and even become emblematic of concepts. Teachers' verbal declarations addressing the same concepts are more than often ignored by the novice students.

The issue of pictorial visualization is far from being trivial. In a way, it is similar to using analogies commonly employed in physics (an electrical current as a stream of water or gas flow, light as water waves, etc.). Analogies provide an idea, but may also mislead. Nevertheless, they establish scaffolding for understanding. Artistic images may do the same by providing a clue to a complex subject. In the course of learning, cycles of interpretation may remove the foreign and encourage the desired or convey a novel idea not even considered otherwise. After being debated, such images will be remembered together with their interpretations and the related content will equip mental manipulation in problems and thought experiments. There are, however, other images that appeal to the spiritual aspects of science, beauty, for example. In science, this is, however, a different beauty, which still causes various kinds of satisfaction, for instance, of logical reduction, coherence, the ability to prove, obeying universal law, etc. We will now expand on only one of its aspects, which can be directly related to imagery: the pleasure of understanding.

For centuries, the concept of beauty has been considered emblematic of Italian culture. Beauty attracts people always and universally, but in the time of the Renaissance, beauty transgressed mere appearance as was frequent in the classical art of Greece. It expanded to the beauty of spiritual virtues. In particular, it reached the beauty of understanding so much required in doing and appreciation of science. Here, refinement is noteworthy.

---

[79] Levrini et al. (2014).
[80] Goren and Galili (2018).

(a)

Fig. 10.21a The face of David enlarged (Photo by Jörg Bittner Unna)

Fig. 10.21 The statue of David (a copy) by Michelangelo in the center of Florence - an ideal hero manifesting the balanced complementarity of spiritual and physical perfection. (Photo by the author)

Consider Florence, the place where people especially venerated beauty. In the sixteenth century, they placed the statue of David by Michelangelo in the central square of their city as a symbol of beauty (Fig. 10.21). The statue is emblematic of its beauty, but it was a beauty of a specific kind. The people of Florence choose David to represent their ideal of man. They placed the statue in the very center of their city. It was the act of definition. At face value, David defined the beauty of form, youth, force, and courage. It stands there now—a lovely young warrior of perfect proportions. Yet, the youth was neither Apollo nor, say, Alexander the Great, but the Biblical hero, King David. David, the vanquisher of Goliath, was also a poet and that fact introduced the new type of the harmony venerated in the Renaissance. He symbolized the fusion of a warrior, the power of force, courage, and devotion to his people, with something very different. David created the book of Psalms, which talks to and about God and has been in continuous use for 3000 years by myriads

(b)   (c)

Fig. 10.21b, c (**b**) Student at a lecture. The fragment from the relief by the Masegne brothers, Jacobello and Pierpaolo (1383) in Bologna. Compare with Fig. 10.21a. (Photo by the author) (**c**) The sculpture of a scholar in contemplation, in the National Museum of Korea (The national treasure #78), Seoul

around the world in their everyday prayers. In this sense, David apparently symbolized the symbiosis of internal and external beauty, the ultimate harmony that seemingly left no place for anything else to be added. Was it so?

It was not. The people of Bologna did not agree with Florence and pointed to another dimension of beauty missing in the Florencian set—*the beauty and pleasure of understanding*, not less, and possibly more divine in its nature. In the eleventh century, the people of Bologna established a new type of temple, the temple of knowledge, the university—Alma Mater Studiorum. Their heroes became people of knowledge and understanding: students and professors. Within the national tradition of artistic visualization, Bolognian artists produced images of students. In particular, let us look at one student depicted at a lecture (Fig. 10.21b). In parallel to David, a young warrior–poet, an emblem of internal and external beauty, in Bologna, there stands the figure of the young student who is delighted by the knowledge revealed to him. For its extraordinarily expressive power, this image became emblematic of Bologna. Half-smiling, silent, with a look of intellectual admiration, this image visualized the new idea—*the beauty and pleasure of understanding*.

Juxtaposition of the two pieces of art, (Fig. 10.21a, b), from Bologna and Florence, could be revealing, emphasizing the specific beauty of science. This beauty is much beyond being "fun," as science is often advertised in public, but instead, it appeals to the genus of science itself, venerated by scientists ever since its foundation.

This idea was much beyond Italy, it was universal and characterized humankind globally (David was above ethnicity…). To see this, consider the image of Pensive Bodhisattva (Fig. 10.21c). It was produced in Korea in the sixth century. Its meaning is similar to that of the student in Fig. 10.21b. The half-smile is just the same.

The admiring glance upward, however, is replaced by closed eyes, and the face slightly inclined downward. It depicts the same moment of the deeply experienced *pleasure of understanding*.

Obviously, artistic images affect the observer not only by their aesthetics but also through their meaning perceived. Observing the students' faces in classes when we infused artistic images into lecturing, we could not miss their empathy and compassion and enthusiasm expressed on their faces, suggesting a kind of emotional resonance caused by abstract ideas becoming concrete and pleasing in form. Meaning and appearance matched creating a special pleasure.

We talk about spiritual impact. Yet, since this resonance sometimes expands to the improved understanding of content, some scholars infer that understanding, wrapped in emotion, mobilizes additional cognitive resources, thus providing also pragmatic benefit.[81] We may therefore argue that using such images may transform the *pressure for understanding,* so much familiar to students, to the *pleasure of understanding*. It creates a bridge between science (related to objective pragmatic benefits) and the humanities (related to subjective and spiritual values), often perceived in unrelated opposition. Understanding the formal disciplinary content does not exclude, but is often enriched by images. It is much more than *fun*; it is *pleasure*. It is a serious business, because it reveals us the *genus* of science, and this is truly exciting.

Additional Perspective on the Rationale of Using Art Images

It was asserted that pictorial images can illustrate abstract concepts of science and can facilitate interpretive discussions, which would make science teaching more appealing to a wider population than traditionally addressed in science class. This strategy involves the dialogue between art and science, which is usually outside regular science classes. The reason for this tradition is the common belief that art and science are essentially different cognitive areas and is supported by the fact that they are governed by different hemispheres of the brain.[82] This aspect is already more profound than a mere call for an additional tool in pedagogical practice using a superficial appeal. Neuropsychology allows us to peek into the world of cognition that is behind the behavioristic manifestations in the form "I just like it...." In a simplified schematic understanding, the affiliation of cognitive activities relates art to the right hemisphere and science to the left one (Fig. 10.22).

Leaving aside the obvious recognition of the necessary complementarity of the items from both columns in cognitive practice, we draw attention to the fact that Art belongs to the group, which includes the holistic perspective, creativity, imagination, and all requirements in learning and applying scientific knowledge. Apparently, tradition relates science to logic, analytic, and rational thought. Indeed, who oppose this claim? In fact, this explains the historical origin of the rationale behind the traditional science curriculum adopted in a straightforward application. It however

---

[81] Scheffler (2009, p. 132)

[82] For example, Orenstein (1975), Harrison (2015) and https://en.wikipedia.org/wiki/Lateralization_of_brain_function

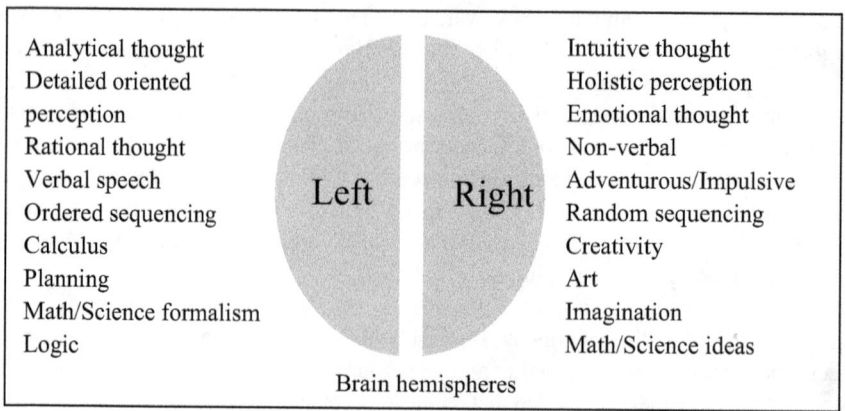

Analytical thought	Intuitive thought
Detailed oriented perception	Holistic perception
Rational thought	Emotional thought
Verbal speech	Non-verbal
Ordered sequencing	Adventurous/Impulsive
Calculus	Random sequencing
Planning	Creativity
Math/Science formalism	Art
Logic	Imagination
	Math/Science ideas

Brain hemispheres

Fig. 10.22 Common affiliation of cognitive activities to brain hemispheres as developed in neuropsychology

appears oversimplified and actually a diminution. From a series of the implied deficiencies of such policy, Einstein referred only to the one that looked to him decisive—imagination. In response to the tendency to make science a Machian economy of thought, an algorithmic application of the well-ordered sequential activities of problem solving and thus to separate science from imagination, Einstein reflected with the comment "We have created a society that honors the servant and has forgotten the gift,"[83] thus defining this approach as erroneous. This issue deserves much more attention in science education. From all possible aspects, we have expanded here the same claim to art with relation to science education, perhaps, because of its being the most salient for possessing an affective nature.

In a series of images, we have illustrated the close relationship of the relevant image with the idea to be addressed in science education. Indeed, artistic images were not produced for teaching science. This makes the role of the teacher pivotal: to draw attention; to provide an initial interpretation; to create a bridge to the particular context, curriculum, and cultural environment; and to the knowledge background of the audience.

Artistic images can introduce the abstract ideas of the nature of scientific knowledge, its concepts, and method. Their appeal and aesthetic impact enhance students' learning. The investigation of such impact requires more research; yet, already the initial experience informs of a positive resonance in teachers and students, which is inspiring.

---

[83] Quoted in Harrison (2015, p. 336).

Picture Credits

- Illustration picture in A. Kircher, *Ars Magna lucis et umbr,* 1650. Public domain. https://commons.wikimedia.org/w/index.php?curid=7672837
- Fig. 10.3a Illustration picture *Typus Logice*, in Reisch, *Margarita Philosophica,* 1503, Public domain. https://commons.wikimedia.org/w/index.php?curid=454467
- Fig. 10.3b Illustration picture *The Tower of Learning*, in Reisch, *Margarita Philosophica,* 1503, Public domain. https://commons.wikimedia.org/w/index.php?curid=41479386
- Fig. 10.6a Fresco *Stigmatization of St Francis,* Giotto, Assisi, 1300. Public domain. https://commons.wikimedia.org/wiki/File:Giotto_-_Legend_of_St_Francis_-_19-_-_Stigmatization_of_St_Francis.jpg
- Fig. 10.6b Painting *Stigmatization of St Francis*, Giotto, Louvre, Paris, 1300. Public domain. https://commons.wikimedia.org/wiki/File:Giotto_di_Bondone_-_Stigmatization_of_St_Francis_-_WGA09329.jpg
- Fig. 10.6c Fresco *Stigmatization of St Francis,* Giotto, Santa Croce, Florence, 1325. Public domain. http://www.travelingintuscany.com/art/giotto/santacrocebardichapel.htm#1
- Fig. 10.6d Painting *Stigmatization of St Catharine*, Manetti, 1630, Siena, Santa Caterina in Fontebranda, Siena. Public domain. https://www.europeana.eu/portal/en/record/08562/D4F21114684ACDEC0CE3ADF58C769071EBF1D1A9.html
- Fig. 10.7b Fresco Fig. 10.6c after renovation in 2014, private communication
- Fig. 10.10a Fragment from the fresco *School of Athens,* Rafael, 1511. The Vatican. Public domain. https://en.wikipedia.org/wiki/The_School_of_Athens#/media/File:Sanzio_01_Plato_Aristotle.jpg
- Fig. 10.10b Portrait of Newton. Godfrey Kneller, 1689. Public domain. https://commons.wikimedia.org/wiki/File:Sir_Isaac_Newton_(1643-1727).jpg Portrait of Descartes. After Frans Hals, 1785, Crédit communal de Belgique. Public domain. https://commons.wikimedia.org/w/index.php?curid=2774313
- Fig. 10.10c Fragment from fresco *The Disputation of the Holy Sacrament.* Raphael, 1510. The Vatican. Public domain. https://commons.wikimedia.org/w/index.php?curid=15460950
- Fig. 10.11a Fresco *School of Athens,* Rafael, 1511. The Vatican. Public domain. https://commons.wikimedia.org/w/index.php?curid=4406048
- Fig. 10.11b Fresco *The Disputation of the Holy Sacrament*, Rafael, 1511. The Vatican. Public domain. https://commons.wikimedia.org/w/index.php?curid=15460950
- Fig. 10.12a Fragment of painting *Calumny of Apelles* Botticelli, 1497. The Yorck Project (2002). Public domain. https://commons.wikimedia.org/w/index.php?curid=60359867
- Fig. 10.12b Painting *The Incredulity of Saint Thomas,* Caravaggio, 1602. Public domain. https://commons.wikimedia.org/w/index.php?curid=6804893
- Fig. 10.13a The Nobel Prize medal of Albert Einstein. With permission: Albert Einstein Archives, the Hebrew University of Jerusalem
- Fig. 10.13b Mosaic in Capella Palatina, Chapel Palatino, Palermo. Public domain. Illustration photo in P. Toesca, La Chapelle Palatine de Palerme, 1955. Sidera Milano: Edizioni d'Arte
- Fig. 10.13c Schematic drawing of Pygmalion casting Galatea. Sketch by unknown author. Public domain. https://lh3.googleusercontent.com/proxy/m4QqhD7iWfKTUtv2otWbtDDrMNljS2q-9HZ_dFeuh_1IJ0VA_nM-4s7OI7XgVl7qMhvtp1FYD25RGXiZWBDCIMTAI8L5cWFgELcXVHoz=w1200-h630-p-k-no-nu
- Fig. 10.14a Painting *The Last Judgement*, S. Lochner, 1435, Wallraf-Richartz Museum, Cologne, Germany. Photo credit: José Luiz Bernardes Ribeiro, CC Attribution-Share 4.0 International license. https://commons.wikimedia.org/w/index.php?curid=61455800
- Fig. 10.14b Painting *The Last Supper*, Rosselli, 1480, Web Gallery of Art: Public domain. https://commons.wikimedia.org/w/index.php?curid=4539502
- Fig. 10.14c Fragments from *The Last Supper,* Leonardo da Vinci, 1490. Public domain. https://en.wikipedia.org/wiki/The_Last_Supper_(Leonardo)#/media/File:The_Last_Supper_-_Leonardo_Da_Vinci_-_High_Resolution_32x16.jpg

- Fig. 10.15 Stained glass windows, 1230, the Notre-Dame Cathedral, Chartres, France. Photo by PtrQs, the Creative Commons Attribution-Share Alike 4.0 International license. https://commons.wikimedia.org/w/index.php?curid=58332525
- Fig. 10.16a Fragment from the Ghent Altarpiece, 1429. Van Eyck, St. Bavo Cathedral, Gent. Public domain. https://commons.wikimedia.org/wiki/File:Jan_van_Eyck_-_The_Ghent_Altarpiece_-_Singing_Angels_(detail)_-_WGA07642.jpg
- Fig. 10.16b Gravure of Chartres Cathedral, Jean Baptiste Rigaud, 1750. Picture credit: Bildforyou7, the CC Attribution-Share Alike 3.0 Unported license. https://commons.wikimedia.org/wiki/File:Inneres_der_Kathedrale.jpg
- Fig. 10.16c Fragment of fresco *God the Father*, Michelangelo, 1509, Capella Sistine, the Vatican. Public domain. https://commons.wikimedia.org/w/index.php?curid=1551130
- Fig. 10.17a Icon *The Ladder of Divine Ascent* by anonymous author. Monastery of St. Catherine, Sinai 12 c. Public domain. https://commons.wikimedia.org/w/index.php?curid=2386101
- Fig. 10.17b Roman mosaic from Pompeii, *The Plato's Academy,* 1st c. National Archaeological Museum, Naples. Public Domain. Photo credit: Jebulon, the CC0 1.0 Universal Public Domain Dedication. https://en.wikipedia.org/wiki/Plato%27s_Academy_mosaic#/media/File:MANNapoli_124545_plato's_academy_mosaic.jpg
- Fig. 10.18b *Portrait of Picasso* by **Juan Gris**, 1912. Public domain. https://commons.wikimedia.org/wiki/File:Juan_Gris_-_Portrait_of_Pablo_Picasso_-_Google_Art_Project.jpg
- Fig. 10.19b Statue of *Unique Forms of Continuity in Space,* U. Boccioni, 1913. Public domain. https://commons.wikimedia.org/wiki/File:%27Unique_Forms_of_Continuity_in_Space%27,_1913_bronze_by_Umberto_Boccioni.jpg
- Fig. 10.20 *The Ancient of Days,* by William Blake, 1794. Public domain. https://commons.wikimedia.org/w/index.php?curid=27197029
- Fig. 10.20 Einstein with Bohr in 1930. Photo by Paul Ehrenfest, Public domain.
- Einstein and Bohr, photo by P. Ehrenfest, 1930. Public domain. https://commons.wikimedia.org/wiki/File:Niels_Bohr_Albert_Einstein2_by_Ehrenfest.jpg
- Fig. 10.21a Face of David. CC Attribution 3.0 licensed and allowed for copy by Jörg Bittner Unna https://commons.wikimedia.org/wiki/File:%27David%27_by_Michelangelo_Fir_JBU013.jpg
- Fig. 10.21b *Fragment from the mediaeval tomb, Masegne brothers,* 1383, the *Medieval* Civic *Museum,* Bologna. Public domain. https://commons.wikimedia.org/w/index.php?curid=4942241
- Fig. 10.21c The sculpture of *Pensive Bodhisattva* (National Treasure No. 78). The collection of the National Museum of Korea. Picture credit: the Korea Open Government License Type I: Attribution. © the National Museum of Korea https://en.wikipedia.org/wiki/File:Pensive_Bodhisattva_(National_Treasure_No._78)_01.jpg

References

Andersen, E. (2007). *The geometry of art. The history of the mathematical theory of perspective from Alberti to Monge.* Springer.

Argentieri, D. (1956). Leonardo's optics. In E. Vollmer (Ed.), *Leonardo da Vinci* (pp. 405–436). Reynal.

Arnheim, R. (1974). *Art and visual perception. A psychology of the creative eye.* The University of California.

Arnold, V. I. (1998). The antiscientific revolution and mathematics. Talk at the meeting of the Pontifical Academy at Vatican, 26 October, 1998, *Changing concepts of nature at the turn of the millennium.* https://www.math.ru.nl/~mueger/arnold.pdf. Retrieved 17.09.2019.

Arnold, V. I. (2002). Математическая Дуэль Вокруг Бурбаки. *Вестник Российской Академии Наук, 72*(3), 245–250.

Baigrie, B. (1996a). Descartes's Scientific Illustrations and "la grande mechaniquede la nature". In B. Baigrie (Ed.), *Picturing knowledge: Historical and philosophical problems concerning the use of art in science* (pp. 86–134). University of Toronto Press.

Baigrie, B. (Ed.). (1996b). *Picturing knowledge: Historical and philosophical problems concerning the use of art in science*. University of Toronto Press.

Beller, M. (1999). *Quantun dialogue. The making of a revolution*. The University of Chicago Press.

Bonaventure, B. (1235). Pescia, San Francesco. Retrieved on September 9, 2013, http://www.flickriver.com/photos/renzodionigi/4589204591/

Born, M. (1962). *Einstein's theory of relativity*. Dover.

Calvi, I. (1956). Military engineering and arms. In E. Vollmer (Ed.), *Leonardo da Vinci* (pp. 275–306). Reynal.

Canestrini, G. (1956). Leonardo's machines. In *Leonardo da Vinci* (pp. 493–507). Collected Studies. Reynal.

Cellarius, A. (1661). *Harmonia Macrocosmica*. Apud Joannem Janssonium.

Cohen, R. M., & Drabkin, E. I. (1948). *A source book in Greek science*. McGraw-Hill.

Crombie, A. C. (1959). *Medieval and early modern science*. Doubleday Anchor Books.

De Broglie, L. (1953). *Pier Duhem's life and work*. Forward in Duhem, P. (1906/1982). *The aim and structure of physical theory*. Princeton University Press.

Dekker, E. (2004). *Catalogue of orbs, spheres and globes*. Instituto e Museo di Storia della Scienza, Giunti.

Descartes, R. (1637/1965). *Discourse on method, optics, geometry and meteorology. Second discourse – Of refraction*. Bobbs-Merrill.

Descartes, R. (1644/1982). *Principles of philosophy*. D. Reidel.

Einstein, A. (1905/1923). On the electrodynamics of moving bodies. 1. Definition of simultaneity. In *The principle of relativity* (pp. 38–40). Dover.

Evers, B. (Ed.). (2006). *Architectural theory from renaissance to the present*. Taschen.

Fludd, R. (1617). *Utriusque Cosmi, Maioris scilicet et Minoris, metaphysica, physica, atque technica Historia*. Johan Theodore de Bry.

Foucault, M. (1972). *Archaeology of knowledge*. Routledge.

Galilei, G. (1623/1957). *The Assayer*. Translated by Stillman Drake. In *Discoveries and opinions of Galileo* (pp. 237–238). Anchor Books.

Galili, I., & Zinn, B. (2007). Physics and art – A cultural symbiosis in physics education. *Science & Education, 16*(3–5), 441–460.

Galili, I. (2001). Weight versus gravitational force: Historical and educational perspectives. *International Journal of Science Education, 23*(10), 1073–1093.

Galili, I., & Goldberg, F. (1993). Left-right conversions in a plane mirror. *The Physics Teacher, 31*(8), 463–466.

Galili, I., & Kaplan, D. (1997). Changing approach in teaching electromagnetism in a conceptually oriented introductory physics course. *American Journal of Physics, 65* (7), 657–668.

Gliozzi, M. (1965). Storia della Fisica, Vol. II. Storia della Scienze. Torino: Italy.

Gombrich, E. H. (1972). *Art and illusion. A study in the psychology of pictorial representation*. Phaidon.

Goodman, N. (1968/1976). *Languages of art. An approach to a theory of symbols*. The Bobbs-Merrill Company.

Goodwin, J. (1979). *Robert Fludd. Hermetic philosopher and surveyor of two worlds*. Thames & Hudson.

Goren, E. & Galili, I. (2018). A summary lecture as a delay organizer of students' knowledge of mechanics – A discipline-culture approach. *Proceedings of the 11th conference of the European science education research association (ESERA)*. Dublin, Ireland.

Gutman, H. B. (1941). The medieval content of Raphael's "school of Athens". *Journal of the History of Ideas, 2*(4), 420–429.

Haas, R. (2012). Raphael's school of Athens: A theorem in a painting? *Journal of Humanistic Mathematics, 2*(2). Available at: http://scholarship.claremont.edu/jhm/vol2/iss2/3

Hall, B. S. (1996). The didactic and the elegant: Some thoughts on scientific and technological illustrations in the middle ages and renaissance. In B. Baigrie (Ed.), *Picturing knowl-*

edge. Historical and philosophical problems concerning the use of art in science (pp. 3–39). University of Toronto Press.

Harrison, D. (2015). *Brain asymmetry and neural systems*. Springer.

Heidegger, M. (1972). *On time and being*. Harper.

Heidegger, M. (1977). Science and reflection. In M. Heidegger (Ed.), *Concerning technology and other essays*. Harper.

Heilbronn, J. L. (1998). *Geometry civilized. History, culture, and technique*. Claredon Press.

Heisenberg, W. (1959/1971). *Physics and philosophy. The revolution in modern science*. Harper.

Herz-Fischler, R. (1987). *A mathematical history of the golden number*. Dover.

Huxley, T. H. (1882). On science and art in relation to education. In Huxley, T. H. (1897) *Science and education. Essays*. D. Appleton and Company.

Irzik, G., & Nola, R. (2011). A family resemblance approach to the nature of science for science education. *Science & Education, 20*, 591–607.

Jung, C. G. (1944/1993). *Psychology and alchemy*. Princeton University Press.

Landau, L. D., & Lifshitz, E. M. (1960). *Course of theoretical physics* (9 vols). Pergamon Press.

Lawlor, R. (1982). *Sacred geometry: Philosophy and practice*. Thames & Hudson.

Leo, B. (1227/2010). *The mirror of perfection: To Wit the blessed Francis of Assisi*. Burns & Oates. http://www.archive.org

Levrini, O., Bertozzi, E., Gagliardi, M., Grimellini-Tomasini, N., Pecori, B., Tasquier, G., & Galili, I. (2014). Meeting the discipline-culture framework of physics knowledge: An experiment in Italian secondary school. *Science & Education, 23*, 1701–1731.

Livio, M. (2002). *The golden ratio. The story of phi, the most astonishing number*. Broadway Books.

Lopes, D. (2004). *Understanding pictures*. Clarendon Press.

Losee, J. (1993). *A historical introduction to the philosophy of science*. Oxford University Press.

Mach, E. (1883/1919/1989). *The science of mechanics, a critical and historical account of its development*. Open Court.

Margotta, R. (1968). *The story of medicine*. Golden Press.

Maxwell, J. C. (1890/1965). In W. D. Niven (Ed.), *The scientific papers of James Clerk Maxwell* (2 vols.). Cambridge University Press/Dover.

Micheletti, E. (2007). *Santa Croce*. Becocci Editore.

Miller, A. I. (1984). *Imagery in scientific thought: Creating 20th-century physics*. Birkhauser.

Miniati, M. (1991). *Museo di Storia della Scienza. Catalogo*. Instituto e Museo di Storia della Scienza, Giunti.

Muray, P., & Muray, L. (1996). *The Oxford companion to Christian art and architecture*. Oxford University Press.

Nersessian, N. J. (1992). How do scientists think? Capturing the dynamics of conceptual change in science. In R. Giere (Ed.), *Cognitive Models of Science. Minnesota studies in the philosophy of science* (pp. 3–45). University of Minnesota Press.

Orenstein, R. E. (1975). *The psychology of consciousness*. Penguin Books.

Pais, A. (1982/2005). *Subtle is the Lord... the science and life of Albert Einstein*. Oxford University Press.

Park, D. (1997). *The fire within the eye. A historical essay on the nature and meaning of light*. Princeton University Press.

Phelan, J. (2002). The philosopher as hero: Raphael's the school of Athens. In *Artcyclopedia*. Retrieved on October 14, 2021, http://www.artcyclopedia.com/feature-2002-09.html

Piltz, A. (1981). *The world of medieval learning*. Basil Blackwell.

Poincare, H. (1903). *Science and method*. Dover.

Reisch, G. (1517). *Margarita philosophica, cum additionibus novis: ab auctore suo studiosissima revisione quarto supper additis*. Johann Schott.

Reti, L. (Ed.). (1974). *The unknown Leonardo*. McGraw-Hill.

Roob, A. (2001). *Alchemy & Mysticism*. Taschen.

Scheffler, I. (2009). *Words of truth. A philosophy of knowledge*. Willey-Blackwell.

Shlain, L. (1991). *Art and physics*. William Morrow.

Stein, B., & Galili, I. (2018). *Introduction of observer dependent concepts into physics teaching of middle school*. Proceedings of the 11th Conference of the European Science Education Research Association (ESERA). Dublin.

Tseitlin, M., & Galili, I. (2005). Teaching physics in looking for its self: From a physics-discipline to a physics-culture. *Science & Education, 14*(3–5), 235–261.

Vasari, G. (1991). *The lives of the artists*. Oxford University Press.

Vincent, R. (2001). *Geometrie du Nombre d'Or*. Chalagam.

Wallace, R. (1966). *The world of Leonardo*. Time-Life.

Whitfield, P. (1995). *The mapping of the heavens*. Pomegranate Artbooks, British Museum.

Whittaker, E. (1910/1951). *A history of the theories of aether and electricity*. Happer & Brothers.

Wind, G. D. (2002). *Correggio. Hero of dome*. Silvana Editoiale.

Wittgenstein, L. (1953/2001). *Philosophical investigations*. Blackwell.

Chapter 11
Epilogue: Discipline-Culture for the Pleasure of Understanding

The noblest pleasure is the joy of understanding

Leonardo da Vinci

Here is the place for making summarizing comments regarding the subject considered - science education. Becoming "educated in science" can be understood in a very broad sense, and this rather vague phrase often expresses people's feeling of satisfaction in the mere idea that they possess something very valuable – scientific knowledge. A spontaneous self-evaluation in this regard is often not indicative because the subject is far from trivial. Therefore, in light of what was discussed in this book, we may suggest to agree, at least, that in this area a supervisor's guidance is essential. We tried to address several aspects of complexity of the subject which we consider being of central importance. Among the vast number of components comprising the knowledge under examination, we brought to the fore its organization, relationships between components, their nature, and hierarchical structure.

To appreciate the problem, consider the view familiar to the numerous tourists who patiently stand in a long queue to enter the library of Trinity College in Dublin. Reaching the huge central hall, they look around being clearly excited (Fig. 11.1). When I asked people about this experience, they mentioned the awe and pride of human knowledge (as if they saw the knowledge itself…) created by our civilization and preserved in the library.[1]

Indeed, this magnificent panorama has become emblematic of our heritage of culture and knowledge. Yet, besides admiration, I felt a kind of frustration… I

---

[1] When my friend knew that I was going to visit Dublin, he had a single recommendation – visit the library at Trinity College. From then on, I have always done the same, and have received many messages of gratitude for my suggestion as people have shared my appreciation of that special place. I have this picture on the wall of my room for its pleasing effect as well as being an auxiliary tool in answering people about what we, teachers, are doing in class besides reciting the well-known content of textbooks. Seemingly, the picture helps us to realize some of the problems faced in education, what is exactly to be chosen and recounted to students…

© Springer Nature Switzerland AG 2021
I. Galili, *Scientific Knowledge as a Culture*, Science: Philosophy, History and Education, https://doi.org/10.1007/978-3-030-80201-1_11

realized, even if I concentrate solely on the books in a specific domain, there is no chance of mastering them all. This is apart from realizing that such an enterprise would leave no time for doing anything new and original of my own. This is exactly the situation in science. But if this is so, does learning science have any chance of success? Are we destined to remain ignorant in the face of an insurmountable abundance of the already accumulated knowledge?[2] We, teachers, often observe that students believe: the more you learn – the more you know... This simple and frustrating truth, leaving no chance to ever "truly" know... Fortunately, it is not so. The learning is possible and can be effective. The presented perspective of frustration is essentially deficient in missing the idea of science as an *organized* body of knowledge. This means that scientific knowledge in each of its domains can be reduced to a *few* fundamental theories, each containing a *hierarchy*. This feature allows us to grasp the whole without reading too many books... How can this be?

Fig. 11.1 The long room in the library of Trinity College in Dublin

---

[2] This might look as a reverberation of the feelings expressed by Socrates, Plato, Nicolas of Cusa, Newton and many other brilliant scholars whom many of us would like to resemble in capability of learning and understanding... After much learning over many years, they claimed to remain "ignorant"... This expressed a peculiar feature of education – revealing a whole new world in which those who persist find much more to learn and investigate than they could have imagined prior to their learning... That "ignorance" was perceived from a much more advanced position in no way similar to what they knew before learning.

We have presented and discussed a special structure of organization. It addressed two fundamental aspects of the nature of science: *science* being *an objective theory of nature*, and *science* being *a specific culture*. Discipline-culture (DC) is a framework which represents scientific disciplines as drawing on fundamental theories which comprise a family and know each other. Familiarizing ourselves with such a structure provides us with a great advantage: awareness of the major ideas of the theories in each discipline, their ontological and epistemological basis. Being *cultural* implies not ignoring, but recognizing the partners of the conceptual dialogue. Cultural knowledge of a theory presumes awareness of its open problems and the alternative accounts of the same subject matter by other theories. This approach makes explicit the areas of validity of each theory and refines the frequently misinterpreted claim of multiple (incommensurable) theories of science, implying the misleading perception of a mere tentativeness of any scientific knowledge.

DC organizes the knowledge elements in a way that clarifies their status through specifying their affiliation to the nucleus, body, or periphery of a certain theory. The mere effort of such classification is beneficial as stimulating meaningful learning for the students and providing competence to the teachers. We illustrated this structure in the domains of mechanics and optics.

The DC approach provides much required guidance in choosing the content of history and philosophy relevant for science education, aiming at the balance in presenting the three types of scientific content in a scientific theory. The conceptual dialogue upgrades a mere accretion of knowledge elements by awareness of their status and affiliation in the body of knowledge, facilitating meaningful learning. Commitment to the triadic structure of disciplinary content essentially discharges the frustrating "the more, the better" and refines the otherwise obscurant "less is more."

Further upgrading of meaningful learning of science requires addressing the features often labeled as the nature of science. Within the dispute with alternative views, the following features of science were argued by us here:

- Science deals with constructing objective theories of actual reality. This program outlines natural philosophy and modern science throughout history.
- Scientific knowledge is comprised of a small number of fundamental theories which include a multitude of knowledge elements – concepts, principles, laws, problems, experiments, as well as tools and apparatus designed in accordance with specific theories.
- Scientific knowledge presents episteme – the rational knowledge which must be distinguished from other types of knowledge such as experience, tradition, faith, magic, etc. It is open to refinement, to critique, and it varies in tentativeness from the hypothetical to the highly certain and accomplished.
- Scientific knowledge is a cultural product. It presents objective knowledge reached through a specific procedure of production and accumulation. It includes theoretical analysis, experimentation, observation, verification, and prediction, all interwoven in logical patterns. Scientific knowledge emerges in numerous stages of interaction with reality causing refinement and self-correction.

- Scientific activity is immersed in society, which influences scientists in their *inquiry* in form and thematic. However, at the stage of knowledge *justification*, scientific products are subject to objective tests and verification, acquiring meaning and existence independent of creators and their intentions.
- Scientific knowledge is symbiotic with scientific method. The latter incorporates modeling, logical analysis, mathematical account, experimentation with control variables, and statistical analysis of the accumulated data, all complementing in knowledge production.

This nature opposes the claims of science being merely subjective, tentative, never fully proved, or precise, lacking historical continuity or a specific method of construction and verification. Such superficial claims threaten with another "cultural revolution" tried in the past, causing a devastating setback in science and science education. To prevent it, we need to learn and teach about such, being aware of the cases of pseudoscience and brutal intrusion into science.

Beyond declaration, the nature of science needs epistemological comments in the course of regular teaching, sometimes special discussions, drawing on enculturation rather than indoctrination. The DC-based curriculum addresses the subject matter including epistemology. This implies, for instance, emphasizing dual concept definitions, nominal and operational. The history of weight concept may serve as an exemplar of such treatment (Chap. 5). Currently, it is often missed in science classes and programs for prospective teachers, frequently lacking pertinent exposure to history or the philosophy of science.

Within the DC approach, teaching about the periphery of, say, Quantum Mechanics, addresses Classical Mechanics, and this strategy leads to the new understanding of the classical theory, correct in a regular environment but invalid in the atomic world. Learning Classical Mechanics as a DC touches on the ideas of Aristotelian, Hellenistic, and Medieval mechanics (Chap. 2). The new teaching of Optics (Chap. 7) is especially attractive for its inclusion of three theories (the theory of rays – Geometrical Optics, the theory of waves – Physical Optics, and the theory of photons – Modern Optics). All correct in different areas of validity, they powerfully illustrate the nature of science.

Finally, our approach calls for acknowledging that culture can be expressed in a variety of ways. Art is one of them. Its images may effectively express the ideas otherwise requiring long, verbal descriptions (Chap. 10). Subscribing to the metaphoric artistic images significantly reinforces the appealing power of teaching. Art requires interpretation, but the impact rewards with amazing aesthetic benefits. Such teaching engages a far wider audience, inspires many students, enables teachers to reach those "humanists" otherwise refraining from science class. It expands the community of students who experience a special kind of pleasure - the pleasure of understanding, "the noblest joy," in the words of Leonardo da Vinci, providing deep satisfaction and spiritual maturation.

One may explain this impact through employing the perspective of dichotomy of brain activity (dual cognitive affiliation to the two brain hemispheres). The fusion of holistic, imagery, emotional with analytic, rational, sequential logic produces science education of broad appeal. It can upgrade the common disciplinary science curriculum which petrifies the dichotomy of students, "physicists" and "poets," good and bad in science. By applying a solely disciplinary curriculum, our society wastes the original minds which can essentially enrich us all.

Combining the cultural approach in its internal (in science) and external (in society) senses will radically change science education in the way warranted by the genus of science.

***

Picture Credit

Correction to: Scientific Knowledge as a Culture

Correction to:
I. Galili, *Scientific Knowledge as a Culture*,
Science: Philosophy, History and Education,
https://doi.org/10.1007/978-3-030-80201-1

This book was inadvertently published without updating the following corrections:

Corrections:

p. ix, About the Book by Prof. Michael R. Matthews

p. xv, Replace with "The Hebrew University of Jerusalem"

p. 184, Fig 4.18: Insert "(sketch by Guy Galili)" after "in a rotating cabin of a spaceship"

p. 463, Fig 10.21a: (i) Change Fig 10.21a to Fig 10.21 in citation and caption, (ii) update caption as "The statue of David (a copy) by Michelangelo in the center of Florence - an ideal hero manifesting the balanced complementarity of spiritual and physical perfection. (Photo by the author)"

p. 463, Insert ",say," in between "nor" and "Alexander"

The updated versions of the chapters can be found at
https://doi.org/10.1007/978-3-030-80201-1_4
https://doi.org/10.1007/978-3-030-80201-1_10

The updated version of this book can be found at
https://doi.org/10.1007/978-3-030-80201-1

© Springer Nature Switzerland AG 2022
I. Galili, *Scientific Knowledge as a Culture*, Science: Philosophy, History and Education, https://doi.org/10.1007/978-3-030-80201-1_12

p. 463, Insert Fig. 10.21a and caption "The face of David enlarged (Photo by Jörg Bittner Unna)"

p. 464, Insert "Compare with Fig. 10.21a" after "Bologna" in the caption

p. 468, Insert picture credit for Fig 10.21a "Fig.10.21a Face of David. CC Attribution 3.0 licensed and allowed for copy by Jörg Bittner Unna https://commons.wikime-dia.org/wiki/File:%27David%27_by_Michelangelo_Fir_JBU013.jpg"

The book and the affected chapters have been updated with these corrections.

CPSIA information can be obtained
at www.ICGtesting.com
Printed in the USA
BVHW052039080223
658147BV00008B/107